D0073283

Photovoltaic Systems Engineering

Third Edition

Roger A. Messenger • Jerry Ventre

CRC Press
Taylor & Francis Group
Boca Raton London New York

CRC Press is an imprint of the
Taylor & Francis Group, an **informa** business

CRC Press
Taylor & Francis Group
6000 Broken Sound Parkway NW, Suite 300
Boca Raton, FL 33487-2742

© 2010 by Taylor and Francis Group, LLC
CRC Press is an imprint of Taylor & Francis Group, an Informa business

No claim to original U.S. Government works

Printed in the United States of America on acid-free paper
10 9 8 7 6 5 4 3

International Standard Book Number: 978-1-4398-0292-2 (Hardback)

This book contains information obtained from authentic and highly regarded sources. Reasonable efforts have been made to publish reliable data and information, but the author and publisher cannot assume responsibility for the validity of all materials or the consequences of their use. The authors and publishers have attempted to trace the copyright holders of all material reproduced in this publication and apologize to copyright holders if permission to publish in this form has not been obtained. If any copyright material has not been acknowledged please write and let us know so we may rectify in any future reprint.

Except as permitted under U.S. Copyright Law, no part of this book may be reprinted, reproduced, transmitted, or utilized in any form by any electronic, mechanical, or other means, now known or hereafter invented, including photocopying, microfilming, and recording, or in any information storage or retrieval system, without written permission from the publishers.

For permission to photocopy or use material electronically from this work, please access www.copyright.com (http://www.copyright.com/) or contact the Copyright Clearance Center, Inc. (CCC), 222 Rosewood Drive, Danvers, MA 01923, 978-750-8400. CCC is a not-for-profit organization that provides licenses and registration for a variety of users. For organizations that have been granted a photocopy license by the CCC, a separate system of payment has been arranged.

Trademark Notice: Product or corporate names may be trademarks or registered trademarks, and are used only for identification and explanation without intent to infringe.

Library of Congress Cataloging-in-Publication Data

Messenger, Roger A.
Photovoltaic systems engineering / Roger A. Messenger and Jerry Ventre. -- 3rd ed.
p. cm.
"A CRC title."
Includes bibliographical references and index.
ISBN 978-1-4398-0292-2
1. Photovoltaic power systems. 2. Dwellings--Power supply. 3. Building-integrated photovoltaic systems. I. Title.

TK1087.M47 2010
621.31'244--dc22 2009032962

Visit the Taylor & Francis Web site at
http://www.taylorandfrancis.com

and the CRC Press Web site at
http://www.crcpress.com

Dedication

We did not inherit this world from our parents …
We are borrowing it from our children.
(Author unknown)

It is our fervent hope that the engineers who read this book will dedicate themselves to the creation of a world where children and grandchildren will be left with air they can breathe and water they can drink, where humans and the rest of nature will nurture one another.

Contents

Preface xvii
Acknowledgments xxi
About the Authors xxiii

Chapter 1
Background 1

1.1 Introduction 1
1.2 Population and Energy Demand 2
1.3 Energy Units 2
1.4 Current World Energy Use Patterns 3
1.5 Exponential Growth 7
1.5.1 Introduction 7
1.5.2 Compound Interest 7
1.5.3 Doubling Time 8
1.5.4 Accumulation 9
1.5.5 Resource Lifetime in an Exponential Environment 10
1.6 Hubbert's Gaussian Model 11
1.7 Net Energy, Btu Economics, and the Test for Sustainability 13
1.8 Direct Conversion of Sunlight to Electricity with Photovoltaics 14
References 18
Suggested Reading 19

Chapter 2
The Sun 21

2.1 Introduction 21
2.2 The Solar Spectrum 21
2.3 The Effect of Atmosphere on Sunlight 23
2.4 Sunlight Specifics 25
2.4.1 Introduction 25
2.4.2 Definitions 26
2.4.3 The Orbit and Rotation of the Earth 27
2.4.4 Tracking the Sun 29
2.4.5 Measuring Sunlight 32
2.4.5.1 Precision Measurements 32
2.4.5.2 Less Precise Measurements 35
2.5 Capturing Sunlight 35
2.5.1 Maximizing Irradiation on the Collector 35
2.5.2 Shading 38
2.5.2.1 Field Measurement of Shading Objects 38
2.5.2.2 Computational Methods of Determining Shading 39

2.5.3 Special Orientation Considerations 40
2.5.3.1 Horizontal Mounting 41
2.5.3.2 Non-South-Facing Mounting 41
References 46
Suggested Reading 46

Chapter 3
Introduction to PV Systems 47

3.1 Introduction 47
3.2 The PV Cell 49
3.3 The PV Module 52
3.4 The PV Array 55
3.5 Energy Storage 55
3.5.1 Introduction 55
3.5.2 The Lead-Acid Storage Battery 56
3.5.2.1. Chemistry of the Lead-Acid Cell 56
3.5.2.2 Properties of the Lead-Acid Storage Battery 57
3.5.3 The Nickel Cadmium Storage Battery 62
3.5.3.1 Chemistry of the Nickel Cadmium Storage Battery 62
3.5.3.2 Properties of the Nickel Cadmium System 63
3.5.4 Other Battery Systems 64
3.5.5 Hydrogen Storage 65
3.5.6 The Fuel Cell 66
3.5.7 Other Storage Options 67
3.6 PV System Loads 67
3.7 PV System Availability 69
3.8 Associated System Electronic Components 72
3.8.1 Introduction 72
3.8.2 Charge Controllers 72
3.8.2.1 Charging Considerations 73
3.8.2.2 Discharging Considerations 76
3.8.3 Maximum Power Point Trackers and Linear Current Boosters 76
3.8.4 Inverters 80
3.8.4.1 Square Wave Inverters 81
3.8.4.2 Modified Sine Wave Inverters 82
3.8.4.3 Pulse-Width-Modulated Inverters 85
3.8.4.4 Other Desirable Inverter Features 88
3.9 Generators 89
3.9.1 Introduction 89
3.9.2 Types and Sizes of Generators 90
3.9.3 Generator Operating Characteristics 91
3.9.3.1 Rotation Speed 91
3.9.3.2 Efficiency 91
3.9.3.3 Fuel Types 92

3.9.3.4 Altitude Effects 93
3.9.3.5 Waveform Harmonic Content 93
3.9.3.6 Frequency Stability 93
3.9.3.7 Amplitude Stability 94
3.9.3.8 Noise Level 94
3.9.3.9 Type of Starting 94
3.9.3.10 Overload Characteristics 94
3.9.3.11 Power Factor Considerations 94
3.9.4 Generator Maintenance 95
3.9.5 Generator Selection 95
3.10 Balance of System (BOS) Components 95
3.10.1 Switches, Circuit Breakers, Fuses, and Receptacles 96
3.10.2 Ground Fault, Surge, and Lightning Protection 97
3.10.3 Inverter Bypass Switches and Source Circuit Combiner Boxes 98
3.10.4 Grounding Devices 98
References 101
Suggested Reading 102

Chapter 4
Grid-Connected Utility-Interactive PV Systems 103

4.1 Introduction 103
4.2 Applicable Codes and Standards 104
4.2.1 The National Electrical Code 104
4.2.1.1 Introduction 104
4.2.1.2 Voltage Drop and Wire Sizing 106
4.2.2 IEEE Standard 1547-2003 109
4.2.2.1 Introduction 109
4.2.2.2 Specific Requirements 109
4.2.2.3 Comparison of PV Inverters to Mechanically Rotating Generators 112
4.2.2.4 Islanding Analysis 112
4.2.3 Other Issues 115
4.2.3.1 Aesthetics 115
4.2.3.2 Electromagnetic Interference 115
4.2.3.3 Surge Protection 115
4.2.3.4 Structural Considerations 116
4.3 Design Considerations for Straight Grid-Connected PV Systems 117
4.3.1 Determining System Energy Output 117
4.3.2 Array Installation 119
4.3.3 Inverter Selection and Mounting 119
4.3.4 Other Installation Considerations 120
4.4 Design of a System Based on Desired Annual System Performance 121
4.4.1 Array Sizing 121
4.4.2 Inverter Selection 122

4.4.3 Module Selection 123
4.4.4 Balance of System 125
4.4.4.1 Module Extension Wire, Rooftop Junction Box, and Module and Array Frame Grounding 126
4.4.4.2 Wire and Conduit from Rooftop Junction Box to Inverter 127
4.4.4.3 DC and AC Disconnects and Overcurrent Protection 130
4.4.4.4 Point of Utility Connection 130
4.4.4.5 Final System Electrical Schematic Diagram 131
4.5 Design of a System Based on Available Roof Space 132
4.5.1 Array Selection 132
4.5.2 Inverter Selection 136
4.5.3 Balance of System 136
4.5.3.1 Module Extension Wire, Rooftop Junction Box, and Module and Array Frame Grounding 136
4.5.3.2 Wire and Conduit from Rooftop Junction Box to Inverter 137
4.5.3.3 DC and AC Disconnects and Overcurrent Protection 137
4.5.3.4 Point of Utility Connection 138
4.5.3.5 Estimating System Annual Performance 138
4.5.3.6 Final System Electrical Schematic Diagram 138
4.5.4 Extension of Design to Lower Wind Speed Region 139
4.6 Design of a Microinverter-Based System 140
4.6.1 Introduction 140
4.6.2 System Design 140
4.6.3 Bells and Whistles (i.e., Monitoring Possibilities) 142
4.7 Design of a Nominal 21 kW System That Feeds a Three-Phase Distribution Panel 142
4.7.1 Introduction 142
4.7.2 Array Configuration 143
4.7.3 System DC Wiring 144
4.7.4 System AC Wiring 146
4.7.4.1 Wire and Overcurrent Protection Sizing 146
4.7.4.2 Voltage Drop Calculations 148
4.7.4.3 Connecting to a 277/480 V Y-Connected Source 149
4.8 Design of a Nominal 250 kW System 150
4.8.1 Introduction 150
4.8.2 Configuring the Array 150
4.8.3 Source Circuit Voltage Drop Calculations 151
4.8.4 DC Disconnects and Overcurrent Protection 155
4.8.5 AC Wire Sizing, Disconnects, and Overcurrent Protection 155
4.9 System Performance Monitoring 158
References 161

Chapter 5
Mechanical Considerations 163

5.1 Introduction 163
5.2 Important Properties of Materials 163
5.2.1 Introduction 163
5.2.2 Mechanical Properties 164
5.2.3 Stress and Strain 166
5.2.4 Strength of Materials 169
5.2.5 Column Buckling 170
5.2.6 Thermal Expansion and Contraction 170
5.2.7 Chemical Corrosion and Ultraviolet Degradation 172
5.2.8 Properties of Steel 174
5.2.9 Properties of Aluminum 176
5.3 Establishing Mechanical System Requirements 177
5.3.1 Mechanical System Design Process 177
5.3.2 Functional Requirements 178
5.3.3 Operational Requirements 178
5.3.4 Constraints 179
5.3.5 Trade-Offs 180
5.4 Design and Installation Guidelines 180
5.4.1 Standards and Codes 180
5.4.2 Building Code Requirements 181
5.5 Forces Acting on PV Arrays 182
5.5.1 Structural Loading Considerations 182
5.5.2 Dead Loads 183
5.5.3 Live Loads 183
5.5.4 Wind Loads 183
5.5.5 Snow Loads 191
5.5.6 Other Loads 192
5.6 Array Mounting System Design 192
5.6.1 Introduction 192
5.6.2 Objectives in Designing the Array Mounting System 192
5.6.2.1 Minimizing Installation Costs 192
5.6.2.2 Building Integration Considerations 194
5.6.2.3 Costs and Durability of Array-Roof Configurations 195
5.6.3 Enhancing Array Performance 195
5.6.3.1 Irradiance-Enhancement 195
5.6.3.2 Shading 196
5.6.3.3 Array-Cooling 196
5.6.3.4 Protection from Vandalism 196
5.6.4 Roof-Mounted Arrays 196
5.6.4.1 Stand-Off Mounting 196
5.6.4.2 Rack Mounting 198

5.6.4.3 Integrated Mounting 198
5.6.4.4 Direct Mounting 199
5.6.5 Ground-Mounted Arrays 199
5.6.5.1 Rack Mounting 200
5.6.5.2 Pole Mounting 200
5.6.5.3 Tracking-Stand Mounting 200
5.6.6 Aesthetics 201
5.7 Computing Mechanical Loads and Stresses 202
5.7.1 Introduction 202
5.7.2 Withdrawal Loads 202
5.7.3 Tensile Stresses 203
5.7.4 Buckling 203
5.8 Stand-Off, Roof Mount Examples 205
5.8.1 Introduction to ASCE-7 Wind Load Analysis Tabular Method 205
5.8.2 Array Mount Design: High Wind Speed Case 207
5.8.3 Array Mount Design: Low Wind Speed Case 209
5.8.4 Exposure C, Exposure D, and Other Correction Factors 210
References 212
Suggested Reading 213

Chapter 6
Battery-Backup Grid-Connected PV Systems 215

6.1 Introduction 215
6.2 Battery-Backup Design Basics 216
6.2.1 Introduction 216
6.2.2 Load Determination 217
6.2.3 Inverter Sizing 217
6.2.4 Battery Sizing 217
6.2.5 Sizing the Array 219
6.3 A Single-Inverter 120 V Battery-Backup System Based on Standby Loads 220
6.3.1 Determination of Standby Loads 220
6.3.2 Inverter Selection 221
6.3.3 Battery Selection 221
6.3.4 Array Sizing 223
6.3.5 Charge Controller and Module Selection 224
6.3.6 BOS Selection and Completion of the Design 226
6.3.6.1 Array Mounting Equipment 226
6.3.6.2 Rooftop Junction Box 227
6.3.6.3 Source Circuit Combiner Box and Surge Arrester 227
6.3.6.4 Wire and Circuit Breaker Sizing: dc Side 227
6.3.6.5 Wire and Circuit Breaker Sizing: ac Side 229
6.3.6.6 Wiring of Standby Loads 230

6.3.6.7 Equipment Grounding Conductor and Grounding Electrode Conductor Sizing 231
6.3.7 Programming the Inverter and Charge Controller 232
6.3.8 Fossil Fuel Generator Connection Options 234
6.4 A 120/240 V Battery-Backup System Based on Available Roof Space 235
6.4.1 Introduction 235
6.4.2 Module Selection and Source Circuit Design 235
6.4.3 Source Circuit Combiner Box and Charge Controller Selection ... 236
6.4.4 Inverter Selection 238
6.4.5 Determination of Standby Loads and Battery Selection 238
6.4.6 BOS Selection and Completion of Design 239
6.4.6.1 Rooftop Junction Box 239
6.4.6.2 Source Circuit Combiner Box and Surge Arresters 240
6.4.6.3 Wire and Circuit Breaker Sizing: dc Side 241
6.4.6.4 Wire and Circuit Breaker Sizing: ac Side 242
6.4.6.5 Wiring of Standby Loads 242
6.4.6.6 Equipment Grounding Conductor and Grounding Electrode Conductor Sizing 243
6.5 An 18 kW Battery-Backup System Using Inverters in Tandem 244
6.5.1 Introduction 244
6.5.2 Inverter and Charge Controller Selection 244
6.5.2.1 Inverter Selection 244
6.5.2.2 Charge Controllers 246
6.5.3 Module Selection and Array Layout 246
6.5.3.1 Module Selection 247
6.5.3.2 Array Layout 248
6.5.3.3 Array Performance 250
6.5.4 Battery and BOS Selection 251
6.5.5 Wire Sizing 251
6.5.6 Final Design 254
6.6 AC-Coupled Battery-Backup Systems 257
6.6.1 Introduction 257
6.6.2 A 120/240 V Battery-Backup Inverter with 240 V Straight Grid-Connected Inverter 258
6.6.3 A 120 V Battery-Backup Inverter with a 240 V Straight Grid-Connected Inverter 260
6.6.4 A 120/208 V 3-Phase AC-Coupled System 261
6.7 Battery Connections 263
References 269

Chapter 7
Stand-Alone PV Systems 271

7.1 Introduction 271
7.2 The Simplest Configuration: Module and Fan 272

7.3 A PV-Powered Water-Pumping System 274
7.3.1 Introduction 274
7.3.2 Selection of System Components 275
7.3.3 Design Approach for Simple Pumping System 277
7.3.3.1 Determination of Pump Operating Current and Voltage 277
7.3.3.2 Module Selection 278
7.4 A PV-Powered Parking Lot Lighting System 280
7.4.1 Determination of the Lighting Load 280
7.4.2 Parking Lot Lighting Design 282
7.4.2.1 Determination of Lamp Wattage and Daily Load Presented by the Fixture 283
7.4.2.2 Determination of Battery Storage Requirements 284
7.4.2.3 Determination of Array Size 284
7.4.2.4 Final System Schematic 285
7.4.2.5 Structural Comment 286
7.5 A Cathodic Protection System 287
7.5.1 Introduction 287
7.5.2 System Design 288
7.6 A Portable Highway Advisory Sign 291
7.6.1 Introduction 291
7.6.2 Determination of Available Average Power 292
7.7 A Critical-Need Refrigeration System 293
7.7.1 Introduction 293
7.7.2 Load Determination 294
7.7.3 Battery Sizing 294
7.7.4 Array Sizing 295
7.7.4.1 Fixed Array 295
7.7.4.2 Tracking Array Mount 297
7.7.4.3 Special Cases 297
7.7.5 BOS Component Selection 298
7.7.6 Overall System Design 299
7.8 A PV-Powered Mountain Cabin 300
7.8.1 Introduction 300
7.8.2 Load Determination 301
7.8.3 Battery Selection 303
7.8.4 Array Sizing and Tilt 304
7.8.5 Controller and Inverter Selection 308
7.8.5.1 Charge Controller 308
7.8.5.2 Inverter 308
7.8.6 BOS Component Selection 309
7.8.6.1 Wire, Circuit Breaker, and Switch Selection 309
7.8.6.2 Other Items 310
7.9 A Hybrid-Powered, Off-Grid Residence 311
7.9.1 Introduction 311

7.9.2 Summary of Loads ... 313
7.9.3 Battery Selection ... 314
7.9.4 Array Design ... 315
7.9.5 Generator Selection ... 318
7.9.6 Generator Operating Hours and Operating Cost ... 319
7.9.7 Charge Controller and Inverter Selection ... 320
7.9.8 Wire, Circuit Breaker, and Disconnect Selection ... 322
7.9.9 BOS Component Selection ... 324
7.9.10 Total System Design ... 324
7.10 Summary of Design Procedures ... 325
References ... 330
Suggested Reading ... 330

Chapter 8
Economic Considerations ... 331

8.1 Introduction ... 331
8.2 Life-Cycle Costing ... 332
8.2.1 The Time Value of Money ... 332
8.2.2 Present Worth Factors and Present Worth ... 333
8.2.3 Life-Cycle Cost ... 336
8.2.4 Annualized Life-Cycle Cost ... 340
8.2.5 Unit Electrical Cost ... 341
8.3 Borrowing Money ... 341
8.3.1 Introduction ... 341
8.3.2 Determination of Annual Payments on Borrowed Money ... 341
8.3.3 The Effect of Borrowing on Life-Cycle Cost ... 343
8.4 Payback Analysis ... 344
8.5 Externalities ... 347
8.5.1 Introduction ... 347
8.5.2 Subsidies ... 348
8.5.3 Externalities and Photovoltaics ... 348
References ... 350
Suggested Reading ... 350

Chapter 9
Externalities and Photovoltaics ... 351

9.1 Introduction ... 351
9.2 Externalities ... 352
9.3 Environmental Effects of Energy Sources ... 353
9.3.1 Introduction ... 353
9.3.2 Air Pollution ... 353
9.3.2.1 The Clean Air Act and the U.S. Environmental Protection Agency ... 353
9.3.2.2 Greenhouse Gases ... 356

9.3.3 Water and Soil Pollution 357
9.3.4 Infrastructure Degradation 358
9.3.5 Quantifying the Cost of Externalities 358
9.3.5.1 The Cost of CO_2 358
9.3.5.2 Sequestering CO_2 with Trees 359
9.3.5.3 The Clean Power Estimator 360
9.3.5.4 Attainment Levels as Commodities 361
9.3.5.5 Subsidies 361
9.3.6 Health and Safety as Externalities 362
9.4 Externalities Associated with PV Systems 363
9.4.1 Environmental Effects of PV System Production 363
9.4.2 Environmental Effects of PV System Deployment and Operation 364
9.4.3 Environmental Effects of PV System Decommissioning 365
References 366

Chapter 10
The Physics of Photovoltaic Cells 369

10.1 Introduction 369
10.2 Optical Absorption 369
10.2.1 Introduction 369
10.2.2 Semiconductor Materials 370
10.2.3 Generation of EHP by Photon Absorption 371
10.2.4 Photoconductors 374
10.3 Extrinsic Semiconductors and the pn Junction 376
10.3.1 Extrinsic Semiconductors 376
10.3.2 The pn Junction 377
10.3.2.1 Drift and Diffusion 377
10.3.2.2 Junction Formation and Built-In Potential 379
10.3.2.3 The Illuminated pn Junction 382
10.3.2.4 The Externally Biased pn Junction 383
10.4 Maximizing PV Cell Performance 386
10.4.1 Introduction 386
10.4.2 Minimizing the Reverse Saturation Current 386
10.4.3 Optimizing Photocurrent 387
10.4.3.1 Minimizing Reflection of Incident Photons 387
10.4.3.2 Maximizing Minority Carrier Diffusion Lengths 388
10.4.3.3 Maximizing Junction Width 390
10.4.3.4 Minimizing Surface Recombination Velocity 393
10.4.3.5 A Final Expression for the Photocurrent 394
10.4.4 Minimizing Cell Resistance Losses 396
10.5 Exotic Junctions 398
10.5.1 Introduction 398
10.5.2 Graded Junctions 398

10.5.3 Heterojunctions 399
10.5.4 Schottky Junctions 400
10.5.5 Multijunctions 402
10.5.6 Tunnel Junctions 403
References 405

Chapter 11
Present and Proposed PV Cells and Systems 407

11.1 Introduction 407
11.2 Silicon PV Cells 409
11.2.1 Production of Pure Silicon 409
11.2.2 Single-Crystal Silicon Cells 410
11.2.2.1 Fabrication of the Wafer 410
11.2.2.2 Fabrication of the Junction 411
11.2.2.3 Contacts 413
11.2.2.4 Antireflective Coating 416
11.2.2.5 Modules 416
11.2.3 A High-Efficiency Si Cell with All Contacts on the Back 417
11.2.4 Multicrystalline Silicon Cells 418
11.2.5 Other Thin Silicon Cells 419
11.2.6 Amorphous Silicon Cells 420
11.2.6.1 Fabrication 420
11.2.6.2 Cell Performance 422
11.3 Gallium Arsenide Cells 423
11.3.1 Introduction 423
11.3.2 Production of Pure Cell Components 423
11.3.2.1 Gallium 423
11.3.2.2 Arsenic 424
11.3.2.3 Germanium 424
11.3.3 Fabrication of the Gallium Arsenide Cell 425
11.3.4 Cell Performance 427
11.4 Copper Indium (Gallium) Diselenide Cells 428
11.4.1 Introduction 428
11.4.2 Production of Pure Cell Components 428
11.4.2.1 Copper 428
11.4.2.2 Indium 429
11.4.2.3 Selenium 430
11.4.2.4 Cadmium 431
11.4.2.5 Sulfur 431
11.4.2.6 Molybdenum 432
11.4.3 Fabrication of the CIS Cell 432
11.4.4 Cell Performance 433
11.4.4.1 Transient Effects 434

11.5 Cadmium Telluride Cells 435
11.5.1 Introduction 435
11.5.2 Production of Pure Tellurium 435
11.5.3 Production of the CdTe Cell 436
11.5.4 Cell Performance 437
11.6 Emerging Technologies 438
11.6.1 New Developments in Silicon Technology 438
11.6.2 CIS-Family-Based Absorbers 440
11.6.3 Other III-V and II-VI Emerging Technologies 441
11.6.4 Other Technologies 442
11.6.4.1 Thermophotovoltaic Cells 442
11.6.4.2 Intermediate Band Solar Cells 442
11.6.4.3 Supertandem Cells 442
11.6.4.4 Hot Carrier Cells 442
11.6.4.5 Optical Up- and Down-Conversion 443
11.6.4.6 Organic PV Cells 443
11.7 New Developments in System Design 443
11.7.1 Micro Grids 443
11.7.2 Smart Grids 444
11.7.3 Inverter Performance Enhancement 445
11.8 Summary 445
References 446

Appendix A
Average Daily Irradiation for Selected Cities 449

Appendix B
Design Review Checklist 461

Index 463

Preface

The goal of this textbook is to present a comprehensive engineering basis for photovoltaic (PV) system design so engineers can understand the *what*, *why*, and *how* associated with the electrical, mechanical, economic, and aesthetic aspects of PV system design. The book is intended to educate the engineer in the design of PV systems so that when engineering judgment is needed, the engineer will be able to make intelligent decisions based upon a clear understanding of the parameters involved. This goal differentiates this textbook from the many design and installation manuals currently available that train the reader how to do it, but not why.

Widespread acceptance of the first two editions, coupled with significant growth and new ideas in the PV industry, along with five additional years of experience with PV system design and installation for the authors, and, for that matter, a bit of nudging from the publisher, have led to the publication of this third edition. It should be noted that the U. S. Department of Energy now estimates a factor of 14 increase in grid-connected systems between 2009 and 2017, depending upon various factors such as incentives for renewables and availability and price of conventional fuels. So, who knows? There might even be jobs waiting out there for many who learn the material in this book.

The *what* question is addressed in the first three chapters, which present an updated background of energy production and consumption, some mathematical background for understanding energy supply and demand, a summary of the solar spectrum, how to locate the sun, and how to optimize the capture of its energy, as well as the various components that are used in PV systems. A section on the analemma has been added to Chapter 2, and Chapter 3 has been shortened a bit so actual design work can begin by about the sixth week of a three-credit course.

The *why* and *how* questions are dealt with in the remaining chapters in which every effort is made to explain why certain PV designs are done in certain ways, as well as how the design process is implemented. Included in the *why* part of the PV design criteria are economic and environmental issues that are discussed in Chapters 8 and 9. Chapter 5 has been embellished with additional practical considerations added to the theoretical background associated with mechanical and structural design. Chapters 4, 6, and 7 have been nearly completely reworked and rearranged to incorporate the most recently available technology and design and installation practice.

The rearranging of chapters acknowledges that straight grid-connected systems now dominate the market. Coincidentally, straight grid-connected systems are also easier to design, so Chapter 4 is now dedicated entirely to these. It is also designed so instructors can assign relevant design problems earlier in the course to avoid having designs due at final exam time. We have found that when significant design projects are due just before final exams the students tend to spend time on design projects that might better be spent studying for the final exam, especially if the final exam counts more toward the overall course grade. Paradoxically, the better designers end up with lower course grades.

Appendix A has been retained, even though Web references are now available that will yield data based on any specified PV array orientation, rather than the five orientations listed. We have opted to leave out the appendix on Web sites, because we have found that they become obsolete even more quickly than equipment. Appendix B presents a recommended format for submittal of a PV design package for permitting or for design review that has been widely used in Florida by the authors.

A modified top-down approach is used in the presentation of the material. The material is organized to present a relatively quick exposure to all of the building blocks of the PV system, followed by design, design, and design. Even the physics of PV cells of Chapter 10 and the material on present and future cells of Chapter 11 are presented with a design flavor. The focus is on adjusting the parameters of PV cells to optimize their performance, as well as on presenting the physical basis of PV cell operation.

Homework problems are incorporated that require both analysis and design, since the ability to perform analysis is the precursor to being able to understand how to implement good design. Many of the problems have multiple answers, such as "Calculate the number of daylight hours on the day you were born in the city of your birth." We have eliminated a few homework problems based on old technology and added a number of new problems based on contemporary technology. Hopefully, there is a sufficient number to enable students to test their understanding of the material.

We recommend that the course be presented so that by the end of Chapter 4, students will be able to engage in a comprehensive design project, and by the end of Chapter 7, they will be able to design relatively complex systems. We like to assign two design projects—a straight grid-connected system based on Chapter 4 material and a battery-backup grid-connected system based on Chapter 6 material. At the discretion of the instructor, an additional design of a stand-alone system might be considered at the end of Chapter 7, or the stand-alone design might be assigned rather than the battery-backup grid-connected system.

While it is possible to cover all the material in this textbook in a three-credit semester course, it may be necessary to skim over some of the topics. This is where the discretion of the instructor enters the picture. For example, each of the design examples of Chapter 4 introduces something new, but a few examples might be left as exercises for the reader with a preface by the instructor as to what is new in the example. Alternatively, by summarizing the old material in each example and then focusing on the new material, the *why* of the new concepts can be emphasized.

The order of presentation of the material actually seems to foster a genuine reader interest in the relevance and importance of the material. Subject matter covers a wide range of topics, from chemistry to circuit analysis to electronics, solid state device theory, and economics. The material is presented at a level that can best be understood by those who have reached upper division at the engineering undergraduate level and have also completed coursework in circuits and in electronics.

We recognize that the movement to reduce credit toward the bachelor's degree has left many programs with less flexibility in the selection of undergraduate elective courses, and note that the material in this textbook can also be used for a beginning graduate level course.

While the primary purpose of this material is for classroom use, with an emphasis on the electrical components of PV systems, we have endeavored to present the material in a manner sufficiently comprehensive that it will also serve the practicing engineer as a useful reference book.

One of the authors has twice taught the course as an Internet course using the first edition of the book. Those students who were actually sufficiently motivated to keep up with the course generally reported that they found the text to be very readable and a reasonable replacement for lectures. We highly recommend that if the Internet is tried, that quizzes be given frequently to coerce the student into feeling that this course is just as important as her or his linear systems analysis course. Informal discussion sessions can also be useful in this regard.

The photovoltaic field is evolving rapidly. While every effort has been made to present contemporary material in this work, the fact that it has evolved over a period of a year almost guarantees that by the time it is adopted, some of the material will be outdated. For the engineer who wishes to remain current in the field, many of the references and Web sites listed will keep him/her up to date. Proceedings of the many PV conferences, symposia, and workshops, along with manufacturers' data, are especially helpful.

This textbook should provide the engineer with the intellectual tools needed for understanding new technologies and new ideas in this rapidly emerging field. The authors hope that at least one in every 4.6837 students will make his or her own contribution to the PV knowledge pool.

We apologize at the outset for the occasional presentation of information that may be considered to be practical or, perhaps, even interesting or useful. We fully recognize that engineering students expect the material in engineering courses to be of a highly theoretical nature with little apparent practical application. We have made every effort to incorporate heavy theory to satisfy this appetite whenever possible.

Additional materials (updates) will be available at: http://www.crcpress.com/product/isbn/9781439802922.

Acknowledgments

We are convinced that it is virtually impossible to undertake and complete a project such as this without the encouragement, guidance, and assistance from a host of friends, family, and colleagues.

For this edition, we again asked many questions of many people as we rounded up information for the wide range of topics contained herein. A wealth of information flowed our way from the National Renewable Energy Laboratory (NREL) and Sandia National Laboratories (SNL), as well as from many manufacturers and distributors of a diverse range of PV system components, including representatives of SolarWorld, SunPower, Xantrex, Outback Power Systems, SatCon, and Direct Power and Water. We feel confident that had we asked, we would have been able to obtain additional cooperation and materials from many more. And, of course, student feedback on the second edition provided significant insight to the authors on how to make the material easier to understand. We hope we have accomplished this goal.

Special thanks to Kimandy Lawrence of VB Engineering for the many discussions held during and after normal work hours on various nuances associated with certain systems, including, but not limited to, ac-coupled battery-backup systems, microinverter systems, and other experimental designs. Also, special thanks to Christian Smith and Tom Bennett for their work on solving many of the new homework problems.

Last, but not least, we thank Nancy Ventre and Jane Caputi for encouraging us to work on this manuscript rather than play golf or tennis or watch TV.

Roger Messenger
Jerry Ventre

About the Authors

Roger Messenger, Ph.D., is professor emeritus of electrical engineering at Florida Atlantic University in Boca Raton. He received his Ph.D. in electrical engineering from the University of Minnesota and is a Registered Professional Engineer, a Certified Electrical Contractor, and an NABCEP Certified PV Installer, who enjoys working on a field installation as much as he enjoys teaching a class or working on the design of a system or contemplating the theory of operation of a system. His research work has ranged from electrical noise in gas discharge tubes to deep impurities in silicon to energy conservation. He worked on the development and promulgation of the original Code for Energy Efficiency in Building Construction in Florida and has conducted extensive field studies of energy consumption and conservation in buildings and swimming pools.

Since his retirement from Florida Atlantic University in 2005, Dr. Messenger has worked as vice president for engineering at VB Engineering, Inc., in Boca Raton. The VB Engineering team has completed several hundred PV designs to date, including the design of the 5808-module, 4-acre, 1 MW system on the roof of the Orange County Convention Center in Orlando, Florida. He has also been active in the Florida Solar Energy Industries Association and the Florida Alliance for Renewable Energy, has served as a peer reviewer for the U.S. Department of Energy, and has sat on the Florida Solar Energy Center Policy Advisory Board. He has conducted numerous seminars on designing, installing, and inspecting PV systems.

Jerry Ventre, Ph.D., is a consultant in photovoltaic systems engineering, specializing in system design, product certification, and workforce development. For 25 years, he led the photovoltaic and distributed power programs at the Florida Solar Energy Center, a research institute of the University of Central Florida. During that time, he also managed the Photovoltaic Southeast Regional Experimentation Station for the U.S. Department of Energy. He received his B.S., M.S., and Ph.D. in aerospace engineering from the University of Cincinnati, and has more than 35 years of experience in research, development, design, and systems analysis. He served on the aerospace engineering faculties of both the University of Cincinnati and the University of Central Florida, is a Registered Professional Engineer, and, among many courses, he has taught photovoltaic systems at the graduate level. He has designed solid rocket motors and jet engines for the Advanced Engine Technology Department of the General Electric Company, and has performed research for numerous agencies, including NASA, the National Renewable Energy Laboratory, Sandia National Laboratories, Oak Ridge National Laboratory, the U.S. Navy, the FAA, and the U.S. Department of Energy. He has over 150 technical publications, has been active in technical societies, and has been the recipient of a number of awards for contributions to engineering and engineering education.

CHAPTER 1

Background

1.1 INTRODUCTION

On June 26, 1997, the United States Million Solar Roofs Initiative (MSRI) was announced at the United Nations Special Session on Environment and Development in New York [1]. It was proclaimed that "now we will work with businesses and communities to use the sun's energy to reduce our reliance on fossil fuels by installing solar panels on 1 million more roofs around our nation by 2010. Capturing the sun's warmth can help us to turn down the Earth's temperature."

In 1997, little concern was being expressed in the public sector about energy problems and there was perhaps even less discussion related to global warming. Now, 12 years later as this chapter goes to press, both topics are receiving significant attention. By the end of 2007, it is estimated that more than 600,000 solar systems had been installed around the country [2], and the State of California had introduced its own Million Solar Roofs Initiative [3]. The 600,000 installations included photovoltaic, solar water heating, and solar pool heating systems. But installations have not been limited to the United States. Germany leads the world in installed solar and wind energy systems, and Japan, Spain, and many other countries also have aggressive renewable energy programs [2,3].

More and more states are now adopting renewable portfolio standards (RPS), which set goals for the percentage of the electrical energy mix to be provided by renewable sources by a certain date. Implementation of these goals will require engineers who understand photovoltaic systems in order to be able to make intelligent system design choices. Renewable portfolio standards, coupled with carefully devised incentive programs, should provide the momentum for a sustained effort in the deployment of solar technologies well beyond the year 2010. In fact, the MSRI may need to be extended to a 100 Million Solar Roofs Initiative to meet the sustainable energy needs of future generations. This book is dedicated to the engineers and technicians who have been, and may become, involved in turning this dream into reality.

1.2 POPULATION AND ENERGY DEMAND

The human population of the earth has now passed 6.7 billion [4], and all of these inhabitants want the energy necessary to sustain their lives. Exactly how much energy is required to meet these needs and exactly what sources of energy will meet these needs will be questions to be addressed by the present and future generations. One certainty, however, is that developing nations will be attempting to increase their per capita energy use significantly. For example, in 1997, the People's Republic of China was building electrical generating plants at the rate of 300 MW per week. These plants used relatively inexpensive, old, inefficient, coal-fired technology and provided electricity to predominantly inefficient end uses [5]. By 2008, pollution from factories and power plants in China had become so intense that some Olympic athletes had expressed reluctance to compete in the 2008 Olympic Games in Beijing for health reasons. The potential consequences to the planet of continued increase in the use of fossil fuels are profound [6,7]. Before proceeding with the details of photovoltaic power systems, a promising source of energy for the future, it is instructive to look at the current technical and economic energy picture. This look will enable the reader to better assess the contributions that engineers will need to make toward a sustainable energy future for the planet.

1.3 ENERGY UNITS

Energy is measured in a number of ways, including the calorie, the Btu, the quad, the foot-pound (ft-lb), and the kilowatt hour (kWh). For the benefit of those who may not have memorized the appendices of their freshman physics books, we repeat the definitions of these quantities for an earth-based system at or about a temperature of 27°C [8]:

1 calorie is the heat needed to raise the temperature of 1 mL of water by 1°C.
1 Btu is the heat needed to raise the temperature of 1 lb of water by 1°F.
1 quad is 1 quadrillion (10^{15}) Btu.
1 ft-lb is the energy expended in raising 1 lb through a distance of 1 ft.
1 kWh is the energy expended by 1 kW operating for 1 h.

With these definitions, the following equivalencies can be determined:

1 Btu = 252 calories
1 kWh = 3413 Btu = 2,655,000 ft-lb
1 ft-lb = 0.001285 Btu
1 quad = 10^{15} Btu = 2.930×10^{11} kWh

Because the emphasis of this text will be on electrical generation, and because the kWh is the common unit for electrical energy, the equivalence between kWh and ft-lb is especially noteworthy. For example, suppose a 150 lb person wished to generate 1 kWh, assuming a system with 100% efficiency. One way would be to climb to

Figure 1.1 Three ways to produce 1 kWh of electricity.

the top of a 17,700 ft mountain to create 1 kWh of potential energy. Then, by returning to sea level by way of a chair, connected via a pulley system to a generator, the person's potential energy could be converted to electrical energy. This kWh could then be sold at wholesale for about 6 cents. Another somewhat simpler method is to burn approximately 11 fluid ounces of petroleum to produce steam to turn a steam turbine as shown in Figure 1.1.

Still another method is to deploy about 2 m^2 of photovoltaic (PV) cells. This system will produce about 1 kWh per day for 25 years or more with no stops for refueling, no noise, minimal maintenance, and no release of CO_2, SO_2, or NO_2 while the electricity is being produced.

1.4 CURRENT WORLD ENERGY USE PATTERNS

Figure 1.2 shows the increase in worldwide energy production by source since 1980, with projections through 2020 [9]. In 2000, worldwide annual primary energy consumption was 397.40 quads [10]. The developed countries of the world consumed approximately 75% of this energy, while nearly 2 billion people in developing countries, mostly within the tropics, remained without electricity. In 2005, worldwide primary energy consumption increased to 462 quads [9], and 2 billion people still lived without electricity.

The petroleum curve in Figure 1.2 shows how price can affect energy consumption. It also shows that there can be a time delay between market forces and market responses. Note that the production of petroleum continues upward after the 1973 oil

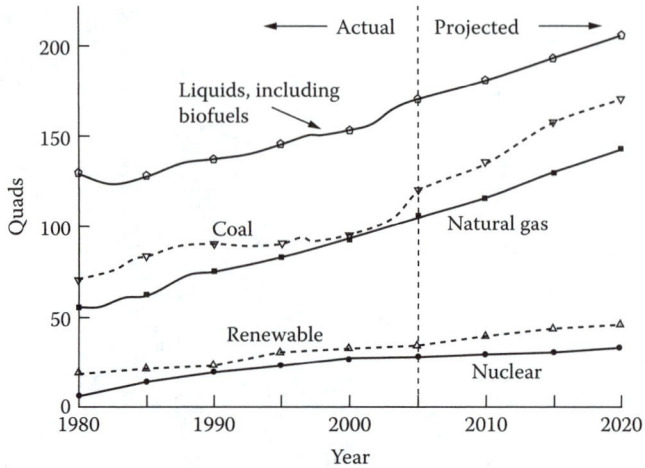

Figure 1.2 Growth of worldwide energy production since 1980, by source. (Based on data from U.S. Department of Energy, Energy Information Administration, Washington, D.C., International Energy Outlook 2008, September 2008, www.eia.doe.gov/oaif/ieo/index/html.)

embargo and the subsequent significant petroleum price increases during the remaining 1970s and early 1980s. During this period, high petroleum prices spurred the development of energy efficiency legislation, such as the National Energy Conservation and Policy Act, codes for energy efficiency in building construction, and increased vehicle fleet mileage requirements. Consumers also responded by reducing energy use by lowering thermostats and installing insulation and other energy conservation measures. The result was lower petroleum production for a period in the mid-1980s, because the demand was lower. During this same period, more efficient use of electricity led to the cancellation of nuclear plant construction, resulting in a significant decrease in the growth rate of nuclear-produced electricity. Finally, concern over oil price control and embargoes prompted a switch from petroleum to coal and natural gas for use in fossil-fired electrical generation. Figure 1.3 shows the global mix of energy sources in the year 2005 [9]. The "renewables" category includes sources such as hydroelectric, wind, biomass, geothermal, and photovoltaics.

Figures 1.4 and 1.5 illustrate that the world faces a challenge of mammoth proportions as developing countries strive to achieve energy equity with the developed countries [9,11]. Note that energy equity is simply another term for the attempt to achieve comparable standards of living. But achieving a higher standard of living can carry with it a price. The price includes not only monetary obligations, but also the potential for significant environmental degradation if energy equity is pursued via the least expensive, first-cost options. Regrettably, this is the most probable scenario, because it is already under way in regions such as Eastern Europe and Asia. In fact, it is probably more likely that use of least-cost energy options may lead to comparable per capita energy use but may simultaneously degrade the standard of

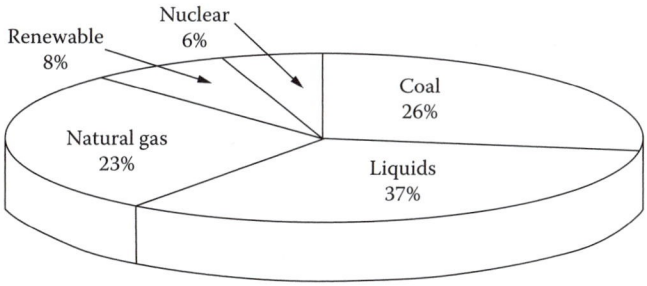

Figure 1.3 Global energy production mix by fuel type in the year 2005. (Based on data from U.S. Department of Energy, Energy Information Administration, Washington, D.C., International Energy Outlook 2008, September 2008, www.eia.doe.gov/oaif/ieo/index/html.)

living by producing air not suitable for breathing and water not suitable for drinking. These issues will be dealt with in more detail in Chapter 9.

Figure 1.4 shows that North America, with slightly more than 5% of the world population, consumes 26% of the world's energy, while China and India, with nearly one third of the world's population, consume 18% of the world's energy [9]. What is missing in Figure 1.4 is the efficiency with which the energy is consumed in these regions.

Figure 1.5 is particularly interesting because it clearly shows examples of countries where energy is used least and most efficiently [11]. Those countries with the highest ratios of per capita GDP to per capita kW have developed more energy-efficient economies than those with low ratios. Some have shifted from predominantly manufacturing economies to predominantly information economies, with the result that a smaller fraction of energy is used for manufacturing. Countries at the low end of the scale tend to be farming oriented, with the use of mostly manual labor, or manufacturing oriented, with very low wages for workers and low first-cost energy sources.

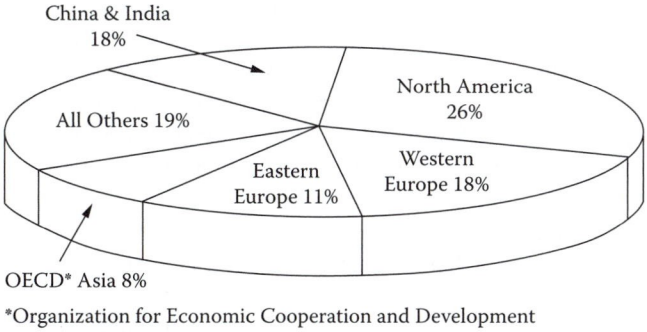

Figure 1.4 Distribution of energy users in the year 2005. (Based on data from Van Mierlo, Energy Consumption vs. GDP, in World Energy Resources and Consumption, *Wikipedia, the free encyclopedia*, http://en.wikipedia.org/wiki/World_energy_resources_and_consumption.)

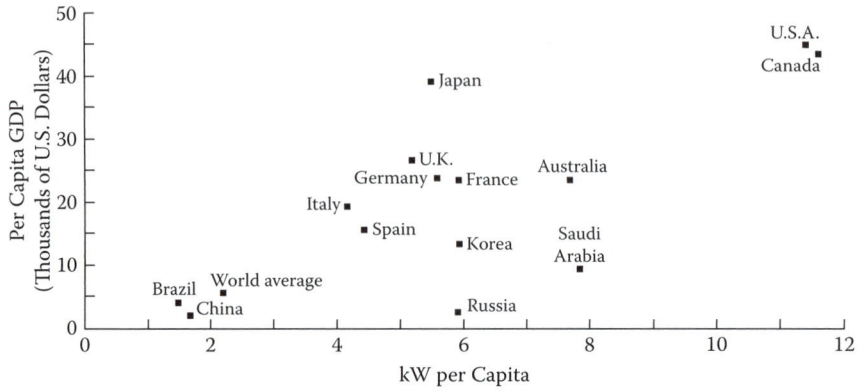

Figure 1.5 Worldwide per capita GDP versus per capita kW for the year 2005. (Based on data from Van Mierlo, Energy Consumption vs. GDP, in World Energy Resources and Consumption, *Wikipedia, the free encyclopedia*, http://en.wikipedia.org/wiki/World_energy_resources_and_consumption.)

As developing countries increase their manufacturing capabilities, using cheap but polluting local energy sources, there may be pressure to relax pollution control standards within the developed world in order to maintain the ability to compete with production from developing countries. Opposition to free trade agreements has been partially based on such environmental concerns, because such agreements forbid any country to impose import tariffs on goods produced in countries having weak environmental standards. So the developed countries are challenged to export efficiency to the developing countries. In fact, one of the arguments the United States has made against the signing of the Kyoto Treaty on the reduction of greenhouse gases is that it does not require equal restrictions on greenhouse gas emissions in developing countries. The reasoning here is that it puts the United States at an economic disadvantage, but in Chapter 9, the argument will be made that efficient use of energy, with minimal environmental degradation, leads to economic advantage.

But what does this discussion have to do with photovoltaic power production? Simple. As will be shown in Chapter 9, photovoltaic energy sources are very clean, but current photovoltaic deployment costs cannot compete with the initial installed direct costs of fossil sources of electrical generation in most cases. It means that the consumer must be familiar with life-cycle costing and that the engineer must be able to create the most cost-effective photovoltaic solution. It also means that a significant amount of research and development must be done to ensure the continuation of the decrease in the price of photovoltaic generation. It also means that work must continue in the effort to put a price tag on environmental degradation caused by energy sources, so this price can be factored into the total cost to society of any energy source.

The bottom line is that there remains a significant amount of work in research, development, and public education to be done in the energy field, and particularly in the field of photovoltaics. And much of this work can best be done by knowledgeable engineers who, for example, understand the concept of exponential growth.

1.5 EXPONENTIAL GROWTH

1.5.1 Introduction

Exponential growth is probably most familiar to the electrical engineer in the diode equation, in which I and V are the diode current and voltage, I_0 is the reverse saturation current, and kT/q is 26 mV at $T = 300$ K, or, specifically,

$$I = I_0(e^{\frac{qV}{KT}} - 1). \tag{1.1}$$

While this equation is fundamental to the performance of photovoltaic cells, many other physical processes are also characterized by exponential growth.

Exponential growth is commonly referred to as compound interest. Almost everyone has heard about it, but few understand the ramifications of constant annual percentage increase in a quantity, whether it be money, population, or energy supply or demand. One of the first to warn of the dangers of exponential growth was Malthus, in 1798 [12]. He warned that population growth would exceed the ability to feed the population. The Malthusian theory is often subject to ridicule by growth enthusiasts [13]. The intent of this discussion is neither to support nor discount the predictions of Malthus, but merely to illustrate an important mathematical principle that engineers often overlook. The application of the principles of exponential growth is widespread in society, so the principles of exponential growth should be just as important to a well-informed engineer as the second law of thermodynamics. For those who may have missed the second law of thermodynamics in either chemistry or thermodynamics class, it states that in every process, less energy comes out than what is put in. In other words, there is no free lunch.

1.5.2 Compound Interest

Compound interest is the process of compounding simple interest. If a quantity N_0 is subject to an interest rate i, with i expressed as a fraction (i.e., $i = \%/100$), the quantity will increase (or decrease, if $i < 0$) to a value of

$$N(1) = N_o\,(1 + i) \tag{1.2}$$

after one prescribed time period has elapsed. If the quantity present after the prescribed time period is allowed to remain and to continue to accumulate at the same rate, then the quantity is subject to compound interest and the amount present after n time periods will be

$$N(n) = N_o\,(1 + i)^n. \tag{1.3}$$

To show that this formula is a form of the exponential function, one need only recall that

$$y^X = e^{x \ln y}.$$

Hence,

$$N(n) = N_o\, e^{n \ln(1+i)}. \tag{1.4}$$

Some special properties of the exponential function can now be considered.

1.5.3 Doubling Time

To determine the time, D, it will take for the original quantity to double, one need only set $N(n) = 2N_o$ and solve for n. The result (and the answer to Problem 1.1) is

$$n = \frac{\ln 2}{\ln(1+i)} = 2 \tag{1.5}$$

For small values of i, $\ln(1 + i)$ can be approximated as $\ln(1 + i) \cong i$. Noting that $\ln 2 = 0.693$ leads to the formula so popular in the financial world, that is,

$$D = n \cong 0.7/i. \tag{1.6}$$

Hence, for an interest rate of 7% per year ($i = 0.07$), the doubling time will be 10 years. For an interest rate of 10%, the doubling time will be approximately 7 years. However, as the interest rate exceeds 10%, the approximation becomes less valid, and the exact expression should be used for accurate results. In the case where the interest rate is negative, it should be obvious that no doubling can occur. The authors have proved this to be the case with various investments in the stock market.

An important property of the exponential function is that the doubling process continues for all time. Hence, if the doubling time is 10 years, then the quantity will double again in another 10 years, so it will now be 4 times its original value. In another 10 years it will double again to 8 times its original value. After 40 years, the quantity will be 16 times its original value. Figure 1.6 shows this exponential increase.

Note that if the function $y = Ae^{bx} = A10^{bx\log e}$ is plotted with linear coordinates, the familiar exponential curve appears, as in Figure 1.6a. If the logarithm of each side is taken, the result is

$$\log y = \log A + (b \log e)x. \tag{1.7}$$

Hence, if $\log y$ is plotted as a function of x, the graph will be linear with slope $b \log e$, as shown in Figure 1.6b. Figure 1.6b shows that plotting $\log y$ versus x is a convenient

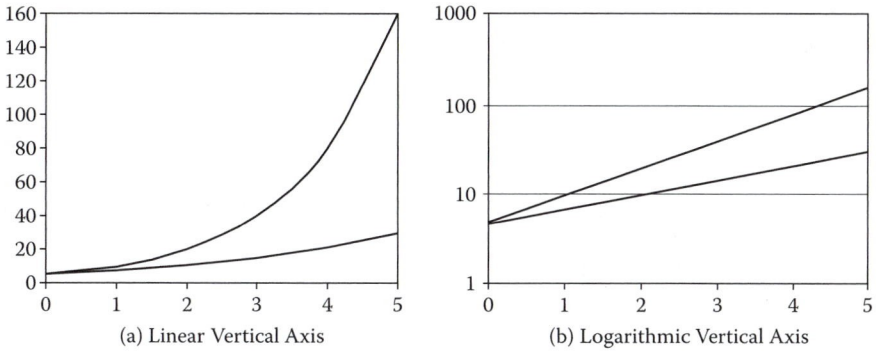

Figure 1.6 Examples of exponential functions plotted on linear and on semilogarithmic coordinates.

way to check for an exponential relationship between two variables. Note also that the vertical axis can be labeled either in terms of log y or, simply, in terms of y. When the axis is labeled in terms of y, it is understood that the axis is linear in the logarithm of y. For the person who prefers to let computers do the work, either Excel® or MATLAB® can conveniently plot data and produce a least mean square curve fit to the data along with an estimate of the goodness of the fit. When a set of data is available, all one needs to do is to list the independent and dependent variables side by side on an Excel spreadsheet and then use the Chart Wizard to plot an x-y scatter plot of the data. Selecting the chart option with the plot of the data but no connecting lines and following through to the placement of the chart on the spreadsheet, one can then click on the data points, go to "Chart" on the pull-down menu, and select "Add Trendline." This opens a dialog box that offers a choice of six trend/regression types. To obtain the best fit, select a regression type, select "Options," and choose to add the equation and the R^2 value to the graph. An R^2 value of 1 indicates a perfect fit of the data to the curve by the equation that is displayed. Low R^2 values suggest choosing a different regression type. Several good user guides for Excel are available for those who would like to further explore its use as a convenient analysis tool for use in photovoltaic system design or other technical endeavors [14,15]. MATLAB and other math programs are also powerful tools.

For the reader who has not had the pleasure of using Excel to find the least mean square fit to a data set, several problems are given at the end of this chapter. It is anticipated that the reader will use Excel frequently as a tool for the design of photovoltaic systems in later chapters.

1.5.4 Accumulation

Another important property of the exponential function is the amount of accumulation (or depletion) of a quantity in a doubling time. It is a straightforward math exercise to show that the amount accumulated (or depleted) in a doubling time is equal to the amount accumulated (or depleted) in all previous history (see problem 1).

The implications of this result are far-reaching. For example, a prominent political figure once noted that there was still as much oil underground in the United States as what had been pumped out since pumping first started more than 140 years ago. This was at a time when oil extraction was increasing at a rate of approximately 7% per year. If the extraction had continued to increase at 7% per year, which has a doubling time of approximately 10 years, in the next doubling time all of the remaining petroleum would have been extracted. Many other important public figures have made similar statements that tend to assign a linear nature to the exponential function [16]. Could this be an argument for engineers to run for public office?

In fact, extraction did not continue to increase at this rate. With regard to the use of resources, Hubbert [17] developed a model that incorporates a Gaussian function for depletion, which seems to have more validity than the exponential model. The rising edge of the Gaussian function, however, is conveniently approximated by an exponential.

The accumulation formula, of course, may also apply to the deployment of new technology. Since the early 1990s, photovoltaic power production has been increasing at a very impressive rate. Problems 1.15 and 1.16 offer an opportunity to explore the relevance of this observation if this rate of increase continues.

1.5.5 Resource Lifetime in an Exponential Environment

The previous discussion of exponential growth has been based on total amounts of a quantity at any given time. If the time derivative of the exponential expression for quantity is taken, the rate of change of the quantity is obtained. Because the derivative of an exponential is also an exponential, the same rules apply to the derivative as to the function. Distinguishing between amount present and rate of use (or increase) is important when determining the lifetime of a resource. Hence, when considering an exponential expression, one needs to establish whether it refers to barrels or barrels per day, or, perhaps, megawatts or megawatts per year of photovoltaic deployment.

The final concept to explore relating to the exponential function is the lifetime of a resource under conditions of exponential increase. It is common to predict the lifetime of a resource under the current rate of consumption. This involves simple arithmetic, because if there are Z (quantity) widgets left to use and if we use X (rate of use) of them per year, then the widgets will last for Y years, where $Y = Z/X$. But what happens to the expected lifetime of the widget if people decide that they really like widgets and they decide to use them at an increasing rate of $100i\%$ per year? This problem can be solved by assuming that C_0 represents the present rate of consumption of a resource and Y_0 represents the estimated lifetime of the resource at the present rate of consumption. Then, if consumption increases by $100i\%$ per year, the rate of consumption at any point in time, x, is given by

$$C(x) = C_o\,(1 + i)^x. \tag{1.8}$$

The accumulated consumption over a period of m years, TOT, can be found from previous formulas, or, more formally, by evaluating

$$TOT = \int_0^m C(x)dx = \int_0^m C_o e^{x\ln(1+i)}dx$$

$$= \frac{C_o}{\ln(1+i)}(e^{m\ln(1+i)} - 1).$$

Next, set the total to equal the estimated amount remaining (C_oY_o) and solve for m, because this will yield the number of years for the total consumption to equal the amount remaining. The result is

$$m = \frac{\ln[Y_o\ln(1+i)+1]}{\ln(1+i)}. \tag{1.9}$$

As an example of the use of this result, assume that a resource is estimated to last for another 100 years at present consumption rates ($Y_o = 100$), but that consumption will increase at the rate of 5% per year ($i = .05$). Using these numbers in the above formula gives $m = 36.31$ years. If the estimate is off by a factor of 10, and there is really a 1000-year supply left at current consumption rates, then $m = 80.09$ years.

As a perhaps more reassuring example of the use of this result, it is also possible that the consumption of a resource might decline at a constant percentage per year. This could happen if the resource were to be replaced by another resource, for example. For the previous example, with $Y_o = 100$ years and an annual decrease of 0.5% (i.e., $i = -0.005$), the new lifetime becomes 139 years, and if $i = -0.01$, the resource will last forever.

Hence, two important observations emerge from the lifetime formula:

1. If annual consumption of a resource increases exponentially, it is not important how much is thought to remain; it will be consumed much faster than one can imagine.
2. If annual consumption decreases exponentially, it is possible to extend the lifetime of a resource to forever.

1.6 HUBBERT'S GAUSSIAN MODEL

In 1956, M. King Hubbert, who was employed by the Shell Oil Company, published his now acclaimed theory of resource depletion [17]. Simply put, Hubbert reasoned that the life of a finite resource follows a Gaussian curve described by the equation, often referred to as the error function or normal curve,

$$R(t) = R_m e^{\frac{-(t-t_o)^2}{2s^2}}, \tag{1.10}$$

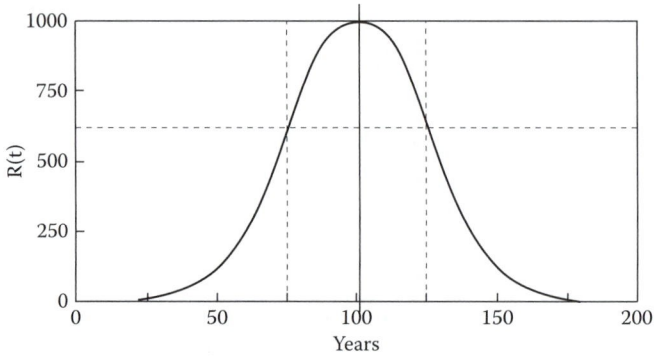

Figure 1.7 Gaussian function with $R_m = 1000$, $t_0 = 100$, and $s = 25$.

where

$R(t)$ represents the consumption rate at a given time, t,
R_m represents the maximum consumption rate, and
s represents a shape factor for the curve, commonly known as the standard deviation.

He also observed that production of petroleum lagged discovery by approximately 40 years. Figure 1.7 shows a plot of Equation (1.10), and Figure 1.8 shows Hubbert's 1971 curves relating to petroleum production. The rising edge represents a nearly exponential function, until it nears the peak of the curve, where leveling occurs, followed by nearly exponential decay. Because the curve of Figure 1.7 plots consumption rate, note that according to Hubbert's theory, when half a resource is consumed, the consumption will have leveled at its maximum value and then will begin its nearly exponential decline. According to Hubbert's model, domestic petroleum production in the United States would peak near 1970, and world petroleum production would peak near the turn of the century. Indeed, it appears that the domestic prediction has

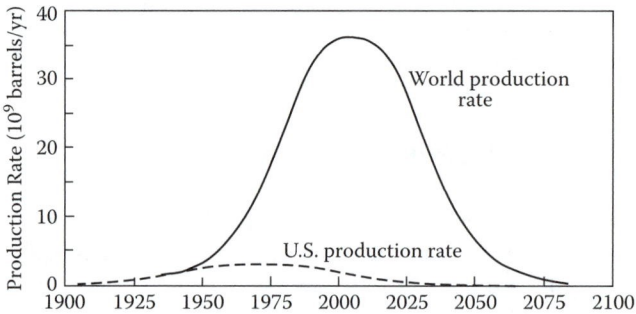

Figure 1.8 Hubbert's predictions for United States and world petroleum production. (Adapted from Hubbert, M.K. *Scientific Am.*, Vol. 225, No. 3, September 1971, 60–70. With permission.)

been confirmed. According to projections of the U.S. Department of Energy, Energy Information Agency, both domestic and Alaskan petroleum production will continue to decline through the year 2015 [18].

Interestingly enough, it also turns out with the Hubbert theory that if an error is made in estimating the amount of a resource present, the peak of the curve will not be shifted by a significant amount. Perhaps the most significant conclusion of the Hubbert theory, however, is that the consumption of a resource follows a smooth curve rather than an abrupt one that involves unabated consumption until suddenly no more is left. Hopefully, approaching the peak sends a message to find a replacement.

1.7 NET ENERGY, BTU ECONOMICS, AND THE TEST FOR SUSTAINABILITY

The net energy associated with an energy source is simply the difference between the energy required to obtain and convert the source to useful energy and the actual useful energy obtained from the source. For example, in order to be able to burn a barrel of oil, it is necessary to find the oil, extract it, transport it, refine it, and construct the facility for burning it. The refined oil must then be transported to the burning site and, presumably, after the oil is burned, any environmental damage resulting from the extraction, transportation, refining, and burning should be repaired. Energy is involved in all of these steps. The bottom line is that if it takes more than a barrel of oil worth of energy to obtain and convert the energy available in the barrel, one should question seriously whether it makes sense to burn the oil in the first place.

In some cases, it may make sense to expend the energy to get the resource. Suppose, for example, that another use were discovered for oil, such as providing an essential chemical for the cure of cancer. Then it would make sense to expend energy from sources other than oil, even if the energy exceeded the energy value of the oil, in order to make the oil available for the more important use. Another situation would be to use a lower-grade or lower-quality form of energy to produce a higher-grade or higher-quality form of energy. Sometimes such an action might make energy sense.

For example, burning coal to produce electricity takes about three units of coal energy to produce one unit of electrical energy. Until television sets that run directly from coal are invented, this inefficient process of converting coal energy to electrical energy will probably continue, in spite of the known environmental problems with coal.

The concept of net energy was introduced by Odum and Odum in 1976 [19,20]. They incorporated the net energy concept into a new standard for economics that they felt made better sense than the gold standard. They called it the Btu standard. The Btu standard recognizes that everything has an energy content. Henderson [21] has written extensively on the concept of Btu economics. The reader is encouraged to read both Odum and Odum and Henderson during a term break for enlightening discussions of how the economic system might be changed to an energy-based standard.

For the purposes of this book, the test for sustainability for an energy source will include two factors. The first will be whether the source is finite. A finite source is generally termed *nonrenewable*, while an infinite source is termed *renewable*. The second will be whether the source has positive net energy. That is, once energy is expended to produce the source, will the source then generate more energy than was required for its production?

The idea that a source can produce more energy than was used to create the source may seem inconsistent with the second law of thermodynamics. However, if we allow the use of energy from a very large reservoir as a supply of energy to be converted by the source, then the source becomes nearly infinite. In the case of the sun, which is expected to survive for another 4 billion years or so [22], we have such a reservoir. Thus, for example, if a photovoltaic cell can generate more electrical energy over its lifetime than was expended in its production and deployment and ultimately in its disposal, including environmental energy costs, then the cell would be considered to have positive net energy.

The concept of net energy will be considered in the context of photovoltaic cell production and in discussion of environmental effects of energy sources.

1.8 DIRECT CONVERSION OF SUNLIGHT TO ELECTRICITY WITH PHOTOVOLTAICS

Becquerel [23] first discovered that sunlight can be converted directly into electricity in 1839, when he observed the photogalvanic effect. Then, in 1876, Adams and Day found that selenium has photovoltaic properties. When Planck postulated the quantum nature of light in 1900, the door was opened for other scientists to build on this theory. It was in 1930 that Wilson proposed the quantum theory of solids, providing a theoretical linkage between the photon and the properties of solids. Ten years later, Mott and Schottky developed the theory of the solid-state diode, and in 1949, Bardeen, Brattain, and Shockley invented the bipolar transistor. This invention, of course, revolutionized the world of solid-state devices. The first solar cell, developed by Chapin, Fuller, and Pearson, followed in 1954. It had an efficiency of 6%. Only 4 years later, the first solar cells were used on the Vanguard I orbiting satellite.

One might wonder why it took so long to develop the photovoltaic cell. The answer lies in the difficulty in producing sufficiently pure materials to obtain a reasonable level of cell efficiency. Prior to the development of the bipolar transistor and the advent of the space program, there was little impetus for concentrating on preparing highly pure semiconductor materials. Coal and oil were meeting the world's need for electricity, and vacuum tubes were meeting the needs of the electronics industry. However, because vacuum tubes and conventional power sources were impractical for space use, solid-state technology gained its foothold.

Photovoltaic cells are made of semiconductor materials and are assembled into modules of 36 or more cells. This observation is significant, because it means the same industry that has, in the past 50 years, progressed from the development of the bipolar transistor to integrated circuits containing millions of transistors, is also involved in the development of photovoltaic cells. Figure 1.9 shows the decline in cost of photovoltaic modules since 1975. Much of the initial cost reduction has been due to process improvement in the production of the cells. At this point, the limiting

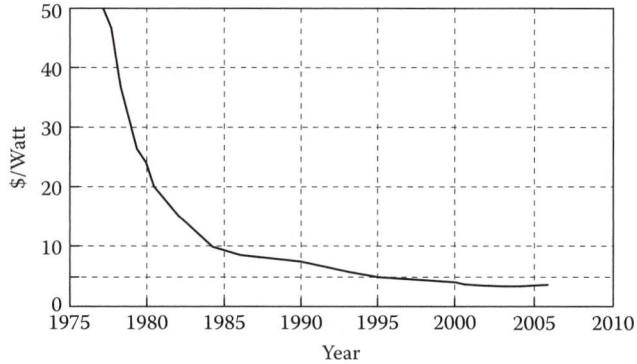

Figure 1.9 The decline in cost per watt for photovoltaic modules. (Based on data from Maycock, Ed., *Photovoltaic News*, Vol. 18, No 1, January 1999; *International Marketing Data and Statistics,* 22nd ed., Euromonitor PIC, London, 1998; Maycock, *Renewable Energy World*, July–August 2002, 147–161; Earth Policy Institute, Eco-Economy Indicators: SOLAR POWER—Data www.earth-policy.org/Indicators/Solar/2007_data.htm.)

factor is becoming the energy cost of the cells combined with a worldwide shortage of refined silicon. Hence, the challenge of the future will be to reduce the energy content of the cell production process while maintaining or increasing cell performance, efficiency, and reliability.

Figure 1.10 shows world photovoltaic shipments in megawatts from 1975 to 2006. Note that the data are plotted on semilogarithmic coordinates. The actual data since

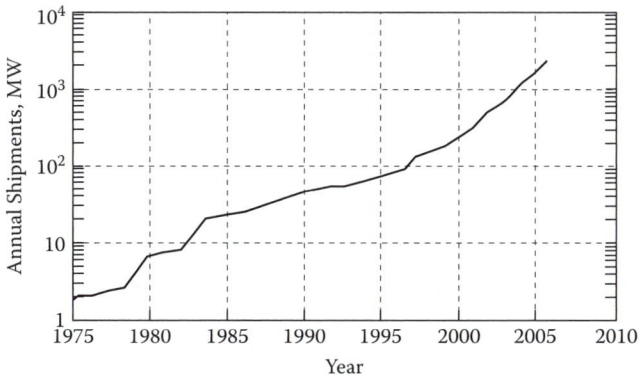

Figure 1.10 Worldwide photovoltaic shipments, 1975–2006. (Based on data from *International Marketing Data and Statistics,* 22nd ed., Euromonitor PIC, London, 1998; Maycock, *Renewable Energy World*, July–August 2002, 147–161; Earth Policy Institute, Eco-Economy Indicators: SOLAR POWER—Data www.earth-policy.org/Indicators/Solar/2007_data.htm; Maycock, *PV News*, Vol. 22, No. 3, March 2003.)

1995 are given in Problem 1.17 so the reader can generate a plot and determine the goodness of fit and the rate of growth over this period. A further important observation is that in 1995, 45% of the world's photovoltaic modules were manufactured in the United States, while Europe, Japan, and the rest of the world manufactured 80% of the world's photovoltaic modules in 2002 [26] and 93% of the world's photovoltaic cells in 2007 [29]. Just as the United States has allowed the manufacture of consumer electronics to transfer to other countries, it appears that it is also allowing the same to happen with the photovoltaic industry. It will be interesting to observe this trend in the future.

As an initial test as to whether to continue with this book, it is useful to determine whether the net energy associated with photovoltaic cells is positive. Indeed, current photovoltaic cells can produce more than a 10:1 return on the energy invested in their production, and future improvements in production technology and practice may result in values up to 45:1 [30]. Hence, it appears to be worth investigating this technology in more detail.

Problems

1.1 Prove Equation (1.5).

1.2 Calculate the approximate and exact doubling times for annual percentage increases of 5%, 10%, 15%, and 20%.

1.3 The human population of the earth is approximately 7 billion and is increasing at approximately 2% per year. The diameter of the earth is approximately 8000 miles and the surface of the earth is approximately two-thirds water. Calculate the population doubling time, and then set up a spreadsheet that will show (a) the population, (b) the number of square feet of land area per person, and (c) the length of the side of a square that will produce the required area per person. Carry out the spreadsheet for 15 doubling times, assuming that the rate of population increase remains constant. What conclusions can you draw from this exercise?

1.4 Show that in an exponential growth scenario, the amount accumulated in a doubling time equals the amount accumulated in all previous history.

1.5 Assume there is enough coal left to last for another 300 years at current consumption rates.

a. Determine how long the coal will last if its use is increased at a rate of 5% per year.

b. If there is enough coal to last for 10,000 years at current consumption rates, then how long will it last if its use increases by 5% per year?

c. Can you predict any other possible consequences if coal burning increases at 5% per year for the short or long term?

d. Determine the annual percentage reduction in coal consumption to ensure that coal will last forever, assuming the 300-year lifetime at present consumption rates.

1.6 The half-life is the time it takes a substance to decay exponentially to half the original amount. If the half-life of a radioactive isotope is 500 years, how many years will it take for an amount of the isotope to decay to 1% of its original value? Assume the isotope decays exponentially.

1.7 If a colony of bacteria lives in a jar and doubles in number every day, and it takes 30 days to fill the jar with bacteria:

a. How long does it take for the jar to be half full?

b. How long before the bacteria notice they have a problem? (You may want to pretend you are a bacterium.)

c. If on the 30th day, three more jars are found, how much longer will the colony be able to continue to multiply at its present rate?

1.8 An enterprising young engineer enters an interesting salary agreement with an employer. She agrees to work for a penny the first day, 2 cents the second, 4 cents the third, and so on, each day doubling the amount of the previous day. Set up a spreadsheet that will show her daily and cumulative earnings for her first 30 days of employment.

1.9 Burning a gallon of petroleum produces approximately 25 lb of carbon dioxide, and burning a ton of coal produces approximately 7000 lb of carbon dioxide.

a. If a barrel of petroleum contains 42 gal, if the world consumes 80 million barrels of petroleum per day and if the atmosphere weighs 14.7 psi of earth surface area, calculate the weight of carbon dioxide generated each year from burning petroleum and compare this amount with the weight of the atmosphere.

b. If a total of 16 million tons of coal are burned every day on the earth, calculate the weight of carbon dioxide generated each year from coal burning, and compare it with the weight of the atmosphere.

1.10 Assume a world population of 7 billion and a U.S. population of 270 million.

a. Using the data in Figures 1.2–1.4, determine the total world energy consumption in quads if the rest of the world were to use the same per capita energy as in the United States.

b. If the energy source mix were to remain the same as the present mix in achieving the scenario of part a, what would be the percentage increase in CO_2 emissions?

1.11 Obtain data on worldwide energy consumption by sector from the U.S. Department of Energy, Energy Information Administration Web site. Plot the data, and estimate annual percentage growth rates for the seven regions reported and then for the world.

1.12 The following measurements of $x(t)$ are made:

t	0	1	2.3	3.0	4.5	5.2	6.5	8.0
x	2.1	8.4	31	65	360	850	3700	20,000

Construct a semilog plot of $x(t)$ either manually or with a computer, and determine whether the function appears to have an exponential dependence. If so, determine $x(t)$.

1.13 a. What does the area under the Gaussian curve represent?

b. Show that 68% of the area under the Gaussian curve lies within one standard deviation, s, of the maximum value of the function.

c. What percentage of the area lies within $2s$?

1.14 Determine R_m, t_o, and s for the worldwide graph of Figure 1.8.

1.15 Look up actual U.S. and world petroleum production figures, and plot them on Hubbert's curves to compare the actual production with the theoretical production.

1.16 Based on the data of Figure 1.10,

a. Estimate the year when PV shipments will reach 10 GW.

b. Estimate the year when PV shipments will reach 50 GW.

c. Estimate the year when PV shipments will reach 2700 GW.

1.17 The Earth Policy Institute [27] reports the following worldwide PV production figures. Plot the data on an Excel graph, establish an equation to represent the data, and then answer the three questions posed in Problem 1.16. Compare the results of the two problems.

Year	1995	1996	1997	1998	1999	2000	2001	2002	2003	2004	2005	2006
MW	77.6	88.6	126	155	201	277	386	547	748	1194	1786	2521

REFERENCES

1. Remarks by President William Clinton in Address to the United Nations Special Session on Environment and Development, June 26, 1997.
2. Sherwood, L., U.S. Solar Market Trends 2007, Interstate Renewable Energy Council, August 2008 www.irecusa.org/fileadmin/user_upload/NationalOutreachPubs/IRECSolarMarketTrendsAugust2008_2.pdf.
3. Dorn, Jonathan G., Solar Cell Production Jumps 50 Percent in 2007, Earth Policy Institute, December 2007 www.earth-policy.org/Indicators/Solar/2007.htm.
4. U.S. Census Bureau, U.S. and World Population Clocks—POPClocks www.census.gov/main/www/popclock.html.
5. Lindley, D., An overview of renewable energy in China, *Renewable Energy World*, November 1998, 65–69.
6. Gore, Al, *An Inconvenient Truth: The Planetary Emergency of Global Warming and What We Can Do About It*, Rodale, Emmaus, PA, 2006.
7. Intergovernmental Panel on Climate Change, Fourth Assessment Report, Climate Change 2007, www.ipcc.ch/.
8. Fishbane, P. M., Gasiorowicz, S., and Thornton, S. T., *Physics for Scientists and Engineers,* 2nd ed., Prentice Hall, Upper Saddle River, NJ, 1996.
9. U.S. Department of Energy, Energy Information Administration, Washington, D.C., International Energy Outlook 2008, September 2008, www.eia.doe.gov/oaif/ieo/index/html.
10. Annual Energy Review, 2001, U.S. Department of Energy, Energy Information Administration, Washington, D.C. www.eia.doe.gov/emeu/aer.
11. Van Mierlo, F., Energy Consumption vs. GDP, in World Energy Resources and Consumption, *Wikipedia, the free encyclopedia*, http://en.wikipedia.org/wiki/World_energy_resources_and_consumption.
12. Malthus, T. R., *An Essay on the Principle of Population, as It Affects the Future Improvement of Society, with Remarks on the Speculations of Mr. Godwin, M. Condorcet, and Other Writers,* printed for J. Johnson in St. Paul's Church-Yard, London, 1798.
13. Bahr, H. M., Chadwick, B. A., and Thomas, D. L., Eds., *Population, Resources, and the Future: Non-Malthusian Perspectives*, Brigham Young University Press, Provo, UT, 1974.
14. Bloch, S. C., *Excel for Engineers and Scientists*, 2nd ed., John Wiley & Sons, Hoboken, NJ, 2003.
15. Gottfried, B., *Spreadsheet Tools for Engineers Using Excel, Including Excel 2002*, McGraw-Hill, New York, 2003.
16. Bartlett, A. A., Forgotten fundamentals of the energy crisis, *Am. J. Phys.*, Vol. 46, No. 9, September 1978, 876–888.
17. Hubbert, M. K., The energy resources of the earth, *Scientific Am.*, Vol. 225, No. 3, September 1971, 60–70.
18. Annual Energy Review, 2001, U.S. Department of Energy, Energy Information Administration, Washington, D.C. www.eia.doe.gov/emeu/aer.
19. Odum, H. T. and Odum, E. C., *Energy Basis for Man and Nature,* McGraw-Hill, New York, 1976.
20. Odum, H. T. and Odum, E. C., *Energy Basis for Man and Nature,* 2nd ed., McGraw-Hill, New York, 1981.
21. Henderson, H., *The Politics of the Solar Age: Alternatives to Economics*, Anchor Press/Doubleday, Garden City, NY, 1981.
22. Foukal, P., *Solar Astrophysics*, John Wiley & Sons, New York, 1990.

23. Markvart, T., Ed., *Solar Electricity*, John Wiley & Sons, Chichester, U.K., 1994.
24. Maycock, P. D., Ed., *Photovoltaic News*, Vol. 18, No. 1, January 1999.
25. *International Marketing Data and Statistics,* 22nd ed., Euromonitor PIC, London, 1998.
26. Maycock, P. D., The world PV market, production increases 36%, *Renewable Energy World*, July–August 2002, 147–161.
27. Earth Policy Institute, Eco-Economy Indicators: SOLAR POWER—Data www.earth-policy.org/Indicators/Solar/2007_data.htm.
28. Maycock, P. D., *PV News*, Vol. 22, No. 3, March 2003.
29. International Energy Agency: PV Power Systems Programme, Solar Power International 08, San Diego, CA, October 2008.
30. Hagens, Nate, The Energy Return of (Industrial) Solar—Passive Solar, PV, Wind and Hydro, The Oil Drum: Discussion about energy and our future, April 29, 2008 www.theoildrum.com/files/PV_table.PNG.

SUGGESTED READING

Aubrecht, G. J., *Energy,* 2nd ed., Prentice-Hall, Upper Saddle River, NJ, 1995.

Gore, Al, *Earth in the Balance: Ecology and the Human Spirit*, Rodale, New York, 1992.

Kerr, R. A., The next oil crisis looms large and perhaps close, *Science*, Vol. 281, August 21, 1998, 1128–1131.

Krupp, Fred and Horn, Miriam, *Earth: The Sequel*, W. W. Norton & Co., New York, London, 2008.

Lovins, Amory B. et al., Winning the Oil Endgame: Innovation for Profits, Jobs, and Security, Rocky Mountain Institute, Snowmass, CO, 2004.

"President Clinton Announces Million Solar Roofs by 2010," *FEMP Focus*, August/September 1997, 7.

Silberberg, M., *Chemistry: The Molecular Nature of Matter and Change*, Mosby, St. Louis, 1996.

Starke, L., Ed., *Vital Signs 1998: The Environmental Trends That Are Shaping Our Future*, W. W. Norton, New York, London, 1998.

CHAPTER 2

The Sun

2.1 INTRODUCTION

Optimization of the performance of photovoltaic and other systems that convert sunlight into other useful forms of energy depends on knowledge of the properties of sunlight. This chapter provides a synopsis of important solar phenomena, including the solar spectrum, atmospheric effects, solar radiation components, determination of sun position, measurement of solar parameters, and positioning of the solar collector. In this chapter, an attempt is made to quantify the obvious and, perhaps, some not-so-obvious observations: Why is it light during the day and dark at night? Why are there more daylight hours in summer than in winter? Why is the sun higher in the sky in summer than in winter? Why are there more summer sunlight hours in northern latitudes than in latitudes closer to the equator? Why does less energy reach the surface of the earth on cloudy days? Why is it better for a solar collector to face the sun? What happens if a solar collector does not face the sun directly? How much energy is available from the sun? Why is the sky blue?

2.2 THE SOLAR SPECTRUM

The sun provides the energy needed to sustain life in our solar system. In one hour, the earth receives enough energy from the sun to meet its energy needs for nearly a year. In other words, this is about 5000 times the input to the earth's energy budget from all other sources. In order to maximize the utilization of this important energy resource, it is useful to understand some of the properties of this "ball of fire" in the sky.

The sun is composed of a mixture of gases, with a predominance of hydrogen. As the sun converts hydrogen to helium in a massive thermonuclear fusion reaction, mass is converted to energy according to Einstein's famous formula, $E = mc^2$. As a result of this reaction, the surface of the sun is maintained at a temperature of approximately 5800 K. This energy is radiated away from the sun uniformly in all directions, in close agreement with Planck's blackbody radiation formula. The energy density per unit area, w_λ, as a function of wavelength, λ, is given by

$$w_\lambda = \frac{2\pi hc^2\lambda^{-5}}{e^{\frac{hc}{\lambda kT}} - 1} \text{ (W/m}^2\text{/unit wavelength in meters)}, \tag{2.1}$$

where

$h = 6.63 \times 10^{-34}$ W s^2 (Planck's constant),
$c = 3.00 \times 10^8$ m/s (speed of light in a vacuum),
$k = 1.38 \times 10^{-23}$ J/K (Boltzmann's constant), and
T = temperature of blackbody in K.

Equation (2.1) yields the energy density at the *surface* of the sun in W/m^2 per unit wavelength in m. By the time this energy has traveled 150 million km to the earth, the total extraterrestrial energy density decreases to 1367 W/m^2 and is often referred to as the solar constant (see Problem 2.1) [1].

Figure 2.1 shows plots of Planck's blackbody radiation formula for several different temperatures, along with the extraterrestrial solar spectrum. Note that at lower temperatures, nearly all of the spectrum lies outside the visible range in the infrared range, whereas at 5800 K, the characteristic white color of the sun is obtained due to the mix of wavelengths in the visible spectral range. At even higher temperatures, the color shifts toward blue, and at lower temperatures, the color shifts toward red. Incandescent lamp filaments, for example, typically are operated at temperatures of approximately 2700 K and, hence, emit an approximate blackbody spectrum characteristic of 2700 K. Depending on the color temperature of the light source, photographic equipment must be compensated to obtain true colors, unless appropriate

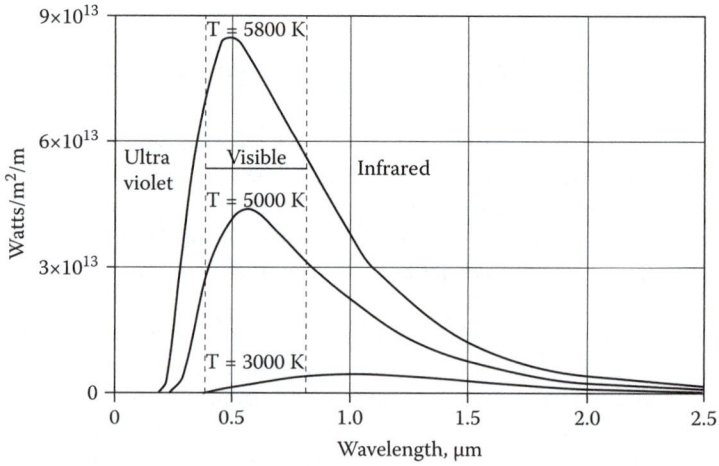

Figure 2.1 Blackbody radiation spectra for temperatures of 3000, 5000, and 5800 K.

filters are used. The extraterrestrial solar spectrum indicates that the sun can be reasonably approximated as a blackbody radiator.

Most sources of light are not perfect blackbody sources. Reasonable approximations of blackbody spectra are obtained from sources that emit light as a result of heating a filament to a high temperature. But to be a perfect radiator, the object must also be a perfect absorber of light, which is not the case for common light sources. Nonfilament light sources, such as gas discharge lamps, for example, emit either the discrete spectral characteristic of the specific gas, or, alternatively, emit light containing many discrete spectral lines from emission from phosphorescent materials. Thus, fluorescent, high-pressure sodium and high-intensity discharge lamps have spectral characteristics of gas discharge systems. Light-emitting diodes (LEDs), depending on their design, also may have discrete spectra or a combination of spectral lines. These are important facts to recognize, because in the testing of photovoltaic systems, it is important to be able to produce standardized spectral testing conditions.

Knowledge of the spectral composition of the sun is important for understanding the effects of the atmosphere on the radiation from the sun and for understanding which materials should offer the best performance in the conversion of sunlight to electricity.

2.3 THE EFFECT OF ATMOSPHERE ON SUNLIGHT

As sunlight enters the earth's atmosphere, some is absorbed, some is scattered, and some passes through unaffected by the molecules in the atmosphere and is either absorbed or reflected by objects at ground level.

Different molecules do different things. Water vapor, carbon dioxide, and ozone, for example, absorb significant amounts of sunlight at certain wavelengths. Ozone plays an important role by absorbing a significant amount of radiation in the ultraviolet region of the spectrum, while water vapor and carbon dioxide absorb primarily in the visible and infrared parts of the spectrum. These absorption lines are shown in Figure 2.2.

Absorbed sunlight increases the energy of the absorbing molecules, thus raising their temperature. Scattered sunlight is responsible for light entering north-facing windows when the sun is in the south. Scattered sunlight, in fact, is what makes the sky blue. Without the atmosphere and its ability to scatter sunlight, the sky would appear black, as it does on the moon. Direct sunlight consists of parallel rays, which are necessary if the light is to be focused. The reader has probably experimented with a magnifying glass and found that it is not possible to burn holes in paper when a cloud covers the sun. This is because the diffuse light present under these conditions is coming from all directions and cannot be focused.

All of these components of sunlight have been given names. Sunlight that reaches the earth's surface without scattering is called direct or beam radiation. Scattered sunlight is called diffuse radiation. Sunlight that is reflected from the ground is called albedo radiation, and the sum of all three components of sunlight is called global radiation.

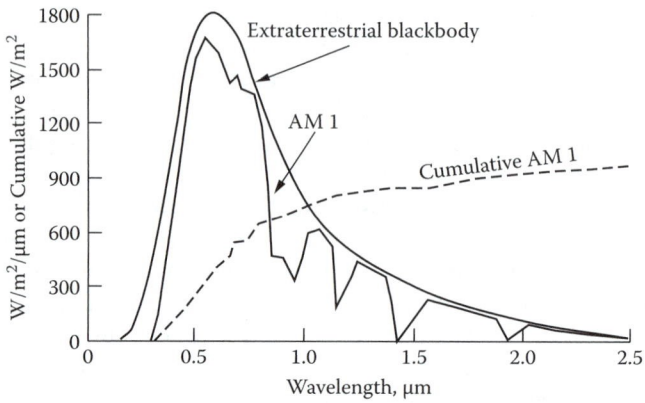

Figure 2.2 AM 1 solar spectrum after atmospheric absorption effects, including plots of extraterrestrial blackbody spectrum plus cumulative incident energy versus wavelength.

The amount of sunlight either absorbed or scattered depends on the length of the path through the atmosphere. This path length is generally compared with a vertical path directly to sea level, which is designated as air mass = 1 (AM 1). Hence, the air mass at a higher altitude will be less than unity for sun directly overhead, and the air mass generally will be more than unity for nonvertical sun angles. In general, the air mass through which sunlight passes is proportional to the secant of the zenith angle, θ_z, which is the angle measured between the direct beam and the vertical. At AM 1, after absorption has been accounted for, the intensity of the global radiation is generally reduced from 1367 W/m^2 at the top of the atmosphere to just over 1000 W/m^2 at sea level. Hence, for an AM 1 path length, the intensity of sunlight is reduced to 70% of its original AM 0 value. In equation form, assuming that the absorption constant depends directly on air mass, this observation can be expressed as

$$I = 1367(0.7)^{AM}. \tag{2.2}$$

This equation is, of course, obvious for AM 1. But does it hold for air masses different from unity? According to Meinel and Meinel [2], a better fit to observed data is given by

$$I = 1367(0.7)^{(AM)^{0.678}}. \tag{2.3}$$

On the average, over the surface of the earth, an amount of heat is reradiated into space at night that is just equal to the amount absorbed from the sun during the day. As long as this steady-state condition persists, the average temperature of the earth will remain constant. However, if for any reason the amount of heat absorbed is not equal to the amount reradiated, the planet will either cool down or heat up. This delicate balance can be upset by events such as volcanoes that fill the atmosphere with fine ash that reflects the sunlight away from the earth, thus reducing the amount of incident sunlight. The balance can also be upset by gases such as carbon dioxide and methane, which are mostly transparent to short-wavelength (visible) radiation, but more absorbing to long-wavelength (infrared) radiation. Because incident sunlight is dominated by short wavelengths characteristic of the 5800 K sun surface temperature, and because the reradiated energy is dominated by long wavelengths characteristic of the earth surface temperature of approximately 300 K, these greenhouse gases tend to prevent the earth from reradiating heat at night. In order to reach a new balance, it is necessary for the earth to increase its temperature, because radiation is proportional to T^4, where T is the temperature in K.

The natural compensation mechanism is green plants on land and under water. Through the process of photosynthesis, they use sunlight and carbon dioxide to produce plant fiber and oxygen, which is released to the atmosphere. Hence, replacement of green plants with concrete and asphalt and adding carbon dioxide to the atmosphere by burning fossil fuels have a combined negative effect on the stability of the concentration of carbon dioxide in the atmosphere. Global warming has been discussed extensively in other literature, some of which is listed in the chapter references.

2.4 SUNLIGHT SPECIFICS

2.4.1 Introduction

Nearly everyone has noticed that the sun shines longer in the summer than in the winter. Nearly everyone also knows that the Sahara Desert receives more sunshine than does London. Another obvious observation is that cloudy places receive less sunlight than sunny places. It may be less evident, however, that the hours of sunlight over a year are the same for every point on the earth, provided that only the hours between sunrise and sunset are counted, regardless of cloud cover. Those regions of the earth closer to the poles that have long winter nights also have long summer days. However, because the sun, on average, is lower in the sky in the polar regions than in the tropics, sunlight must traverse greater air mass in the polar regions than in the tropics. As a result, polar sunlight carries less energy to the surface than tropical sunlight. Not surprisingly, the polar regions are colder than the tropics.

In this section, quantitative formulas will be presented that will enable the reader to determine exactly how long the sun shines in any particular place on any particular day, and to determine how much sunlight can be expected, on average, during any month at various locations. Methods will also be presented for determining the position of the sun at any time on any day at any location. Finally, the effects of varying the orientation of a photovoltaic array on the power and energy produced by the array will be discussed.

2.4.2 Definitions

Irradiance is the measure of the power density of sunlight and is measured in W/m^2. Irradiance is thus an instantaneous quantity and is often identified as the intensity of sunlight. The solar constant for the earth is the irradiance received by it from the sun at the top of the atmosphere, that is, at AM 0, and is equal to 1367 W/m^2. After passing through the atmosphere with a path length of AM 1, the irradiance is reduced to approximately 1000 W/m^2, and has a modified spectral content owing to atmospheric absorption. The irradiance for AM 1.5 is accepted as the standard calibration spectrum for photovoltaic cells.

Irradiation is the measure of energy density of sunlight and is measured in kWh/m^2. Because energy is power integrated over time, irradiation is the integral of irradiance. Normally, the time frame for integration is 1 day, which, of course, means during daylight hours.

Irradiation is often expressed as *peak sun hours* (psh). The psh is simply the length of time in hours at an irradiance level of 1 kW/m^2 needed to produce the daily irradiation obtained from integration of irradiance over all daylight hours. Figure 2.3 illustrates the result of this integration for an example day of sunshine with a few cloudy moments. Note that the figure plots irradiance versus time in order to determine irradiation.

Irradiance and irradiation both apply to all components of sunlight. Hence, at a given time, or for a given day, these quantities will depend on location, weather conditions, and time of year. They will also depend on whether the surface of interest is shaded by trees or buildings and whether the surface is horizontal or inclined. The daily irradiation is numerically equal to the daily psh.

In order to determine the amount of irradiation available at a given location at a given time for conversion to electricity, it is useful to develop several expressions for the irradiance on surfaces, depending on the angle between the surface and the incident beam. It is also interesting to be able to determine the number of hours of sunlight on a given day at a given location.

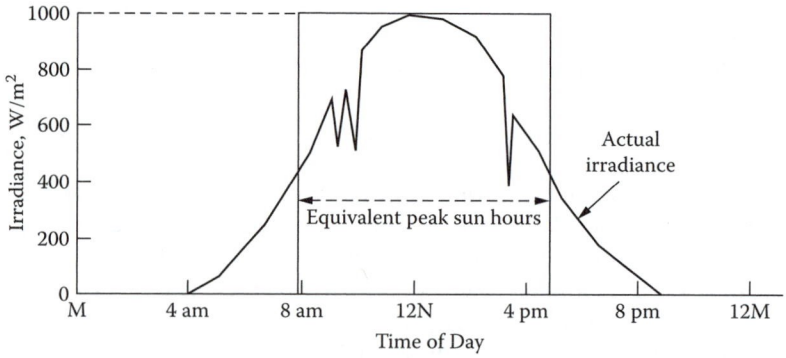

Figure 2.3 Determination of peak sun hours (irradiation) through integration of irradiance.

2.4.3 The Orbit and Rotation of the Earth

The earth revolves around the sun once per year in an elliptical orbit with the sun at one of the foci, such that the distance from sun to the earth is given by [1]

$$d = 1.5 \times 10^{11} \left\{ 1 + 0.017 \sin\left[\frac{360(n-93)}{365} \right] \right\} m, \tag{2.4}$$

where n represents the day of the year with January 1 as day 1. Because the deviation of the orbit from circular is so small, it is normally sufficient to express this distance in terms of its mean value.

The earth also rotates about its own polar axis once per day. The polar axis of the earth is inclined by an angle of 23.45° to the plane of the earth's orbit about the sun. This inclination is what causes the sun to be higher in the sky in the summer than in the winter. It is also the cause of longer summer sunlight hours and shorter winter sunlight hours. Figure 2.4 shows the earth's orbit around the sun with the inclined polar axis. Note that on the first day of Northern Hemisphere summer, the sun appears vertically above the Tropic of Cancer, which is latitude 23.45° N of the equator. On the first day of winter, the sun appears vertically above the Tropic of Capricorn, which is latitude 23.45° S of the equator. On the first day of spring and the first day of fall, the sun is directly above the equator. From fall to spring, the sun is south of the equator and from spring to fall the sun is north of the equator. The angle of deviation of the sun from directly above the equator is called the declination, δ. If angles north of the equator are considered positive and angles south of the equator are considered negative, then at any given day of the year, n, the declination can be found from

$$\delta = 23.45° \sin\left[\frac{360(n-80)}{365} \right]. \tag{2.5}$$

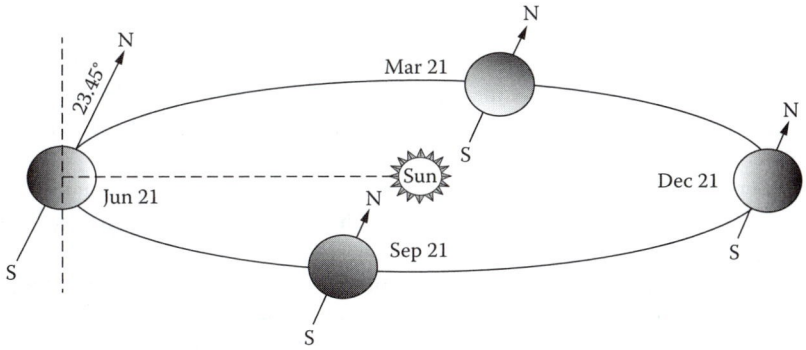

Figure 2.4 The orbit of the earth and the declination at different times of the year.

This formula, of course, is only a good approximation because the year is not exactly 365 days long and the first day of spring is not always the 80th day of the year. In any case, to determine the location of the sun in the sky at any time of day at any time of year at any location on the planet, the declination is an important parameter.

It is also important to be able to determine the clock time at which solar noon occurs. Solar noon occurs at 12 noon clock time at only one longitude, L_1, within any time zone. At longitudes east of L_1, solar noon will occur before 12 noon, and at longitudes west of L_1, solar noon will occur after 12 noon. On a sunny day, solar noon can be determined as that time when a shadow points directly north or directly south, depending on the latitude. Note that in the tropics, part of the year the shadow will point north, and the rest of the year the shadow will point south.

Fortunately, if the longitude is known, it is straightforward to determine the approximate relationship between clock noon and solar noon. Because there are 24 h in a day and because the earth rotates 360° during this period, this means that the earth rotates at the rate of 15°/h. It is also convenient that longitude zero corresponds to clock noon at solar noon. As a result, solar noon occurs at clock noon at multiples of 15° east or west longitude. Furthermore, because it takes the earth 60 min to rotate 15°, it is straightforward to interpolate to find solar noon at intermediate longitudes.

For example, at a longitude of 80° west, solar noon can be found by noting that 80° is between 75° and 90°, where solar noon occurs at clock noon *standard time*. Because 80° is west of 75°, when the sun is directly south at 75°, the sun will be east of south at 80°. The clock time at which the sun will be south at 80° (solar noon for 80°) is thus found by interpolation to be

$$t = 12 + \frac{80-75}{15} \times 60 = 12 + 20\,\text{min} = 12:20 \text{ p.m.}$$

Note that this time is standard time relative to the time zone for which solar noon occurs at 75° W. If 90° W is used as the solar noon reference, then solar noon at 80° will occur 40 min before solar noon occurs at 90°. Note that the answer is still the same. At 80° W, solar noon occurs at 12:20 p.m. Eastern Standard Time or at 11:20 a.m. Central Standard Time.

One glitch in the solar noon argument involves those unique locations on the earth such as Newfoundland, Canada, or India, where there is only half an hour shift between adjacent time zones, or Alaska, where a single clock time zone covers nearly 30° of longitude.

Another glitch in determining solar noon results from the combination of the declination and the elliptical orbit of the earth. During half the year, the earth is moving closer to the sun and during the other half, the earth is moving away from the sun. This results in a variation in solar noon called the analemma [3]. One way of describing the analemma is to plot the position of the tip of a shadow from a fixed object at a specific time each day, each week, or each month. Whichever time period is used, the resulting plot over a year will look like the number 8. The reader familiar with globes may recall seeing the analemma on the globe.

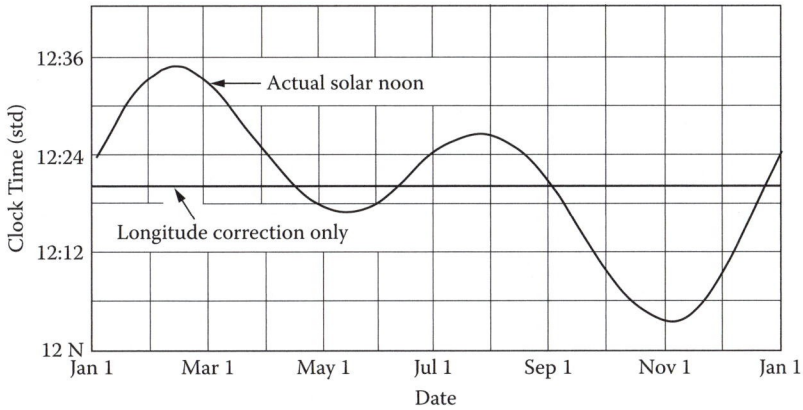

Figure 2.5 Daily variation of solar noon versus clock time in Boca Raton, Florida.

Another convenient way of describing the analemma is to plot the clock time at which solar noon occurs over the period of a year. This plot will refine the solar noon value determined from the longitude correction. Solar noon data can be obtained from http://aa.usno.navy.mil/data/docs [4]. Figure 2.5 shows a plot of solar noon versus clock time for Boca Raton, Florida, which is located at a longitude of approximately 80° W. Note that the greatest variation in solar noon is approximately ±16 min from 12:20 p.m., the value calculated from the longitude correction formula.

2.4.4 Tracking the Sun

To completely specify the position of the sun, it is necessary to specify three coordinates. If one assumes the distance from the sun to the earth to be constant, then the position of the sun can be specified using two angles. Two common choices are the solar altitude and the azimuth.

The **solar altitude**, α, is the angle between the horizon and the incident solar beam in a plane determined by the zenith and the sun, as shown in Figure 2.6.

The angular deviation of the sun from directly south can be described by the **azimuth angle**, ψ, which measures the sun's angular position east or west of south. The azimuth angle is zero at solar noon and increases toward the east. It is the angle between the intersection of the vertical plane determined by the observer and the sun with the horizontal and the horizontal line facing directly south from the observer, assuming the path of the sun to be south of the observer. The reader should note that in many publications, the azimuth angle is referenced to north, such that solar noon appears at $\psi = 180°$. In fact, compass angles are azimuth angles referenced to magnetic north.

Another useful, although redundant, angle in describing the position of the sun is the angular displacement of the sun from solar noon in the plane of apparent travel

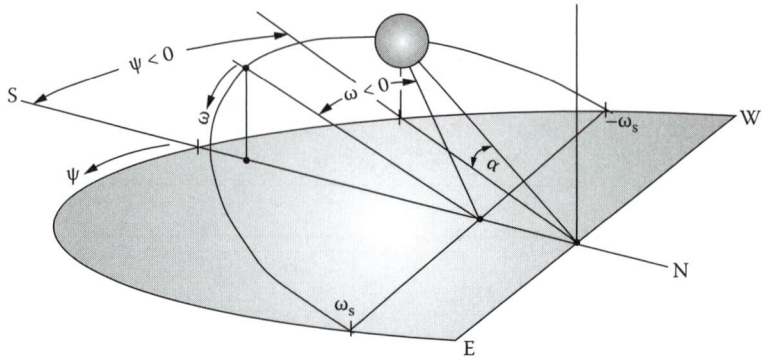

Figure 2.6 Sun angles, showing altitude, azimuth, and hour angle.

of the sun. The hour angle is the difference between noon and the desired time of day in terms of a 360° rotation in 24 h. In other words,

$$\omega = \frac{12 - T}{24} \times 360° = 15(12 - T)°, \tag{2.6}$$

where T is the time of day expressed with respect to solar midnight, on a 24 h clock. For example, for $T = 0$ or 24 (midnight), $\omega = \pm 180°$ and for $T = 9$ a.m., $\omega = 45°$. By relating ω to the other angles previously discussed, it is possible to show [1] that the sunrise angle is given by

$$\omega_s = \cos^{-1}(-\tan\phi \tan\delta), \tag{2.7}$$

which, in turn, implies that the sunset angle is given by $-\omega_s$. This formula is useful because it enables one to determine the number of hours on a specific day at a specific latitude that the sun is above the horizon. Converting the sunrise angle to hours from sunrise to solar noon, and then multiplying by 2 to include the hours from solar noon to sunset, yields the number of hours of daylight to be

$$DH = \frac{48}{360} \times \omega_s = \frac{\cos^{-1}(-\tan\phi \tan\delta)}{7.5} \text{ h} \tag{2.8}$$

Two very important relationships among α and ψ can be determined by the reader who enjoys trigonometry. If δ, ϕ, and ω are known, then the position of the sun, in terms of α and ψ at this location at this date and time, can be determined from [1]:

$$\sin\alpha = \sin\delta \sin\phi + \cos\delta \cos\phi \cos\omega \tag{2.9}$$

and

$$\cos\psi = \frac{\sin\alpha\sin\phi - \sin\delta}{\cos\alpha\cos\phi}. \tag{2.10}$$

Note that in all the foregoing expressions, angles are measured in degrees.

It is interesting to look at the values of Equations (2.9) and (2.10) at solar noon, when $\omega = 0$. The results of Problem 2.9 show that at solar noon, $\theta_Z = \phi - \delta$, where θ_Z, the zenith angle, is the angle between the vertical and the sun position. Thus, θ_Z is the complement of α. During the year, the highest point of the sun in the sky will be at $\theta_z = \phi - 23.45°$, and the lowest point of the solar noon sun in the sky will be at $\theta_z = \phi + 23.45°$, provided that $\phi > 23.45°$. It is particularly interesting to note that if $\phi > 90° - 23.45° = 66.55°$, then the lowest point of the sun in the sky is below the horizon, meaning that the sun does not rise or set that day. This, of course, is the situation in polar regions, which are subject to periods of 24 h of darkness. These same regions, of course, are also subject to equal periods of 24 h of sun 6 months later. If $\phi < 23.45°$, θ_z will at some time during the summer be negative. This simply means that the sun will appear north of directly overhead at solar noon. The relationships among θ_z, ϕ, and δ at solar noon are illustrated in Figure 2.7.

Because Equations (2.9) and (2.10) are somewhat difficult to visualize, it is convenient to plot α versus ψ for specific latitudes and days of the year. Figure 2.8 shows a series of plots of altitude versus azimuth at a latitude of 30°N. The curves show approximately how high in the sky the sun will be at a certain time of day during a particular month, with the azimuth angle determined by the time of day.

Markvart [1] gives detailed formulas for sun position and the components of global irradiance. The interested reader is encouraged to review them. Interestingly enough, when all of the mathematical work is finished, one ends up with an answer that does not account for cloud cover. The most reliable means of accounting for cloud cover is to make measurements over a long period of time in order to determine average figures. Figure 2.9 shows average daily irradiation over the earth for (a) summer and (b) winter. Appendix A shows average daily irradiation for a number of locations on a monthly basis.

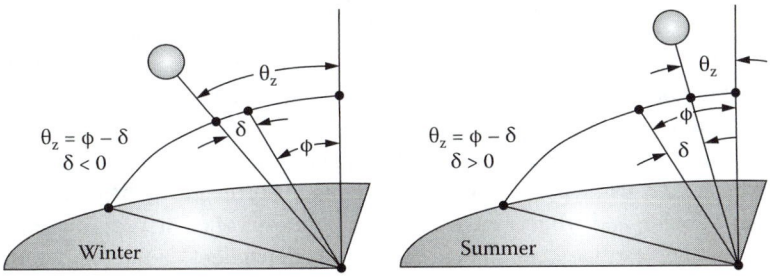

Figure 2.7 Relationships among zenith angle, latitude, and declination at solar noon in winter and summer.

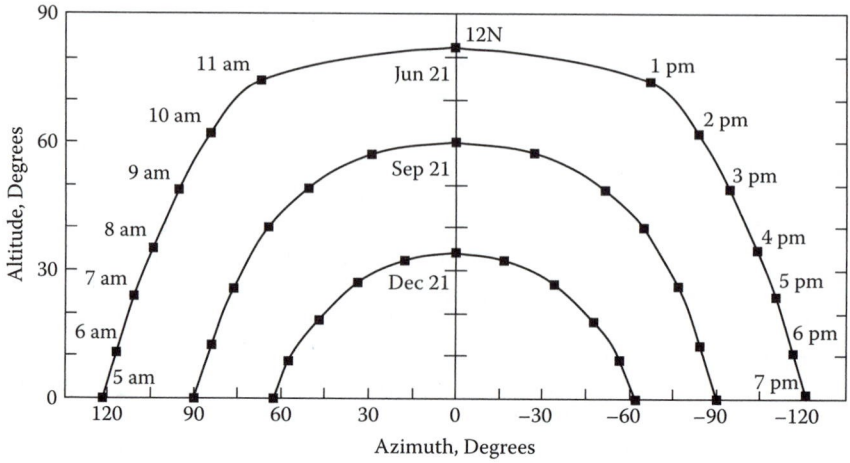

Figure 2.8 Plot of solar altitude versus azimuth for different months of the year at a latitude of 30° N.

2.4.5 Measuring Sunlight

Because all the calculations and approximations in the world cannot yield exact predictions of the amount of sunlight that will fall on a given surface at a given angle at a given time at a given day in a given place, the design of photovoltaic power systems is dependent on the use of data based on measurements averaged over a long time.

Depending on whether it is desired to measure global, beam, diffuse, or albedo components of irradiance, or whether it is simply desired to measure when sunlight exceeds a certain brightness level, different types of instruments are used.

2.4.5.1 Precision Measurements

The pyranometer is designed to measure global radiation. It is normally mounted horizontally to collect general data for global radiation on a horizontal surface. However, it is also often mounted in the plane of a photovoltaic collector in order to measure the global radiation incident on the inclined surface.

The pyranometer is designed to respond to all wavelengths and, hence, it responds accurately to the total power in any incident spectrum.

The black-and-white pyranometer operates on the principle of differential heating of a series of black-and-white wedges. The temperatures of each wedge are measured with thermocouples that yield voltage differences dependent on the temperature differences. This instrument is somewhat less accurate than the precision unit.

The normal incidence pyrheliometer uses a long, narrow tube to collect beam radiation over a narrow beam solid angle, generally about 5.5°. Because the

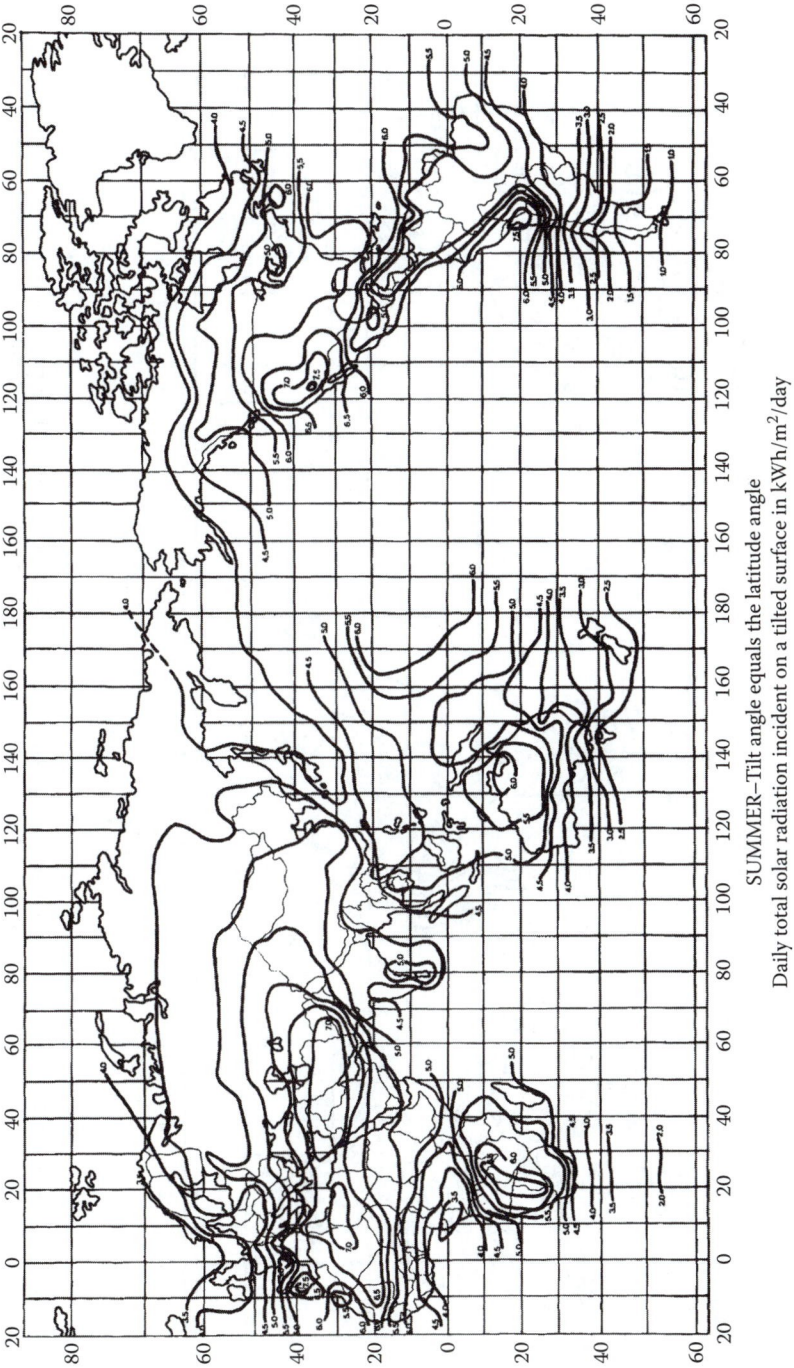

Figure 2.9a Worldwide average daily irradiation (kWh/m²) for summer. (Courtesy National Renewable Energy Laboratory, 30-Year Average of Monthly Solar Radiation, 1961–1990, Spreadsheet Portable Data Files. http://rredc.nrel.gov/solar/old_data/nsrdb/redbook/sum2/)

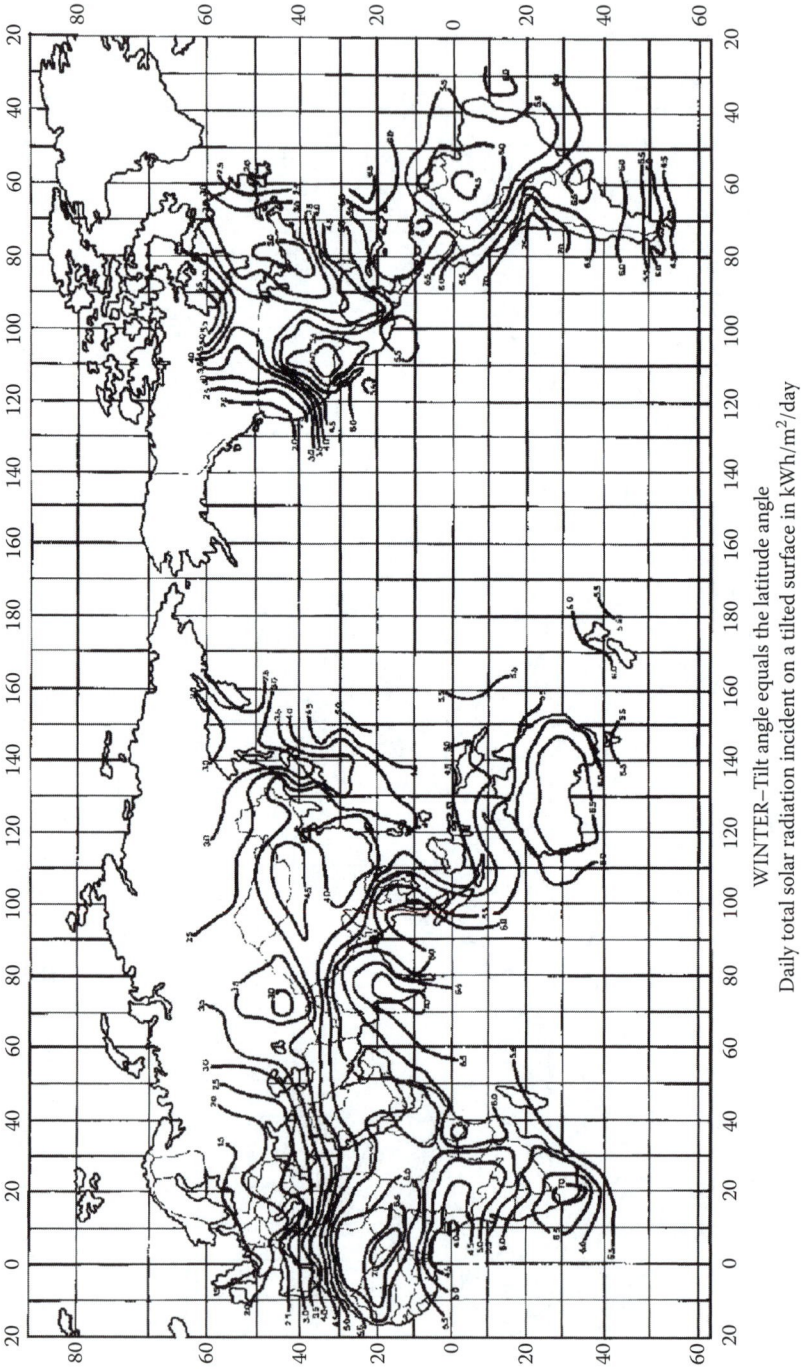

Figure 2.9b Worldwide average daily irradiation (kWh/m²) for winter. (Courtesy National Renewable Energy Laboratory, 30-Year Average of Monthly Solar Radiation, 1961–1990, Spreadsheet Portable Data Files. http://rredc.nrel.gov/solar/old_data/nsrdb/redbook/sum2/)

instrument is only sensitive to beam radiation, a tracker is needed if continuous readings are desired.

The pyranometer can be mounted on a shadow band stand to block out beam radiation so that it will respond only to the diffuse component. The stand is mounted so the path of the sun will be directly above the band during daylight hours. Because δ changes from day to day and because the band blocks out only a few degrees, it is necessary to readjust the band every few days.

2.4.5.2 Less Precise Measurements

Many inexpensive instruments are also available for measuring light intensity, including instruments based on cadmium sulfide photocells and silicon photodiodes. These devices give good indications of relative intensity, but are not sensitive to the total solar spectrum and thus cannot be accurately calibrated to measure total energy. These devices also do not normally have lenses that capture incident radiation from all directions. Devices that capture solid angles from a few degrees to upward of 90° are available.

2.5 CAPTURING SUNLIGHT

2.5.1 Maximizing Irradiation on the Collector

The designer of any system that collects sunlight must decide on a method of mounting the system. Perhaps the easiest mounting of most systems is to mount them horizontally. This orientation, of course, does not optimize collection because the beam radiation component collected is proportional to the cosine of the angle between the incident beam and the normal to the plane of the collector, as shown in Figure 2.10. Depending on the ratio of diffuse to beam irradiance components, the fraction of available energy collected will be between cos γ and unity. Of course, in a highly diffuse environment, the beam irradiance will be only a small fraction of the global irradiance. Several alternatives to horizontal mounting exist. Because

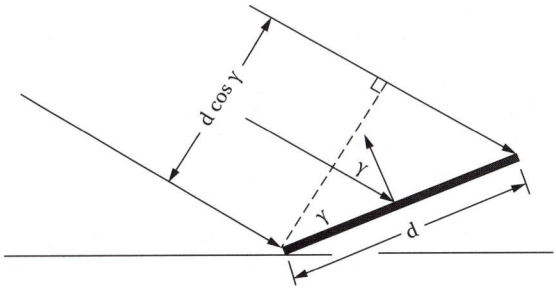

Figure 2.10 Two-dimensional illustration of effect of collector tilt on effective area presented to beam component of radiation.

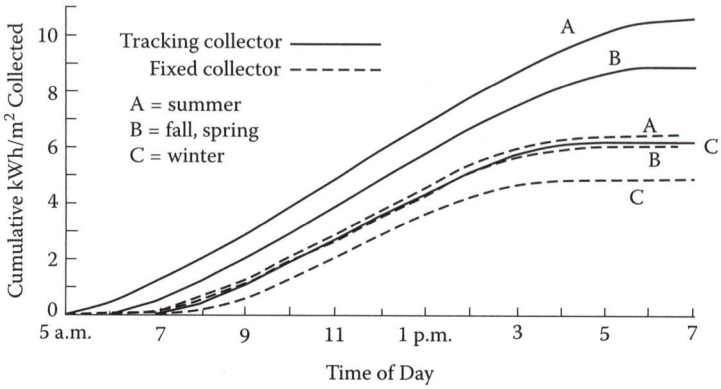

Figure 2.11 Cumulative daily irradiation received by fixed and tracking collectors for different seasons: direct beam contribution only.

$\theta_Z = \phi - \delta$ defines the position of the sun at solar noon, if a collector plane is perpendicular to this angle, it will be perpendicular to the sun at solar noon. This is the point at which the sun is highest in the sky, resulting in its minimum path through the atmosphere and corresponding lowest air mass for the day. Because the sun travels through an angle of 15° per hour, it will be close to perpendicular to the collector for a period of approximately 2 hours. Beyond this time, the intensity of the sunlight decreases because of the increase in air mass, and the angle between incident sunlight and the normal to the collector increases. These two factors cause the energy collected by the collector to decrease relatively rapidly during the hours before 10 a.m. and after 2 p.m. Figure 2.11 shows the approximate cumulative irradiation received by a south-facing collector tilted at the latitude angle in a region where the beam radiation component is significantly stronger than either the diffuse or the albedo components.

If the collector is mounted so that it can track the sun, then the incident irradiance is affected only by the increasing air mass as the sun approaches the horizon. Figure 2.11 also shows the additional cumulative irradiation under direct beam conditions received by a tracking collector. Approximately 50% more energy can be collected in the summer in a dry climate such as that found in Phoenix, Arizona, by using a tracking collector. During winter months, however, only about 20% more energy is collected using a tracker. In Seattle, Washington, which receives somewhat more diffuse sunlight than does Phoenix, a tracking collector will collect about 35% more in the summer but only 9% more energy compared to an optimized fixed collector in the winter (see Appendix A). Whether to use a two-axis tracking collector then becomes an important economic decision for the engineer, because a tracking mount is more costly than a fixed mount.

To make the mounting selection even more interesting, one can consider a single-axis tracker, which rotates about an axis perpendicular to θ_z. One can then also consider mountings that can be adjusted manually several times per day or, perhaps,

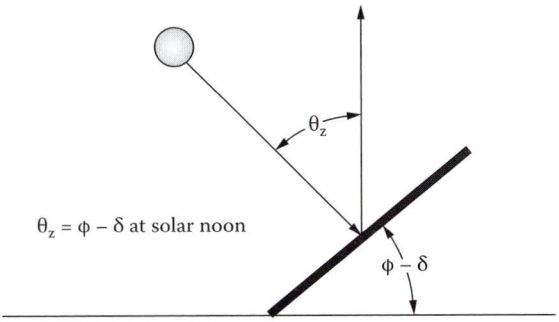

Figure 2.12 Optimizing the mounting angle of a fixed collector.

several times per year. Each of these options will enable the collection of an amount of energy that lies somewhere between the optimized fixed collector and the two-axis tracking collector results.

Collector orientation may also be seasonally dependent. For example, a remote cabin, used only during summer months, will need its collector oriented for optimal summer collection. However, if the cabin is used in the winter as a ski lodge, or in the fall as a hunting base, then the collector may need to be optimized for one of these seasons. For optimal performance on any given day, a fixed collector should be mounted with its plane at an angle of $\phi - \delta$ with respect to the horizontal, as shown in Figure 2.12. This will cause the plane of the collector to be perpendicular to the sun at solar noon.

For optimal seasonal performance, then, one simply chooses the average value of δ for the season. For summer in the northern hemisphere, δ varies sinusoidally between 0° on March 21 to 23.45° on June 21 and back to 0° on September 21. Plot this variation as half a sine wave with amplitude 23.45°. Those who have evaluated the average value of a sine wave in conjunction with the output of a rectifier circuit may recall that the average value of half a sine wave having amplitude A is $2A/\pi$. Hence, the average declination between March 21 and September 21 is 14.93°. Similarly, the average declination for the period from September 21 to March 21 is –14.93°. Hence, for best average summer performance, a collector should be mounted at approximately $\phi - 15°$, and for best average winter performance, it should be mounted at $\phi + 15°$.

For optimum spring or fall performance or optimum annual performance, the collector should be mounted at $\phi°$. Because normally it is acceptable to orient a collector for optimal seasonal or, perhaps, annual performance, many tables of irradiation on collectors are available for collector tilts of either ϕ, $\phi + 15°$, or $\phi - 15°$ at various locations on the globe. Examples of these tables as compiled by the National Renewable Energy Laboratory are presented in Appendix A. Figure 2.13 shows how monthly performance depends on collector tilt angle. Different locations will show different relative monthly performance, depending on local weather seasonal behavior.

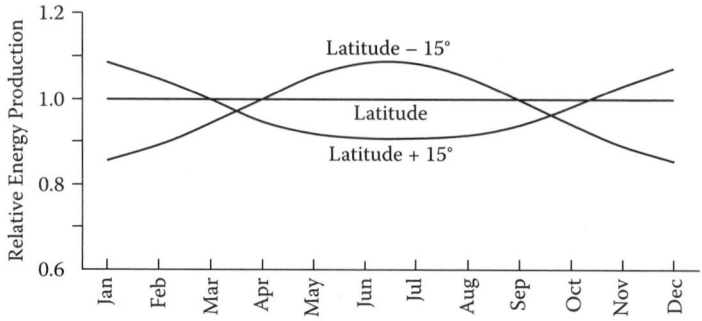

Figure 2.13 Monthly collector performance as a function of collector tilt angle.

2.5.2 Shading

Even a small amount of shade on a PV module can significantly reduce the module output current. It is thus of paramount importance to select a site for a PV system where the PV array will remain unshaded for as much of the day as possible. This is easy if there are no objects that might shade the array, but it is probably more likely that a site will have objects nearby that shade the array at some time of the day on some day of the year. The PV system designer must thus be able to use his or her knowledge of sun position to determine the times at which a PV array might be shaded.

2.5.2.1 Field Measurement of Shading Objects

Figure 2.14 shows a device that incorporates plots of altitude versus azimuth for selected latitudes. The device is used at the proposed site to determine when the array will be shaded by observing the position of reflected objects on the screen of the device. By sketching the outlines of shading objects on the screen with the device at the proposed location of the collector, the user can then determine when the collector will be shaded. For the example in Figure 2.14, the collector will be free of shade during May, June, July, and August. In September, the collector will be unshaded between about 9:15 a.m. and 3:15 p.m. sun time, while in November, December, and January, the collector is shaded most of the day except for around 1 p.m. sun time. This proposed collector location is thus acceptable if the collector is to be used only during the summer months.

The device also provides a measure of the fraction of available solar energy that will be collected under various shading conditions. For example, in September, 7% of the total daily insolation arrives between 11:00 and 11:30 a.m. on a south-facing surface, while in December, 8% of the total daily insolation arrives between 11:00 and 11:30 a.m. sun time. For the example in Figure 2.14, during March about 67% of the daily available sun will arrive at the collector during the unshaded period between 9:30 a.m. and 2:45 p.m. This instrument now also has software available

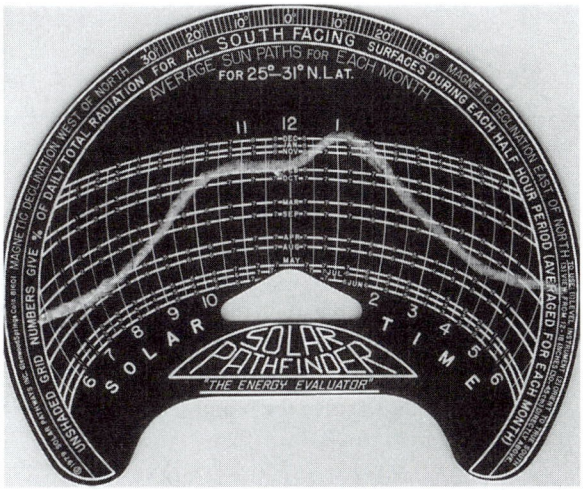

Figure 2.14 Solar Pathfinder showing region of shading. Shading occurs at points above the white line on the pattern. (Florida Solar Energy Center Photo.)

that will analyze the data taken and predict solar harvest taking into account the observed shading [6].

The inclinometer is another device that can be used to determine potential shading problems. The inclinometer simply measures the angle between the horizontal at the height of the PV array and the top of objects that might present shading problems. This angle, along with an azimuth angle measured with a compass, corrected to true north, can then be compared with the altitude versus azimuth chart for the sun for the location of the installation for different months of the year.

For the field person who is willing to spend a bit more on a measuring device, several photographic-based systems are also available that add some automation and a bit more precision to the field measurement of shade process [7].

2.5.2.2 Computational Methods of Determining Shading

Computational methods of determining shading range from solving Equations (2.9) and (2.10) for a specific site to using a variety of computer programs. Google Sketch-Up© is a popular freeware program that can be downloaded from the Web [8]. It can be used with Google Earth© to determine shading at existing sites. AutoCAD has a plug-in that does a very nice job of shading analysis on planned new construction drawings [9]. Any of these methods is useful for any mounting configuration.

As an example, consider a ground-mounted array as shown in Figure 2.15. Assume the array is located at 30° N latitude (ϕ) and that it is desired to space the rows such that rows do not shade each other between the hours of 8:30 a.m. to 3:30 p.m. sun time at any time of the year.

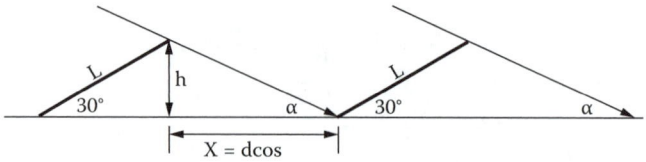

Figure 2.15 Determining the spacing between rows of PV modules.

If a shading program is not available, then the first step is to solve Equations (2.9) and (2.10) for the sun position. Because the worst case will be on December 21, and because the angles will be symmetrical about south, one only needs to know δ, ϕ, and ω for 8:30 a.m. sun time in order to solve for α and ψ. Solving for δ and ω yields

$$\delta = 23.45° \sin\left[\frac{360(n-80)}{365}\right] = 23.45° \sin\left[\frac{360(355-80)}{365}\right] = -23.45°$$

as expected, and

$$\omega = 15(12-T)° = 15(12-8.5) = 52.5°.$$

Thus, from Equation (2.9),

$$\sin\alpha = \sin(-23.45°)\sin 30° + \cos(-23.45°)\cos 30° \cos 52.5° = 0.2847$$

and from Equation (2.10),

$$\cos\Psi = \frac{0.2847 \times \sin 30° - \sin(-23.45°)}{\cos(\sin^{-1} 0.2847)\cos 30°} = 0.6508.$$

Thus, $\alpha = 16.54°$ and $\psi = 49.4°$ east of south.

Next, the length of a shadow, d, can be found using a bit of trigonometry. If the length of the PV modules is L, then $h = L\sin 30°$, and $d = h/\tan\alpha = L\sin 30°/\tan\alpha$. So, for example, if $L = 62$ in., then $d = 104$ in. Because this shadow points 49.4° west of north, the projection of the shadow to the north will give a distance of 104cos (49.4°) = 68 in., which will thus be the minimum row spacing.

2.5.3 Special Orientation Considerations

It should be noted that all previous discussion about collector orientation has assumed fixed collectors to be facing directly south (or north if in the southern hemisphere). Sometimes it is not practical to orient a collector directly to the south, and sometimes it may even be desirable to use a different orientation. In other situations, a horizontal mount is preferable. The question is to what extent collection is sacrificed by a non-south-facing collector, or, alternatively, what may be gained with an alternate orientation.

2.5.3.1 Horizontal Mounting

In some PV installations, notably on floating buoys or watercraft, the orientation of the collector changes as the direction of travel of the watercraft changes. If a buoy has a single anchor, it will rotate. In these or similar cases, the collectors are normally mounted horizontally.

One of the entries in the National Renewable Energy Laboratory tables ([5] and Appendix A) tabulates global irradiation on a horizontal surface, so it is easy to determine the loss in solar harvest over optimal harvest by comparing the horizontal entries with the optimal entries.

The loss in performance with a horizontal mount depends mostly on the latitude of the installation and the season of the year, but also on the specific location and the ratio of beam to diffuse components of irradiance for that location. Because the optimal seasonal performance of collectors is obtained somewhere between $\phi \pm 15°$, if the installation is in the tropics, a relatively small amount of annual collection is lost with a horizontal mount.

2.5.3.2 Non-South-Facing Mounting

Sometimes it is not possible or convenient to install a collector facing directly south. If south-facing is the preferred orientation, Morse and Czarnecki [10] have reported that the collector can be facing up to 22.5° away from south with less than a 2% reduction in annual collection at latitudes up to 45°. In fact, the loss in monthly or annual solar harvest can be determined quite readily for any location and array orientation by using PVWatts [11].

In other cases, maximum PV system output may be desirable at a time of day other than solar noon. For example, in many regions, peak utility electrical generation occurs between the hours of 3 and 6 p.m. If a PV system is connected to the utility grid, it may be desirable to maximize system output during the utility peak. For a fixed mounting, this would require having the collector face the sun at the midpoint of this time period on a date near the middle of the period during which maximum collector output is desired. This orientation will produce peak output when the sun path is at a higher air mass than at solar noon, resulting in a slightly lower peak output than would be obtained with a south-facing collector.

Figure 2.16 shows a comparison of PV output for south-facing and west-facing PV systems. Note the appreciable increase in system output later in the afternoon for the west-facing system. During utility peak time, the system power output is increased by 70% for the west-facing system, but the west-facing system develops 14.6% less total annual energy than the south-facing system. Although the total energy output of the west-facing system is lower than that of the south-facing system, the loss of total energy may be acceptable if the added power output during utility peak hours

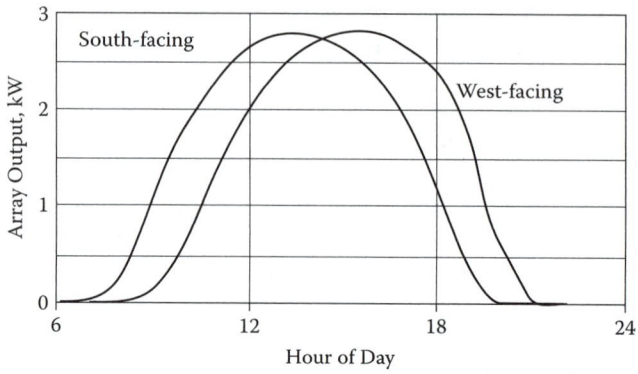

Figure 2.16 Comparison of power and energy output for south-facing and west-facing PV systems in Sacramento, California, in summer. The south-facing system produced 22,417 Wh and the west-facing system produced 22,192 Wh. (Courtesy of Collier, Photovoltaics in the Sacramento Municipal Utility District, Interconnecting Small Photovoltaic Systems to Florida's Utility Grid, A Technical Workshop for Florida's Utilities, Cocoa, FL, October 22, 1998.)

will offset the need to use more expensive peaking generation. This observation holds particularly if it will offset the need to install the peaking generation in the first place.

Computation of the desired orientation for maximum output in a direction other than south is straightforward:

1. Determine the latitude of the location.
2. Calculate the declination for the design day, using Equation (2.5).
3. Determine the time of day (local time) for peak PV system output.
4. Convert the local time to solar time, based upon site longitude.
5. Calculate the hour angle using Equation (2.6).
6. Use Equations (2.9) and (2.10) to determine α and ψ.

Example

Suppose it is desired to determine the position of the sun over Atlanta, Georgia, at 2 p.m. on July 1. Starting with step 1 of the 6 steps outlined earlier, the latitude of Atlanta is found in Appendix A to be 33°39′ = 33.65° N. Then Equation (2.5) is used to determine the declination for July 1. Because July 1 is the 181st day of the year, Equation (2.5) gives

$$\delta = 23.45 \sin\left[\frac{360(181-80)}{365}\right] = 23.12°.$$

With 2 p.m. local time (Eastern Daylight Time) at the longitude of Atlanta (84°36′ W = 84.60° W) as the time of day for peak system output, the solar time is found to be 1 p.m.

(Eastern Standard Time) plus an additional

$$t = \frac{84.6 - 75}{15} \times 60 = 38.4 \text{ min, or 1:38 p.m.}$$

Next, substituting the solar time in Equation (2.8) yields the hour angle as

$$\omega = \frac{12 - 13.64}{24}(360°) = -24.60°.$$

The solar altitude is next found from Equation (2.9) with the result

$$\sin\alpha = \sin 23.12° \sin 33.65° + \cos 23.12° \cos 33.65° \cos(-24.6°) = 0.9161.$$

Next, the azimuth is found from Equation (2.10) to be

$$\cos\psi = \frac{0.9161 \sin 33.65° - \sin 23.12°}{\cos[\sin^{-1} 0.9161] \cos 33.65°} = 0.3445.$$

So, finally, $\alpha = \sin^{-1} 0.9161 = 66.36°$ and $\psi = \cos^{-1} 0.3445 = -69.85°$. Note that $\cos^{-1} 0.3445 = \pm 69.85°$. The minus sign is chosen because at 1:38 p.m. solar time the sun is west of south.

Problems

2.1 Show that, for a surface temperature of 5800 K, the sun will deliver 1367 W/m^2 to the earth. This requires integration of the blackbody radiation formula over all wavelengths to determine the total available energy at the surface of the sun in W/m^2. Numerical integration is recommended. Be careful to note the range of wavelengths that contribute the most to the spectrum. Then note that the energy density decreases as the square of the distance from the source, similar to the behavior of an electric field emanating from a point source. The diameter of the sun is 1.393×10^9 m, and the mean distance from sun to the earth is 1.5×10^{11} m.

2.2 If the diameter of the sun is 1.393×10^9 m, and if the average density of the sun is approximately $1.4 \times$ the density of water, and if 2×10^{19} kg/yr of hydrogen is consumed by fusion, how long will it take for the sun to consume 25% of its mass in the fusion process?

2.3 Calculate the zenith angles needed to produce AM 1.5 and AM 2.0 if AM 1.0 occurs at zero degrees.

2.4 Calculate the zenith angle at solar noon at a latitude of 40° N on May 1.

2.5 Calculate the number of hours the sun was above the horizon on your birthday at your birthplace.

2.6 Calculate the irradiance of sunlight for AM 1.5 and for AM 2.0, assuming no cloud cover, using Equations (2.2) and (2.3). Then write a computer program that will plot irradiance versus AM for $1 \leq \text{AM} \leq 10$ for each equation.

2.7 Assume no cloud cover and, hence, that the solar irradiance is predominantly a beam component. If a nontracking collector is perpendicular to the incident radiation at solar noon, estimate the irradiance on the collector at 1, 2, 3, 4, 5, and 6 hours past solar noon, taking into account air mass and collector orientation. Then estimate the total daily irradiation on the collector in kWh. Assume AM 1.0 at solar noon and a latitude of 20° N.

2.8 Write a computer program using MATLAB, Excel, or a similar program that will plot solar altitude versus azimuth. Plot sets of curves similar to Figure 2.8 for the months of March, June, September, and December for Denver, Colorado, Mexico City, Mexico, and Fairbanks, Alaska.

2.9 Using Equations (2.9) and (2.10), show that at solar noon, $\theta_z = \phi - \delta$ and $\psi = 0$.

2.10 Write a computer program that will generate a plot of solar altitude versus azimuth for the 12 months of the year for your home town. It would be nice if each curve would have time-of-day indicators.

2.11 Neglecting the analemma effect, calculate the time of day at which solar noon would occur at your longitude. Then compare the north indicated by a compass, corrected for compass declination, with the north indicated by a shadow at calculated solar noon and estimate the error in the shadow direction (i.e., solar noon) due to the analemma effect. Estimate actual solar noon time on the basis of the difference between compass north and the shadow "north."

2.12 Noting the dependence of air mass on θ_z,

a. Generate a table that shows α, ψ, and I (using Equation 2.3) versus time for a latitude of your choice on a day of your choice.

b. Assuming a tracking collector and no cloud cover, estimate the total energy available for collection during your chosen day.

2.13 Plot I versus AM, and then plot I versus α.

2.14 Calculate the collector orientation that will produce maximum summer output between 2 and 5 p.m. in Tucson, Arizona. Use July 21 as the assumed midpoint of summer, and correct for the longitude of the site.

2.15 Calculate the collector orientation that will produce maximum summer output at 9 a.m. Daylight Savings Time in Minneapolis, Minnesota, using July 21 as the assumed midpoint of summer and using a longitude correction.

2.16 A collector in Boca Raton, Florida (ϕ = 26.4° N, longitude = 80.1° W), is mounted on a roof with a 5:12 pitch, facing 30° south of west. Determine the time of day and days of the year that the direct beam radiation component is normal to the array.

2.17. Determine the location (latitude and longitude) where, on May 29, sunrise is at 6:30 a.m. and sunset is at 8:08 p.m. EDT. At what time does solar noon occur at this location?

2.18 Using Figure 2.14, make a table of unshaded collector times for each month of the year.

2.19 An array of collectors consists of three rows of south-facing collectors as shown in Figure P2.1. If the array is located at 40° N latitude, determine the spacing, d, between the rows needed to prevent shading of one row by another row.

2.20 The top of a tree is found to be at an angle of 15° between the horizontal and the corner of a collector and at an azimuth angle of –30°. If the site is at a latitude of 30° N, determine the months (if any) when the tree will shade the array at the point from which the measurements are taken.

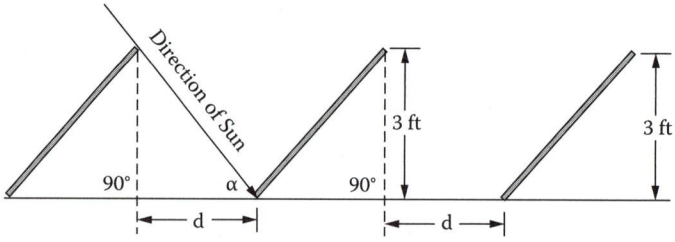

Figure P2.1 Rows of rack-mounted modules.

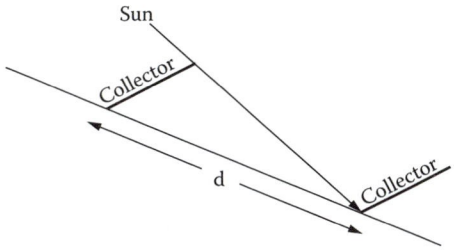

Figure P2.2 Rows of rack-mounted modules on a sloped roof.

2.21 A residence has a 5:12 roof pitch and is located at latitude 27° N and longitude 83° W, facing 15° west of south. If two rows of 66 × 142 cm collectors are to be mounted on the back roof so that they will have a tilt of latitude, as shown in Figure P2.2:

a. How far apart will the rows need to be if collectors are to remain unshaded for 6 hours on December 21?
b. Assuming Eastern Standard Time, over what time period on December 21 will the collectors be unshaded, assuming the collectors are facing 15° west of south?
c. Assuming no other shading objects, over what time period will the collectors be unshaded on March 21?

2.22 A commercial building has a flat roof as shown in Figure P2.3. The roof has a parapet around it that is 3 ft high, and an air conditioner located as shown that is 4 ft above the roof. The building is located at 40° N. A PV system is to be installed using modules that measure 3 ft by 5 ft, in rows that will remain unshaded for at least 6 h every day of the year. The modules are to be mounted at a tilt of 25°, in portrait mode, facing south. The bottom edges of the modules will be 6 in. above the roof. Determine the maximum number of modules that can be installed on the roof so that no module will be shaded during the 6 h period. Pay attention to module rows shading other rows, shading from the air conditioner, and shading from the parapet. Draw a diagram that shows your result.

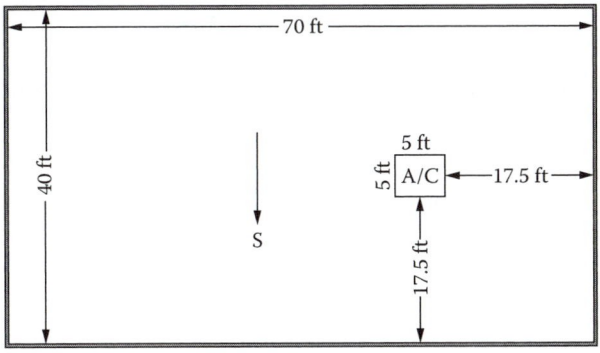

Figure P2.3 Roof layout for Problem 2.22.

2.23 Refer to the tables in Appendix A for Miami, Florida; Sacramento, California; Boston, Massachusetts; and Fairbanks, Arkansas. Create a table that shows the annual collection losses, in percent, for fixed-tilt collectors, compared to maximum collection when the collector is tilted at latitude. Include all of the tilt angles listed in the tables in your table. If all that counts is total annual peak sun hours, what generalizations can you offer with regard to array tilts?

REFERENCES

1. Markvart, T., Ed., *Solar Electricity*, John Wiley & Sons, Chichester, U.K., 1994.
2. Meinel, A. B. and Meinel, M. P., *Applied Solar Energy, an Introduction*, Addison-Wesley, Reading, MA, 1976.
3. Discussion and explanation of analemma, complete with formulas and diagrams. www.analemma.com.
4. "Sun or Moon Rise/Set Table for one year" tabulates sunrise, sunset, moonrise and moonset on a daily basis for a year for a large number of cities. http://aa.usno.navy.mil/data/docs.
5. National Renewable Energy Laboratory, 30-Year Average of Monthly Solar Radiation, 1961–1990, Spreadsheet Portable Data Files. http://rredc.nrel.gov/solar/old_data/nsrdb/redbook/sum2/.
6. Solar Pathfinder Assistant—Latest Version/Updates, www.solarpathfinder.com/spa-updates.html.
7. Wiley electronics Acme Solar Site Evaluation Tool information. http://www.we-llc.com/ASSET.html.
8. Google freeware program for shading analysis. http://sketchup.google.com.
9. A web search for shading analysis programs will yield hundreds of results.
10. Morse, R. N. and Czarnecki, J. T., Report E.E. 6 of Engineering Section, Commonwealth Scientific and Industrial Res. Org., Melbourne, Australia, 1958.
11. PVWATTS on-line calculator for determining PV system performance as a function of location and orientation of the array. http://rredc.nrel.gov/solar/codes_algs/PVWATTS/version2/.
12. Collier, D. E., Photovoltaics in the Sacramento Municipal Utility District, Interconnecting Small Photovoltaic Systems to Florida's Utility Grid, A Technical Workshop for Florida's Utilities, Cocoa, FL, October 22, 1998.

SUGGESTED READING

Howell, J. R., Bannerot, R. B., and Vliet, G. C., *Solar-Thermal Energy Systems, Analysis and Design,* McGraw-Hill, New York, 1982.

Hu, C. and White, R. M., *Solar Cells, From Basics to Advanced Systems*, McGraw-Hill, New York, 1983.

Hubbert, M. K., The energy resources of the Earth, *Scientific Am.*, Vol 225, No 3, September 1971, 60–70.

McCluney, W. R., Sun position in Florida, Design Note, Florida Solar Energy Center, Cocoa, FL (FSEC-DN-4-83), 1985.

Stand-Alone Photovoltaic Systems: A Handbook of Recommended Design Practices, Sandia National Laboratories, Albuquerque, NM, 1995.

Water Pumping: The Solar Alternative, Sandia National Laboratories, Albuquerque, NM, 1996.

CHAPTER 3

Introduction to PV Systems

3.1 INTRODUCTION

Photovoltaic (PV) systems are designed around the PV cell. Because a typical PV cell produces less than 5 W at approximately 0.5 V dc, cells must be connected in series-parallel configurations to produce enough power for high-power applications. Figure 3.1 shows how cells are configured into modules, and how modules are connected as arrays. Modules may have peak output powers ranging from a few watts, depending on the intended application, to more than 300 W. Typical array output power is in the 100 watts to kilowatt range, although megawatt arrays are now becoming more commonplace.

Because PV arrays produce power only when illuminated, PV systems sometimes employ an energy storage mechanism so that the captured electrical energy can be made available at a later time. Most commonly, the storage mechanism consists of rechargeable batteries, but it is also possible to employ more exotic storage mechanisms. In addition to energy storage, storage batteries also provide transient suppression, system voltage regulation, and a source of current that can exceed PV array capabilities.

When a battery storage mechanism is employed, it is common to also incorporate a charge controller into the system, so that the batteries can be prevented from reaching either an overcharged or overdischarged condition. It is also possible that some or all of the loads to be served by the system may be ac loads. If this is the case, an inverter will be needed to convert the dc from the PV array to ac. If a system incorporates a backup system to take over if the PV system does not produce adequate energy, then the system will need a controller to operate the backup system.

It is also possible that the PV system will be interconnected with the utility grid. Such systems may deliver excess PV energy to the grid or use the grid as a backup system in case of insufficient PV generation. These grid-interconnected systems need to incorporate suitable interfacing circuitry so that the PV system will be disconnected from the grid in the event of grid failure. Figure 3.2 shows the components of several types of PV systems. This chapter will emphasize the characteristics of PV system components in order to pave the way for designing systems in the following chapters. The physics of PV cells, with an emphasis on the challenges to the cell innovator, are covered in Chapter 10, and current specific cell technologies are discussed in Chapter 11.

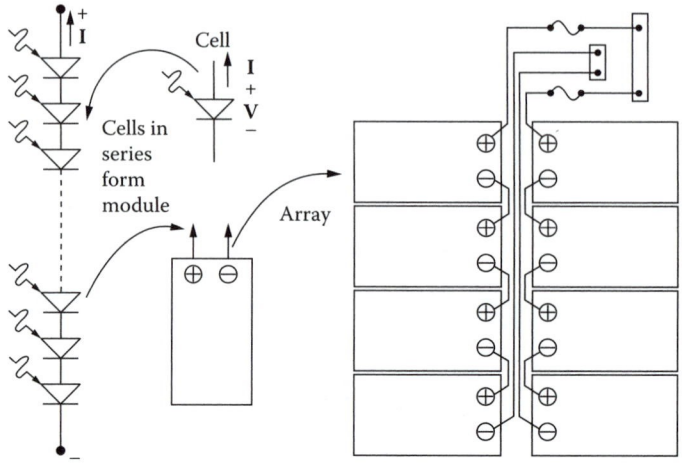

Figure 3.1 Cells, modules, and arrays.

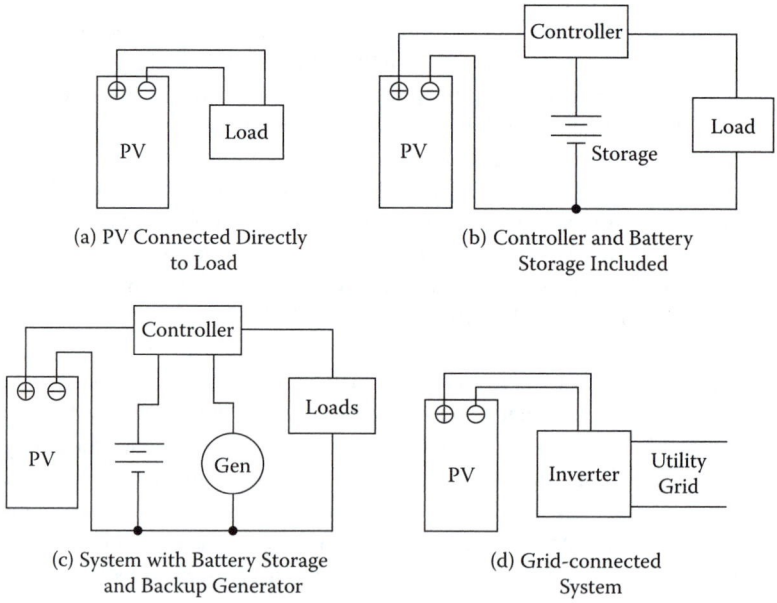

Figure 3.2 Examples of PV systems.

3.2 THE PV CELL

The PV cell is a specially designed pn junction or Schottky barrier device. The well-known diode equation describes the operation of the shaded PV cell.

When the cell is illuminated, electron–hole pairs (EHPs) are produced by the interaction of the incident photons with the atoms of the cell. The electric field created by the cell junction causes the photon-generated-EHPs to separate, with the electrons drifting into the n-region of the cell and the holes drifting into the p-region, provided that the EHPs are generated sufficiently close to the pn junction. This process will be discussed in detail in Chapter 10. For the purposes of this chapter, knowledge of the terminal properties of the PV cell is all that is needed.

Figure 3.3 shows the I-V characteristics of a typical PV cell. Note that the amounts of current and voltage available from the cell depend on the cell illumination level. In the ideal case, the I-V characteristic equation is

$$I = I_\ell - I_o\left(e^{\frac{qV}{kT}} - 1\right) \tag{3.1}$$

where I_ℓ is the component of cell current due to photons, $q = 1.6 \times 10^{-19}$ C, $k = 1.38 \times 10^{-23}$ J/K, and T is the cell temperature in K. While the I-V characteristics of actual PV cells differ somewhat from this ideal version, Equation (3.1) provides a means of determining the ideal performance limits of PV cells.

Figure 3.3 shows that the PV cell has both a limiting voltage and a limiting current. Hence, the cell is not damaged by operating it under either open-circuit or short-circuit conditions. To determine the short-circuit current of a PV cell, simply

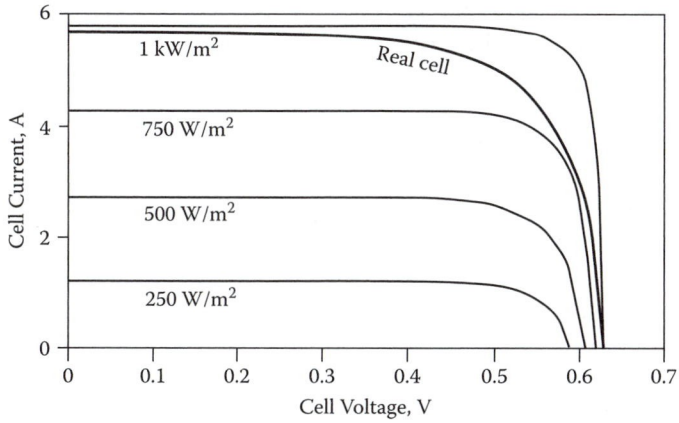

Figure 3.3 *I-V* characteristics of real and ideal PV cells under different illumination levels.

set $V = 0$ in the exponent. This leads to $I_{SC} = I_\ell$. To a very good approximation, the cell current is directly proportional to the cell irradiance. Thus, if the cell current is known under standard test conditions, $G_o = 1$ kW/m^2 at AM 1.5, then the cell current at any other irradiance, G, is given by

$$I_\ell(G) = (G/G_o)I_\ell(G_o). \tag{3.2}$$

To determine the open-circuit voltage of the cell, the cell current is set to zero and Equation (3.1) is solved for V_{OC}, yielding the result

$$V_{OC} = \frac{kT}{q}\ln\frac{I_\ell + I_o}{I_o} \cong \frac{kT}{q}\ln\frac{I_\ell}{I_o}, \tag{3.3}$$

because normally $I_\ell >> I_o$. For example, if the ratio of photocurrent to reverse saturation current is 10^{10}, using a thermal voltage (kT/q) of 26 mV yields $V_{OC} = 0.6$ V. Note that the open-circuit voltage is only logarithmically dependent on the cell illumination, while the short-circuit current is directly proportional to it.

Multiplying the cell current by the cell voltage yields the cell power, as shown in Figure 3.4. Note that at any given illumination level, there is one point on the cell *I-V* characteristic where the cell produces maximum power. Note also that the voltage at which the maximum power points occur is minimally dependent upon the cell illumination level.

If I_m represents the cell current at maximum power, and if V_m represents the cell voltage at maximum power, then the cell maximum power can be expressed as

$$P_{\max} = I_m V_m = FFI_{SC}V_{OC}, \tag{3.4}$$

where FF is defined as the cell fill factor. The fill factor is a measure of the quality of the cell. Cells with large internal resistance will have smaller fill factors, while the ideal cell will have a fill factor of unity. Note that a unity fill factor suggests a

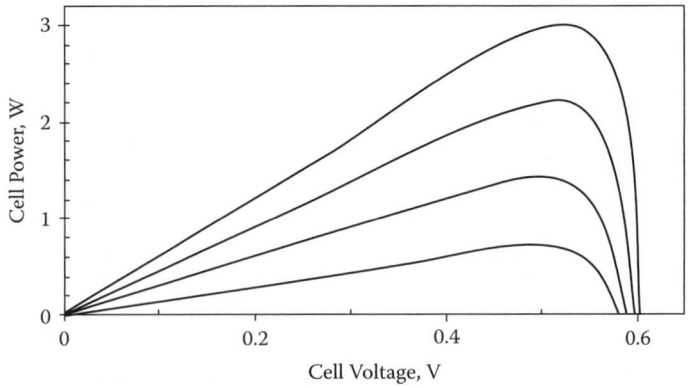

Figure 3.4 Power versus voltage for a PV cell for four illumination levels.

rectangular cell *I-V* characteristic. Such a characteristic implies that the cell operates as either an ideal voltage source or as an ideal current source. Although a real cell does not have a rectangular characteristic, it is clear that it has a region where its operation approximates that of an ideal voltage source and another region where its operation approximates that of an ideal current source.

For the cell having an ideal *I-V* characteristic governed by Equation (3.1), with $V_{OC} = 0.596$ V and $I_{SC} = 2.0$ A, the fill factor will be approximately 0.83. Typical fill factors for real PV cells, depending on the technology, may vary from 0.5 to 0.82. The secret to maximizing the fill factor is to maximize the ratio of photocurrent to reverse saturation current while minimizing series resistance and maximizing shunt resistance within the cell.

The cell power versus cell voltage curve is especially important when considering maximizing power transferred to the cell load. This topic will be investigated in more detail in Section 3.6, where methods of matching the load to the source are discussed.

The PV cell *I-V* curve is also temperature sensitive. A quick look at Equation (3.3) might suggest that the open circuit voltage is directly proportional to the absolute temperature of the cell. A longer look, however, will reveal that the reverse saturation current is also highly temperature dependent. The net result, which will be covered in detail in Chapter 10, is that the open-circuit voltage of an ideal crystalline silicon PV cell decreases by approximately 2.3 mV/°C increase in temperature, which amounts to approximately 0.5%/°C. The short-circuit current, on the other hand, remains nearly constant. As a result, the cell power also decreases by approximately 0.5%/°C. Figure 3.5 shows the temperature dependence of the PV cell power versus voltage characteristic.

It is important to remember that when a cell is illuminated, it will generally convert less than 20% of the irradiance into electricity. The balance is converted to heat, resulting in heating of the cell. As a result, the cell can be expected to operate above

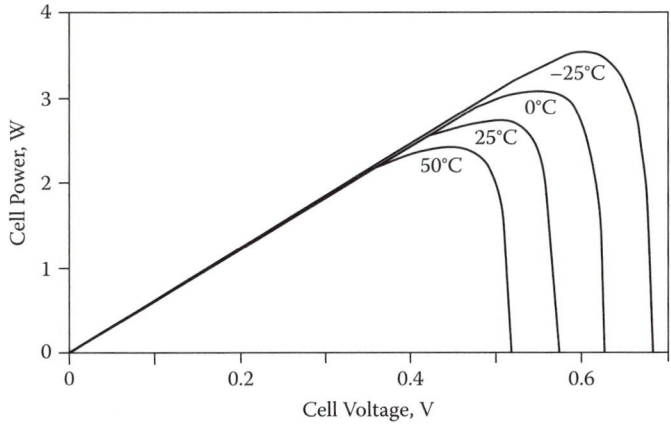

Figure 3.5 Temperature dependence of the power versus voltage curve for a PV cell.

ambient temperature. If the cell is a part of a concentrating system, then it will heat even more, resulting in additional temperature degradation of cell performance.

The photocurrent developed in a PV cell is dependent on the intensity of the light incident on the cell. The photocurrent is also highly dependent on the wavelength of the incident light. In Chapter 2 it was noted that terrestrial sunlight approximates the spectrum of a 5800 K blackbody source. PV cells are made of materials for which conversion to electricity of this spectrum is as efficient as possible. Depending on the cell technology, some cells must be thicker than others to maximize absorption. Cells often have textured surfaces and may also be coated with an antireflective coating to minimize reflection of sunlight away from the cells.

3.3 THE PV MODULE

In order to obtain adequate output voltage, PV cells are connected in series to form a PV module. With silicon, single cell open-circuit voltages are typically close to 0.6 V, and maximum power voltages are close to 0.5 V at 25°C. Historically, when the majority of PV systems were stand-alone systems with battery backup, it was desirable to combine a sufficient number of cells in series to provide effective battery charging. Because nominal 12 V lead-acid battery charging voltages range from 14 to 16 V, allowing for operation of modules at elevated temperatures, 36-cell modules were the norm for many years. These modules operated close to their maximum power points when charging 12 V lead-acid batteries.

Recently, however, two things have changed the design parameters for PV modules. First of all, since 1999, more PV modules have been used in grid-connected systems worldwide than in stand-alone systems [1]. In a straight grid-connected system with no battery backup, it is now common to design the PV array so that the maximum open-circuit voltage is just under 600 V dc. These arrays are connected directly to maximum power point tracking (MPPT) inverter inputs. So many modern modules have 54 to 72 cells, and sometimes even more, with correspondingly higher open-circuit voltages and higher module power ratings.

The other change has been in the technology of charge controllers. When the first edition of this book was published, charge controllers connected the array directly to the batteries at the battery voltage. So the array was designed to operate close to the normal charging voltage of the batteries to approximate maximum power operation. Modern charge controllers, however, incorporate MPPT input circuitry, so that the array can operate at a maximum power voltage that exceeds the battery voltage and still, via the charge controller, deliver the correct charging voltage to the batteries.

An important observation relating to the series connection of PV cells relates to shading of individual cells. If any one of the cells in a module should be shaded, the performance of that cell will be degraded. Because the cells are in series, this means that the cell may become forward biased if other unshaded modules are connected in parallel, resulting in heating of the cell. This phenomenon can cause premature cell failure. To protect the system against such failure, modules are generally protected

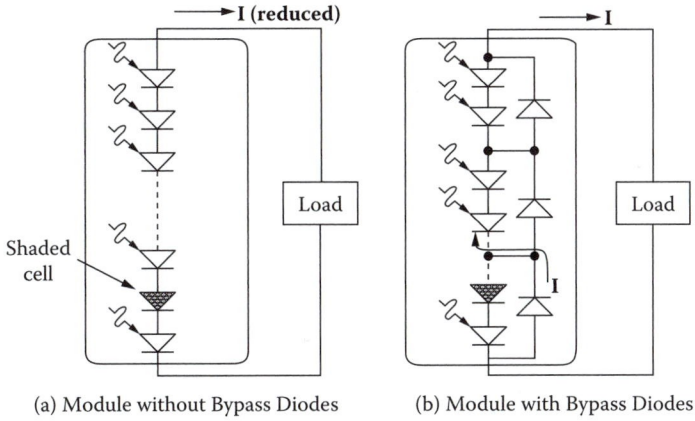

(a) Module without Bypass Diodes (b) Module with Bypass Diodes

Figure 3.6 Use of bypass diodes to protect shaded cells.

with **bypass diodes**, as shown in Figure 3.6. If PV current cannot flow through one or more the PV cells in the module, it will flow through the bypass diode instead.

When cells are mounted into modules, they are often covered with antireflective coating, and then with a special laminate to prevent degradation of the cell contacts. The module housing is generally metal, which provides physical strength to the module. When the PV cells are mounted in the module, they can be characterized as having a nominal operating cell temperature (NOCT). The NOCT is the temperature the cells will reach when operated at open circuit in an ambient temperature of 20°C at AM 1.5 irradiance conditions, $G = 0.8\ \text{kW/m}^2$, and a wind speed less than 1 m/s. For variations in ambient temperature and irradiance, the cell temperature (in °C) can be estimated quite accurately with the linear approximation that

$$T_C = T_A + \left(\frac{NOCT - 20}{0.8}\right)G. \tag{3.5}$$

The combined effects of irradiance and ambient temperature on cell performance merit careful consideration. Because the open-circuit voltage of a silicon cell decreases by 2.3 mV/°C, the open-circuit voltage of a module will decrease by $2.3n$ mV/°C, where n is the number of series cells in the module.

Hence, for example, if a 36-cell module has an NOCT of 46°C with V_{OC} = 19.40 V at 25°C and when G = 0.8 kW/m^2, then the cell temperature will rise to 62.5°C when the ambient temperature rises to 30°C and G increases to 1 kW/m^2. This 37.5°C increase in cell temperature will result in a decrease of the open-circuit voltage to 16.3 V, a 16% decrease. Furthermore, excessive temperature elevation may cause the cell to fail prematurely.

Finally, a word about module efficiency. It is important to note that the efficiency of a module will be determined by its weakest link. Because the cells are series connected, it is important that cells in the module be matched as closely as possible.

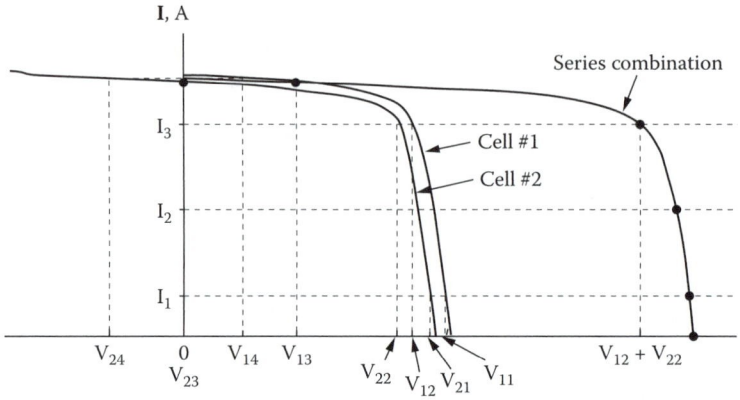

Figure 3.7 Determination of composite operating characteristic for a PV module.

If this is not the case, while some cells are operating at peak efficiency, others may not be optimized. As a result, the power output from the module will be less than the product of the number of cells and the maximum power of a single cell.

Figure 3.7 shows how individual cell operating characteristics must be combined to produce the composite operating characteristic of the module. Note that because all cells carry the same current, the voltages of individual cells must adjust accordingly to meet the current constraint imposed by the external load. The composite *I-V* characteristic for *n* cells in series is thus obtained by adding the individual cell voltages that result for each cell to deliver the required current. The maximum current available from the module is affected by the cell with the lowest current under specific load conditions at the operating irradiance. Hence, it is desirable for all cells in the module to have identical *I* versus *V* versus irradiance curves. It is not difficult to imagine circumstances where the power output of one cell may differ from the power output of other cells, especially if the module is operated in a location that flocks of birds find to be a convenient rest stop.

To arrive at the composite *I-V* curve for two cells in series, simply add the corresponding voltages required to produce the required current, as shown in Figure 3.7. Note that if one cell has a larger I_{sc} than the other cell, as indicated, while the voltage across the combination is zero, the voltage of each cell will not necessarily be zero. In Figure 3.7, cell #2 has a lower I_{sc} than cell #1, and thus its voltage goes negative until its current equals the current of cell #1. This means that cell #1 is generating power, and cell #2 is dissipating this power.

When 30 or more cells are connected in series, if they do not have identical *I-V* curves, then when the module is short-circuited, some of the cells will be generating power while others will be dissipating power. The greater the mismatch among cells in the module, the greater the level of power dissipated in the weaker cells. If all cells are perfectly matched, no power is dissipated within the module under short-circuit conditions, and the overall efficiency of the module will be the same as the efficiency of individual cells.

3.4 THE PV ARRAY

If higher voltages or currents than are available from a single module are required, modules must be connected into arrays. Series connections result in higher voltages, while parallel connections result in higher currents. Just as in the case for cells, when modules are connected in series, it is desirable to have each module's maximum power production occur at the same current. When modules are connected in parallel, it is desirable to have each module's maximum power production occur at the same voltage. Thus, when mounting and connecting modules, the installer should have this information available for each module. In later chapters, criteria will be developed for deciding how many modules to connect in series and how many modules to connect in parallel to optimize the performance of a PV array that provides power to the designer's choice of inverter or charge controller. Which, if any, of the positive or negative output leads of an array should be grounded will also be discussed later.

3.5 ENERGY STORAGE

3.5.1 Introduction

Most electrical engineers have some familiarity with batteries, especially the ones used in flashlights, toys, laptop computers, watches, calculators, video cameras, automobiles, and golf carts. The extent of familiarity normally extends to the decision as to whether to purchase C-Zn batteries or whether to buy the more expensive, but longer-lasting, alkaline cells. Remembering to get the correct size, whether a 9 V or a 1.5 V AA battery, is another of life's energy storage dilemmas. In this section, considerations in battery selection for PV systems are presented. Alternatives to battery energy storage are also presented, so that the PV design engineer will be able to make educated choices, based on sound economic and engineering principles, as to which type of energy storage to incorporate into a specific PV system design.

Many different types of rechargeable batteries suitable for PV applications are currently available. Although several rather exotic technologies are now available, the lead-acid battery is still the most common for relatively economical storage of useful quantities of electrical energy, and will probably remain so for at least the next few years. However, unless improvements in energy density, cost, and lifetime are made with lead-acid technology, other promising technologies may overtake lead-acid technology. For example, Ni-Cd batteries have been in common use in applications that require sealed batteries capable of operating in any position, and still require high energy density. But they are more expensive per joule stored than lead-acid units. Nickel metal hydride and several lithium technologies may also one day provide cost-effective storage.

But, then, batteries are not the only way to store energy. It is also possible to store energy by producing hydrogen, by pumping water uphill, by spinning a flywheel, or by charging a large chemical capacitor. This section explores the applications, advantages, and disadvantages of some of the storage methods currently in use as well as some of the new ideas that have been proposed.

3.5.2 The Lead-Acid Storage Battery

3.5.2.1. Chemistry of the Lead-Acid Cell

Figure 3.8 shows the basics of lead-acid cell operation. In simple terms, the battery consists of a lead cathode and a lead oxide (PbO_2) anode immersed in a sulfuric acid solution. The discharging reaction at the anode consists of the exchange of oxygen ions from the anode with sulfate ions of the electrolyte. At the cathode, the discharge involves sulfate ions from the electrolyte combining with lead ions to form lead sulfate. Removal of sulfate ions from the solution reduces the acidity of the electrolyte. To maintain charge neutrality, two electrons must enter the anode terminal, and two electrons must leave the cathode terminal via the external circuit for each of the two sulfate ions that leave the electrolyte. This corresponds to a positive current leaving the anode terminal, which is consistent with the passive sign convention for power being delivered. Specifically, the chemical process at the anode and cathode during discharge are represented by the respective equations

$$PbO_2 + 4H^+ + SO_4^{2-} + 2e^- \rightarrow PbSO_4 + 2H_2O \tag{3.6a}$$

and

$$Pb + SO_4^{2-} \rightarrow PbSO_4 + 2e^-. \tag{3.6b}$$

The oxygen ion, which enters the solution from the anode, combines with two hydrogen ions from the sulfuric acid electrolyte to form water.

When an external voltage source greater than the voltage produced by the reactions at the anode and cathode is applied across the battery terminals, the discharge

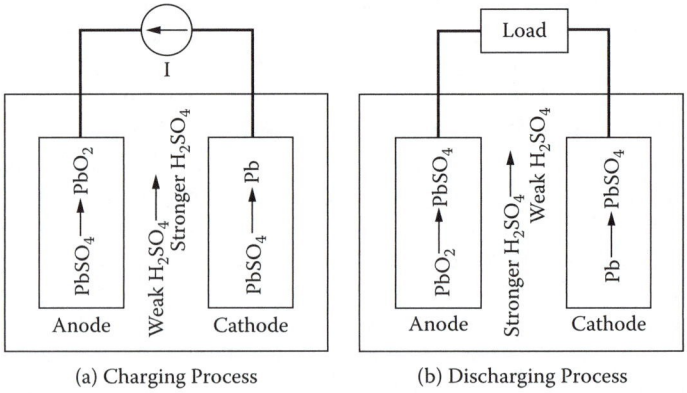

Figure 3.8 Charge and discharge of the lead-acid cell.

process is reversed, and current flows into the anode rather than out of it, thus charging the battery. The chemical processes are then reversed at the anode and cathode to

$$PbSO_4 + 2H_2O \rightarrow PbO_2 + 4H^+ + SO_4^{2-} + 2e^- \tag{3.7a}$$

and

$$PbSO_4 + 2e^- \rightarrow Pb + SO_4^{2-}. \tag{3.7b}$$

To maintain charge balance, the two electrons shown in Equations (3.7a) and (3.7b) must leave the anode via the external connection and travel to the cathode. As the electrons leave the anode, they are replaced by an oxygen ion, which combines with a lead ion in the anode that was created by a sulfate ion entering the electrolyte. As these two electrons are added to the cathode, a sulfate ion is liberated from the cathode to the solution. The two electrons combine with a lead ion, producing a lead atom. The sulfate ions liberated from the cathode and the anode to the solution increase the concentration of sulfuric acid in the electrolyte solution by balancing with hydrogen ions in the solution, while the oxygen ion left behind by the dissociated water molecule is left alone in solution. The positive potential difference between anode and cathode causes the negatively charged oxygen ion to migrate to the anode, where it becomes a part of the anode reaction, ultimately being converted to lead oxide at the anode, as previously indicated.

When the battery discharges, ultimately the surfaces of both electrodes are converted to lead sulfate. If excessive lead sulfate is allowed to build up on the electrodes, the effective surface area of the electrodes is reduced, and cell performance may be affected. It is thus important to avoid fully discharging a lead-acid battery. During the charging process, some of the hydrogen ions combine with free electrons and are converted into gaseous hydrogen. At a certain point during charging, when the cathode is fully converted back to lead, there is no more sulfate at the cathode to maintain continuity in the charging current. At this point, if charging is continued, the electrons entering the cathode can no longer release sulfate ions, so the electrons continue into the electrolyte, combine with hydrogen ions, and produce hydrogen gas. This phenomenon is called gassing, and it is generally not desirable to charge a battery beyond this point because the hydrogen gas can present a safety hazard. Occasional charging of a battery to the gassing stage so the gas bubbles up through the electrolyte can be useful, however, because the bubbling action tends to perform a cleaning action on the electrodes and a mixing action on the electrolyte.

The potential difference between electrodes in the lead-acid system is approximately 2.12 V when the cell is fully charged. As cells are connected in series, multiples of 2.12 V can be achieved. Most commonly, either three or six cells are connected in series, producing nominal voltages of 6 or 12 V.

3.5.2.2 Properties of the Lead-Acid Storage Battery

Ideally, the charging and discharging processes of the lead-acid system should be reversible. In reality, however, they are not. The temperature of operation, the rate of discharge, and the rate of charge all affect the performance of the battery. Because the electrical path of the battery presents ohmic resistance, some of the electrical energy intended for charging, that is, storage, is converted to heat. When hydrogen is lost, it also represents an energy loss. Typically, the charging process is about 95% efficient. The discharge process also results in some losses due to internal resistance of the battery, so only about 95% of the stored energy can be recovered. The overall efficiency of charging and discharging a lead-acid battery is thus about 90%.

Because battery losses to internal resistance are proportional to the square of the current, this means that high-current charging or high-current discharging will tend to result in higher internal losses and less overall performance efficiency. This effect is offset somewhat by the increase in temperature of the battery during charging or discharging at high rates due to the higher I^2R losses. It turns out that a warmer battery can hold more charge. It also turns out that if a battery is too warm for too long, its life expectancy is shortened, so charging and discharging rates need to be carefully observed so that they will not exceed rated values for a specific battery.

The amount of energy stored in a battery is commonly measured in ampere hours. While ampere hours are technically not units of energy, but, rather, units of charge, the amount of charge in a battery is approximately proportional to the energy stored in the battery. If the battery voltage remains constant, then the energy stored is simply the product of the charge and the voltage.

The capacity of a battery is often referred to as C. Thus, if a load is connected to a battery such that the battery will discharge in x hours, the discharge rate is referred to as C/x. The charging of a battery is measured in a similar fashion. Figure 3.9 indicates the effect of discharging rates on the relative amount of charge that can be obtained from a lead-acid battery. Note that higher discharge rates result in less charge being available as energy to a load. At higher charging rates, a smaller fraction of the charging energy is used for charging, and a larger fraction is used to heat up the battery. The battery can be fully charged at higher charging rates, but it takes more energy at higher charging rates to obtain full charge.

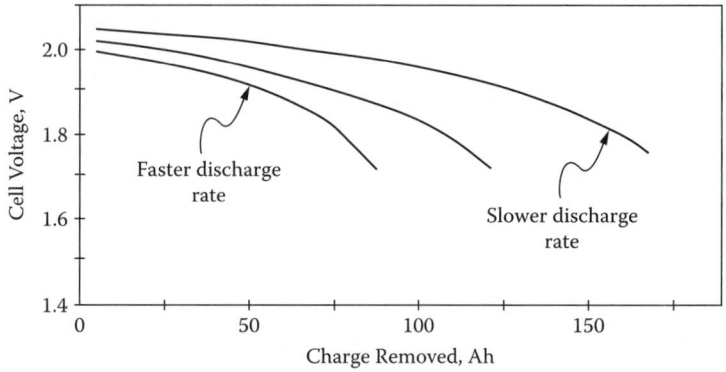

Figure 3.9 Effect of discharge rate on available energy from a lead-acid battery.

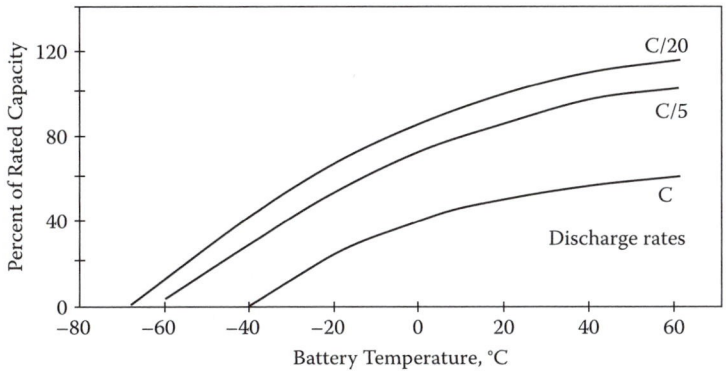

Figure 3.10 Effect of temperature and discharge rate on available energy from a lead-acid battery.

These phenomena can be explained on the basis of the Thevenin equivalent circuit of the battery, which includes a resistance in series with the open-circuit cell voltages. Because the charging or discharging current must pass through the internal (Thevenin) resistance of the battery, a power loss equal to I^2R occurs. For constant current charging or discharging, the charge delivered to or removed from the battery is given by $Q = It$. Because Q is proportional to energy storage, the energy stored is thus proportional to I, whereas the energy lost is proportional to I^2. Hence, at higher charging or discharging rates, a larger fraction of available charging energy is lost to resistive heating. This effect is compensated for somewhat by the fact that warm batteries are capable of storing more charge than cold batteries, as shown in Figure 3.10.

Figure 3.10 shows the effect of discharge rates and temperature on the relative amount of charge that a battery can deliver. Again, slower discharge rates result in a higher overall amount of charge being delivered by the battery. This figure is especially notable to those readers who live in northern latitudes and find it difficult to start their automobiles in subzero weather.

Depending on specific composition of the electrodes, lead-acid batteries may be optimized for shallow discharge or for deep-discharge operation. The shallow discharge units typically have a small amount of calcium combined with the lead to impart greater strength to the otherwise pure lead. The plates can then be made thinner with greater area to produce higher starting currents. These units should not be discharged to less than 75% of their capacity. In automobile applications, these are satisfactory operating conditions, because the battery is needed primarily for operation of the starter motor until the engine starts. After this point, the alternator takes over, recharging the battery and operating the automobile electrical systems. The shallow discharge units have a smaller quantity of lead and are correspondingly less expensive.

Deep-discharge lead-acid batteries use antimony to strengthen the lead and can be cycled down to 20% of their initial capacity. The plates are thicker, with less

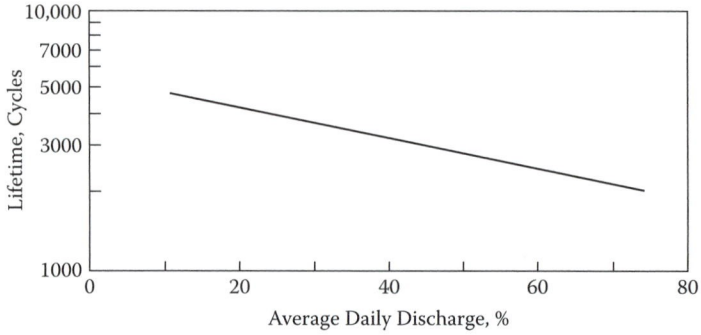

Figure 3.11 Lead-acid battery lifetime in cycles versus depth of discharge per cycle. (Adapted from *Maintenance and Operation of Stand-Alone Photovoltaic Systems*, Sandia National Laboratories, Albuquerque, NM, 1991.)

area and are hence designed for sustained lower-level currents. These batteries are designed for use in golf carts, marine applications, and electric forklifts, as well as for PV system use. Although deep-discharge batteries are designed for deep-discharge applications, their lifetime in cycles depends on the depth of discharge during normal operation. Figure 3.11 shows how the depth of discharge affects the number of operating cycles of a deep-discharge battery. The PV system designer must carefully consider the trade-off between using more batteries operating at shallower discharge rates to extend the overall life of the batteries versus using fewer batteries with deeper discharge rates and the correspondingly lower initial cost. Life-cycle costing of systems will be considered in Chapter 8.

Lead-acid batteries are available in vented (flooded) and nonvented (sealed) enclosures. There is a trade-off between lead-antimony and lead-calcium batteries when venting is considered. In certain lead-calcium batteries, minimal hydrogen and oxygen are lost during charging. This means minimal water is lost from the electrolyte. As a result, it is possible to seal off the cells of these batteries, making them essentially maintenance free. The trade-off, however, is that if these batteries are either purposely or inadvertently discharged to less than 75% of their maximum charge rating, their expected lifetime may be significantly shortened.

Lead-antimony electrodes, on the other hand, may be discharged to 20% of their maximum charge rating. This means that a 100 Ah lead-calcium battery has only 25 Ah available for use, while a 100 Ah lead-antimony battery has 80 Ah available for use, or more than 3 times the availability of the lead-calcium unit. However, the lead-antimony unit produces significantly more hydrogen and oxygen gas from dissociation of water in the electrolyte, and thus water must be added to the battery relatively often to prevent the electrolyte level from falling below the top of the electrodes.

Water loss can be reduced somewhat by the use of cell caps that catalyze the recombination of hydrogen and oxygen back into water, which returns to the cell. Other cell caps have a flame-retardant structure that prevents any flame external to the battery from entering the cell. In any case, the bottom line is that nonsealed,

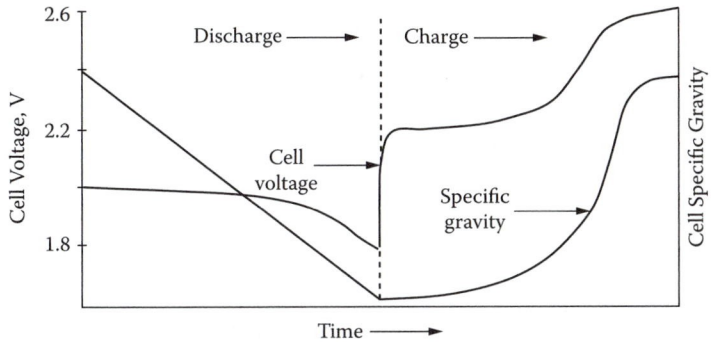

Figure 3.12 Variation of cell electrolyte specific gravity and cell voltage during charge and discharge at constant rate. (Adapted from *Maintenance and Operation of Stand-Alone Photovoltaic Systems*, Sandia National Laboratories, Albuquerque, NM, 1991.)

deep-discharge batteries have open vents and require relatively frequent addition of water for maintenance, depending on cycle time and depth.

An advantage of removable cell caps on batteries is that it is possible to measure the specific gravity of the cells of these batteries. Because one of the most accurate measures of the state of charge of a battery is the specific gravity of the electrolyte, this provides a convenient means of finding faulty battery cells. Figure 3.12 shows how the electrolyte specific gravity and voltage vary during charging and discharging of the battery. Note particularly the effect of the voltage drop across the internal cell resistance during charging and discharging.

For applications where maintenance of batteries is inconvenient, maintenance-free, sealed deep-cycle batteries exist, but are generally at least double the price of equivalent capacity nonsealed lead-acid units. These batteries are valve regulated to recombine gases, with additional treatment to immobilize the electrolyte. *Gel cells* are sealed, lead-acid, deep-cycle batteries that have silica gel added to the electrolyte. Absorbed Glass Mat (AGM) batteries have highly absorbent glass mat separators between plates to bind the electrolyte. In addition to being maintenance free, sealed deep-discharge units do not leak and have long life and low self-discharge. Charging requirements for sealed lead-acid batteries differ somewhat from nonsealed units. While nonsealed units are designed for occasional overcharge, sealed units should not be overcharged.

Once a battery type has been selected, a suitable enclosure must be selected or designed, and appropriate safety precautions must be taken during installation, operation, and maintenance of lead-acid batteries. Two important references are the *National Electrical Code* (*NEC*) [3] and the *IEEE Recommended Practice for Installation Design and Installation of Valve-Regulated Lead-Acid Storage Batteries for Stationary Applications* (IEEE 1187-2002 [4]). Annex C of IEEE 1187-2002 provides information on the degree to which the life of lead-acid batteries is shortened if the batteries are operated at temperatures above 25°C. For example, at a temperature of 35°C, the battery lifetime will be shortened to 44% of its expected lifetime

Table 3.1 Typical Deep-Cycle Lead-Acid Battery Cell Properties Versus State of Charge

State of Charge (%)	Specific Gravity	Cell Voltage (V)	Voltage of 12 V Battery	Freezing Point (°F)
100	1.265	2.12	12.70	−71
75	1.225	2.10	12.60	−35
50	1.190	2.08	12.45	−10
25	1.155	2.03	12.20	>3
0	1.120	1.95	11.70	>17

Source: Based on data from *Maintenance and Operation of Stand-Alone Photovoltaic Systems*, Sandia National Laboratories, Albuquerque, NM, 1991.

at 25°C. This means that it is important to provide adequate ventilation for batteries, not only to vent any hydrogen or other gases, but also to maintain a battery operating temperature close to 25°C. The serious PV system designer should be sure to have access to the *NEC* and IEEE 1187.

As a final note on lead-acid batteries, it is important to be aware of the temperatures at which the electrolytes of batteries will freeze. As the electrolyte becomes more acidic, the freezing temperature is depressed. Hence, fully charged batteries may be operated at low temperatures, while batteries in some state of discharge will need to be kept warmer. Table 3.1 summarizes the relationships between state of charge, specific gravity, cell voltage, and electrolyte freezing point.

When a PV system is designed that uses lead-acid batteries for storage, the designer needs to determine proper charging rates for the batteries and also should determine the ventilation needs of the battery system. These considerations will be covered in Chapter 6.

3.5.3 The Nickel Cadmium Storage Battery

3.5.3.1 Chemistry of the Nickel Cadmium Storage Battery

Ni-Cd batteries use nickel hydroxide for the anode plates and cadmium oxide for the cathode plates, in a structure similar to that of the lead-acid system. The electrolyte in the Ni-Cd system is potassium hydroxide. The NiOH anode is generally made of nickel fibers mixed with graphite- or nickel-coated plastic fibers. Small quantities of other materials such as barium and cobalt compounds are also added to improve performance. The cathode is also frequently made of a cadmium-coated plastic fiber. If the cathode is not a coated plastic, then it is commonly mixed with iron or nickel. The fiber structures of anode and cathode maximize the surface area while minimizing the amount of relatively expensive nickel and cadmium required for the electrodes.

The overall discharge reaction at the electrodes is given by

$$2\mathrm{NiOOH} + 2\mathrm{H_2O} + \mathrm{Cd} \rightarrow 2\mathrm{Ni(OH)_2} + \mathrm{Cd(OH)_2}. \tag{3.8}$$

The reaction is reversed in the charging direction. The voltage of the fully charged cell is 1.29 V. Unlike the lead-acid system, where the specific gravity of the electrolyte changes measurably during discharge or charge, the KOH electrolyte of the Ni-Cd system changes very little during battery operation. In some batteries, LiOH is also added to the electrolyte to improve cycle life and also to improve higher-temperature operation.

3.5.3.2 Properties of the Nickel Cadmium System

Ni-Cd batteries are more robust than lead-acid batteries. They can survive freezing and high temperatures, they can be fully discharged, and they are less affected by overcharging. As a result, in some applications, Ni-Cd batteries may be a better choice because their robust nature may enable the elimination of the system charge controller. If the batteries are to be used in a location where access for maintenance is difficult, the higher cost of these batteries can also often be justified.

The most common industrial Ni-Cd batteries are of the vented pocket plate type. These batteries are in use in industrial, military, and space applications where the previously mentioned properties are important. Capacities range from small sizes up to more than 1200 Ah. Three different thicknesses are used for electrodes, depending on whether the battery is designed for high, medium, or low discharge rates.

Energy densities range from 20 Wh/kg for the pocket plate batteries to more than 50 Wh/kg when cells are made of the plastic-bonded plates.

Unlike the lead-acid system, which loses capacity under conditions of heavy discharge, the Ni-Cd system can be discharged at rates up to C over a wide temperature range, while still providing more than 90% of its capacity to the load. This is particularly attributable to the very low internal resistance of the cells, which, depending on cell area, can be less than a milliohm.

If Ni-Cd batteries are charged and then left unused, they will lose charge at the rate of approximately 2% per day for the first few days, but then stabilize to a relatively low loss rate. Over a 6-month period, the total loss is typically about 20%, depending on whether the battery is a high, medium, or low discharge rate battery. The higher the discharge rating of the battery, the greater the loss of charge over time. Loss of charge is also temperature dependent. The loss rate is greater at higher temperatures, but at a temperature of –20°C, there is almost no loss at all.

The lifetime of a Ni-Cd battery depends on how it is used, but is less dependent on depth of discharge than that of lead-acid batteries. A lifetime of at least 2000 cycles can be expected for a battery when it is not used extensively at elevated temperatures. As a result, under certain applications and operating conditions, a battery may last as long as 25 years, while under more frequent cycling, the lifetime may be reduced to 8 years. It is not unreasonable to expect a Ni-Cd battery to last twice as long as its lead-acid counterpart.

Ni-Cd batteries may be charged at either constant current, constant voltage, or somewhere in between. This makes them especially useful in PV applications, because a PV array tends to act as a constant current source, depending on the cloud cover at the moment. If Ni-Cd batteries are overcharged, gassing will occur, and the water in the electrolyte will decompose, creating a need for more maintenance. Normally, charging from full discharge to full charge should be accomplished in 5 to 7 hours, with a charging voltage between 1.50 and 1.65 V per cell. The ampere-hour charging efficiency of the most efficient Ni-Cd batteries approaches 85%.

Sealed Ni-Cd batteries are also available, but their cost is higher. Their advantages over the vented units include maintenance-free operation, high discharge rates and high charge rates, along with wide temperature range. However, presently they are available only in smaller sizes.

Disadvantages of Ni-Cd batteries include difficulty in determining the state of charge of the batteries and the toxicity of the cadmium, which creates an environmental concern during production and disposal. They are also more expensive than most other rechargeable battery technologies.

As electric and hybrid automobiles become more popular, the demand for high-energy-density batteries will likely lead to the replacement of nickel cadmium technology with lower-cost, higher-energy-density technologies.

3.5.4 Other Battery Systems [5]

A number of other battery systems are currently in use, and others seem to show promise for future use. The nickel-zinc battery is a combination of the Ni-Cd system and the Cu-Zn system, which provides some attractive features, including long life and a capacity advantage. The specific energy of the Ni-Zn system is double the specific energy of the Ni-Cd system. The overall discharge chemical reaction is given by

$$2\mathrm{NiOOH} + \mathrm{Zn} + 2\mathrm{H_2O} \rightarrow 2\mathrm{Ni(OH)_2} + \mathrm{Zn(OH)_2}. \tag{3.9}$$

If the battery is allowed to overcharge, it dissociates the water in the KOH electrolyte, resulting in gassing and the release of hydrogen and oxygen. Hence, charging must be controlled to avoid overcharging. The fully charged cell voltage is approximately 1.73 V.

Ni-Zn battery systems can be made in sizes in excess of 200 Ah. At normal operating temperatures ($\cong 20°C$) and reasonable discharge rates ($\cong C/3$), the cell voltage remains quite stable at about 1.6 V over most of the discharge range.

Another technology that is becoming very popular, particularly in smaller applications such as camcorders and laptop computers, is the nickel-metal hydride (NIMH) battery. This battery replaces the cadmium cathode with an environmentally benign metal hydride cathode, allowing for higher energy density at the cathode and a correspondingly longer lifetime or higher capacity, depending on the design goal. The anode is the same as in the Ni-Cd cell, and KOH is used as the electrolyte. The overall discharge reaction is

$$\mathrm{MH} + \mathrm{NiOOH} \rightarrow \mathrm{M} + \mathrm{Ni(OH)_2}. \tag{3.10}$$

The NIMH cell requires clever use of an oxygen recombination system to prevent loss of oxygen during the charging cycle. Electrodes are also carefully sized to ensure that the useful capacity of the battery is determined by the anode electrode to ensure that the cathode will moderate any overcharge or overdischarge condition.

A number of additional battery types are currently under investigation for PV and similar uses, such as electric vehicles. A partial listing of these batteries includes zinc/silver oxide, metal/air, iron/air, zinc/air, aluminum/air, lithium/air, zinc/bromine, lithium-aluminum/iron sulfide, lithium-aluminum/iron disulfide, sodium/sulfur, sodium/metal chloride, and several variations of lithium ambient-temperature batteries.

The lithium technologies appear quite attractive due to energy densities up to 150 Wh/kg, cell voltages up to 4 V, and long charge retention and shelf life. Other problems must be overcome, however, before use of these batteries becomes widespread. For example, they have low cycle life, relatively poor high-rate performance, relatively poor low-temperature performance, capacity fading, and potential safety problems. It should be remembered that metallic lithium is highly reactive and presents a danger if it contacts water or humans.

The young engineer can expect to see progress in these new technologies during his or her career, especially now that public demand for electric vehicles, hybrid vehicles, and PV systems with battery storage is increasing.

In addition to battery storage, which is presently the most common method of energy storage for PV systems, a number of other energy storage methods are either in use or have been proposed. Many of these methods will also require a significant research effort to bring them into cost-effective, production-scale use.

3.5.5 Hydrogen Storage [6]

Aside from using batteries to store energy produced by PV systems, it is also possible to use hydrogen as a storage medium. The advantage of using hydrogen for energy storage is that when it is desired to recover the stored energy, the hydrogen is reacted with oxygen to form water, which is an exothermic reaction that does not have any carbon byproducts. If the hydrogen is burned in air to produce steam to turn a turbine to generate electricity, water is still produced as the by-product, with a relatively minimal amount of nitrogen oxides. In a fuel cell, hydrogen combines with oxygen to produce water, electricity, and heat. Of course, the hydrogen does involve need to be used to produce electricity. It can also be used as an engine fuel to power a vehicle, such as an automobile, bus, or space shuttle. Indirectly, the result is a solar-powered vehicle that stores energy in hydrogen rather than in batteries. The engineering community has not yet developed a battery-powered, space shuttle booster engine.

Perhaps the most important advantage of hydrogen storage is the energy density of hydrogen. One of the highest energy densities in conventional fuels is that of gasoline, which contains approximately 1,047,000 Btu/ft^3. To store the same amount of energy in high-capacity storage batteries would require approximately 120 batteries having approximately 2000 Btu/ft^3 storage capacity. The storage capacity of liquid hydrogen is close to 240,000 Btu/ft^3, and gaseous hydrogen can store approximately 47,000 Btu/ft^3. Because water cannot be decomposed into gasoline, however, and because gasoline burning produces CO_2, among other undesirable combustion products, clean-burning hydrogen is a more attractive alternative for combustion.

Aside from liquid and gaseous hydrogen storage, other means of obtaining relatively high energy density, including various hydrides, are also available.

Electrolysis of water is a common method of producing hydrogen. The output of a PV system is conveniently dc at a level consistent with that needed for relatively efficient electrolysis. Production of hydrogen by electrolysis may seem to be simply a matter of passing a current through water to produce hydrogen and oxygen, just as many readers have already done in the chemistry lab. The challenge is to produce hydrogen efficiently with a system whose power output varies significantly whenever a cloud passes over the PV array.

Once the hydrogen is produced, it can be used on site, or it can be transported to other locations. Hydrogen proponents have suggested that hydrogen be produced in tropical latitudes having greater annual insolation and shipped to regions having less sun. Thus, after the production of hydrogen, storage and transport of the gas also present engineering challenges.

Recent interest in fuel cells for use in automotive applications has once again piqued interest in the efficient and convenient production of hydrogen. The same technology that may emerge for automotive fuel cell applications may very well spin off into PV energy storage applications as well.

3.5.6 The Fuel Cell

Fuel cells provide a convenient means of reacting hydrogen and oxygen directly to produce electricity and water. They utilize essentially the inverse of the electrolysis process. The first fuel cell was built in 1839 by Sir William Grove [7]. The high cost of fuel cells, however, had kept them out of practical use until NASA decided to use them on space flights to provide power and water.

Fuel cells are being constructed in a wide range of power output capabilities. The smaller cells are sized to power video cameras and laptop computers. On a larger scale, The Southern California Gas Company had ten 200 kW plants on line in 1993, and experimental units in the megawatt range are now in operation [7]. The fuel cell reaction is exothermic, so if cleverly applied, the cells can supply both heat and electricity to a building. By capturing the heat, the overall system efficiency is increased, provided that the heat can be put to good use, such as for district heating or other space or water heating application.

The alkaline cell (AFC) contains a 30% solution of KOH as the electrolyte. The proton exchange (or polymer electrolyte) membrane cell (PEMFC) incorporates a semipermeable membrane that will pass hydrogen ions, resulting in a proton exchange. The phosphoric acid cell (PAFC) is based on a 103% phosphoric acid solution. The molten carbonate cell (MCFC) is based on a eutectic mixture of molten lithium carbonate and potassium carbonate. The solid oxide cell (SOFC) is based on a stabilized zirconium oxide electrolyte [8].

The low-temperature cells are approximately twice as efficient as conventional heat engines in the conversion of energy stored in hydrogen to electricity. This makes fuel cells a very attractive possibility for energy storage and conversion to electricity in an electric vehicle because electric motors can be operated at very high efficiency, especially with modern solid-state controllers. The present disadvantage, of course, is that fuel cells remain quite costly and will require further development to reduce their cost. Fuel cells also require a source of hydrogen, so a means for providing fuel for vehicular applications of fuel cells will need to be created.

3.5.7 Other Storage Options

Perhaps one of the simplest means of storing energy from the sun is to fill a water storage tank so the water can be used after sundown. Maybe the use will be for drinking or irrigation, or maybe the water will be used to turn a turbine to generate electricity before it is used for some other purpose. It is not difficult to calculate the amount of potential energy stored in a gallon of water raised to a height of 10 ft above ground.

Other possible storage mechanisms include compressed air, flywheels, superconducting magnets, and chemical capacitors. All of these can be shown to be useful for certain end uses, but, in general, the cost per kWh stored is quite high at present.

3.6 PV SYSTEM LOADS

The importance of operating PV modules near their maximum power points has already been discussed. Maximum power operation is a challenging problem because it requires that the system load be capable of using all power available from the PV system at all times. Not only does this mean that the system load needs to be maximum when PV system output is maximum, but it also means the system load must adjust itself rather quickly at the onset or dispersion of cloud cover. In any case, the *I-V* characteristic of the ideal load will intersect the locus of maximum power points on the *I-V* characteristics of the PV array for varying illumination levels. The *I-V* characteristic for this ideal load is shown in Figure 3.13 along with the *I-V* characteristics of several other common PV system loads.

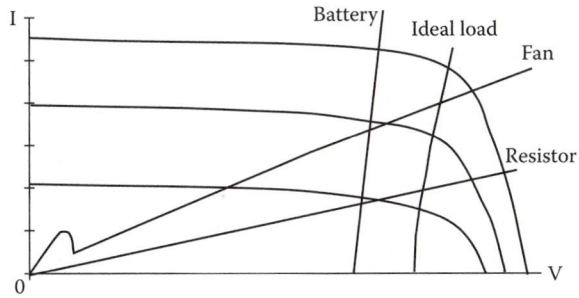

Figure 3.13 *I-V* characteristics for several common loads along with ideal load *I-V* characteristic for maximum power operation of a PV system.

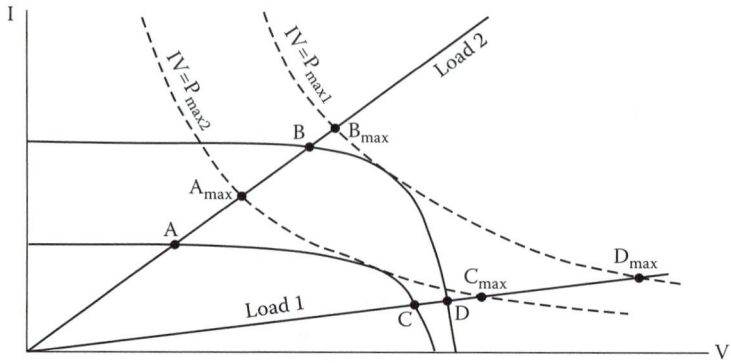

Figure 3.14 Operation of the maximum power point tracker (MPPT).

If the intersection of the *I-V* characteristic of a load with that of a PV source departs significantly from the maximum power point of the source, it may be desirable to employ an electronic maximum power point tracker between the array and the load. Maximum power point trackers (MPPTs) generally employ pulse-width modulation techniques to switch from an input dc voltage to an output dc voltage at a different level, similar to a switching dc power supply. The MPPT employs a feedback loop to sense the PV array output power and change the array output voltage accordingly until the output power is maximized.

Figure 3.14 shows how the MPPT electronically moves the operating point along the maximum power hyperbola (IV = constant) associated with the PV array until it intersects the load *I-V* characteristic. The price paid for this MPPT is not just the dollar price because there is also some power loss in the MPPT. It is thus necessary to justify that the expense of the MPPT will be recovered by the value of the additional energy made available by the device. In Chapter 8, this will be shown to be a straightforward analysis.

Another consideration for loads on PV systems is the trade-off between lifetime operating cost of a load and initial cost of the load. Sometimes a more efficient load is available at a higher initial cost, and sometimes a more efficient load is available at a lower initial cost. If the initial cost of a load is lower, and there are no hidden maintenance costs, then it is a no-brainer to use the more efficient load, unless aesthetic or functional considerations rule it out.

On the other hand, if the initial cost of a more efficient load is higher, then it is important for the PV system designer to perform an economic analysis of both system options, to determine which load is a better choice. In particular, the choice between relatively inexpensive first-cost incandescent lamps and more expensive but more efficient fluorescent lamps involves an interesting cost analysis. Now that LED lamps are becoming commonplace, but at relatively higher cost and correspondingly longer lifetimes, these devices add yet another interesting dimension to the choice of lighting sources.

Also, the cost difference between dc refrigerators and ac refrigerators is considerable, but the efficiency of the dc units was once significantly higher than the ac units, so the purchase of a $2500 dc refrigerator may have been justified. These considerations will be accounted for in Chapter 8.

3.7 PV SYSTEM AVAILABILITY

Critical loads are defined as loads for which power is required at least 99% of the time. Noncritical loads require power for at least 95% of the time. It is important to recognize that these definitions involve a statistical distribution of downtime over the expected lifetime of the system. In other words, a critical system may have availability during any given month or year of its operation that may be less than 99%, while it will have availability during other years in excess of 99%. The 99% figure is considered to be an average over the lifetime of the system.

The reader is probably familiar with causes of downtime on conventional grid-connected systems. Sometimes it is a failure of an electrical generator. Sometimes it is a system overload that requires parts of the system to be shut down selectively to prevent generator overload. Sometimes a tree falls and breaks a power line. About every 100 years, an ice storm of the century causes long-term power outages for millions of utility customers, and sometimes a short circuit occurs and causes a circuit breaker to trip. Even hurricanes have been known to cause an occasional power outage. Clearly, many other situations may occur that will result in the loss of power.

PV systems are subject to similar failure modes. Loose or corroded connections, battery failure, controller failure, and module failure represent a few of the things that might go wrong in a PV system. However, a good preventive maintenance program can keep these failure modes to a minimum. The warrantees on most major PV system equipment are now 10 years or longer.

PV systems do have a factor that affects system performance to which conventional systems are not subjected: unpredictable cloud cover. Unless a PV system is grid-connected, so that grid power will be available at night and when it is cloudy, it is necessary to provide battery backup not only for hours of nighttime operation, but also for those days when the sun either does not shine at all or when it shines too little to make sufficient electricity to meet the required daily need. Most readers are aware that different geographical locations have different seasons when cloudy days are more common. The reader in the north will be familiar with the winter weeks when the sun almost never peeks through the cloud cover. The reader in the south will be familiar with the rainy summer days when little sun is available. The bottom line is that for different geographical locations, different amounts of battery backup are required for critical and noncritical loads if the PV system is not connected to the grid.

Figure 3.15 shows the statistical distribution of downtimes due to weather over a 23-year system lifetime for critical and noncritical systems. Note that a 95% available system is allowed 438 hours of downtime per year, but that for 3.5 years of the 23 years, the system will be down for less than 240 hours and for 8.3 years the

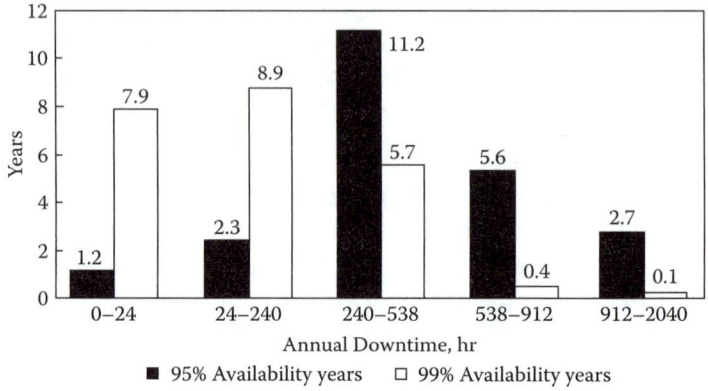

Figure 3.15 Statistical distribution of annual downtime for critical and noncritical PV systems over a 23-year system lifetime. (Adapted from *Stand-Alone Photovoltaic Systems: A Handbook of Recommended Design Practices*, Sandia National Laboratories, Albuquerque, NM, 1995.)

system will be down for somewhere between 538 and 2040 hours per year. The critical system is allowed to be down for 88 hours per year, but will be down for less than 24 hours per year for 7.9 years and will be down for more than 240 hours per year for 6.2 years out of the 23-year lifetime.

Figure 3.16 shows a comparison of necessary days of battery backup for critical and *noncritical* operation as a function of available peak sun hours, assuming a stand-alone system. Days of battery backup are also known as storage days

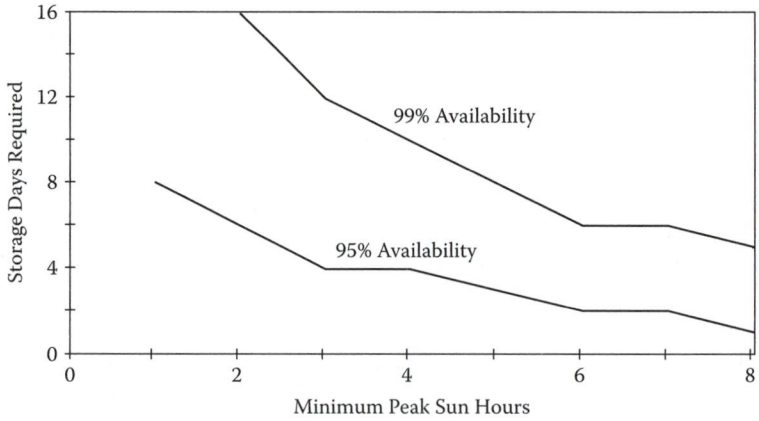

Figure 3.16 Necessary days of autonomy for critical and noncritical PV systems operation versus minimum available peak sun hours. (Adapted from *Stand-Alone Photovoltaic Systems: A Handbook of Recommended Design Practices,* Sandia National Laboratories, Albuquerque, NM, 1995.)

or days of autonomy. Using Excel, second-order polynomial curve fits can be obtained for the data in Figure 3.16. For those familiar with least mean square curve fitting to data, it will be recognized that an R^2 value of 1.0 indicates a perfect fit of the curve to the data, while an R^2 value that approaches zero indicates random scatter of the data. The best fits to the data of Figure 3.16 yield the following equations for estimating necessary storage days, based on minimum average peak sun hours over the year, T_{min}, as determined from insolation data for the listed sites.

$$D_{crit} = 0.2976T_{min}^2 - 4.7262T_{min} + 24 \qquad (R^2 = 0.9914) \tag{3.11a}$$

for critical applications, and

$$D_{non} = 0.1071T_{min}^2 - 1.869T_{min} + 9.4286 \qquad (R^2 = 0.9683) \tag{3.11b}$$

for *noncritical* applications, provided that $T_{min} > 1$ h. These equations should be used only if the critical and *noncritical* storage times have not already been determined for a site, because site-specific cloud cover or sunlight availability may differ from the averages assumed in arriving at Equation (3.11).

Extending the availability of a PV system from 95% to 99% may at first appear to be a simple, linear extension of the 95% system. Such a linear extension would involve only an additional 4% in cost. However, this is far from the case, and should obviously be so considering that essentially no system can provide 100% availability at any cost. Figure 3.17 shows the sharp increase in system cost as the availability of the system approaches 100%.

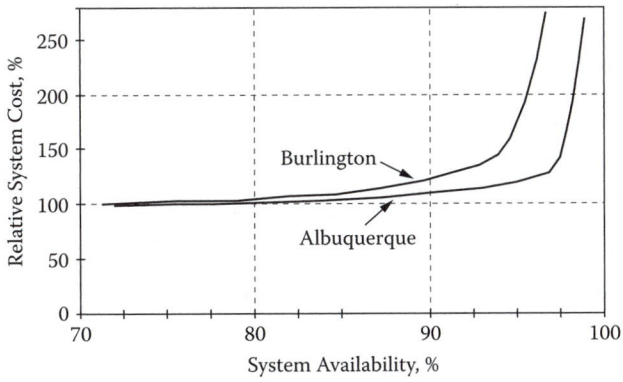

Figure 3.17 Relative costs versus availability for PV systems in Burlington, Vermont, and Albuquerque, New Mexico. (Adapted from *Stand-Alone Photovoltaic Systems: A Handbook of Recommended Design Practices,* Sandia National Laboratories, Albuquerque, NM, 1995.)

The cost of extending the availability of a system toward 100% depends on the ratio of available sun during the worst time of the year to available sun during the best time of the year. The extreme of this situation is represented by the polar latitudes, where no sun at all is available during the winter. Thus, near-100% availability would require enough batteries to store enough energy to meet all the no-sun load needs, plus a sufficient number of PV modules to charge up the batteries. This situation is further complicated by the likelihood that the winter loads would be larger than the summer loads. Clearly, a system with 180 days of autonomy should be more costly than a system with 10 days.

In the less extreme situation, such as found in Seattle, to provide the winter system needs, the system must be overdesigned with respect to the summer needs, resulting in excess power from the array during the summer. There is a good chance that this excess availability will be wasted unless a creative engineer figures out a way to put it to use. One way to do so, of course, is to connect it to the local electrical grid if the grid is available. However, in remote locations, the system may be standalone, and the creativity of the engineer will be challenged.

3.8 ASSOCIATED SYSTEM ELECTRONIC COMPONENTS

3.8.1 Introduction

This section introduces the basic electronic components of PV systems. These components include charge controllers, MPPTs, linear current boosters, and inverters. All of these components handle relatively large amounts of power, and are thus classified under the realm of power electronics. In each case, a simplistic explanation of the operation of the component will be given, with an emphasis on system performance requirements and how to best achieve them. Readers are encouraged to extend their understanding of these systems by consulting a power electronics text.

3.8.2 Charge Controllers

In nearly all systems with battery storage, a charge controller is an essential component. The charge controller must shut down the PV array when the battery is fully charged and may also shut down the load when the battery reaches a prescribed state of discharge. When the "battery" is really a system of batteries connected in series and parallel as needed to meet system needs, the control process becomes somewhat of a challenge. The controller should be adjustable to ensure optimal battery system performance under various charging, discharging, and temperature conditions.

Battery terminal voltage under various conditions of charge, discharge, and temperature has been presented in Figures 3.9 and 3.10 and Table 3.1. These results can be used to determine a Thevenin equivalent circuit for the battery system, as shown in Figure 3.18. The key is that during charging, the battery terminal voltage, V_T, will exceed the battery cell voltage, V_B, because $V_T = V_B + \mathrm{IR}_B$. During discharge, $V_T < V_B$, because under discharge conditions, $V_T = V_B - \mathrm{IR}_B$ as the

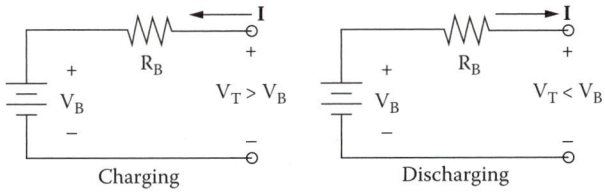

Figure 3.18 Thevenin equivalent circuit for a battery under charging and discharging conditions.

current direction is now reversed. The battery cell voltage is simply the battery open-circuit voltage.

The requirements for charging and discharging are made more complicated by the fact that the Thevenin equivalent circuit for the battery system is temperature dependent for both the open-circuit voltage and for the resistance. As temperature decreases, open-circuit voltage decreases [10] and resistance increases. Furthermore, the Thevenin equivalent circuit for an old battery is different from that of a new battery of the same type. Hence, for a charge controller to handle all of these parameters, it should incorporate several important features. Depending on the specific application, it may be possible to omit one or more of the following features.

3.8.2.1 Charging Considerations

First, consider the charging part of the process. Assume that the battery is fully charged when the terminal voltage reaches 15 V with a specific charging current. Assume also that when the terminal voltage reaches 15 V, the array will be disconnected somehow from the batteries and that when the terminal voltage falls below 15 V, the array will be reconnected. Now note that when the array is disconnected from the terminals, the terminal voltage will drop below 15 V, because there is no further voltage drop across the battery internal resistance. The controller thus assumes that the battery is not yet charged and the battery is once again connected to the PV array, which causes the terminal voltage to exceed 15 V, which in turn causes the array to be disconnected. This oscillatory process may continue until ultimately the battery becomes overcharged or until additional circuitry in the controller senses the oscillation and decreases the charging current or voltage. It should be noted that if the charging current pulses on and off, in fact, its average value will decrease to the duty cycle of the pulsing times the peak value of the current pulses.

Figure 3.19 shows how the terminal voltage of a battery depends on the charge or discharge rate and the state of charge for a typical battery. Note, for example, that if the battery is charged at a *C*/5 rate, full charge will be reached at a terminal voltage of 16 V, whereas if the battery is charged at *C*/20, then the battery will reach full charge at a terminal voltage of 14.1 V. If the charging current is then reduced to zero, the terminal voltage will drop to below 13 V.

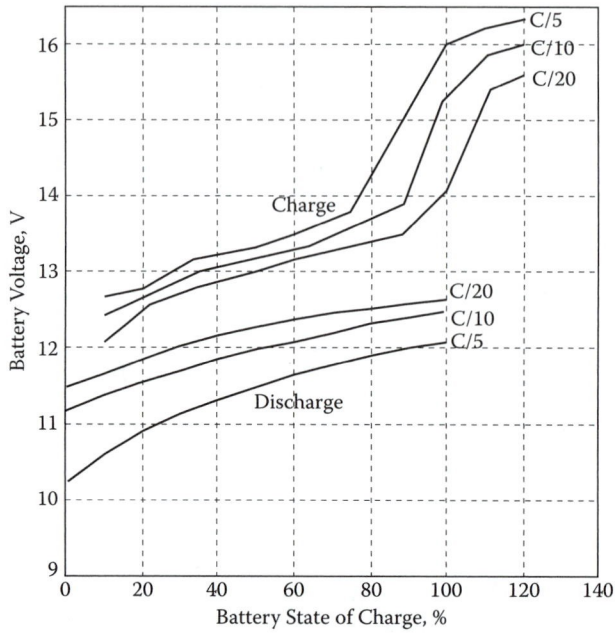

Figure 3.19 Terminal voltage as a function of charging or discharging rate and state of charge. (Adapted from *Maintenance and Operation of Stand-Alone Photovoltaic Systems*, Sandia National Laboratories, Albuquerque, NM, 1991.)

One way to eliminate overcharging resulting from the oscillatory process would be to reduce the turnoff set point of the controller. This, however, may result in insufficient charging of the battery. Another method is to introduce hysteresis into the circuit, as shown in Figure 3.20, so that the array will not reconnect to the batteries until the batteries have discharged somewhat. The reader who has been wondering what to do with the regenerative comparator circuit that was presented in an electronics course now may have a better idea of a use for this circuit.

A more careful examination of Figure 3.19 suggests that an even better charging algorithm might be to initially charge at a relatively high rate, such as *C*/5. When the

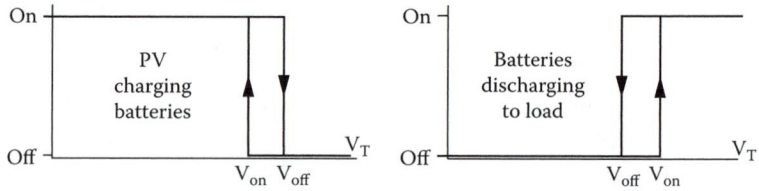

Figure 3.20 Hysteresis loops in charge controller using regenerative comparator for voltage sensing.

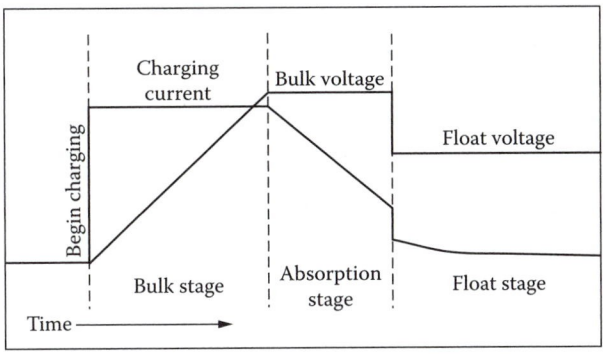

Figure 3.21 Three-stage battery charge control.

terminal voltage reaches about 15 V, indicating approximately 85% of full charge, the charging rate is then decreased, taking temperature into account, until the battery ultimately reaches 100% charge at a very low charging rate and a correspondingly lower voltage. This method is employed in many of the charge controllers currently being marketed for use with PV systems.

Figure 3.21 shows the regions of charging associated with the algorithm suggested in the previous paragraph. Initially, the charge controller acts as a current source. If the charging mechanism is a PV array, then presumably full array current will be used for charging. This is the bulk stage. When the charging voltage reaches a preset level, the *bulk voltage*, the charging mode is switched to constant voltage, during which the charging current decreases nearly linearly. This is called the *absorption* stage. Note that sometimes the bulk voltage is referred to as the absorption voltage because it is the voltage at which constant voltage charging begins. The absorption mode is continued for a time preprogrammed into the controller, after which the charging voltage is decreased to the *float* voltage. The float voltage is then maintained by the charge controller. The float voltage must be set to a level that will not result in damage to the battery.

In fact, because battery temperature affects battery terminal voltage and state of charge, modern charge controllers incorporate battery temperature sensor probes that provide temperature information to the controller that results in automatic adjustment of charging set points for the charging modes. Because battery cell voltage drops slightly with temperature decrease, but battery internal resistance increases with temperature decrease, the battery charging voltage is increased as temperature decreases to overcome the increased resistance and maintain the desired charging current levels.

A further mode that is available in modern chargers is the equalization mode. The equalization mode involves application of a voltage higher than the bulk voltage for a relatively short time after the batteries are fully charged. This interval of overcharging causes gassing, which mixes the electrolyte as a result of the turbulence caused by the escaping gases. This mixing helps prevent sulfate buildup on the plates and brings all individual cells to a full state of charge.

Only unsealed or vented batteries need equalization. For specific equalization recommendations, manufacturers' literature on the battery should be consulted. Some charge controllers allow for automatic equalization every month or so, but often it is also possible to set the controllers for manual equalization. Electrolyte levels should be checked before and after equalization because the gassing process results from decomposition of the electrolyte.

The battery disconnect may result in the array's being short-circuited, open-circuited, or perhaps connected to an auxiliary load that will use excess array energy. If the array is short-circuited to disconnect it from the batteries, the controller is called a shunt controller. Open-circuiting the array is performed by a series controller. One advantage of the shunt controller is that it maintains a constant battery terminal voltage at an acceptable level by bypassing enough charging current to achieve this result. The disadvantage is the amount of power that must be dissipated by the shunt and the heat sinking necessary to remove this heat from the shunt device.

3.8.2.2 Discharging Considerations

Now consider the discharge part of the cycle. Assume the battery terminal voltage drops below the prescribed minimum level. If the controller disconnects the load, the battery terminal voltage will rise above the minimum and the load will turn on again, and once again an oscillatory condition may exist. Thus, once again an application for hysteresis is identified, and another regenerative comparator circuit is justified for the output of the controller.

At this point, all that remains to be done is to make the set points of the charging regenerative comparator temperature sensitive with the correct temperature correction coefficient, and the controller is complete. Of course, if the controller is designed to reduce charging current in order to bring the batteries up to exactly full charge before shutting down, and if the controller selectively shuts down loads to ensure the battery is optimally discharged, then the overall system efficiency will be improved over the strictly hysteresis-controlled system.

New designs continue to emerge for controllers as engineers continue their quest for the optimal design. Ideally, a charge controller will make full use of the output power of the PV array, charge the batteries completely, and stop the discharge of the batteries at exactly the prescribed set point, without using any power itself.

3.8.3 Maximum Power Point Trackers and Linear Current Boosters

Electronic maximum power point trackers (MPPTs) have already been mentioned in Section 3.6. Linear current boosters (LCBs) are special-purpose MPPTs designed for matching the PV array characteristic to the characteristic of dc motors designed for daytime operation, such as in pumping applications. In particular, a pump motor must overcome a relatively large starting torque. If a good match between array characteristic and pump characteristic is not made, it may result in the pump operating under locked rotor conditions and shortening of the life of the pump motor due to input electrical energy being converted to heat rather than to mechanical output.

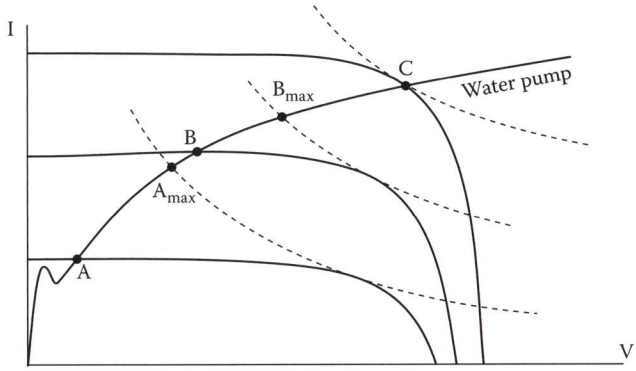

Figure 3.22 Pump and PV *I-V* characteristics, showing the need for use of MPPT.

Figure 3.22 shows a typical pump *I-V* characteristic, along with a set of PV array *I-V* curves for different illumination levels. The fact that the pump characteristic is relatively far from the array characteristic maximum power point for lower illumination levels shows why an LCB can enable the pump to deliver up to 20% more fluid. The LCB input voltage and current track V_{mp} and I_{mp} of the PV array. The LCB output voltage and current levels maintain the same power level as the input, except for relatively small conversion losses, but at reduced voltage and increased current levels to satisfy the pump motor characteristic. The fact that the LCB increases current to the load accounts for the name of the device. In effect, the LCB acts as an electronic transformer for dc currents and voltages.

MPPTs and linear current boosters are generally adaptations of dc-to-dc switching voltage regulators, as indicated in Section 3.6. Coupling to the load for maximum power transfer may require providing either a higher voltage at a lower current or a lower voltage at a higher current. Either a buck-boost or a boost-buck conversion scheme is commonly used in conjunction with load voltage and current sensors tied into a feedback loop using a microcontroller to vary the switching times on the switching device to produce optimal output voltage.

The LCB is used in special cases when only a boost of current is needed. This means a decrease in voltage will accompany the current boost in order to keep output power equal to input power. Because only a decrease in voltage is required, no boost is needed in the converter. Hence, a simple buck converter with associated tracking and control electronics will meet the design requirements of the device. In fact, this is also the case for most MPPTs. They typically convert a higher array voltage and lower array current to a lower voltage and a higher current, with minimal loss of power in the conversion process. The efficiency of good MPPT devices is typically 98% or better.

Figure 3.23a shows a simplified diagram of a buck converter circuit. When the MOSFET is switched on, current from the PV array can only flow through the inductor into the parallel RC combination, where the capacitor voltage increases. When the MOSFET is off, current must remain flowing in the inductor, so the inductor

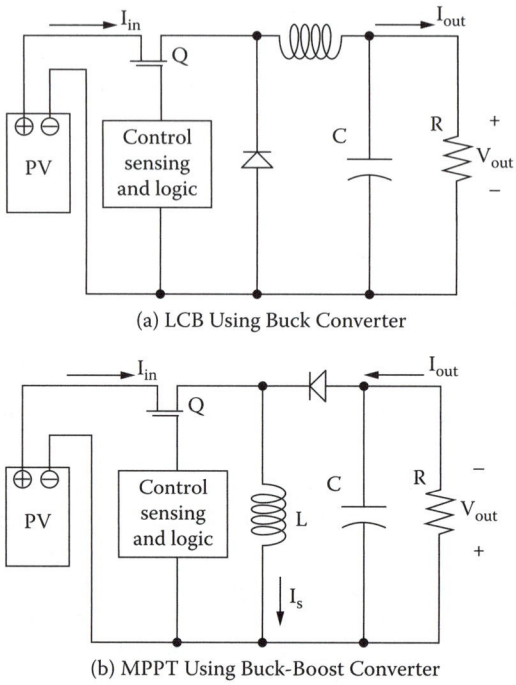

Figure 3.23 Maximum power point tracker and linear current booster.

current is now supplied by the capacitor through the diode, causing the capacitor to discharge. The extent to which the capacitor charges or discharges depends on the duty cycle of the MOSFET. If the MOSFET is on continuously, the capacitor will charge to the array voltage. If the MOSFET is not on at all, the capacitor will not charge at all. In general, the output voltage and current of an ideal buck converter are given by

$$V_{out} = DV_{in} \tag{3.12a}$$

and

$$I_{out} = I_{in}/D \tag{3.12b}$$

where D is the duty cycle of the MOSFET, expressed as a fraction ($0 < D < 1$). Note that the polarity of the output voltage is the same as the polarity of the input voltage, so there is no problem keeping the same grounded conductor at the input and the output of the converter.

Figure 3.23b shows the basic elements of a buck-boost converter that might be used in an MPPT. Note that in this system, the polarity of the output voltage is opposite the polarity of the input voltage. This can be a problem if a common negative (grounded) conductor is required for the PV circuit and the output of the converter. In this case, the circuit needs to be somewhat more sophisticated. On the other hand, if, for example, the PV input required a grounded positive conductor and the controller output required a grounded negative conductor, this switch of grounds can be achieved with this design.

The output voltage and current of the ideal buck-boost converter are given by

$$V_{out} = \frac{D}{1-D} V_{in}, \tag{3.12c}$$

$$I_{out} = \frac{1-D}{D} I_{in} \tag{3.12d}$$

where again D is the duty cycle of the MOSFET. Note that if $D < 0.5$, $V_{out} < V_{in}$, and if $D > 0.5$, $V_{out} > V_{in}$. Hence, the MPPT is capable of either increasing or decreasing its output voltage in order to track an array maximum power point. Of course, if the MPPT output voltage decreases, then its output current will increase, and vice versa, such that, in the ideal case, the output power will equal the input power.

During the time Q is on, energy is stored in the inductor. When Q turns off, the inductor current must continue to flow, so it then flows through R and C and the diode, charging C to V_{out}, since the capacitor voltage equals the output voltage. When Q turns on again, the diode becomes reverse-biased, and current is built up again in the inductor while the capacitor discharges through the resistor. Finally, when Q turns off, the cycle repeats itself. The values of L and C and the switching frequency determine the amount of ripple in the output voltage. Because no energy is lost in ideal inductors and capacitors, and because Q and the diode approximate ideal switches, essentially all power extracted from V_{in} must be transferred to the load. Of course, in reality, these components will have some losses, and the efficiency of the MPPT will be less than 100%. However, a well-designed MPPT will have an overall efficiency greater than 95%, with many units currently being marketed having advertised efficiencies that are close to 98% [11,12].

Another application of the MPPT is to ensure optimal charging of batteries. The MPPT charge controller electronically tracks the PV array maximum power point to ensure that maximum charging current is delivered to the battery bank. The result is a charge controller with high input voltage, low input current, lower output voltage, and higher output current. Charge controllers without MPPT input circuitry connect the PV array directly to the battery voltage, while MPPT charge controllers operate the array at its maximum power point for charging the batteries at the proper charging voltage, as shown in Figure 3.24.

By designing the array to have a higher maximum power voltage, smaller wiring can be used between the array and charge controller, with lower power loss in the wiring and lower wiring cost. Note that with the PV array of Figure 3.24, the MPPT charge controller will deliver about 2100 W to the batteries, while the conventional controller will deliver only about 1300 W to the batteries.

In terms of charging current, the MPPT charge controller delivers approximately 2100 ÷ 56 = 37.5 A, while the conventional charge controller delivers only 1300 ÷ 56 = 23.2 A. In fairness to the conventional charge controller, however, one

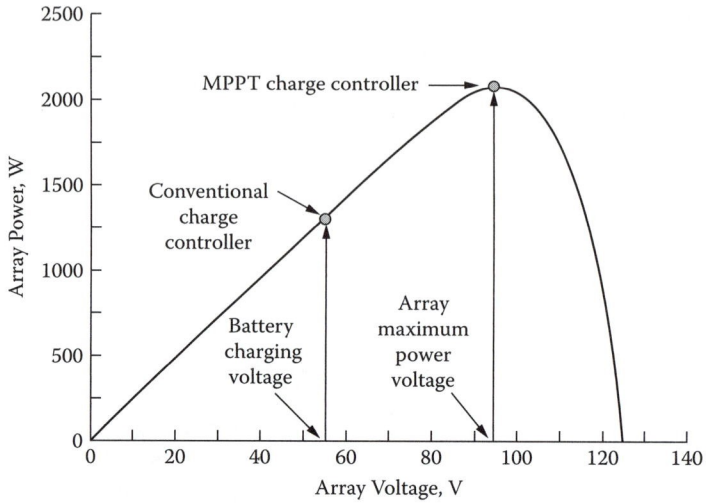

Figure 3.24 Comparison of PV array operating points for conventional and MPPT charge controllers.

should admit that normally with a conventional charge controller, the array would be designed with its maximum power point closer to 60 V so that the battery charging voltage will be closer to the array maximum power point voltage.

At the time of this writing, essentially all charge controllers and inverters used in grid-connected PV systems used maximum power point tracking. The array-to-battery efficiency of good MPPT charge controllers is typically close to 98%. Problem 3.14 explores this concept further.

3.8.4 Inverters

Depending on the requirements of the load, a number of different types of inverters are available. Selection of the proper inverter for a particular application depends on the waveform requirements of the load and on the efficiency of the inverter. Inverter selection will also depend on whether the inverter will be a part of a grid-connected system or a stand-alone system. Even though warrantees are now typically 10 years on PV inverters, many opportunities still exist for the design engineer to improve on inverters.

Table 3.2 summarizes the presently available inverters. Inverter performance is generally characterized in terms of the rated power output, the surge capacity, the efficiency, and the harmonic distortion. Because maximum efficiency may be achieved near rated output, it is important to consider the efficiency versus output power curve for the inverter when selecting the inverter. Certain loads have significant starting currents, so it is important to provide adequate surge current capacity in the inverter to meet the load surge requirements. Other loads will either overheat or introduce unwanted noise if the harmonic distortion of their power supply is not below a specific level.

Table 3.2 Summary of Inverter Performance Parameters

Parameter	Square Wave	Modified Sine Wave	Pulse Width Modulated	Sine Wave[a]
Output power range (W)	Up to 1,000,000	Up to 5,000	Up to 20,000	Up to 500,000
Surge capacity (multiple of rated output power)	Up to 20 ×	Up to 4×	Up to 2.5×	Up to 4×
Typical efficiency over output range	70%–98%	>90%	>90%	>90%
Harmonic distortion	Up to 40%	>5%	<5%	<5%

Sources: Based on data from *Maintenance and Operation of Stand-Alone Photovoltaic Systems*, Sandia National Laboratories, Albuquerque, NM, 1991; http://www.xantrex.com/ (Information on Xantrex Power Electronics Equipment); http://www.satcon.com/ (Information on Satcon PV Inverters).

[a] Multilevel H-bridge or similar technology to yield utility-grade sine wave output.

In general, the square wave inverter is the least expensive and is relatively efficient, but is limited in its applications. It has the best surge capacity but the highest harmonic distortion. The modified sine inverter is more complicated, but still relatively efficient. The pulse-width-modulated inverter has higher cost, high efficiency, and minimal distortion. The pure sine inverter, with digitally generated waveform, has the least distortion, and can have efficiencies in excess of 96%.

3.8.4.1 Square Wave Inverters

The simplest inverters are square wave inverters. These inverters employ solid-state switches connected as either astable multivibrators or as externally controlled switches. Use of astable multivibrators enables the inverter to be used in a self-contained PV system, whereas externally synchronized switches are used when the inverter is to be synchronized with an external ac power source. Figure 3.25 shows how a single source can be switched alternately in positive and negative directions to produce a square wave.

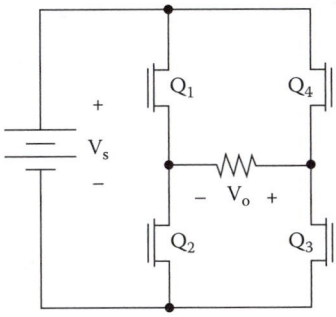

Figure 3.25 Converting a dc source to a square wave.

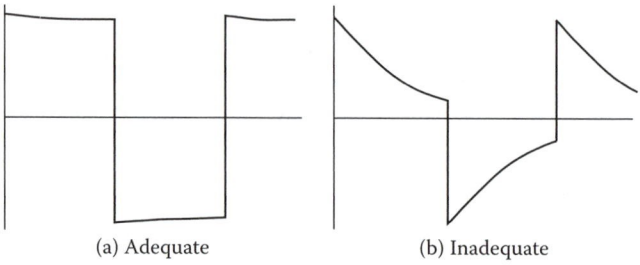

Figure 3.26 Output waveforms for a square wave inverter with adequate and with inadequate transformer turns.

The key to efficient switching is to have current flowing from the array at all times while maintaining zero volts across the switching element when it passes current, and zero current through the switching element when it has voltage across it. This can be approximated reasonably well either with insulated gate bipolar transistors, power MOSFETs, or silicon-controlled rectifiers. Once the dc is converted into a square wave, its amplitude normally needs to be increased to produce a 120 V rms ac waveform. Because the rms value of a square wave is simply the amplitude of a square wave, a system with a 12 V dc input will need a transformer with a 10:1 turns ratio.

It is important, however, to realize that the transformer must be designed with a sufficient number of turns so that the time constant determined by the magnetizing inductance of the transformer and the source resistance will be long enough to maintain the square wave. Too few turns on the transformer will cause the output waveform to droop as shown in Figure 3.26. When specifying the transformer for this application, it is easy to assume that transformers are ideal and to simply try to use a 120:12 V transformer backward. The 120:12 V transformer, however, probably would have been designed for use with a sinusoidal signal, and thus would not have enough turns to handle the square wave effectively. A square wave inverter with a good transformer and efficient switching can operate at efficiencies in the 90% range.

3.8.4.2 Modified Sine Wave Inverters

For a number of applications, a square wave is inadequate for meeting the harmonic distortion requirements of the load. For example, because square waves have significant harmonic content, and because hysteresis and eddy current losses in magnetic materials increase significantly with increase in frequency, square wave excitation may cause some motors or magnetic fluorescent ballasts to overheat.

Square wave harmonics can also introduce noise into a system. Thus, before selecting an inverter, it is important to verify that the proposed load will operate with square wave excitation.

If square wave excitation is not suitable for a load, it is possible that a modified sine wave will work. A number of methods are available to convert the output of a

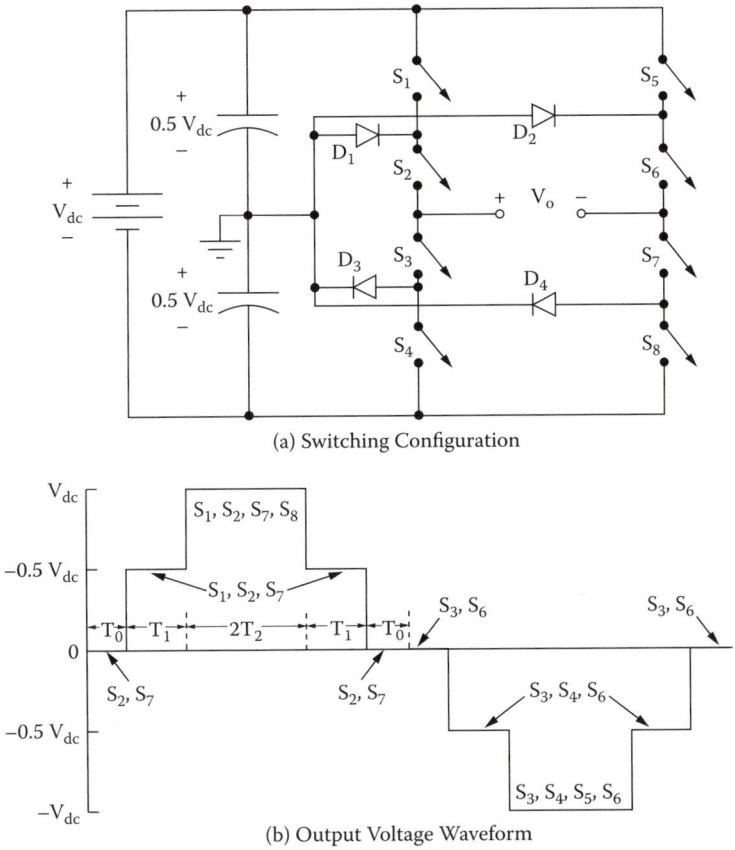

Figure 3.27 Multilevel H-bridge modified sine wave inverter.

dc source into some approximation of a sine wave. One such method involves using a multilevel H-bridge as shown in Figure 3.27a to generate a waveform such as the one shown in Figure 3.27b. The idea is to time the individual voltage levels so that harmonic distortion will be minimized and the rms value of the output voltage will remain constant in the event that the input dc voltage should vary.

The sequence for closing switches to obtain portions of the waveform in Figure 3.27b is shown on the waveform diagram. The actual switches may be bipolar transistors, MOS transistors, silicon-controlled rectifiers (SCRs), or insulated gate bipolar transistors. As an example of how the various voltage levels in v_o are obtained, consider the interval T_1, where, according to the output voltage waveform, $v_o = 0.5V_{dc}$. If S_1 is closed, diode D_1 becomes reverse biased and appears as an open circuit. If S_2 is closed, then the positive terminal of v_o is connected to $0.5V_{dc}$. Closing S_7 provides a path to ground from the negative terminal of v_o through forward-biased D_4. Hence, if voltage drops across diodes and electronic switches are neglected, $v_o = 0.5V_{dc}$. Next, consider the interval where v_o is shown as $-V_{dc}$. During this interval, when S_3 and S_4

are closed, D_3 becomes reverse biased and appears as an open circuit, and the positive terminal of v_o is thus connected to $-0.5V_{dc}$. Closing S_5 and S_6 reverse-biases D_2 and connects the negative terminal of v_o to $>0.5V_{dc}$. Hence, $v_o = -V_{dc}$. It is left as an exercise for the reader to verify the remaining switching combinations.

Noting the symmetry of v_o, and recognizing that $T_0 > T_1 > T_2 = 0.25T$, where T is the period of the waveform, it is possible to determine the rms value of v_o from the first quarter cycle by solving

$$V_{rms} = \left[\frac{4}{T}\int_0^{\frac{T}{4}} v_o^2(t)dt\right]^{\frac{1}{2}} = \left[\frac{4}{T}\left\{\int_0^{T_0} 0dt + \int_{T_0}^{T_0+T_1} (0.5V_{dc})^2 dt + \int_{T_0+T_1}^{T_0+T_1+T_2} V_{dc}^2 dt\right\}\right]^{\frac{1}{2}}$$

$$= V_{dc}\left[\frac{T_1}{T} + \frac{4T_2}{T}\right]^{\frac{1}{2}}. \tag{3.13}$$

As an example, assume that $V_{dc} = 160$ V, and it is desired to produce $V_{rms} = 120$ V by keeping $T_0 = T_1$ and solving for T_2. The result is that $T_0 = T_1 = T/16$ and $T_2 = T/8$. Problem 3.18 offers the reader a chance to solve for T_0, T_1, and T_2 for different values of V_{dc}. Note that because $T_1 + 4T_2 < T$, if $V_{dc} < V_{rms}$, there is no solution for T_0, T_1, and T_2.

H-bridges can be designed with $4n + 1$ levels, where n is an integer. The greater the number of levels, the closer v_o can be made to approximate a sine wave. In addition, the voltage drop across any individual switch is reduced as the number of levels is increased. Problem 3.21 involves the design of a 9-level H-bridge.

The design challenge in modern inverters is to maximize the range of input voltages over which the output of the multilevel H-bridge maintains a constant rms value with minimal distortion. The next design constraint, of course, is to ensure that the inverter input will track the PV array maximum power point and, finally, the inverter should have an efficiency close to 100%. Modern inverters generally have MPPT input voltage ranges of nearly 300 V between maximum and minimum tracking voltages.

Most computer uninterruptible power supplies produce a 5-level modified sine wave, so it is safe to say that this waveform is acceptable for use in providing backup ac power to a computer. However, IEEE Standard 1547-2003 [15] requires that any source connected to the utility line must have less than 5% total harmonic distortion (THD). For those who have not memorized the formula for THD from their electronics textbook, recall that THD is the percentage ratio of the sum of the rms values of all the harmonics above the fundamental frequency to the rms value of the fundamental frequency. It is unlikely that a 5-level modified sine wave inverter will meet the 5% THD rule, especially if V_{dc} is allowed to vary while keeping V_{rms} constant. This is the reason for the existence of pure sine inverters. Using higher-level H-bridges to approximate pure sine waves is one method. Another method is to use pulse width modulation techniques.

3.8.4.3 Pulse-Width-Modulated Inverters

The pulse-width-modulated (PWM) inverter produces a waveform that has an average value at any instant equivalent to the level of a selected wave at that instant. PWM inverters are perhaps among the most versatile of the family of inverters. They are similar to the PWM described in the discussion of the MPPT, with the exception that the MPPT PWM signal is designed to have a constant average value to produce a regulated dc output, while the inverter PWM signal is designed to have a time-dependent average value that can have any arbitrary waveform at any arbitrary frequency at any arbitrary amplitude. For use in PV applications, it is generally desirable to have a sinusoidal waveform with a predictable amplitude and frequency.

Figure 3.28 shows how a PWM waveform can produce waveforms of differing amplitudes and frequencies by controlling the on-and-off time of a pulse waveform. The waveform is controlled by controlling the relative duty cycle of successive pulses. The amplitude is controlled by controlling the overall duty cycle, and the frequency is determined by controlling the repetition time for the pulse sequence.

By switching the pulse between a positive level and a negative level, it is possible to construct waveforms with zero average value, which is particularly important when driving loads for which a dc component in the excitation may cause losses in the load. This includes applying dc to an ac motor or to a transformer. Application of

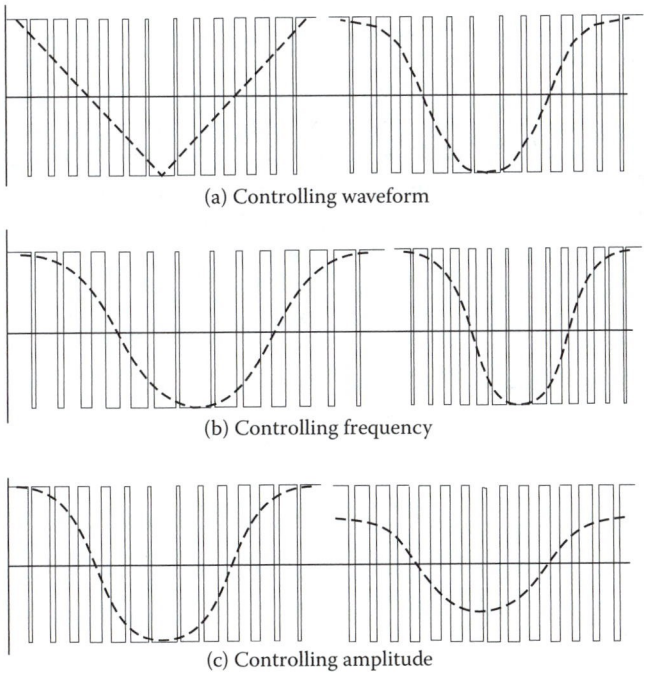

Figure 3.28 PWM control of waveform, frequency, and amplitude.

dc to such loads can result in significant I^2R heating with possible failure of the winding insulation and subsequent catastrophic failure of the motor or the transformer, including the possibility of fire. For this reason, elimination of any dc component in the output of an inverter is important regardless of the type of inverter.

PWM inverters are particularly useful when used as ac motor controllers, because the speed of an ac motor can be controlled by adjustment of the frequency of the motor excitation. It must be recalled, however, that the flux developed in the motor is inversely proportional to the excitation frequency, and that every motor has a saturation flux above which the motor draws significant current without further flux increase. Hence, as a motor excitation frequency is decreased, the peak value of the applied voltage must also decrease proportionally to keep the motor out of saturation. A PWM controller coupled with MPPT capability can be a very efficient means of maximizing the efficiency of a PV pumping system.

Figure 3.29 shows one means of generating a PWM waveform. The process begins with a triangle wave at the frequency of the pulse waveform applied to one input of a voltage comparator. Next, the desired waveform is applied as a modulating waveform at the other input of the comparator. Whenever the modulating signal exceeds the triangle waveform, the comparator output goes high, and whenever the modulating signal is less than the triangle waveform, the comparator output goes low. The comparator output is then used to switch the output pulse on and off. The result is an output PWM waveform that has a moving average value proportional to the modulating signal. The average value of the PWM output is obtained in a manner similar to the detection process used to pick off the modulation signal of an amplitude-modulated communications signal. A filter needs to be incorporated that will not allow the voltage (or current) of the load to change instantaneously, but will allow it to change quickly enough to follow the moving average of the PWM waveform. This can be achieved with inductive and/or capacitive filtering, depending on whether it is desired to smooth the current, the voltage, or both.

The harmonic content of the two-level PWM waveform can be significantly reduced through the use of a three-level PWM waveform. A three-level PWM

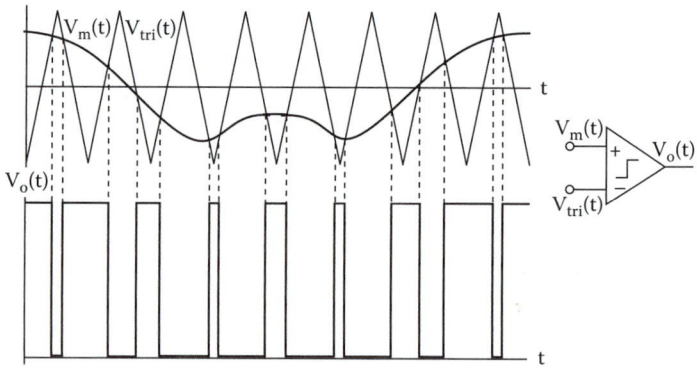

Figure 3.29 Generation of a PWM waveform.

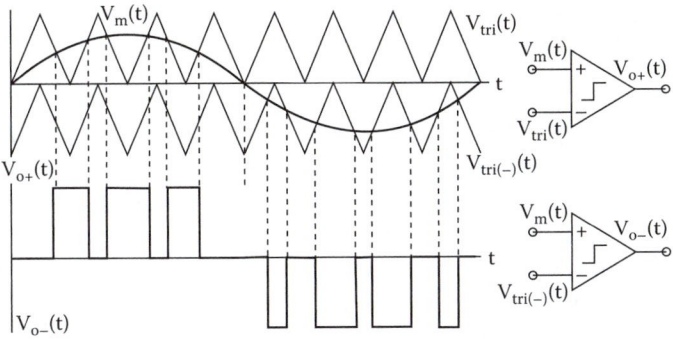

Figure 3.30 Three-level PWM inverter configuration and output signal.

waveform can be generated by incorporating a second comparator into the circuit, so that one comparator will control the positive-going pulse when the modulating signal is positive and the other comparator will control the negative-going pulse when the modulating signal is negative. The challenge here, however, is to switch the PV input so the array operates at maximum power voltage if the inverter is powered directly from the array. If a storage mechanism is used for array current, so that the array is continuously supplying dc power to the storage mechanism, then the storage mechanism can be switched on and off to meet the load requirements without interrupting the power flow from the array. Alternatively, if the array voltage is switched between V_{OC} and some voltage less than V_{mp}, then the duty cycle of the connection to the array can be varied such that the average value of the array voltage seen by the inverter is V_{mp}. Figure 3.30 shows a three-level PWM inverter output signal derived from dc input from battery storage.

Figure 3.31 shows a PWM inverter circuit using a full-bridge switching network for the dc source, including MPPT. This is an adaptation of the full-bridge that was shown for the generation of a square wave. The inverter can be fixed frequency, line commutated (synchronized), or variable frequency, depending on the source of the

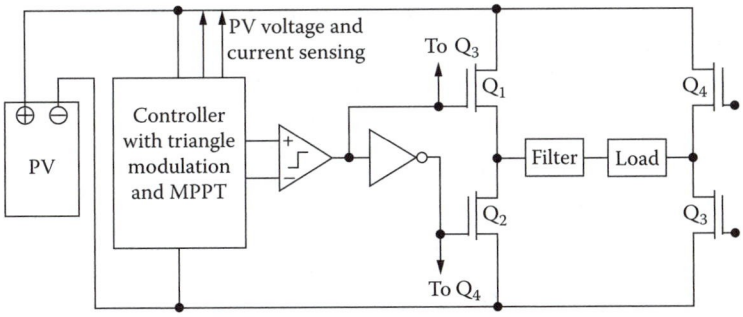

Figure 3.31 Full-bridge PWM converter showing controller and PV connections.

modulation signal. The controller generates the triangle waveform and the modulating waveform, v_m, with the amplitude and/or frequency of v_m determined by the PV array output voltage and current as sensed by the MPPT. Switches Q_1 and Q_3 of the bridge are switched on during the positive pulse excursions, and Q_2 and Q_4 are switched on during the negative excursions. This provides a continuous power drain on the PV array, thus enabling maximum power point tracking.

Keeping the switching elements as close to ideal as possible, along with minimal loss in the filtering elements, can yield very high overall efficiencies for these units. When a switching element has either zero voltage or zero current, it does not dissipate power. During a switching transition, however, switching elements have both voltage and current and, hence, dissipate power. The trade-off in a PWM inverter, then, is to have the frequency as high as possible to give adequate waveform reproduction, but not so high as to increase switching losses to unacceptable levels. Most manufacturers now advertise PWM inverters with efficiencies above 90% over a wide range of inverter output power.

3.8.4.4 Other Desirable Inverter Features

Modern inverters for PV applications often incorporate very sophisticated features to optimize inverter performance. In fact, the instructions and installation manuals for some of these inverters are nearly as lengthy as this textbook [16,17].

For stand-alone inverters and for grid-connected inverters with stand-alone option, a "search" mode is often incorporated into the design. In the search mode, the inverter uses minimal energy to keep its electronics operational when no loads are connected. In the search mode, approximately once every second, the inverter sends out an ac voltage pulse a few cycles long. If the inverter senses that no current flows as a result of the pulse, it concludes that no load is connected and waits to send another pulse. On the other hand, if current flows at a magnitude greater than a preprogrammed value, the inverter recognizes that a load has been connected and supplies a continuous output voltage. If the inverter is not producing an output voltage on a continuous basis, power losses in the inverter are minimized.

In some cases, however, an inverter in the search mode can be "confused." For example, if the inverter is set to provide power as long as more than 25 W is connected, it can be confused by a 20 W incandescent light bulb. The reason is that when the lamp filament is cold, it has a lower resistance and a surge of current flows until the filament is heated. Hence, a 20 W incandescent light bulb will appear to be a higher wattage load when it first turns on, and then will drop back to 20 W when the filament is hot. This, then, will cause the inverter to turn off and resume the search mode, where it will find the same light bulb and will turn it on again until it gets hot, resulting in a light that flashes on and off. Other confusing loads include the remote receivers in electronic equipment such as televisions, VCRs, and CD players. When they are on, they draw very small amounts of power, so when an inverter is in the search mode, it is quite possible that the remote receivers will shut down.

For stand-alone inverters, it is necessary that the output appear as a voltage source, preferably as close to ideal as possible. For utility interactive inverters, however, it is more convenient to let the grid voltage fulfill the voltage source role and to have the inverter act as a current source that feeds current into the utility grid. Utility interactive inverters with battery backup that are capable of supplying power to standby loads if the utility grid is disconnected must be capable of switching over from grid-synchronized current source to internally synchronized voltage source to power standby loads.

Utility interactive inverters must be designed so that if the utility goes down, the inverter output to the utility also shuts down. This safety feature is relatively easy to design into an inverter if no other distributed electrical source is connected to it. However, if another nonutility source is connected to the line, it is important that the PV system inverter not recognize this source as a utility source and continue to supply power to the line. This is called *islanding* and will be discussed in detail in Chapter 4.

Many utility-interactive inverters that have a standby backup feature also incorporate a battery charger in the inverter so that the utility can be used to charge the batteries in the event that the PV system has not fully charged the batteries. This feature only works, of course, when the utility grid is energized. Fortunately, this is generally most of the time.

Another convenient feature found in many inverters is a generator start option. This allows for a separate ac generator to be used as a backup to the PV system in either a utility interactive or a stand-alone system. When the battery level of charge drops to a prescribed level, the inverter will send a starting signal to the generator. If the inverter has a real-time clock, the time of day for generator starting can also be controlled, so the generator will not come on just before the sun is about to charge the batteries.

Finally, most high-quality inverters that are connected directly to a PV array will incorporate maximum power point tracking circuitry at their dc input so that the PV array will deliver power at its maximum power level to the inverter.

3.9 GENERATORS

3.9.1 Introduction

In some cases, when either a large discrepancy between seasonal loads exists or when seasonal sun availability varies greatly, a stand-alone system designed completely around PV components will result in the deployment of a large PV array to meet the needs of one season. Meanwhile, during other seasons, much of the energy available from the array is not used. This is similar to the problem of meeting critical system needs with PV, where the cost generally increases rapidly as the system availability exceeds 95%. In such cases, it is often more cost-effective to employ an alternate source of electricity to be available when the PV array is not meeting system needs. While it is conceivable that the backup source for a stand-alone system may be wind or other renewable source, it is more common to employ either a gasoline, diesel, or propane generator as a system backup. In a grid-connected system, the utility grid provides any needed backup power, unless, of course, the grid goes down.

With the relatively low acquisition costs of small and portable generators, it may appear that it would make better economic sense to simply use the generator without the PV array. However, life-cycle cost analysis, as is discussed in Chapter 8, sometimes shows the use of a mix of PV and conventional generation to be more economical than an engine-powered generator. Furthermore, the fact that PV generation is quiet and clean adds further appeal to using a maximum practical amount of PV generation in the system.

When it makes economic sense to use a generator, it is important for the engineer to have some knowledge of generator options. The next section deals with some of the considerations in selecting a generator.

3.9.2 Types and Sizes of Generators

While most small electrical generators use a gasoline engine as the mechanical prime mover, methane-, propane-, and diesel-powered engines are also available. A number of factors will enter the selection process, including initial cost, power requirements, fuel availability, and maintenance requirements.

While it is possible to obtain dc generators for use in battery-charging applications, dc generators are generally not recommended because of maintenance requirements. They have brushes and commutators that require frequent replacement and cleaning. Many years ago, automotive batteries were charged with dc generators, but now all batteries are charged with alternators, which are simply ac generators that are connected to rectifiers to convert their output to dc. Similarly, ac generators are used in PV systems, even if the systems are dc systems, to minimize maintenance costs.

Table 3.3 lists a few common generator sizes and approximate costs, based on 2008 data. The smaller units are normally used as portable units, while larger sizes are intended for use as fixed units. Rated output is generally about 90% of maximum output.

Table 3.3 Common Generator Sizes, Features, and Approximate Costs

Max KW	Engine (HP)	Duty Rating	Voltage Regulation	Noise Level (dBA)	≈ kWh/gal @ Rated Load	Approx. Cost ($)
2.5	5.5	Economy	Yes	74	≈4.0	510
5.0	8.0	Economy	Yes	77	≈4.0	670
5.5	8.0	Deluxe	Yes	82	≈5.4	1425
8.0	16.0	Deluxe	Yes	76.5	≈4.6	1890
10.0	20.0	Deluxe	Yes	76	≈4.8	2475
5.5	10.0	Industrial	Yes	77	≈5.6	2455
6.0	10.0	Industrial	Yes	77	≈6.9	2475
13.0	19.0	Industrial	Yes	81	≈7.2	5140

Source: Based on data from www.generatorjoe.net/store.asp (a wealth of information on a wide variety of fossil fuel generators).

3.9.3 Generator Operating Characteristics

Factors other than peak power output will enter the selection process if a generator is carefully chosen. Generator specifications also include rotation speed, efficiency, fuel type, altitude effects, waveform harmonic content, frequency stability, noise levels, type of starting, and overload characteristics.

3.9.3.1 Rotation Speed

Normally, 60 Hz generators operate at either 3600 or 1800 rpm. Corresponding speeds for 50 Hz generators are 3000 and 1500 rpm. The 3600 rpm units are two-pole machines and are of simpler construction, resulting in lower acquisition cost. The 1800 rpm machines are 4-pole machines and are somewhat more expensive, but more common in the larger sizes or heavy-duty units.

In general, the higher the rpm, the more the wear and tear on the bearings, which means more frequent maintenance requirements. Two-pole generators are thus most convenient for use in relatively light-duty applications that require less than 400 h per year of operation. Four-pole generators are recommended when more than 400 h of operation per year are anticipated.

Since the last edition of this book, some generator manufacturers [19] have incorporated an inverter into some of their generators. The advantage of this design is that the load on the inverter determines the amount of mechanical power needed to meet the electrical load requirements. As a result, the generator is controlled to operate at variable engine speeds up to maximum engine horsepower. This significantly reduces fuel consumption at low load levels and, as a result, increases the overall efficiency of the generator. In addition, these models tend to run more quietly than conventional fossil fuel generators.

3.9.3.2 Efficiency

Electrical and mechanical losses are present in all generators. However, the greatest losses in a generator system are attributable to the prime mover engine. Because the prime mover and electrical generator will each generally have a particular load at which they will operate at maximum efficiency, manufacturers endeavor to carefully match the two components to produce maximum efficiency at somewhere between 80% and 90% of rated full load. Figure 3.32 shows approximate plots of efficiency versus percent of rated electrical load and kWh per gallon versus percent of rated electrical load for typical small generators compared to small inverter/generator models [18,19].

As generator size increases, the overall maximum efficiency also increases. Figure 3.33 shows maximum efficiency versus generator size [18]. Whether or not to use a larger, more expensive generator, operating at higher efficiency for shorter intervals presents an interesting design challenge to the design engineer.

Maximum generator size is limited in most cases by maximum allowable charging rates for the system batteries, assuming the generator is incorporated into a

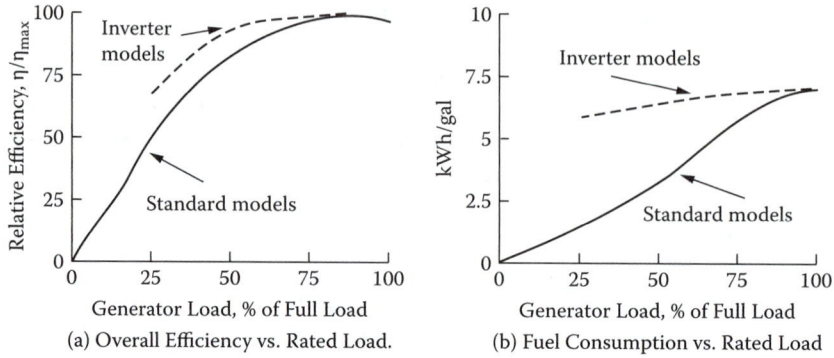

Figure 3.32 Two means of characterizing generator efficiency.

system with batteries. For systems with highly variable electrical loads, it is particularly inefficient to incorporate a generator unless battery storage is provided to present a nearly constant load on the generator or a generator/inverter is used rather than a conventional generator. In either case, it is more efficient to charge batteries at maximum generator efficiency and then turn off the generator.

3.9.3.3 Fuel Types

Choice of fuel for a generator will depend on several factors. Although diesel generators tend to be the most efficient, diesel fuel is sometimes more difficult to obtain and is more likely to become contaminated with water or other contaminants than other fuel choices. In cold temperatures, diesel engines tend to be more difficult to start, so they need to be used under adequately controlled temperature conditions if they are to be started and stopped. Diesel is probably not the best choice for remote locations for these reasons, especially if the site is unattended and reliability is important.

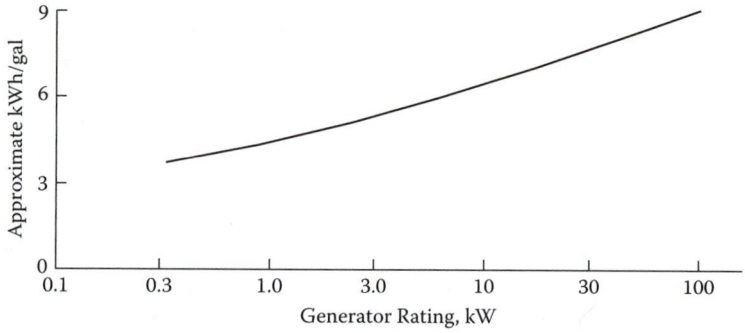

Figure 3.33 Approximate efficiency versus generator size at rated output.

Gasoline is probably the easiest fuel to obtain and will be the fuel of choice for many applications, because gasoline engines can be started and run under relatively adverse conditions. In some situations, however, when either natural gas or propane is used on site for other purposes, such as cooking or space heating, it is convenient to also use it as a fuel for backup electrical generation. This eliminates the need to keep track of more than one supply of fuel, and, in standby operations, such as during utility outages, it is often very inconvenient to obtain replacement supplies of gasoline. However, as will be noted in the next section, natural gas efficiency decreases more rapidly with increase in altitude than gasoline or propane efficiency, so this observation may need to be taken into account.

3.9.3.4 Altitude Effects

Many engineers have observed that their automobiles do not respond as well at high altitudes as they do at lower altitudes. As altitude increases, less air is available for the combustion process, and combustion engines thus degrade in performance with increase in altitude.

The design engineer should first check manufacturers' specifications for altitude derating on generators. If manufacturers' specifications do not list derating factors, then it is reasonable to derate gasoline, diesel, and propane generators by 3% per 1000 ft of altitude above sea level and to derate natural gas engines by 5% per 1000 ft of altitude above sea level.

3.9.3.5 Waveform Harmonic Content

Generally, the output waveform of an electrical generator is adequate for nearly all applications. For battery charging, almost any waveform is satisfactory, depending on whether the battery charger contains a transformer.

If the charger contains a transformer, there is a remote possibility that excessive dc or harmonic content in the generator waveform may damage the transformer. If the generator is connected directly to the PV system ac loads at any time, then it is a good idea to be sure the generator output waveform is sufficiently "clean" to meet the requirements of the loads. Normally, the only loads with possible sensitivity to waveform quality will be electronic systems, magnetic fluorescent light ballasts, and some motors.

3.9.3.6 Frequency Stability

Again, depending on load requirements and whether the generator is connected to ac loads with critical power frequency requirements, the frequency stability of a generator may need to be taken into account. It is generally desirable to maintain frequency fluctuations at less than ±0.5 Hz for ac loads, but this degree of frequency stability is not necessary for many PV system loads. Conventional generators often have specified frequency stability of 5%.

3.9.3.7 Amplitude Stability

It is generally desirable to have the generator output voltage amplitude remain within 5% of its no-load value. This implies a voltage regulation figure of 5%. Depending on whether the load is inductive or capacitive, the load voltage under full-load conditions may actually exceed the no-load voltage. The voltage regulation of the generator is defined by

$$VR = \frac{V_{nl} - V_{fl}}{V_{fl}}, \tag{3.14}$$

where V_{nl} and V_{fl} are the no-load and full-load voltages, respectively. Hence, if $V_{fl} > V_{nl}$, VR will be negative.

Again, voltage regulation requirements will depend on the specific application. If the PV system is designed to operate the generator at maximum efficiency, the load will remain nearly constant, and the output amplitude and frequency will remain quite stable.

3.9.3.8 Noise Level

Some generators are noisy, and others are less noisy. Local ordinances should be checked, but normally the noise level demands of the user will be more stringent than local ordinances. National parks, for example, have relatively strict noise regulations.

3.9.3.9 Type of Starting

If the generator is to be controlled by the system controller, then it will need to be equipped with electric starting. Otherwise, it may be satisfactory to use a generator with manual starting.

3.9.3.10 Overload Characteristics

A synchronous generator can suffer serious consequences if it is overloaded. Overloads can occur with the application of excessive steady-state loads, but also can occur as the result of transient loads, such as motor starting.

For example, if the generator is to operate an electric motor, the starting current of the motor can place the generator in an overload condition that can cause the generator to slow down, resulting in reduced generator output. This in turn can result in insufficient starting current for the motor and slow the generator even more. The result could be damage to the prime mover and possibly to the generator armature winding, leading to an expensive repair bill.

3.9.3.11 Power Factor Considerations

The generator must be able to meet the real power requirements of the load as well as its reactive power requirements. Because I^2R copper losses in the generator

Table 3.4 Cost and Maintenance Information on Small Electrical Generators

Type	Size Range (kW)	Application	Cost ($/W)	Maintenance Intervals		Engine Rebuild (h)
				Oil Change (h)	Tune-Up (h)	
3600 rpm gas	1–20	Light use	.50	25	300	2000–5000
1800 rpm gas	5–20	Heavy use	.75	50	300	2000–5000
Diesel	3–100	Industrial	1.00	125–750	500–1500	6000

Source: Based on data from *Stand-Alone Photovoltaic Systems: A Handbook of Recommended Design Practices, Sandia National Laboratories, Albuquerque, NM, 1995.*

armature winding depend on the magnitude of the armature current, generators are rated in kVA rather than in watts. It is also possible that the generator will have a minimum power factor rating under full-load conditions. If so, the minimum power factor anticipated for the load must also be considered when the generator is selected.

3.9.4 Generator Maintenance

Table 3.4 shows cost and maintenance information on 3600 rpm gasoline generators, 1800 rpm gasoline generators, and diesel generators. Note that gas includes gasoline, natural gas, and propane. Note also that *small* is defined as sizes less than 100 kW. Diesel generators up to 5 MW are not uncommon, but are not typically intended for home or recreational vehicle use.

3.9.5 Generator Selection

Selection of a generator for a system may involve all the previous considerations. If the generator is not expected to be used very much, a light-duty or portable unit may suffice. If it is to be an essential, reliable system component, capable of running for long periods and for a significant number of hours per year, then a heavy-duty unit is preferable. The interesting challenge for the engineer is to determine the dividing line between light duty and heavy duty. After this choice is made, then all the other items previously listed may be considered. Generally, the unit with the lowest life-cycle cost will be the best unit for the application, provided that all operating parameters have been incorporated into the analysis.

3.10 BALANCE OF SYSTEM (BOS) COMPONENTS

Aside from the major components of a PV system, that is, PV modules, charge controllers, batteries, inverters and, possibly, generators, there are numerous other smaller/less expensive components necessary to complete a PV system installation. They include overcurrent protection devices, disconnects, surge protectors, array mounts, receptacles, ground fault detection and interruption devices, wiring, connectors, wiring enclosures, inverter bypass switches, source circuit combiner boxes, grounding connections, and battery containers. In some cases, the connected loads are considered to be part of the BOS, and in other cases the loads are not considered to be a part of the system. The distinction is generally made when the system is installed to operate a specific load, as opposed to being installed to be one contributor to the operation of any load.

Certain BOS components are regulated by codes or standards. Array mounts, for example, must meet any wind-loading requirements of applicable building codes. Battery compartments are covered in the *NEC*.

In certain environments, BOS components may need to be resistant to corrosion from exposure to salt air or may need to be appropriate for other environmental considerations. If a PV system is used as a structural component of a building, then a number of other codes and standards may become applicable.

The *National Electrical Code®* (*NEC*) [3] specifies the requirements for choosing most of the BOS components. *NEC* requirements will be described and used in detail in the next few chapters.

3.10.1 Switches, Circuit Breakers, Fuses, and Receptacles

All switches, circuit breakers, and fuses used in the dc sections of PV systems must be rated for use with dc. Switches, circuit breakers, and fuses used in ac circuits must be rated for ac use. Switches have both current and voltage ratings. If a switch is used to control a motor, it must be rated to handle the horsepower of the motor at the operating voltage of the motor.

Circuit breakers must be sized in accordance with *NEC* requirements. For example, although the maximum current ratings for #14, #12, and #10 THHN wire are 25, 30, and 40 A, respectively, the maximum fuse or circuit breaker sizes allowed for use with these wire sizes are 15, 20, and 30 A, respectively. Larger wire sizes may be fused at their rated ampacities (i.e., I_{max}).

For motors, it may be necessary to install a fuse or a circuit breaker with a rating that exceeds the circuit ampacity in order to accommodate the starting current of a motor. When this is the case, the motor must have a form of overload protection that will disconnect the motor if the motor current exceeds approximately 125% of its rated running current, depending on the size and type of motor. Details of wiring for motors and motor controllers are covered in great detail in *NEC* Article 430.

Different voltages require different receptacles. While it is not very likely that 12 VDC will damage a piece of 120 VAC equipment except possibly a motor or transformer, it is almost certainly true that 120 VAC will damage a piece of 12 VDC equipment. It is thus necessary to use different attachment cap and receptacle configurations for different voltages.

Because the *NEC* will be referenced in nearly all the design examples in later chapters, specific applications of *NEC* requirements will become more clearly evident as the examples are developed. Further discussion of overcurrent protection, disconnects, and wire sizing in accordance with *NEC* requirements will be covered in detail in these examples.

3.10.2 Ground Fault, Surge, and Lightning Protection

Presumably, current will leave the PV array via the positive conductor, and the same amount of current will return to the array via the negative conductor. This will be the case, provided that no alternate return paths are present.

The *NEC* requires that metallic frames and other metal parts of PV systems be connected to ground. The conductors used for this purpose are called grounding conductors. Generally, the positive or negative conductor (but not both) is also connected to ground at some point along the system. If so, the conductor connected to ground is called the grounded conductor. Because the grounded conductor is connected to ground at only one point, current will flow in the grounded conductor, but will not flow in any of the grounding conductors, as there is no closed circuit in which the current can flow.

However, if for some reason the ungrounded conductor were connected to ground, then there would be an alternate closed path in which current would be able to flow through the grounding conductors. If this is the case, then Kirchhoff's current law requires that the current in the ungrounded conductor equal the sum of the currents in the grounded and grounding conductors. The net result is that no longer will the currents in the positive and negative conductors be equal. The greatest danger in dc ground fault currents in a PV system is the means by which they are established. If the ground fault results from a loose connection, the connection may begin to arc and become a fire hazard.

The *NEC* requires ground fault protection to be provided to disconnect the array in the event of a difference in current between the positive and negative array conductors leading to the controller/inverter/loads.

Another important component used to provide protection to the array and inverter is the surge protector. Surge protectors are similar to zener diodes in that they are made of material that is essentially insulating until a predetermined voltage appears across the material. At this point, avalanche breakdown occurs and the surge protector acts as a current shunt. Metal oxide varistor (MOV) types of surge protector respond in nanoseconds and can bypass many joules of surge energy. However, a disadvantage of MOV surge protectors is that they draw a small current at all times and, generally, the failure mode of a MOV device is a short circuit. SiO (SOV) surge protectors are an improvement in terms of the MOV disadvantages. It is good practice to incorporate surge protection into system design at a location close to the common system ground point.

Additional protection from lightning strikes can be obtained with lightning rod systems. Contrary to popular belief, lightning rods do not attract lightning. Rather, they present a sharp point (or points) that is connected to ground. If a potential difference should appear between the rod and the surrounding atmosphere, a very high electric field builds up around the point to the extent that a corona discharge takes place, equalizing the charge between the air and the ground. Corona discharge is also known as glow discharge. Lightning is known as arc discharge. Arc discharge is violent, but glow discharge is essentially harmless because the charge is equalized over a much longer period of time.

3.10.3 Inverter Bypass Switches and Source Circuit Combiner Boxes

Inverter bypass switches are used in battery-backup systems to bypass the inverter in the event that maintenance is needed on the inverter and an alternate connection between grid and standby loads is required. Inverter bypass switches will be described in Chapter 6.

Source circuit combiner boxes are used to combine the outputs of individual source circuits of a PV array into a single PV output circuit. The combiner boxes include either fuses or circuit breakers in each source circuit to protect the source circuit from backfed current from other source circuits in the event that a problem should occur in a source circuit that either lowers the current or voltage in the circuit. At the time of this writing, PV systems that are directly grid connected with no battery backup generally operate at array voltages greater than 150 V. Also, at the time of this writing, no economical dc-rated circuit breakers are available with voltage ratings greater than 150 V. As a result, dc-rated fuses are used in source circuit combiner boxes designed for operating at voltages between 150 and 600 V dc and dc-rated circuit breakers are generally used in source circuit combiner boxes that are designed to operate at voltages below 150 V. Figure 3.34 shows a collection of BOS components.

3.10.4 Grounding Devices

All non-current-carrying metallic parts of the system, such as array mounts, PV module frames, equipment enclosures, junction boxes, and conduit, must be grounded. Grounded means that these items must be at ground potential. The *NEC* has established the sizes of conductors and the types of connectors that must be used for grounding equipment. As system designs are treated in later chapters, careful attention will be paid to the selection of proper grounding equipment.

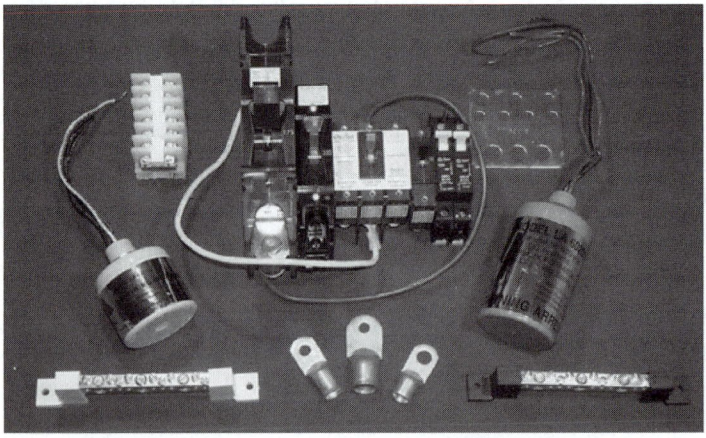

Figure 3.34 Examples of balance of system (BOS) components.

Problems

3.1 If $I_l = 2$ A, $I_o = 10^{-10}$ A, and $T = 300$ K for a PV cell, determine the maximum power point of the cell by differentiating the expression obtained by multiplying Equation (3.1) by the cell voltage.

3.2 Determine the range of operating voltages for which a 36-cell module will have power output within 90% of maximum power. You may assume $I_o = 10^{-10}$ A and $V_{OC} = 0.596$ V at an operating temperature of 300 K. Assume all cells in the module to be identical.

3.3 A PV module is found to operate at a temperature of 60°C under conditions of $T_A = 30$°C and G = 980 W/m^2. Determine the nominal operating cell temperature (NOCT) of the module.

3.4 Plot V_m versus T for $-25 < T < +75$°C for a 36-cell Si module for which each cell has $V_m = 0.5$ V at 25°C.

3.5 Plot P_m versus T for the module of Problem 3.4 if $I_m = 5.85$ A at 25°C.

3.6 Two 36-cell PV modules are connected in series. One is shaded, and the other is fully illuminated, such that the *I-V* characteristics of each module are as shown in Figure P3.1.

a. If the output of the two series modules is shorted, estimate the power dissipated in the shaded module.

b. If the two modules are equipped with bypass diodes across each 12 series cells, estimate the power dissipated in the shaded module.

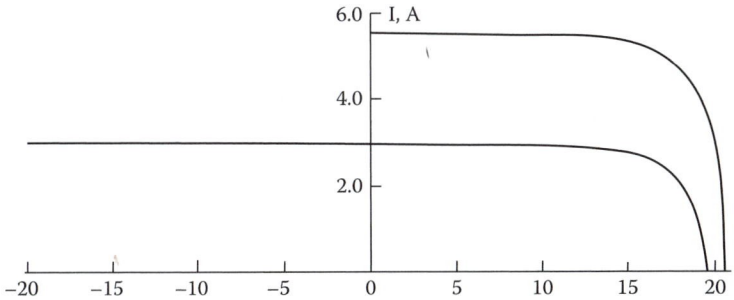

Figure P3.1

3.7 Rewrite the chemical equations of the Ni-Cd and NIMH systems showing the half-reactions at the anodes and at the cathodes. Determine the charge transfer mechanism.

3.8 How many gallons would have to be pumped into a tank raised 10 ft from the ground in order to be able to recover 1 kWh of electricity at a conversion efficiency of 100%, assuming the water is allowed to fall the entire 10 ft to operate the electrical generator?

3.9 Determine the capacitance of a capacitor needed to store 1 kWh if it is charged to 500 V.

3.10 Determine the inductance needed to store 1 kWh if the inductor is carrying a reasonable current, to be determined by the problem solver.

3.11 Determine the size and rotation speed of a flywheel designed to store 1 kWh of energy.

3.12 Describe two additional charging algorithms that will result in a fully charged battery and minimal waste of PV array energy.

3.13 If a maximum power point tracker operates at 98% efficiency with a PV array that has the characteristics shown in Figure P3.2, using the load curves as shown in the figure:

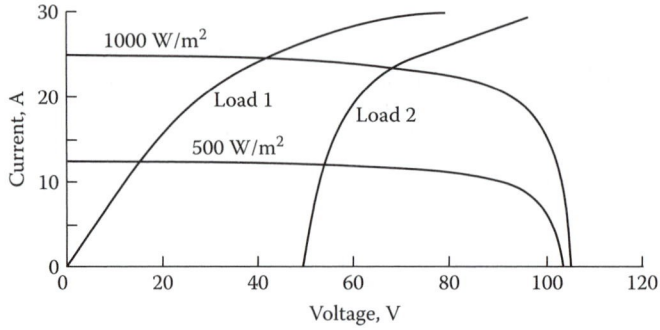

Figure P3.2

a. Determine the additional power available to each load when the array operates at 1000 W/m^2.
b. For each load, determine the charge controller input current and output current at 1000 W/m^2.
c. For each load, how long would the system need to operate at 1000 W/m^2 in order to recover the cost of the MPPT if the MPPT cost \$500 and the PV-generated electricity has a value of \$.40/kWh?
d. Repeat a–c if the array operates at 500 W/m^2.

3.14 A battery charge controller incorporates an MPPT to optimize charging of the batteries. Assume the maximum power voltage of a PV array to be 99 V and the bulk charge voltage level of a 48 V vented lead-acid battery bank to be 56 V. If the conversion efficiency of the MPPT is 98%, estimate the percentage increase in charge delivered to the battery under these conditions over that which would be delivered by a controller that causes the array to operate at 56 V rather than 96 V. You may assume the PV array consists of three modules connected in series, that V_{mp} is independent of irradiance levels, and each module has V_{OC} = 42 V, I_{SC} = 6.25 A, V_{mp} = 33 V, and I_{mp} = 5.76 A.

3.15 For the circuit of Figure 3.25, explain the on-off sequencing of the MOSFET switches to produce a symmetrical square wave output. Sketch the result.

3.16 Determine whether the dc source in Figure 3.25 remains on constantly during the switching. If not, explore modifications to the circuitry to ensure that the dc source is delivering constant dc current.

3.17 Show how a dc-dc converter can be used as a part of an inverter designed to have a square wave output of 120 V with a 12 V dc input. The idea is to design the inverter without a transformer. Sketch a block diagram, showing some components to clearly express your design.

3.18 For the 5-level H-bridge of Figure 3.27, determine T_2 if $T_1 = T_0$, V_{rms} = 120 V, and
a. V_{dc} = 150 V
b. V_{dc} = 155 V
c. V_{dc} = 160 V
d. V_{dc} = 165 V
Note that Excel or MATLAB can be very useful for this problem.

3.19 For the 5-level H-bridge of Figure 3.27, determine a set of values of T_0, T_1, and T_2 that will minimize the total harmonic distortion of the waveform if V_{dc} = 160 V and V_{rms} = 120 V. A convenient tool is to construct the waveform in PSPICE using series pulse voltage sources and then to run a fast Fourier transform of the resulting waveform.

3.20 For the 5-level H-bridge of Figure 3.27, determine the maximum value for V_{dc} if the design requires $T_2 > 0.025T$ and $T_0 > 0.025T$.

3.21 Referring to Figure 3.27a, design a 9-level H-bridge, using four capacitors and an appropriate number of series switches in each string. Show the waveform and the switching needed to obtain each of the nine levels of the output voltage.

3.22 For the 9-level H-bridge output waveform, derive a formula for the rms value of the output voltage.

3.23 Construct the 9-level H-bridge output waveform using series pulse voltage sources in PSPICE and perform a fast Fourier transform on the waveform to explore the harmonic distortion of the waveform. Keep V_{dc} and V_{rms} constant, while varying the duration of the different levels of the waveform in order to minimize the total harmonic distortion (THD) of the waveform.

3.24 Make a list of loads that might confuse an inverter that is in "sleep" mode. Explain why each load causes a problem.

3.25 The following data are given for a series of gasoline-powered electrical generators:

Rated output	1500 W	2300 W	3000 W	4500 W
Fuel tank size	2.9 gal	2.9 gal	4.5 gal	4.5 gal
Run time/tank	9 h	9 h	8.3 h	5.6 h

Calculate the kWh/gal for each of these generators under rated load conditions.

REFERENCES

1. RISE Information Portal, The Future of Grid-Connected Systems, www.rise.org.au/info/Applic/Gridconnect/indix.html (accessed November 2008).
2. *Maintenance and Operation of Stand-Alone Photovoltaic Systems*, Sandia National Laboratories, Albuquerque, NM, 1991.
3. *NFPA 70 National Electrical Code, 2008 Ed.*
4. IEEE Standard 1187-2002, *IEEE Recommended Practice for Installation Design and Installation of Valve-Regulated Lead-Acid Storage Batteries for Stationary Applications*, Institute of Electrical and Electronics Engineers, Inc., New York, 2002.
5. Linden, D., Ed., *Handbook of Batteries,* 2nd ed., McGraw-Hill, New York, 1994.
6. Hancock, O. G., Jr., Hydrogen storage: the hydrogen alternative, *Photovoltaics International*, March 1986.
7. Skerrett, P. J., Fuel UPD, *Popular Science*, June 1993.
8. Roland, B., Nitsch, J., and Wendt, H., *Hydrogen and Fuel Cells: The Clean Energy System*, Elsevier Sequoia, Oxford, U.K., 1992.
9. *Stand-Alone Photovoltaic Systems: A Handbook of Recommended Design Practices,* Sandia National Laboratories, Albuquerque, NM, 1995.
10. Buchanan, J. R., Lead-Acid Batteries Intro, www.buchanan1.net (accessed November 2008).
11. UL 1741: 2005, Standard for Static Inverters and Charge Controllers for Use in Independent Power Production Systems, Underwriters Laboratories, Inc., Northbrook, IL, May 1999.
12. ANSI C84.1–1995, Electric Power Systems and Equipment—Voltage Ratings (60 Hertz), American National Standards Institute, New York, 1995.
13. http://www.xantrex.com/ (Information on Xantrex Power Electronics Equipment.)
14 http://www.satcon.com/ (Information on Satcon PV Inverters.)

15. IEEE 1547-2003, *Standard for Interconnecting Distributed Resources with Electric Power Systems*, IEEE Standards Coordinating Committee 21, Photovoltaics, 2003.
16. SMA America, Inc., *Solar Inverter Sunny Boy 5000US, 6000US, 7000US Installation Guide, V2.2.*
17. Outback Power Systems, FX Series Inverter/Charger FX/VFX/GTFX/GVFX/Mobile Installation Manual, 2008 (www.outbackpower.com).
18. Welcome to Generator Joe home page, www.generatorjoe.net/store.asp (a wealth of information on a wide variety of fossil fuel generators).
19. www.hondapowerequipment.com/products/modeldetail.aspx?page=modeldetail&sec

SUGGESTED READING

ASCE Standard 7-05, Minimum Design Loads for Buildings and Other Structures, American Society of Civil Engineers, Reston, VA, 2005.

Hoogers, G., Ed., *Fuel Cell Technology Handbook*, CRC Press, Boca Raton, FL, 2002.

Hu, C. and White, R., *Solar Cells: From Basics to Advanced Systems*, McGraw-Hill, New York, 1983.

Krein, Philip T., *Elements of Power Electronics*, Oxford University Press, New York, 1998.

Markvart, T., Ed., *Solar Electricity*, John Wiley & Sons, Chichester, U.K., 1994.

Roden, M. S. and Carpenter, G. L., *Electronic Design: From Concept to Reality*, Discovery Press, Burbank, CA, 1997.

Sedra, A. S. and Smith, K. C., *Microelectronic Circuits,* 4th ed., Oxford University Press, New York, 1998.

Silberberg, M., *Chemistry: The Molecular Nature of Matter and Change*, Mosby, St. Louis, 1996.

Skvarenina, T. L., Ed., *The Power Electronics Handbook*, CRC Press, Boca Raton, FL, 2002.

Vithayathil, J., *Power Electronics: Principles and Applications*, McGraw-Hill, New York, 1995.

CHAPTER 4

Grid-Connected Utility-Interactive PV Systems

4.1 INTRODUCTION

As the cost of photovoltaic (PV) systems continues to decrease, utility-interactive systems are becoming more economically viable. Furthermore, increases in consumer awareness often result in a willingness to pay a premium price for clean electrical energy. This feedback loop, coupled with the increased demand for stand-alone systems, as well as the financial incentives offered by a number of states and countries, has resulted in a healthy demand for PV system components. And this increased demand has enabled PV module and balance of system (BOS) component manufacturers to scale up manufacturing facilities to take advantage of economies of scale to further reduce system costs.

In addition to cost reductions, the increased demand for PV systems has led to significant efforts to improve the reliability of PV system components, designs, and installations. So, not only are PV systems decreasing in cost, but they are also increasing in reliability.

By 1999, the technical issues associated with connecting PV systems to the utility grid had essentially been solved. In 2000, IEEE adopted Standard 929-2000 [1]. Any PV system meeting the performance criteria of IEEE Standard 929, using power-conditioning units/inverters (inverters) listed under UL 1741 and installed in accordance with the current *National Electrical Code*, automatically met all established technical performance criteria. Since then, IEEE has adopted further refinements to Standard 929-2000, which are reflected in IEEE 1547-2003 [2].

Grid-connected utility-interactive PV systems are capable of supplying power to the utility grid. There are essentially two classes of grid-connected systems: straight grid-connected with no battery backup and grid-connected with battery backup. Grid-connected systems with battery backup can be further separated into dc-coupled systems and ac-coupled systems, and in some cases, battery-backup systems do not sell back to the grid, but only purchase power from the grid for the purpose of charging the system batteries.

This chapter will focus on the design of straight grid-connected PV systems. Straight grid-connected systems are the most straightforward systems to understand

and are currently the most commonly installed systems. Battery-backup systems will be covered in Chapter 6, and stand-alone PV systems will be covered in Chapter 7.

Before beginning the design process, it will be helpful to take a closer look at some of the codes and standards that govern the design and installation of PV systems.

4.2 APPLICABLE CODES AND STANDARDS

A number of codes and standards have been created to ensure the safety of electrical systems. These codes and standards also generally address the efficiency and reliability of systems. Because a PV system is capable of generating sufficiently high voltages to present a potential electrical safety hazard, PV systems are included in these codes. Perhaps the two most common codes and standards that deal with PV systems are the *National Electrical Code®* [3] and IEEE 1547-2003 [2]. The PV engineer should be familiar with both. But the list does not end here. PV systems have a mechanical component as well as an electrical component, and, hence, a number of mechanical and structural codes also apply. If the PV system is to be integrated into the construction of a building, the list of codes and standards becomes even longer.

Table 4.1 provides a partial listing of the many codes and standards that may be applicable to any particular PV installation.

4.2.1 The National Electrical Code

4.2.1.1 Introduction

The *National Electrical Code* (*NEC*) [3] is published by the National Fire Protection Association and is updated approximately every 3 years. This text will use the 2008 edition of the *NEC* as a reference. The *NEC* consists of a collection of articles that apply to considerations such as wiring methods, grounding, motor circuits, and nearly every conceivable topic in which electrical safety and efficient utilization is a consideration, including Article 690, which deals specifically with PV systems.

The *NEC* specifies the sizes and types of switches, fuses, and wire to be used, and specifies where these items must be located in the system. These components are necessary, not only to protect the end user, but also to protect the maintenance technician. As specific PV system design examples are discussed, compliance with the *NEC* will be incorporated into the design process discussion. In particular, proper wire sizing to limit voltage drop in connecting wires to acceptable limits, proper use of switches, circuit breakers and fuses, types of insulation and conductors, types of electrical conduit, and proper grounding will be considered.

Normally, copper wire will be used in PV system wiring. Aluminum wire is allowed by the *NEC*, but generally it is used only over longer distances for carrying higher currents when the cost of using copper would be prohibitive. For example,

Table 4.1 Partial Listing of Codes and Standards That May Apply to PV Systems

Reference #	Title/Contents
NEC 2008	*National Electrical Code*/Wiring methods (comprehensive)
IEEE 928	IEEE Recommended Criteria for Terrestrial Photovoltaic Power Systems
IEEE 929	IEEE Recommended Practice for Utility Interface of Residential and Intermediate Photovoltaic Systems
IEEE 937	IEEE Recommended Practice for Installation and Maintenance of Lead-Acid Batteries for Photovoltaic Systems
IEEE 1013	IEEE Recommended Practice for Sizing Lead-Acid Batteries for Photovoltaic Systems
IEEE 1144	Sizing of Industrial Nickel Cadmium Batteries for PV Systems
IEEE 1145	IEEE Recommended Practice for Installation and Maintenance of Nickel Cadmium Batteries for Photovoltaic Systems
IEEE 1187	Recommended Practice for Design and Installation of Valve-Regulated Lead-Acid (VRLA) Storage Batteries for Stationary Applications
IEEE 1262	Recommended Practice for Qualification of Photovoltaic Modules
IEEE 1361	Recommended Practice for Determining Performance Characteristics and Suitability of Batteries in Photovoltaic Systems
IEEE 1373	Recommended Practice for Field Test Methods and Procedures for Grid-Connected Photovoltaic Systems
IEEE 1374	Guide for Terrestrial Photovoltaic Power System Safety
IEEE 1479	Recommended Practice for the Evaluation of Photovoltaic Module Energy Production
IEEE 1513	Recommended Practice for Qualification of Concentrator Photovoltaic Receiver Sections and Modules
IEEE 1526	Recommended Practice for Testing the Performance of Stand-Alone Photovoltaic Systems
IEEE 1547	IEEE Standard for Interconnecting Distributed Resources with Electric Power Systems
IEC TC-82	A compendium of 25 standards relating to the electrical and mechanical performance testing and measurement of PV systems
ISO 9001	An international quality standard, comprising of 20 segments, dealing with all aspects of design, manufacturing, and delivery of service
UL 1741	Standard for Static Inverters and Charge Controllers for Use in Photovoltaic Power Systems
ANSI Z97.1	Relates to safety relating to potential glass breakage
ASCE 7-05	Minimum Design Loads for Buildings and Other Structures
ASTM	A compendium of tests and standards that may apply to building integrated photovoltaic systems

Sources: Based on data from IEEE 929-2000, IEEE Recommended Practice for Utility Interface of Residential and Intermediate Photovoltaic (PV) Systems, IEEE Standards Coordinating Committee 21, Photovoltaics, 2000; IEEE 1547-2003, IEEE Standard for Connecting Distributed Resources to Electric Power Systems, IEEE Standards Coordinating Committee 21, Fuel Cells, Photovoltaics, Distributed Generation and Energy Storage, 2003; *NFPA 70 National Electrical Code, 2008 ed.*, National Fire Protection Association, Quincy, MA, 2007; UL 1741: 2001, Inverters, Converters and Controllers for Use in Independent Power Systems, Underwriters Laboratories, Inc., Northbrook, IL, 2005; ANSI C84.1—1995, Electric Power Systems and Equipment—Voltage Ratings (60 Hertz), American National Standards Institute, New York, 1995. Wohlgemuth, Testing for BIPV, *Proc. Photovoltaic Performance and Reliability Workshop*, Cocoa Beach, FL, November 1998, 113. DeBlasio, Status of IEEE Standards Coordinating Committee (SCC 21) and IEC Standards Technical Committee 82 (TC82), *Proc. Photovoltaic Performance and Reliability Workshop*, Cocoa Beach, FL, November 1998, 236; IEEE Standard 519–1992, *Recommended Practices and Requirements for Harmonic Control in Electrical Power Systems*, Institute of Electrical and Electronics Engineers, Inc., New York, 1992; Zgonena, T., Safety Considerations for Static Grid-Tied Photovoltaic Inverters, Interconnecting Small PV Systems to Florida's Utility Grid, A Technical Workshop for Florida's Utilities, Cocoa, FL, October 22, 1998; ASCE Standard 7-05, Minimum Design Loads for Buildings and Other Structures, American Society of Civil Engineers, Reston, VA, 2005.

because copper is significantly heavier than aluminum, the insulators on power poles would need to be considerably stronger if copper wire were used instead of aluminum. Aluminum has a lower conductivity than copper, it oxidizes faster than copper, and unless terminals are tightened properly, it is more likely to loosen in a connector. Furthermore, aluminum oxide is not readily visually distinguishable from aluminum.

Because aluminum oxide is a good insulator, special care must be taken when terminating aluminum to ensure that the exposed aluminum is free of oxidation. If aluminum wire is used in a system, devices with terminations approved for use with aluminum must be used.

The purpose of this section is to highlight several *NEC* requirements. The reader who goes beyond the walls of the classroom into the design of PV systems should invest in the latest edition of the *NEC* to be sure to have all the latest requirements on items such as ground fault detection and interruption, source and output circuit maximum voltage ratings, acceptable conductor placement, and storage batteries.

4.2.1.2 Voltage Drop and Wire Sizing

Table 4.2 shows the dc resistance per 1000 ft and rated current (ampacity) for copper conductors with insulation rated at 90°C. The ampacities are for no more than three current-carrying conductors in a conduit at a temperature of 30°C or less. If more than three current-carrying conductors are in a conduit, or if the conductors are operated in an ambient that exceeds 30°C, their ampacity must be derated. Ampacities for single conductors in free air are higher than those listed in Table 4.2. While many different types of insulation exist, as shown in *NEC* Article 310, type THWN-2 is a high-temperature (90°C), moisture- and oil-resistant thermoplastic insulation that is commonly in use for wiring protected by conduit. Type THHN insulation is almost always also rated as THWN-2.

The *NEC* requires that the total voltage drop in feeder and branch circuits be less than 5%, with the drop in either feeder or branch circuits limited to no more than 3%. A feeder circuit is a circuit that provides power to an electrical distribution panel. In a common residential electrical service, the feeder circuit is the wiring between

Table 4.2 Properties of Copper Conductors with 90°C Insulation

Size	18	16	14	12	10	8	6	4
dc Ω/kft	7.77	4.89	3.07	1.93	1.21	0.764	0.491	0.308
I_{max} (A)	14	18	25	30	40	55	75	95
Size	3	2	1	0	00	000	0000	250 kcm
dc Ω/kft	0.245	0.194	0.154	0.122	0.0967	0.0766	0.0608	0.0515
I_{max} (A)	110	130	150	170	195	225	260	290

Note: Solid conductor wire sizes 18–8. Stranded for larger sizes.
Source: Based on data from *NFPA 70 National Electrical Code, 2008 ed.*, National Fire Protection Association, Quincy, MA, 2007.

the electric meter and the circuit breaker panel. The branch circuits are the circuits that provide power to the individual electrical loads, such as lighting, refrigerators, dishwashers, air conditioners, etc., that are connected to the distribution panel. The PV equivalence of branch circuits can be considered to be the PV source circuits that connect the PV array to the power conditioning equipment, and the PV equivalence of feeder circuits can be considered to be the PV inverter output circuits that connect to the utility. In any case, good PV system design generally requires voltage drop in any PV circuits to be less than 2%.

Many tables exist in various design manuals that list the maximum distance a certain size conductor can be run with a given current and still not produce excessive voltage drop. For the engineer with a calculator, however, all one needs to do is to recognize that a circuit consists of wire in both directions, so that a load located 50 ft from a voltage source will need 100 ft of wire to carry the current to and from the load. If d is the distance from source to load and V_s is the source voltage, the percent voltage drop in the wire is thus given by

$$\%VD = 100\frac{I}{V_s}\left(\frac{\Omega}{kft}\right)\left(\frac{2d}{1000}\right) = \frac{0.2Id}{V_s}\left(\frac{\Omega}{kft}\right) \tag{4.1}$$

as long as the circuit is a dc circuit or a single-phase, two-wire ac circuit. Three-phase voltage drop calculations will be introduced in Section 4.7.

Example 4.1

A 20 W, 12 V dc fluorescent lamp is located 50 ft from a 12 V battery. Specify the wire size needed to keep the voltage drop between battery and lamp under 2%.

SOLUTION

First solve Equation (4.1) for (Ω/kft) to obtain

$$\Omega/kft = \frac{(\%VD)V_s}{0.2Id}.$$

Then, determine the load current from $P = IV$, assuming the load voltage to be essentially equal to the supply voltage. Substituting the known values on the right-hand side yields Ω/kft $= (2 \times 12)/(0.2 \times 1.67 \times 50) = 1.437$. Table 4.2 shows that #12 wire has too much resistance, so it is necessary to use #10. The actual voltage drop with #10 wire can now be found from Equation (4.1) to be 1.68%.

Note that although #10 wire will carry 40 A, the current in the circuit is limited to 1.67 A due to the voltage drop problem. It is very important to be aware of the need for larger wire in low-voltage systems. Equation (4.1) and Table 4.2 will find considerable use in examples to follow in this and later chapters.

Table 4.3 Summary of Contents of *NEC*® Article 690

Section	Contents	*NEC* Cross References
I	**General**: Scope, Definitions, Installation, Ground-fault protection, ac modules	Article 240
II	**Circuit Requirements**: Maximum voltage, Circuit sizing and current, Overcurrent protection, Stand-alone systems	Articles 110, 210, 240
III	**Disconnecting Means**: Conductors, Additional provisions, PV equipment, Fuses, Switches and circuit breakers, Installation and service	Article 230
IV	**Wiring Methods**: Methods permitted, Component interconnections, Connectors, Access to boxes	Articles 310, 339, 400
V	**Grounding**: System grounding, Point of system grounding connection, Equipment grounding, Size of equipment grounding conductor, Grounding electrode system	Article 250
VI	**Marking**: Modules, ac modules, PV power source, Point of common connection	
VII	**Connection to Other Sources**: Identified interactive equipment, Loss of interactive system power, Ampacity of neutral conductor, Unbalanced interconnections, Point of connection	Article 230
VIII	**Storage Batteries**: Installation, Charge control, Battery interconnections	Articles 400, 480
IX	**Systems Over 600 V**: General, Definitions	Article 490

Because of the problem with voltage drop at low voltages, and the correspondingly larger wire sizes necessary, it is generally desirable to operate PV systems that deliver any significant amounts of power over any reasonable distances at voltages higher than 12 V. If the previous 20 W load were connected to a 24 V system, the load current would be halved, and the load voltage would be doubled. This allows the resistance of the wiring to be four times higher, or as much as 5.648 Ω/kft, which means that #16 wire is now adequate. From a total cost standpoint, however, the cost of the additional 12 V battery will exceed the difference in cost between #10 wire and #16 wire. Presumably, the cost of the PV modules will be the same, because the amount of power required has not changed.

For concealed wiring, the minimum wire size is #14. Smaller wire sizes are normally used only for portable cords or for attaching single loads.

The *National Electrical Code* is an important source of information for the PV design engineer, because it clearly defines acceptable PV system design practice. Article 690 deals exclusively with PV systems, but refers to other articles such as Article 240 on overcurrent devices, Article 250 on grounding, and Article 310 on conductor ampacities. Other parts of the *NEC* also apply to specific installations or installation methods. Table 4.3 summarizes the components of *NEC* Article 690 and lists some of the other *NEC* articles that apply to PV system installations.

The design examples that follow in this chapter as well as in Chapters 6 and 7 will introduce specific sections of the *NEC* that apply to the specific designs as the need arises.

4.2.2 IEEE Standard 1547-2003

4.2.2.1 Introduction

Prior to the adoption of IEEE Standard 1547-2003 [2], grid-connected PV inverters were required to comply with IEEE Standard 929. IEEE Standard 929 was developed specifically for the purpose of addressing concerns of utilities regarding quality of power delivered to the grid and the need to disconnect the PV system from the utility grid in the event of utility power failure. IEEE Standard 1547 was developed "to provide a uniform standard for interconnection of distributed resources with electric power systems," by providing requirements for performance, operation, testing, safety considerations, and maintenance of the equipment. Thus, in addition to covering electronic inverters as used in PV systems, IEEE Standard 1547 also covers distributed resources that use synchronous machines and induction machines, such as wind and low-head hydro power. If more than one type of distributed resource, up to an aggregate capacity of 10 MVA, is connected at a common point of utility connection, then all must meet the requirements of this standard at the connection point.

While the PV system designer only needs to specify that the inverter must be listed to UL 1741, which is based on IEEE 1547, to be assured that it will meet all utility interconnect requirements, it is interesting to explore some of the requirements of IEEE 1547 in order to appreciate the amount of creative engineering that was involved in creating the standard and the subsequent amount of creative engineering that is necessary to meet the standard with an inverter design. In fact, the engineer who loves to speak in acronyms should probably obtain a copy of IEEE 1547 in order to add another dozen or so acronyms to her vocabulary.

4.2.2.2 Specific Requirements

IEEE 1547 is essentially a functional standard as opposed to a prescriptive standard. It specifies functions that must be performed by a distributed generation system rather than specify how the function must be implemented. It is up to the designer to figure out how to satisfy the functional requirements. Section 4.1 of the standard delineates the general requirements of voltage regulation, integration of grounding, synchronization with the grid, distributed secondary spot networks, disconnect from the grid when the grid is de-energized, monitoring, isolation, electromagnetic interference, withstanding surges, voltage rating of interconnecting device, and response to abnormal grid conditions. Abnormal grid conditions include overvoltage, undervoltage, overfrequency, and underfrequency, and reconnection to the grid after the abnormal condition is cleared. The following discussion will apply to PV inverter compliance with IEEE 1547.

Table 4.4 defines abnormal voltage conditions and lists the necessary response times for the PV inverter, as listed in Table 1 of IEEE 1547-2003. Because PV inverters may have output voltages of 120, 208, 240, 277, or 480 V, the table lists abnormal limits as a percentage of the "normal" voltage. The listed clearing times are

Table 4.4 Required Clearing Times for PV Inverters under Abnormal Grid Voltage Conditions

Voltage Range in % of Base Voltage	<50%	50% to 88%	110% to 120%	≥120%
Clearing Time (s)	0.16	2.0	1.0	0.16

Source: Based on data from IEEE 1547-2003, IEEE Standard for Connecting Distributed Resources to Electric Power Systems, IEEE Standards Coordinating Committee 21, Fuel Cells, Photovoltaics, Distributed Generation and Energy Storage, 2003.

maximum limits for inverters rated at less than 30 kW and are default clearing times for larger inverters.

Table 4.5 defines clearing times for abnormal grid frequencies for inverters that are rated either more than 30 kW or less than 30 kW. The clearing times on the smaller units can either be fixed or adjusted in the field. The clearing times on the larger units must be field adjustable. As in the case of voltage clearing times, the times for the smaller units are maximum values, while the times for the larger units are default values.

If a PV inverter shuts down due to a voltage or frequency disturbance, it must monitor the voltage and frequency of the utility for 5 minutes upon restoration of the grid. If voltage and frequency remain stable for this time period, then the inverter will restart. This same situation, of course, applies to the initial start-up of the inverter. When the ac side of the inverter is switched on, the inverter will not deliver any power for 5 min. For the installer who turns on the switch and waits for the inverter to start, this is the longest 5 min of the day.

IEEE 1547 also defines necessary power quality requirements for PV inverters. The maximum dc component of an inverter output must be less than 0.5% of rated inverter output current. The inverter must not cause objectionable flicker for other grid users, generally defined as not causing their lights to flicker. The final power quality measure is the accuracy of the sine wave output, defined in terms of the

Table 4.5 Required Clearing Times for PV Inverters under Abnormal Grid Frequency Conditions

Inverter Size (kW)	Frequency Range (Hz)	Clearing Time (s)
≤30 kW	>60.5	0.16
	<59.3	0.16
>30 kW	>60.5	0.16
	<(59.8–57.0) (adjustable)	Adjustable 0.16–300
	<57.0	0.16

Source: Based on data from IEEE 1547-2003, IEEE Standard for Connecting Distributed Resources to Electric Power Systems, IEEE Standards Coordinating Committee 21, Fuel Cells, Photovoltaics, Distributed Generation and Energy Storage, 2003.

Table 4.6 Maximum Allowable Harmonic Amplitudes for PV Inverter Output in Percent of Maximum Load Current

Harmonic Range	$n < 11$	$11 \leq n < 17$	$17 \leq n < 23$	$23 \leq n < 35$	$35 \leq n$
% of Rated Current	4.0	2.0	1.5	0.6	0.3

Source: Based on data from IEEE 1547-2003, IEEE Standard for Connecting Distributed Resources to Electric Power Systems, IEEE Standards Coordinating Committee 21, Fuel Cells, Photovoltaics, Distributed Generation and Energy Storage, 2003.

harmonic content of the waveform. Table 4.6 is a summary of the maximum allowed harmonic amplitudes. Note that the overall harmonic distortion must not exceed 5%. The table entries apply to odd harmonics. Even harmonics must be less than 25% of the odd harmonic limits.

The final requirement of IEEE 1547 is that if an unintentional island should occur in which a collection of inverters is connected to the grid and grid power is lost, all of the inverters must be able to distinguish the grid from other inverters such that all inverters in the island shut down within 2 seconds of loss of grid power. The standard also recognizes that there may be an instance where, under the condition of intentional islanding, the inverters remain connected, but this part of the standard is still under development. One such example might be a military base that is connected to the grid, but also has PV systems in operation. If the grid goes down, it would still be desirable to maintain as much power supply as possible for critical base operations.

It is relatively straightforward to detect islanding, because PV inverters that are synchronized with the grid voltage act as nearly ideal current sources. Figure 4.1 shows how a typical inverter will generate a current waveform that is slightly off grid frequency. As long as the grid voltage is present, the inverter adjusts the current waveform frequency at zero crossings of the grid voltage. If the grid is lost, then the inverter current drifts in frequency until it exceeds either the upper or the lower

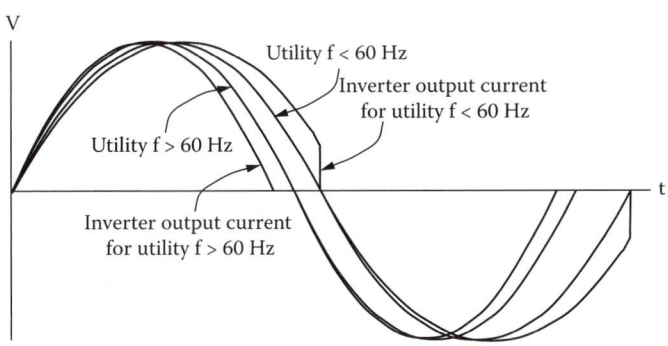

Figure 4.1 Comparison of utility voltage and inverter current for under-frequency and over-frequency utility voltage.

frequency limit. This algorithm, developed at Sandia National Laboratories, is called the Sandia frequency shift (SFS) [11,12]. A similar algorithm, the Sandia voltage shift (SVS) [11,12], is used to detect the output voltage of the inverter going out of range, and shutting down the inverter until the grid is restored.

While IEEE 929 required that the voltage at the inverter output should not be more than 5% higher than the voltage at the point of utility connection, IEEE 1547 no longer contains this requirement. IEEE 1547 is only concerned with conditions at the point of utility connection. To this end, IEEE 1547 prescribes test procedures to verify compliance. These procedures are used by Underwriters Laboratories when they perform these tests prior to listing an inverter to UL 1741. Thus, any inverter listed under UL 1741 has been tested to meet the criteria established in IEEE 1547.

4.2.2.3 Comparison of PV Inverters to Mechanically Rotating Generators

When considering the connection of a PV source to the grid, it is important to distinguish between the electrical characteristics of an inverter and a conventional rotating generator. First of all, most utility-interactive inverters are best modeled as dependent current sources, while rotating generators appear as voltage sources. In the event of a short-circuit fault, a rotating generator can deliver a very large current, limited only by the ability of the prime mover to keep the generator rotating. Any energy stored as rotational energy can be dissipated into a short circuit as electrical energy. On the other hand, if a short circuit occurs at the output of an inverter, little more current than full-load value will flow from the inverter.

Because the inverter acts as a current source, it is easier to ensure that the inverter will meet the standards for utility interconnection. The reason is that the utility is close to being an ideal voltage source. Hence, the inverter can sense the utility voltage and frequency and inject current only if the voltage and frequency fall within prescribed limits. This same circuitry can be used to ensure that the current is injected in phase with the utility voltage. This ensures a high power factor for the inverter output. The sensing circuitry has high impedance inputs and can remain connected to the utility at all times in order to monitor the voltage and frequency stability of the utility.

It should be noted that the voltage values apply to the point of utility connection, also known as the point of common coupling (PCC) for the inverter. If the inverter is located some distance from the PCC, there may be voltage drop on the line between the inverter and the PCC. If so, compensation can be made at the inverter output, because the inverter output voltage in these cases will be higher than the voltage at the PCC in order for the inverter to deliver power to the PCC.

4.2.2.4 Islanding Analysis

The greatest concern over islanding occurs when more than one inverter is in operation within the island. In this case, it is possible that they will support a feedback situation in which each inverter, sensing the combined output of the other inverters, thinks the output of the other inverters constitutes the grid. This possibility is enhanced under worst-case load conditions. It is thus useful to consider what might constitute worst-case load conditions.

Worst-case load conditions occur if, when the island is created by the utility fault, the island voltage does not change quickly to the steady-state fault value. This will occur under two conditions: resonance at the utility frequency and island motor loads with low damping, such as grinding wheels that continue to rotate even though power is removed from the motor.

If a circuit is at resonance with a relatively high Q, then it will continue to oscillate at its natural resonant frequency until the oscillation is ultimately damped out. As long as a motor continues to rotate, it will act as a generator and return its back emf to the grid. Except for those used in wind turbines, induction motors are normally inefficient generators, but synchronous motors are very efficient generators. The generation frequency, of course, depends on the rotation speed, so as the motor slows down, the generated frequency changes.

Circuit analysis textbooks give a general proof that any parallel RLC circuit will be underdamped if it has a $Q > 0.5$. Equation (4.2) represents the response of an underdamped parallel RLC circuit.

$$v(t) = V_m e^{-\alpha t} \cos(\omega_d t + \phi), \tag{4.2}$$

where V_m is the initial amplitude, $\alpha = \frac{1}{2RC}$, $\omega_o = \frac{1}{\sqrt{LC}}$, $\omega_d = \sqrt{\omega_o^2 - \alpha^2}$, and ϕ is a phase angle that depends on initial energy storage in the inductor and capacitor. The Q of the circuit can be found from

$$Q = \frac{\omega_o}{2\alpha} = \omega_o RC. \tag{4.3}$$

It is interesting to use Equations (4.2) and (4.3) to calculate how long it will take for the amplitude of the voltage to decrease to the inverter trip limit as a function of Q.

The two important parameters of the equation are the exponentially decaying amplitude part, $V_m e^{-\alpha t}$, and the natural frequency, ω_d. Assuming the energized grid has a natural resonant frequency outside the trip limits, the inverter will sense the frequency departure and disconnect within the prescribed time. If the natural resonant frequency is within the trip limits, then the inverter will need to disconnect when the voltage falls outside the trip limits. For the case of the motor/flywheel load, again, as soon as the motor speed drops enough to move the motor output frequency outside the trip limits, the inverter will trip. On the other hand, if the inverter keeps the motor running, it may never trip. Given that neither the inverter nor the motor will normally have any provisions for stabilizing the frequency or rotation, however, this occurrence is extremely unlikely.

Table 4.7 tabulates the number of cycles required for the amplitude of the island load voltage to decrease to 50% of its initial value as a function of the Q of the circuit.

Table 4.7 The Number of Cycles for the Voltage to Reach Half the Original Amplitude in an Underdamped, Decaying, Parallel RLC Circuit as a Function of the Q of the Circuit

Q	1	2	3	4	5	6	7	8	9	10
N	0.19	0.43	0.65	0.88	1.10	1.32	1.54	1.76	1.98	2.20

The results are obtained by setting $V_m e^{-\alpha t} = 0.5V_m$. Solving for t yields the result

$$t = \frac{\ln 2}{\alpha}. \tag{4.4}$$

But it can be shown (see Problem 4.3) that $\alpha = \pi f_o/Q$. Thus, the time for the voltage amplitude to fall to half its starting value is given by

$$t = \frac{Q \ln 2}{\pi f_o}. \tag{4.5}$$

It also can be shown that (Problem 4.4) the relationship between resonant frequency, ω_o, and natural resonant frequency, ω_d, is

$$f_o = \frac{\omega_o}{2\pi} = \frac{\omega_d}{2\pi\sqrt{1 - \frac{1}{4Q^2}}} = \frac{f_d}{\sqrt{1 - \frac{1}{4Q^2}}}. \tag{4.6}$$

So, finally, solving for t in terms of the period of the natural frequency,

$$t = \frac{Q \ln 2}{\pi f_d}\sqrt{1 - \frac{1}{4Q^2}} = \left[0.2206\sqrt{Q^2 - \frac{1}{4}}\right] T_d. \tag{4.7}$$

Because T_d represents one cycle at the natural resonant frequency, the coefficient of T_d represents the number of cycles, N, that it takes for the voltage amplitude to decay to half its initial value.

The previous analysis was based on the assumption that the utility island load had only initial stored energy. If energy continues to be added to the load in a synchronous manner, the load will continue to oscillate in a manner not very different from an electronic class C amplifier, depending on the conduction angle of the current source. The time to decay is thus prolonged, perhaps indefinitely, depending on the match between the load and the inverter output. So, this is the worst-case condition that must be overcome, and this is why the SVS and SFS, as previously discussed, were developed.

4.2.3 Other Issues

4.2.3.1 Aesthetics

Although not a part of Article 690, *NEC* 110.12 requires that "electrical equipment shall be installed in a neat and workmanlike manner." This addresses the importance of a professional appearance for an installation and helps remove any potential objections to the aesthetics of the installation. The aesthetics of a PV installation are of particular concern to architects and building owners, who want to make the PV system look like it is an intended part of a structure as opposed to looking like an add-on. The PV community has responded with building integrated PV (BIPV) products such as shingles, tiles, windows, and laminates, as well as framed modules with "more aesthetic appeal" than other framed modules to address the wishes of the design community.

4.2.3.2 Electromagnetic Interference

The U.S. Federal Communications Commission (FCC) requires that any electronic device with an electronic clock that runs at a frequency greater than 9 kHz must meet its standards for electromagnetic interference (EMI) [13]. EMI occurs in two forms: conducted and radiated. Conducted EMI is coupled directly through connections to the power line. Radiated EMI consists of radio frequency signals transmitted by the device directly to the surroundings. Both types of EMI, if sufficiently strong, can interfere with other electronic devices.

Because all modern utility-interactive inverters are microprocessor controlled, all have internal clocks that can be expected to run at frequencies in the MHz range. Hence, all are subject to the radiative and conductive emission standards as set forth in FCC Part 15 of the Code of Federal Regulations. Each unit must have a label that indicates compliance.

Conducted emission will normally not be a problem because of the low-pass filters needed at the output of an inverter to average out the synthesized sinusoidal signal. High-frequency currents that might otherwise be conducted to the utility interconnection are thus suppressed by this filter.

Radiated emission is caused by clock and other high-frequency currents circulating in sufficiently long conductors on the printed circuit boards of the inverter. The long, sometimes looped, conductors act as antennas and radiate the clock pulse signal. These same conductors act as receiving antennas for signals from other devices. Because the clock is not sinusoidal, it contains many harmonics that spread across the spectrum. An important part of suppression of radiated emission is the inverter cover. Hence, it is important to avoid operating the inverter without its cover for prevention of electrical shock, and also for FCC compliance.

4.2.3.3 Surge Protection

PV system surge protection is not directly addressed in the *NEC* or in IEEE 1547, but surge protection in general is covered in *NEC* article 280. Surges may appear on either the utility line or on the PV line. Both are subject to lightning strikes, and the utility line is particularly susceptible to transients from events such as motor starting and stopping, which may cause sharp inductive voltage spikes. A good PV system design will provide MOV or SOV surge protection on both the dc side and the ac side of the inverter. These devices act much like bilateral zener diodes, with response times measured in nanoseconds. They can bypass hundreds of joules of energy that might otherwise enter the inverter or the PV modules and damage the components. Lightning protection is a wise choice for systems installed in areas with a high incidence of lightning. The idea is to dissipate any uneven charge balance between ground and the atmosphere prior to the buildup of sufficient voltage to cause arc discharge.

4.2.3.4 Structural Considerations

The primary consideration in the structural design of utility-interactive PV systems is safety. This includes the safety of individuals as well as the protection of structures and other property that could be damaged as the result of mechanical failure.

Building codes are adopted and enforced by local jurisdictions to help ensure a safe environment in and around buildings. A major difference exists between the installation of utility-interactive versus stand-alone PV systems in that codes are much more likely to be enforced for the former. This is primarily because of the need to electrically connect the interactive system to the grid, which can result in multiple inspections: (a) by the electric utility service provider and (b) by local code officials, which may include both electrical and building inspections. Stand-alone PV systems need only be inspected as required by local building officials.

As will be discussed in Chapter 5, large-area arrays may present a significant wind load on a structure. They are also subject to corrosion and degradation by ultraviolet sunlight components. Furthermore, they may be subjected to other conditions such as rain, snow, ice, and earthquakes. While most arrays are tested for mechanical performance, and while most array mounts have been preengineered to withstand the worst of loading conditions, it is still important for the engineer to determine that the specific array mounting is adequate with respect to existing structural parameters and local weather conditions. The American Society of Civil Engineers (ASCE) standard procedures and formulas for computing the various types of mechanical forces should be followed. In addition to the *National Electrical Code*, the PV engineer should keep abreast of the periodic updates to *Minimum Design Loads for Buildings and Other Structures* (SEI/ASCE 7-05) [10].

Of the various types of forces mentioned earlier, aerodynamic wind loading generally presents the most concern. At most locations around the world, the force effects due to wind loading are much higher than the other forces acting on the structure. For example, both dead- and live-weight loads due to a photovoltaic array on a building are usually less than 5 pounds per square foot (psf). In contrast, the computed wind loads are typically between 24 psf and 55 psf, but sometimes greater than 80 psf, depending on location and the specific array mounting configuration.

In addition to the foregoing considerations, the design engineer must select and configure the array mounting materials such that corrosion and ultraviolet degradation are minimized. Also, if the array is mounted on a building, the structural integrity of the building must not be degraded, and building penetrations must be properly sealed such that the building remains watertight over the life of the photovoltaic array. Structural considerations and calculations are covered in detail in Chapter 5.

4.3 DESIGN CONSIDERATIONS FOR STRAIGHT GRID-CONNECTED PV SYSTEMS

4.3.1 Determining System Energy Output

Depending on the technology, PV modules in common use may have efficiencies ranging from 6% to nearly 20%. Assuming 12% overall module efficiency, this would mean that for irradiance of 1 kW/m^2, each square meter of PV area would generate 120 W if the modules are operating at their maximum power points. Hence, generating 10 kW would require 83.33 m^2 (= 897 ft^2) of roof space, assuming the array to be roof mounted. For most dwellings, this area comes close to being the maximum area that may be south facing, especially if room is provided for access to the modules for maintenance and, perhaps, the array is located on a roof to minimize wind loading. Hence, 10 kW is a reasonably practical size limit for residential PV systems, and systems under 10 kW are considered to be small systems.

The next question is, how many kilowatt hours can the system be expected to generate? The answer depends on the location of the system and on the overall conversion efficiency from dc array output to ac inverter output. While some areas receive upwards of 7 average daily peak sun hours over the year, other areas receive less than 4. A 10 kW PV array can thus be expected to generate somewhere between 40 and 70 kWh/day, or 1200 to 2100 kWh/mo for an annual average array output energy. Loss mechanisms, however, generally result in an overall conversion efficiency between 65% and 80%, depending on whether the system incorporates battery backup and whether a maximum power point tracking charge controller is used for systems with battery backup. Figure 4.2 shows loss mechanisms that occur between the generation of electrical power by the individual modules and final delivery of power to the utility grid.

Because no two modules are exactly the same, when modules are combined into an array, there will be losses due to the fact that the maximum power voltages and currents of all modules will not be exactly the same for each module. Thus, if two modules are in series, they will not both necessarily operate at their maximum power currents because the array will operate at the overall maximum power current if the array feeds a maximum power tracking device. In a similar fashion, if two modules are in parallel, they will not necessarily have exactly the same maximum power voltage, so the overall maximum power voltage of the pair will be tracked by the maximum power tracking device. The net result is that the PV array will have less

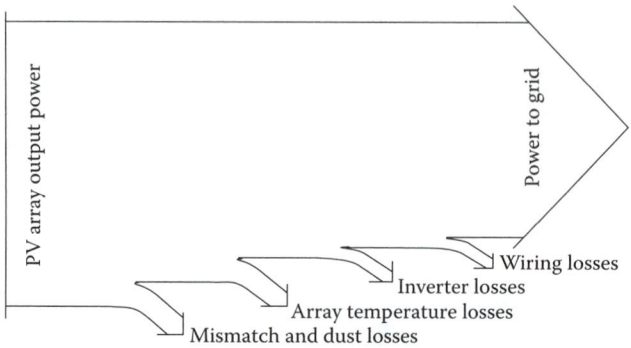

Figure 4.2 Loss mechanisms between module output and inverter output.

than the anticipated amount of power due to these mismatch losses. In any case, the mismatch losses can be grouped with the losses due to dirt on the modules.

Returning to the question of annual kWh delivered to the grid from the array, if the loss mechanisms are accounted for, it is seen that, depending on the specific losses and on the specific irradiation data for the site, a 10 kW array can be expected to deliver somewhere between 700 and 1600 kWh/mo to the utility grid or to PV system owner loads.

A number of computer programs are available for calculating monthly and annual PV system performance. Some are more detailed than others and have a price tag. Others are free (i.e., the government already paid for them). Perhaps the most popular free program for PV system performance analysis is PVWatts [14]. By inputting location and array orientation information, estimated monthly and annual system performance and savings can be obtained.

Because the average residential electrical consumption is in the 1000 kWh/mo range, depending on location once again, it appears that installation of a 10 kW PV system on every rooftop can produce sufficient kWh to exceed the demands of the dwellings on which the PV systems are installed. Of course, the PV systems will be generating electrical energy only while the sun is shining, so if all electricity used is to be PV generated, obviously massive storage capability will be required. One way to reduce the storage needs is to reduce nighttime electrical consumption. This, of course, is contrary to most current utility marketing schemes, because it is economically beneficial to them with present generation to encourage nighttime electrical use to distribute their load more evenly.

One major advantage of distributed electrical generation on rooftops is the elimination of power plant siting concerns. Most consumers tend to be happy to install PV systems on their roofs, whether they own the system themselves or whether the system is owned by the utility. Customer surveys consistently show customer support of cleaner generation, even if it has a premium price. If the choice becomes that of a PV system on their roof versus a coal or nuclear plant nearby, the PV system becomes an even more attractive choice for most consumers. For those who perceive

PV systems as aesthetically unappealing, it must be remembered that beauty is in the eye of the beholder.

4.3.2 Array Installation

The preferred orientation of a PV array will normally be south facing, mounted in a fixed position, tilted at 90% of latitude. This positioning will usually result in maximum annual kWh generation by the array. However, generally little is lost if the array is within 20° of South with a tilt within ±15° of latitude. A check with PVWatts will verify this statement for any specific location.

Depending on the duration of utility peak load and the difference between utility peak load and utility base load, it may be desirable to mount the array with a different exposure. Although the alternate mounting may result in generation at preferred utility times, the alternate mounting will generally result in a reduction of annual kWh production. Thus, unless the utility has a preferred rate for peak electricity, it may not be worth it to the system owner to orient the array to coincide with utility peaking. In some cases a compromise can be had, especially when the roof of the dwelling has more west-facing than south-facing exposure. If the array has multiple exposures, separate inverters will be required for each exposure. Again, PVWatts can be used to estimate the performance of any orientation or orientations of the array.

4.3.3 Inverter Selection and Mounting

It should be remembered that inverters are not noise free. In a manner similar to fluorescent light ballasts and other magnetic loads, the inverters may produce a 120 Hz hum or they may have an internal cooling fan. They should thus be mounted in as practical a location as possible from an electrical and mechanical perspective, but should also not be mounted where noise will be an annoyance. Generally, mounting the inverter outdoors or in a garage or carport is acceptable. If it is mounted near an air conditioner compressor or a swimming pool pump, the inverter noise will be pretty much masked by the other noise.

When selecting an inverter, it is important to remember to specify a utility-interactive unit, because a wide variety of units is available. Some are straight utility-interactive, some can be used as utility-interactive as well as stand-alone, and some are intended for stand-alone use only. While a wide range of performance specifications may be available for any unit, it is important to identify those parameters that are most significant. Features to consider may include any or all of the following:

IEEE 1547 and UL 1741—If the inverter has a UL 1741 listing, then it has been tested to IEEE Standard 1547 and consequently complies with *NEC* requirements. For small systems, all voltage and frequency trip points are factory preset and are not field adjustable.

Power rating—Maximum inverter output power is normally specified at an ambient temperature of 25°C. Surge power rating is less important for a utility-interactive inverter, because any surges encountered in the load can generally be met by the utility. For a stand-alone inverter, surge capability is more important. Utility-interactive units with battery backup act as stand-alone units if the utility disconnects, so in this case, the surge capacity becomes relevant.

Peak efficiency—It is important to note the input power over which the stated peak efficiency is obtained. Because the utility-interactive inverter always is loaded by the utility connection, if it is designed with maximum power point tracking, it will deliver maximum power to the grid over a wide range of PV input power and will operate close to peak efficiency over most of the maximum power point tracking range. Modern straight grid-connected inverters should have peak efficiencies in excess of 95% over most of the operating range.

Maximum power point tracking range—Look for the widest possible range of input voltages over which the unit is capable of tracking maximum power for greatest flexibility in PV array design.

No-load power consumption—No-load power consumption should be very low, although normally a straight grid-connected inverter will operate at maximum power output because the grid always provides a load for all available PV power.

Nighttime losses—If the unit operates on power supplied by the PV system, then there will be no nighttime losses. If the unit has a stand-alone mode of operation, then one can normally expect it to have a "sleep" mode that consumes very little power if no loads are connected. Straight grid-connected units with power ratings below 10 kW will generally have power losses of a few watts.

Warranty—Many quality inverters carry a 10-year warranty, sometimes extendable. Check what the warranty covers.

Code compliance—UL 1741 listing indicates compliance with IEEE 1547. It is possible to purchase inverters that incorporate additional *NEC* compliance components, such as ground fault protection, input and output disconnects, fusible combiners for multiple string inputs so that strings will not need to be fused in a remote or difficult-to-access location, and weatherproof packaging.

Data monitoring—Most grid-connected inverters provide optional data monitoring capability by digitizing system parameters and making them available on an output bus. Some monitoring is available at the inverter, and some units provide for remote data logging via telephone dial-up, Internet, or other communication means.

4.3.4 Other Installation Considerations

Prior to commencing the installation, because the system will very likely be inspected by both the utility and the local municipality, it is important to first complete all the necessary permitting paperwork. This may offer an excellent opportunity for the engineer to perform an educational public service by answering any questions either authority may have about the installation.

Persons not familiar with IEEE 1547, UL 1741, and Article 690 of the *NEC* may be reluctant to accept the straightforward installation allowed by these codes and standards. It is possible that the most time-consuming part of the installation may be convincing the inspectors that the installation will be safe, competent, and code compliant. Fortunately, electrical inspectors tend to be well-versed on the *NEC* and are generally interested in new technology.

It is also conceivable that a zoning restriction on installation of rooftop solar systems on the street side of a building may exist. The permitting process may then involve a trip or two to zoning boards or town council meetings, perhaps with an attorney or a delegation of solar enthusiasts.

4.4 DESIGN OF A SYSTEM BASED ON DESIRED ANNUAL SYSTEM PERFORMANCE

4.4.1 Array Sizing

PV system design may be targeted at the achievement of a variety of objectives. In this example, it will be assumed that the objective is to provide a specified percentage of the electrical consumption of an occupancy. In this case, the building will be a residence in Oklahoma City, Oklahoma that has an average monthly electrical consumption of 1000 kWh. The design goal is to provide 33% of the annual electrical consumption with a grid-connected PV system, which amounts to an average of 333 kWh/mo. Because it is desired to provide 33% of the annual consumption, this amounts to $12 \times 333 = 3996$ kWh/yr.

As luck may have it, the dwelling has 600 ft^2 of unobstructed, unshaded, south-facing roof that has a 5:12 slope. In other words, for every 12 horizontal units, the roof rises 5 vertical units. So the roof slope is $\tan^{-1}(5/12) = 22.6°$.

While a completely formal analysis would involve estimating all the factors that affect dc-ac conversion efficiency, along with looking up the peak sun hours for each month for Oklahoma City, for this example, PVWatts [14] will be used instead. By looking at www.pvwatt.com, and choosing Version 1 and selecting Oklahoma, one notices that there are 2 PVWatts sites in Oklahoma, and, again, coincidentally, one of them is Oklahoma City.

The PVWatts home page shows a number of default values in the dialog boxes. All that is needed is to decide whether to use the default values or to change them. Beginning with the dc rating of the array, one observes that this is what is to be solved for. If the solution is achieved through iterations, then it is acceptable to leave the default value in place. Next, it is reasonable to choose the default value for the conversion efficiency, so this can be left at 0.77. It will be assumed that the array will be mounted parallel to the roof (stand-off mounted) at the slope of the roof, so the array tilt should be changed to 22.6°. Because the roof faces south, the default azimuth can be used. Note that PVWatts uses NORTH = 0° azimuth, with EAST = 90°, SOUTH = 180°, and WEST = 270°. Because the cost of electricity is not relevant in this problem, all the information needed for the calculation has now been entered. So all that is left is to hit the CALCULATE button, go out for coffee, and come back to the answer.

The result comes out at 5624 kWh/yr, which is higher than the design goal. But now, all that is needed is to try an array size that is $(3996 \div 5624) \times 4$ kW = 2.842 kW and see whether the annual estimated ac kWh production will end up at 3996 kWh. Entering 2.842 kW as the array size on the data (home) page yields the not-too-surprising result of 3996 kWh/yr. So, PVWatts has helped to determine the array size of 2.842 kW. Now all that is needed is to find a collection of modules that will produce close to this amount and to find an inverter that will match up to the modules, or vice versa. In this case, a suitable module will be selected.

But before moving on to select an inverter and modules, the reader should experiment with the entries in the PVWatts dialog boxes to explore the variations in monthly and annual system output for various values of array tilt and array azimuth. The first thing to note is that up to 20° either side of south does not make much difference on annual system performance. Then, it is also interesting to note that any tilt within about ±15° of latitude also does not affect annual results much. Because the system is grid connected, normally there will not be concern over monthly performance, as long as the desired annual performance is achieved.

4.4.2 Inverter Selection

Because the array size has been determined to be 2.842 kW, it now makes sense to find an inverter or two that are rated to handle this amount of output power. When looking at inverter specifications, one should note that inverter manufacturers allow PV array sizes up to 10% or more than the rated output power of the inverter. So, for example, a 2.5 kW inverter would not necessarily be ruled out if a suitable one could be found. A slightly more conservative approach, however, would suggest to size the inverter close to the size of the array. Table 4.8 shows the input and output characteristics of two inverters that can handle the array output, provided that the array output does not exceed the maximum allowed inverter input voltage, nor does the array output fall below the lowest maximum power point tracking voltage level.

Other considerations in inverter selection may include whether or not the inverter includes ac and/or dc disconnect, source circuit combining inputs and, of course, the need for the inverter to be listed to UL 1741. Then there are factors such as peak efficiency, CEC (California Energy Commission) rated efficiency, inverter size, weight, and weather rating (outdoor or indoor) that may influence the selection decision.

The next step in the system design is to select a module or modules and then see if the modules can be configured to meet the array power requirements and inverter input limits.

Table 4.8 Input and Output Performance Characteristics of Two Different Inverters

Inverter #	Rated ac Power (W)	Max Array Power (W)	Max dc V_{in} (V)	V_{in} MPPT Range (V)	I_{in} Max (A)	V_{ac} Out (V)	I_{ac} Out Max
1	2800	3100	600	$195 < V_m < 550$	15.4	240	11.7
2	3000	3750	500	$200 < V_m < 400$	17	240	13

Table 4.9 Electrical Characteristics of Four Different PV Modules

Module #	V_{OC} (V)	I_{SC} (A)	V_m (V)	I_m (A)	$\Delta V_{OC}/\Delta T$ (%/°C)	$\Delta V_m/\Delta T$ %/°C	NOCT (°C)
1	32.8	8.05	26.7	7.12	−0.35	−0.49	47
2	44.4	5.30	35.8	4.89	−0.33	−0.47	46
3	68.7	3.83	55.8	3.59	−0.25	−0.29	47
4	48.7	5.99	41.0	5.61	−0.27	−0.38	46

4.4.3 Module Selection

Assume that after searching the Internet for suitable modules, four possibilities have been found, having the performance characteristics shown in Table 4.9. If ambient temperatures range from −20°C to +38°C, then Equation (3.5) can be used to determine the operating temperature range of the module. For the low-temperature limit, solve $T_C = T_A + (\text{NOCT} - 20)G/0.8 = T_A = -20°\text{C}$. Note that this assumes $G = 0$, which would be close to the case at sunrise. The assumption is that the module achieves its lowest temperature just after sunrise. For the high-temperature limit, solve $T_C = 38 + (47 - 20)1/0.8 = 72°\text{C}$ (or 71°C if NOCT = 46°C). These cell/module temperature limits are now used to determine the maximum and minimum number of modules that can be used in a single source circuit. The calculation of the maximum open-circuit voltage of the modules is required by *NEC* 690.7 to ensure that the maximum array open-circuit voltage will not exceed 600 V, but if the selected inverter has a maximum input voltage less than 600 V, then the array voltage must not exceed this value or the inverter may be damaged. The determination of the minimum module operating voltage at high temperatures is needed to ensure that the array output voltage will not fall below the minimum allowable value for the inverter maximum power point tracking range.

Because the module temperature coefficients are expressed in %/°C departure from STC (25°C), it is useful to tabulate the maximum module V_{OC} values and the minimum module V_m values for a single module of each type. When writing the formulas for these determinations, it is important to keep all the minus signs in order so that the resulting values make sense. The formulas become

$$V_{OC}(\max) = V_{OC}(25°\text{C})[1 + (T_{\min} - 25)(\Delta V_{OC}/\Delta T)] \tag{4.8a}$$

and

$$V_m(\min) = V_m(25°\text{C})[1 + (T_{\max} - 25)(\Delta V_m/\Delta T)] \tag{4.8b}$$

where temperatures are expressed in °C and the temperature coefficients are expressed in %/°C, which, in turn, is expressed as a decimal fraction.

Expressing the temperature coefficients as decimal fractions and substituting the known maximum and minimum module temperatures yields the values listed in Table 4.10 for each of the four selected modules. Using the values in Table 4.10, it is

Table 4.10 Tabulation of V_{OC}(max) and V_m(min) for the Modules of Table 4.9

Module #	Max V_{OC}	Min V_m
1	38.0	20.6
2	51.0	28.1
3	76.4	48.2
4	54.6	33.8

now possible to determine the maximum and minimum number of modules of each type that can be used in source circuits for each inverter.

Inverter 1 allows a maximum input voltage of 600 V, so dividing 600 by the maximum module voltages and then *rounding down* gives the maximum number of modules that can be connected in series and still keep the maximum series voltage below 600 V. For example, for module 1, we have 600 ÷ 38.0 = 15.78, which rounds down to 15.

Inverter 2 allows a maximum input voltage of 500 V, so dividing 500 by the maximum module voltages and rounding down gives the maximum number of series source circuit modules for this inverter.

Inverter 1 has a minimum MPPT voltage of 195 V, so dividing 195 by the minimum module voltages and then *rounding up* gives the minimum modules allowed per source circuit. For example, for module 2, we have 195 ÷ 28.1 = 6.93, which rounds up to 7.

Because inverter 2 has a minimum MPPT voltage of 200 V, the minimum number of modules per source circuit is obtained by dividing 200 by the minimum module voltage and rounding up.

Table 4.11 shows the allowable numbers of modules for each inverter for one and two source circuits of each module. Note that when more than one source circuit is used, all must contain equal numbers of modules.

Table 4.11 Summary of Allowed Numbers of Modules That Fall within Inverter Input Voltage Limits

Inverter	Allowed # of Module #1	Allowed # of Module #2	Allowed # of Module #3	Allowed # of Module #4
Inverter #1 with:				
One source circuit	10, 11, 12, 13, 14, or 15	7, 8, 9, 10, or 11	5, 6, or 7	6, 7, 8, 9, or 10
Two source circuits	20, 22, 24, 26, 28, or 30	14, 16, 18, 20, or 22	10, 12, or 14	12, 14, 16, 18, or 20
Inverter #2 with:				
One source circuit	10, 11, 12, or 13	8 or 9	5 or 6	6, 7, 8, or 9
Two source circuits	20, 22, 24, or 26	16 or 18	10 or 12	12, 14, 16, or 18

Note: Inverter input power limits are not considered, so that some of the module combinations may not meet inverter array power restrictions.

The final step in module selection is to determine the number of modules that most closely match the required array power. If the data sheets were available for these modules, one could simply look up the module rated power. But because the powers are not given, it will be necessary to determine the rated powers of the modules by multiplying V_m by I_m for each module. After this rather tedious exercise, the results are: module 1, $P = 190$ W; module 2, $P = 175$ W; module 3, $P = 200$ W; and module 4, $P = 230$ W.

Because the required array power is 2842 W, one can conclude that it will require 14.9579 of module 1, 16.2400 of module 2, 14.2100 of module 3, or 12.3565 of module 4. Noting that fractional modules are not available, and inserting an additional requirement that the *minimum* array size should be 2842 W, the module numbers must be rounded up. So the conclusion is that it will require 15 of module 1, 17 of module 2, 15 of module 3, or 13 of module 4 to achieve the design goal.

Comparing these numbers with the allowed numbers of modules as shown in Table 4.11, one can conclude that it is acceptable to use 15 of module 1 in a single source circuit as long as inverter #1 is used. But 17 of module 2 is not an acceptable number for either inverter, so 18 will need to be used in two 9-module source circuits, using either inverter.

For module 3, because of the high V_{OC}, it will take three source circuits of five modules each for either inverter. So, if this module is selected, then it will be necessary to check the inverter specifications to determine whether the inverter has three combining inputs or whether an external combiner box will need to be used. For module 4, 13 is not an acceptable number for either inverter, so it will require 2 source circuits of 7 modules each with either inverter.

So, for the modules and inverters selected, the array power options are either $15 \times 190 = 2850$ W if module 1 is used, 3150 W if module 2 is used, 3000 W if module 3 is used, and 3220 W if module 4 is used.

So, the module/inverter combination that most closely meets the design requirement in this case is inverter 1 with module 1. Of course, if the owner only wanted approximately 33% of electrical usage to be provided by the PV system, then all of the combinations may be candidates, and other considerations such as price, warranty, or which modules best fit on the roof may be taken into account.

4.4.4 Balance of System

Assuming that inverter 1 and module 1 meet further requirements of cost, availability, warranty, and other features that may be desired by the owner, the rest of the system can now be designed. Selection of the array mount will be saved for Chapter 5, so this means that all that is left is to select module extension wire, a junction box, wire and conduit to connect between the array junction box and the inverter, dc and ac disconnects if they are not already incorporated into the inverter, and wiring from the ac side of the inverter to the point of utility connection.

4.4.4.1 Module Extension Wire, Rooftop Junction Box, and Module and Array Frame Grounding

The modules come with permanently attached positive and negative leads with #10 copper conductors with USE-2 insulation (#10 USE-2) or equivalent. The leads are 36 in. long, which is long enough for the positive lead of one module to connect to the negative lead of the adjacent module. This means that each positive-to-negative connection involves 6 ft of #10 copper wire.

If the 15 modules can be placed in two rows, one with 8 and one with 7, as shown in Figure 4.3, then the modules can be connected to one another in series and to a junction box, without any additional extension wire. In some cases, however, it may be necessary to install the modules in a single row of 15, in 3 rows of 5, or some other combination. If this should be the case, it may be necessary to use one or two additional pieces of USE-2 extension wire to connect to the module wiring to extend to the rooftop junction box.

For the module configuration of Figure 4.3, there is a total of 6 ft of wire per module, so that the total length of open wiring between modules and junction box is 90 ft. This information will be needed in calculating the voltage drop between the array and the inverter.

As an aside, this is a good time to note that the loop of wire among the 15 modules acts as an antenna for electromagnetic disturbances. Because the voltage generated in a loop from a time-varying magnetic flux density is proportional to the area of the loop, it is important to minimize the area of the loop of wire to minimize coupling to nearby lightning strikes. A secondary precaution to protect the equipment is to install a lightning arrestor at the point where the array conductors enter the inverter. This will be shown on the final system circuit diagram.

Note also that a grounding conductor is shown in Figure 4.3. The *NEC* requires that listed grounding techniques be used to ground all metallic components of the PV module frames and the array mounting structure. The size of the grounding conductor is determined by the requirements of *NEC* Section 690.45, which indicates that a system with ground fault detection and interruption is to comply with *NEC* Section

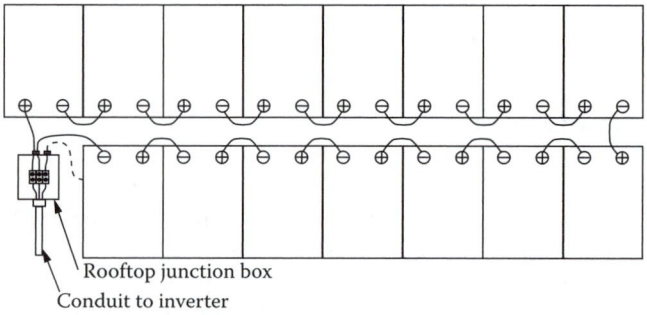

Figure 4.3 Possible rooftop arrangement of PV modules showing transition from open wiring to wiring in conduit via rooftop junction box.

250.122. Because essentially all straight grid-connected inverters that comply with UL-1741 have internal GFDI provisions, this means that *NEC* 250.122 will apply.

Because *NEC* 250.122 specifies the size of the grounding conductor in terms of the circuit breaker or fuse that protects the circuit conductors, it will thus first be necessary to determine the sizes of the circuit conductors and the corresponding fuse size.

4.4.4.2 Wire and Conduit from Rooftop Junction Box to Inverter

To determine the proper wire size for connecting between the rooftop junction box and the inverter, the distance from the junction box to the inverter must be known and the highest anticipated ambient temperature must also be specified. Typically, in a residential PV installation, the distance between the rooftop junction box will be less than 100 ft, unless the dwelling is unusually large. For this example, a distance of 75 ft will be assumed. Also, when the maximum array temperature was determined, it was based on a maximum ambient temperature of 100°F (38°C).

The wire size must be based on satisfying two constraints: ampacity and voltage drop.

The combination of *NEC* 690.8(A) and (B) requires that the minimum ampacity of the array wiring shall be 156% of the source circuit short-circuit current, which, in this case, happens to be the module short-circuit current, which is 8.05 A.

The origin of this seemingly unusual percentage has two components. The first comes from *NEC* 690.8(A), which requires that the *maximum current* of a module is to be considered to be 125% of I_{SC}. The reason for this is that under certain cloud cover conditions, the clouds will act as a magnifying glass and focus the sun onto the PV array with intensity up to 25% higher than full sun. The result is an additional 25% current beyond I_{SC}. The second factor comes from *NEC* 690.8(B), which requires wiring to be selected with the assumption that the maximum PV current will flow continuously. In the *NEC*, continuously means for 3 h or more, and under conditions of continuous operation, an additional 25% must be added to wire and circuit breaker or fuse sizes. So, the final result is that for any PV source circuit or PV output circuit, with the output circuit being the combining of individual source circuit currents, the ampacity of the circuit must be no less than 125% of 125% of I_{SC}, or, multiplying it out, 156% of I_{SC}.

Table 4.2 shows the ampacities of a range of conductor sizes at a temperature of 30°C. These ampacities are excerpts from *NEC* Table 310.16, which applies to conductors in an electrical conduit. If the conductors are exposed to ambient temperatures higher than 30°C, then they must be derated according to the correction factors of *NEC* Table 310.16. Table 4.12 shows the required derating factors for wire with insulation rated at 90°C, which, at present, includes most commonly used insulation types, such as THHN, THWN-2, and USE-2.

An additional derating factor new to the 2008 *NEC* requires increasing the ambient temperature to which conduit is exposed if the conduit is run across a rooftop. The requirement appears in *NEC* Table 310.15(B)(2)(c) and is summarized in Table 4.13. This requirement is not a cumulative correction factor, however, because it is incorporated directly into the ambient temperature correction factor for the conductors.

Table 4.12 Summary of Ambient Temperature Correction Factors for Wire with 90°C Insulation

Temp (°C)	21–25	26–30	31–35	36–40	41–45
Correction	1.04	1.00	0.96	0.91	0.87
Temp (°C)	46–50	51–55	56–60	61–65	66–70
Correction	0.82	0.76	0.71	0.58	0.41

Source: Based on data from *NFPA 70 National Electrical Code, 2008 ed.*, National Fire Protection Association, Quincy, MA, 2007.

The conductors simply must be derated for a higher temperature. For example, if the conduit is spaced 4 in. above the rooftop, then 22°C must be added to the highest expected ambient temperature, which, in this case, is 38°C. So, the temperature used for derating the wire would be 60°C.

Furthermore, if a conduit contains more than three current-carrying conductors, the conductors must be further derated, because with more wires generating I^2R heating, it becomes more difficult to dissipate the heat generated. *NEC* Table 310.15(B)(2)(a) lists the required derating factors for conduit fill. These requirements are shown in Table 4.14. For more than 30 conductors in a conduit, refer to the *NEC* table.

When using Table 4.14, it is important to identify those conductors that are considered to be current carrying versus those that are not considered to be current carrying. In particular, *grounding* conductors are not considered to be current carrying. Grounding conductors must be distinguished from *grounded* conductors. Grounding conductors are for the specific purpose of establishing ground potential on all metal parts of an electrical system. Grounded conductors are current-carrying conductors that are connected to the grounding conductor *at only a single point*. This way the grounded conductor is maintained at a potential close to ground, but no current can flow from the grounded conductor to the grounding conductor because the single connection point does not provide for a return path for current flow. In a properly wired system, the voltage drop in a circuit will not exceed 3%, and will usually be less than 2%. In this case, if half the voltage drop is on the ungrounded conductor and half is on the grounded conductor, then the maximum voltage above ground potential of the grounded conductor will be about 1% of the source voltage.

In summary:

- The 30°C ampacity of source circuit and output circuit conductors must be calculated as 156% of I_{SC}.

Table 4.13 Ambient Temperature Correction Factors for Conduit Run across Rooftops

Height of Conduit above roof (in.)	0–0.5	0.5–3.5	3.5–12	>12
Add to ambient (°C)	33	22	17	14

Source: Based on data from *NFPA 70 National Electrical Code, 2008 ed.*, National Fire Protection Association, Quincy, MA, 2007.

Table 4.14 Derating Factors Based on Conduit Fill

Number of current-carrying conductors	0–3	4–6	7–9	10–20	21–30
Derating factor	1.0	0.8	0.7	0.50	0.45

Source: Based on data from *NFPA 70 National Electrical Code, 2008 ed.*, National Fire Protection Association, Quincy, MA, 2007.

- The 30°C ampacity is then derated for ambient temperatures by multiplying by the appropriate factor.
- A further derating must be applied if more than four current-carrying conductors are run in a conduit.

To keep track of all the ampacity and derating factors that influence the selection of the source circuit and output circuit conductors, it is useful to construct a small table, or, more conveniently, a spreadsheet. The table entries might appear as in Table 4.15. In this table, it is assumed that the conduit leaving the rooftop junction box does not cross the roof, but instead goes directly into an attic space on its journey to the inverter.

For the example in Table 4.15, #10 THWN-2 wire was used to match the wire size of the module leads. It is clear from the table that the ampacity of #10 wire for this application is more than adequate to meet the requirements of the module current (36.40 A > 12.58 A).

The next step is to calculate the voltage drop in the source circuit. For the voltage drop calculation, normally V_m and I_m of the source circuit are used. In some instances, it may be preferable to use the value of V_m at an elevated temperature, but for a small PV system, this is usually not necessary.

There are two components to the voltage drop calculation. The first is the voltage drop in the open module wiring, and the second is the voltage drop in the wiring between rooftop junction box and the inverter. Equation (4.1) along with Table 4.2 can now be used to determine the voltage drops.

For the module open wiring, because there is a total of 90 ft in the complete loop, this is equivalent to a one-way distance of 45 ft. So, using 45 ft for d in Equation (4.1) and using $I_m = 7.12$ A and $V_m = 15 \times 26.7 = 400.5$ V gives

$$\%\mathrm{VD} = 100\frac{\mathrm{I}}{V_S}\left(\frac{\Omega}{\mathrm{kft}}\right)\left(\frac{2\mathrm{d}}{1000}\right) = \frac{0.2 \times 7.12 \times 45}{400.5}(1.21) = 0.194.$$

Next, the voltage drop between rooftop junction box and inverter is calculated. Because #10 wire is still being used, the only item that changes in the voltage drop

Table 4.15 Determination of Conductor Size for PV Source Circuits and/or Output Circuits

I_{SC}	$1.25I_{SC}$	$1.56I_{SC}$	Wire Size	30°C Ampacity	Amb Temp	Amb Derate	Cond Fill	Fill Derate	Derated Ampacity
8.05	10.06	12.58	10	40	38	0.91	2	1.0	36.40

formula is the one-way distance, which, this time, is 75 ft. The corresponding voltage drop is 0.323%. Thus, the overall percent voltage drop of the source circuit is 0.517%, which is well below the allowable 2%. Problem 4.7 gives the reader an opportunity to check on the percent voltage drop if #12 THWN-2 is used between the rooftop junction box and the inverter instead of #10.

4.4.4.3 DC and AC Disconnects and Overcurrent Protection

As luck would have it, the selected inverter comes equipped with a dc disconnect, so the wires from the array are connected to the inverter array input terminals and the dc disconnect is already there. But the inverter does not have an ac disconnect, so one will need to be selected. Several considerations may influence the choice of ac disconnect.

The first consideration is the inverter rated output voltage and current, which is, for inverter 1, 11.7 A @ 240 V. The disconnect must be rated at 125% or greater of the rated output current, which is 14.6 A. This gives several useful pieces of information, such as the minimum wire size to connect the inverter output to the point of utility connection (PUC), the size of the circuit breaker to be used at the PUC if the PUC is to be in a circuit breaker panel, the minimum size of the circuit breaker panel busbars, the size of the ac disconnect, and whether or not the ac disconnect must be fused. The distance from the inverter to the PUC is also needed to confirm that the voltage drop in the selected wire from inverter to PUC is not too large.

Noting that #14 wire is rated at 20 A, but can only have 15 A of overcurrent protection per *NEC*, the first attempt at sizing the wire suggests trying #14. The ac disconnect need only be rated at 15 A, so, in principle, a two-pole 15 A snap switch could be used. The only reasons for not using a simple switch would be if the utility were to require a lockable disconnect with an external handle or if the PUC were to be at the line side of the main disconnect.

The *NEC* does not require a lockable disconnect, but some utilities require a lockable disconnect for redundant assurance that the inverter can be shut down if utility power is lost. If the PUC ends up at the line side of a main disconnect, either because *NEC* 690.64 is not satisfied or because there is no space in the distribution panel for another two-pole, 15 A circuit breaker, then a 15 A fused disconnect will be needed near the PUC.

For this example, it will be assumed that there is room in the distribution panel for another two-pole, 15 A circuit breaker as the PUC. It will also be assumed that the inverter will be located 6 ft from the PUC. The reader is encouraged to use the voltage drop formula to determine the voltage drop for 11.7 A in #14 wire over a distance of 6 ft. If it is less than 2%, then #14 is adequate. If it is more than 2%, then larger wire will need to be used, but it will still have 15 A of overcurrent protection.

4.4.4.4 Point of Utility Connection

NEC 690.64 states that the sum of the ratings of all circuit breakers that supply power to an electric panel shall not exceed 120% of the rating of the panel busbars.

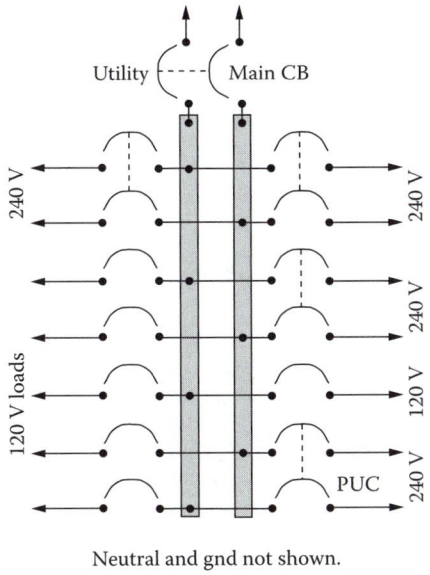

Figure 4.4 PV PUC at distribution panel.

Furthermore, it states that if the sum of the ratings of circuit breakers feeding power to the panel is between 100% and 120% of the busbar rating, then the PUC circuit breaker must be located at the end of the panel opposite the utility feed and labeled so that no one will be tempted to move it. The reason for this rule is that if a busbar is fed from both ends, Kirchhoff's current law requires that for each circuit breaker connected downstream from a feed, if there is a load on the circuit breaker, then the current to that load is no longer carried beyond that point on the busbar. Figure 4.4 shows how feeding the opposite end of a busbar with the PUC helps distribute current more evenly on the busbars.

The 20% rule implies that the busbars to which the 15 A circuit breaker is attached must have a rating of at least 75 A, presumably fed from the utility with a 75 A circuit breaker, because 120% of 75 A is 90 A and 75 A + 15 A = 90 A. Because almost all circuit breaker panels are rated at 100 A or more, the 15 A circuit breaker can be attached to almost any 120/240 V panel that has space for it. In fact, depending on whether the inverter has a setting for an output voltage of 208 V, the circuit breaker could also be connected to a 120/208 V panel. When the PUC circuit breaker is switched on and the inverter output disconnect is switched on, the inverter output automatically will assume the phase of the utility voltage.

4.4.4.5 Final System Electrical Schematic Diagram

Figure 4.5 shows the final electrical schematic diagram of the system. Note that although the diagram shows #10 conductors between the rooftop junction box and

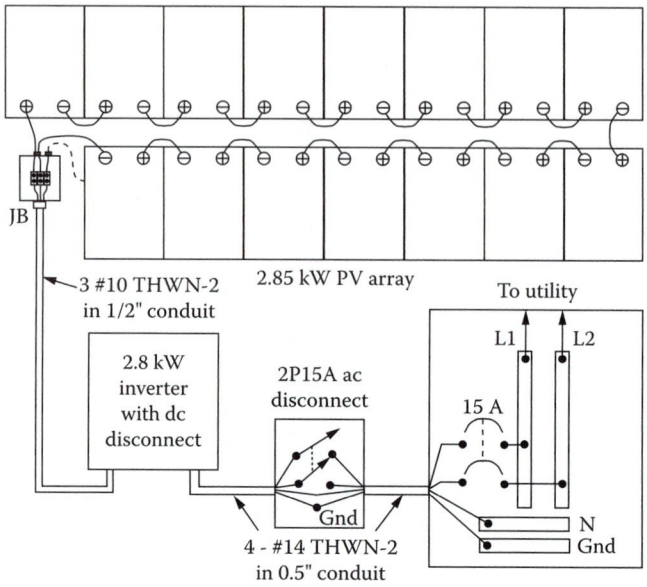

Figure 4.5 Electrical schematic diagram of 2.85 kW PV system.

the inverter, it is likely that homework Problem 4.7 will show that #12 will also be acceptable for this circuit. The 15 A circuit breaker at the PUC determines the size of the equipment grounding conductor that is required. *NEC* 250.122 indicates that for wiring protected by a 15 A circuit breaker, a #14 grounding conductor is adequate. Note that a #10 wire is used for array grounding. The reason for this is simply that a #10 wire has greater physical strength, and, because it will be a bare wire, the larger size should survive the outdoor environment better than a smaller conductor. The inverter installation manual will show where all the wire connections are to be made within the inverter enclosure.

Conduit sizes are determined from tables in Chapter 9 of *NEC* and from additional tables in Annex C of *NEC*.

The only other *NEC* requirement of note is that *NEC* 690.31(E) requires that if source circuit conductors are run inside a structure, then the conductors must be run in metallic conduit up to the first readily accessible disconnecting means.

4.5 DESIGN OF A SYSTEM BASED ON AVAILABLE ROOF SPACE

4.5.1 Array Selection

A common challenge in PV system design is to fit as much power on the roof as possible. With a roof that is just the right size and just the right shape, that is, rectangular, this may not seem to be much of a challenge. But in many areas of the country,

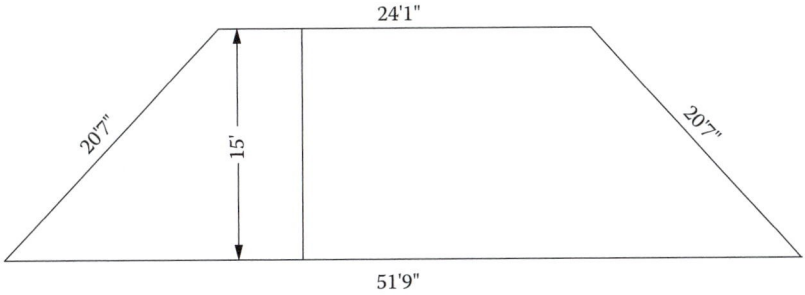

Figure 4.6 A roof in need of a PV array.

finding a south-facing, rectangular roof is like finding a needle in a haystack. Odd roof shapes, shading, and roof protrusions all influence the placement of PV modules. In this example, the roof of Figure 4.6 will be used.

Roof diagrams may come in two forms. One form is a diagram showing the actual measurements of the roof, as in Figure 4.6. The other form is a plan view, where the projection of the roof onto the horizontal is shown. Plan views, believe it or not, are what are found on sets of plans for new construction. They also generally indicate the slope of the roof in terms of rise:run ratios, that is, the ratio of vertical distance to horizontal distance. The slope of the roof is then simply

$$\theta = \tan^{-1}(y/x), \tag{4.9}$$

where y is the vertical rise and x is the horizontal run of a right triangle. The horizontal dimension, x, is the projection of the actual roof dimension, r, in the direction of the slope. The actual dimension in the direction of the slope is the hypotenuse of the right triangle formed by x, y, and r. Thus, if x is known from the plan view, and the ratio y:x is known, then r can be found from

$$r = x/\cos\theta. \tag{4.10}$$

While Equations (4.9) and (4.10) are not exactly rocket science, if the slope is not taken into account in the plan view by the designer, then the installer may end up with some surprises when laying out the array on the actual roof. The roof dimension perpendicular to the slope direction on the plan view is the same as the actual dimension.

Chapter 5 will show that in hurricane-prone regions, there is good reason to keep modules at least 3 ft from the edges or corners of a roof to minimize wind loading. However, in most regions where wind speeds are not expected to exceed 90 mph, as opposed to >125 mph hurricane winds, the designer need not be concerned about the modules ending up too close to the edges of the roof. Just for practice, this example will assume that the installation will be in a hurricane-prone region and the modules will be kept at least 3 ft away from the edge of the roof. So the first step in determining an array layout is to mark the 3 ft boundary on the roof of Figure 4.6, as shown

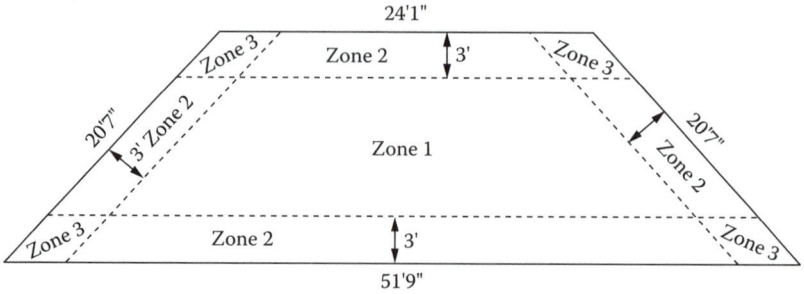

Figure 4.7 Identification of roof zones.

in Figure 4.7. Figure 4.7 also shows the three wind load zones of the roof, as defined in ASCE 7-05 [10]. Keeping the modules 3 ft away from the edges and corners of the roof means keeping them in roof zone 1.

Because Figure 4.6 shows the actual dimensions of the roof, the array layout can be accomplished by drawing the roof to scale and then drawing the proposed modules to scale. The proposed modules are then fitted onto the roof in what looks like a logical configuration to the designer, within any constraints that may be imposed by shading, protrusions, wind loading, or other loads such as snow loads. In fact, there may even be a constraint imposed by the location of the roof trusses. If the designer is willing to try more than one module type, each with different dimensions, then scale model drawings can be made of each module so they can be fitted to the roof. Figure 4.8 shows three different modules drawn to scale showing the module dimensions.

Figure 4.9 shows an attempt to create two rows of each type of module with the modules in landscape position, keeping the rows within roof zone 1. Note that for this roof, only module C can be mounted in two landscape rows and still be kept completely within zone 1. So module C will be used in this design. Figure 4.10 shows that 16 of the C-variety of module can fit within roof zone 1. So the next step is to look at the electrical characteristics of module C.

The data sheet for module C shows $V_{OC} = 68.7$ V, $I_{SC} = 3.83$ A, $V_m = 55.8$ V, $I_m = 3.59$ A, $\Delta V_{OC}/\Delta T = -0.172$ $V/°C$, and $\Delta V_m/\Delta T = -0.29\%/°C$. Note that one temperature coefficient is expressed in $V/°C$ and the other is expressed in $\%/°C$, so the proper units will need to be used, or one set of units will need to be converted to be consistent with the other. For convenience, because Equations (4.8a and 4.8b) are both

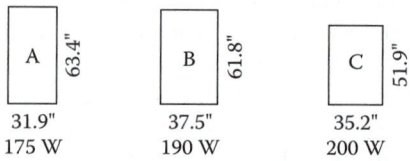

Figure 4.8 Three possible choices for modules.

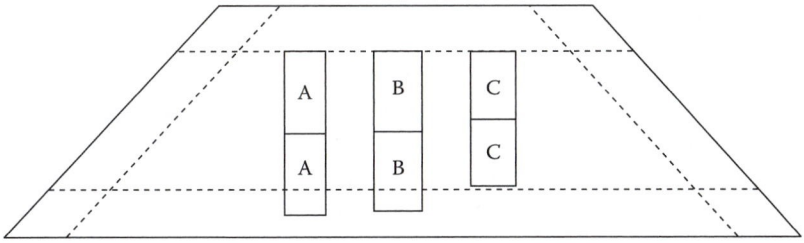

Figure 4.9 Checking module fit in zone 1.

expressed in %/°C, $\Delta V_{OC}/\Delta T$ can be expressed in %/°C by determining what percentage of V_{OC} is represented by 0.172 V. Dividing it out yields 0.172 ÷ 68.7 = 0.25%, so $\Delta V_{OC}/\Delta T = -0.25\%/°C$. The NOCT of the module is 46°C.

With $V_{OC} = 68.7$ V, it should be obvious that it is not possible to connect all 16 modules into a single source circuit, because $16 \times 68.7 \gg 600$ V. So the next logical possibility is two source circuits of eight modules each. Assume that the system will be installed in hurricane-prone South Florida, where the ambient temperature might be expected to fluctuate between –2°C and 35°C. The maximum open-circuit voltage for an 8-module source circuit, using Equation (4.8a), is

$$V_{OC}(\max) = V_{OC}(25°\ C)[1 + (T_{\min} - 25)(\Delta V_{OC}/\Delta T)]$$
$$= 8 \times 68.7[1 + (-2 - 25)(-0.0025)] = 586.7\ V < 600\ V.$$

At the high-temperature end, the maximum module operating temperature can be found from Equation (3.5) to be

$$T_C = T_A + (\text{NOCT-20})G/G_0 = 35 + (46 - 20)/0.8 = 67.5°C.$$

Thus, the minimum value of V_m can be found from Equation (4.8b) to be

$$V_m(\min) = V_m(25°\ C)[1 + (T_{\max} - 25)(\Delta V_m/\Delta T)]$$
$$= 8 \times 55.8[1 + (67.5 - 25)(-0.0029) = 391.4\ V.$$

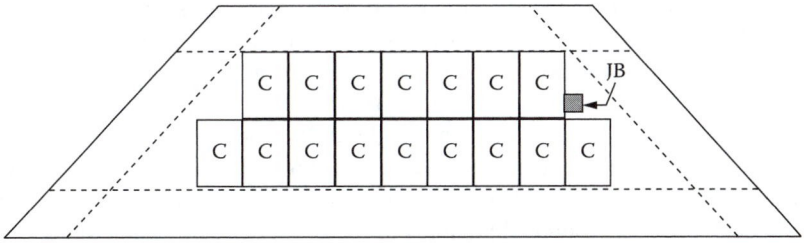

Figure 4.10 Landscape layout of module C in high-wind region.

4.5.2 Inverter Selection

If an inverter is available for which these high and low voltages fall within the inverter input voltage limits and for which the array power meets the inverter array power limits, then the inverter will work in the system. Noting that 16 modules @ 200 W/module gives a 3200 W array power, and that the inverter will need to have a maximum dc input rating of 600 V, with a minimum MPPT voltage of less than 390 V, inverter characteristics can now be checked. After rejecting a few that have adequate power ratings but have too low ratings for input voltage, a 3300 W inverter is found that has $V_{in}(\max) = 600$ V, MPPT range between 195 and 550 V, and $I_{out}(\max) = 15.6$ A @ 240 V. The inverter also has built-in dc and ac disconnect switches, built-in GFDI, includes two combiner inputs, and is rated for outdoor installation. So, the dc wiring of the system can be run directly from array to inverter.

4.5.3 Balance of System

In this example, there are two source circuits that must be run to either a combiner box or directly to the inverter. It will be assumed that a rooftop junction box will be used for a transition from open wiring of the modules to wiring in conduit from rooftop junction box to the inverter. It will also be assumed that the inverter will be mounted to the wall of the occupancy within a few feet of the main disconnect of the occupancy and that the distance from the rooftop junction box to the inverter is 50 ft. It will also be assumed that the roof has a 4:12 pitch.

4.5.3.1 Module Extension Wire, Rooftop Junction Box, and Module and Array Frame Grounding

Figure 4.10 shows the proposed location of the rooftop junction box. Again, modules have 3 ft positive and negative leads of #12 USE-2, so an 8-module source circuit will have 48 ft of module wiring. In the configuration of Figure 4.10, assume that the top row of modules plus the module in the bottom row furthest from the junction box will be one source circuit and the remaining 8 modules in the bottom row will be the other source circuit. In this case, assuming that the modules are installed with the module junction boxes at the top of the bottom row and the bottom of the top row, one lead will be available from each source circuit to be connected directly to the rooftop junction box.

But this means that one lead from the last module in each source circuit will not reach back to the rooftop junction box, so that it will be necessary to add an extension wire to the lead to reach back to the rooftop junction box. Considering that each module is just under 3 ft wide, and that the module junction box is centered near the end of the module, it will require about 21 ft of extension wire for the last module in the top source circuit to reach the rooftop junction box and about 18 ft of extension wire for the last module in the bottom source circuit to reach the rooftop junction box. Thus, the round-trip lengths of the source circuit wiring to the rooftop junction box is 69 ft for source circuit #1 and 66 ft for source circuit #2. These are equivalent to one-way distances of approximately 34 ± 1 ft for each source circuit.

4.5.3.2 Wire and Conduit from Rooftop Junction Box to Inverter

When the grounding conductor for the module frames and array mount frames is included, there will be five conductors running from the rooftop junction box to the inverter. Whether these 5 conductors are #12 or #10, they will fit easily into 0.5 in. conduit. If the conduit between rooftop junction box and inverter runs through the inside of the structure, then the inside section of the conduit must be metallic, and it must be properly grounded. The array conductors are #12, so at this point, the size of the conductors in the conduit can be determined. To determine the highest ambient temperature to which the conductors in conduit are exposed, it will be assumed that the conduit is supported 3.5 in. above the roof for a short distance before it penetrates into the attic of the structure. Referring to Table 4.13, an additional 22°C must be added to the maximum ambient temperature (35°C) to correct for the conduit over the roof. So, the highest ambient wire temperature can be expected to be 57°C. The ambient temperature derating factor for the 90°C wire is 0.71, according to Table 4.12.

In this example, an additional derating will need to be applied because there are now two source circuits, which have a total of four current-carrying wires for the conduit. Thus, a derating factor of 0.8 must also be applied to account for the four current-carrying conductors in the conduit.

Next, 156% of $I_{SC} = 5.97$ A, so the derated ampacity of the conductors from rooftop junction box to inverter must be greater than 5.97 A. If #12 wire is used, with a 30°C ampacity of 30 A, its derated capacity will be $30 \times 0.71 \times 0.8 = 17.04$ A, which is more than adequate. So now the last check is to check the voltage drop in the wiring between source circuits and inverter.

The total equivalent one-way distance between array and inverter is the sum of the equivalent one-way length of the open wiring of the modules plus the one-way length from rooftop junction box to inverter, which happens to be 84 ft. Using Equation (4.1) to evaluate the voltage drop if #12 wire is used the entire distance gives

$$\%VD = 0.2Id(\Omega/\text{kft})/V_s = 0.2 \times 3.59 \times 84 \times 1.93 \div 446 = 0.26.$$

The worst-case %VD occurs when the modules operate at their maximum temperature, causing V_m of the source circuits to drop to 391.4 V. The resulting %VD would then be found by replacing 446 by 391.4 to get 0.297%, which is still well below the allowed 2%. So, the wiring from rooftop junction box to inverter can be #12 THWN-2 and easily meet all ampacity and voltage drop constraints.

4.5.3.3 DC and AC Disconnects and Overcurrent Protection

The inverter comes with a single switch that simultaneously disconnects dc and ac, so the disconnect problem is solved, except for a small discussion of the operation of the switch. The switch incorporates three separate poles in series as the dc disconnect, and two additional poles as the ac disconnect. In addition, the timing of the make-break sequences is such that the ac switch opens before the dc switch opens. The result is that because there is no ac connection, there is no place for dc power to go. As a result, arcing across the dc switch is virtually eliminated, because zero-current switching is taking place. Because dc current has no zero crossings, as does ac current, if a dc current is interrupted instantaneously, resulting in a high d*i*/d*t*, a high voltage appears across the disconnect, causing significant arcing.

Overcurrent protection for the dc side is provided by the inverter. On the ac side, the maximum output current of the inverter is 15.6 A @ 240 V. The output conductors must therefore be sized at 125%, or 19.5 A. Conveniently, this can be achieved with #12 THWN-2 fused at 20 A. So the wiring from inverter to point of utility will consist of three #12 THWN-2 conductors. Note that two of these are phase conductors, and one is the grounding conductor. A neutral conductor is not required with this inverter. So, once again, these three conductors easily fit in a 0.5 in. conduit.

4.5.3.4 Point of Utility Connection

This time, it is assumed that there is no room for an additional two-pole circuit breaker in the distribution panel. This means that the PUC will need to be at the line side of the main disconnect. When this method is used for the PUC, it is necessary to follow *NEC* 240.21(B)(1) on tap conductors. The procedure is to install a fused disconnect within 10 ft of the main disconnect for the occupancy so that *NEC* 240.21(B)(1) will apply. The issue here is that if #12 conductors are connected to the line side of the main disconnect, they will not have overcurrent protection between the fused disconnect and the main disconnect. This is permissible, provided that the fused disconnect is located within 10 ft of the main disconnect, the conductors between the two disconnects are protected by conduit, and the ampacity of the main circuit breaker is no more than 10 times that of the proposed 20 A fuses. Furthermore, regardless of how tempting it may be, it is not acceptable to fasten the #12 conductors under the same lugs as the service conductors in the main disconnect. Either the lugs must be changed to accommodate both conductors or listed tap assemblies must be used on the service entrance conductors.

4.5.3.5 Estimating System Annual Performance

Now that the system design is essentially complete, before proceeding with the final electrical schematic diagram, PVWatts will be consulted to determine the expected system performance. Because the installation is in South Florida, West Palm Beach will be chosen as the site for the PVWatts analysis. Entering the pertinent data on roof slope, roof orientation, and array size, with the default dc-ac conversion factor produces an estimate of 4153 kWh/yr, with estimated maximum production of 394 kWh in the month of May.

4.5.3.6 Final System Electrical Schematic Diagram

Figure 4.11 shows the final system electrical schematic diagram. Note that this diagram is very similar to the previous system with the exception of the PUC at the main disconnect rather than at the distribution panel.

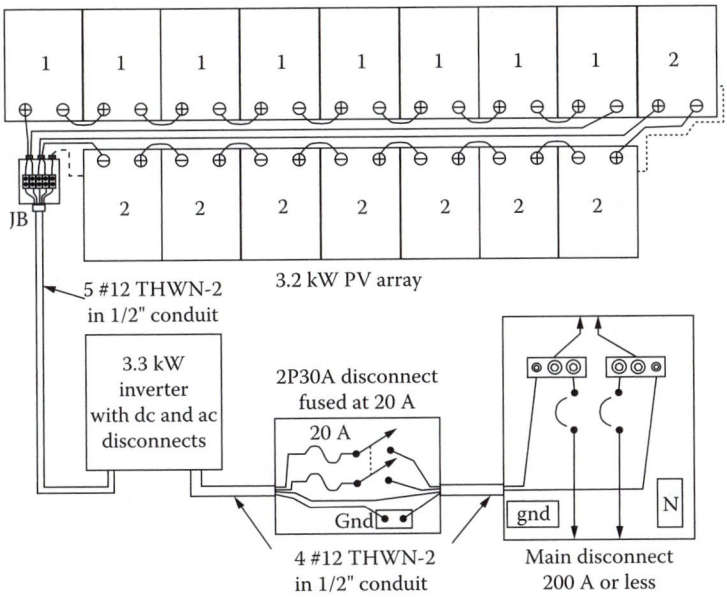

Figure 4.11 Electrical schematic diagram of 3.2 kW PV system.

4.5.4 Extension of Design to Lower Wind Speed Region

Before leaving this example, it is interesting to explore how modules can be fit onto the roof if they are allowed to spill over into roof zones 2 and 3. This is generally the case for lower wind speeds, as will be determined in more detail in the next chapter.

Figure 4.12 shows that 33 modules can be conveniently fit onto the roof. But there is a problem with this number of modules, because 33 is only divisible by 3 and 11. Dividing into 3 source circuits of 11 each results in source circuits having $V_{OC} > 600$ V. Dividing into 11 source circuits of 3 results in a V_m too low for many inverters. So, in this case, a better design would involve 4 source circuits of 8 modules each and a 6000 W inverter. Completion of this design is left as an exercise for the reader in Problem 4.8.

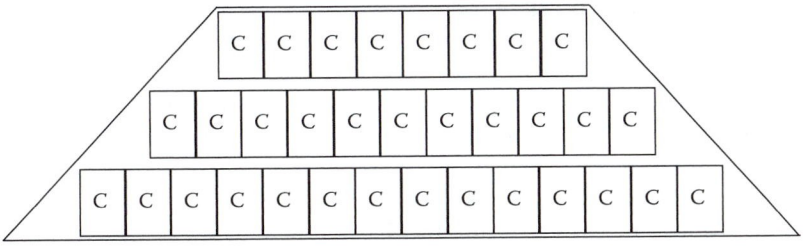

Figure 4.12 Lower wind speed array possibility with module C.

4.6 DESIGN OF A MICROINVERTER-BASED SYSTEM

4.6.1 Introduction

Toward the beginning of the 1990s, the concept of ac modules was introduced by Ascension Technology, Inc. [15]. The idea was to include a small inverter for each PV module, with the inverter attached to the module itself. The inverters were designed to be straight grid-connected inverters that met the requirements for grid connection. At the time, however, the inverter was generally considered to be the least reliable system component, so the PV community met the idea with an element of skepticism. Even though the NEC included a section on ac modules, very few systems were installed.

Proponents of the idea persisted, however, with the argument that if each module had its own inverter, overall efficiency of the entire system would be improved because array mismatch would be eliminated. Furthermore, if one module were shaded, the rest would still operate at maximum efficiency and failure of one module or one inverter would not affect the operation of the rest of the system. In addition, if the array required modules to have a number of different orientations, because each module already had its own inverter, individual module orientation did not affect the performance of any other modules in the system. In any case, after about a decade of neglect, the ac module concept is making a comeback, in the form of microinverters [16]. This time, however, rather than mounting the inverter directly to the module, the inverters are mounted close to the module so that a variety of modules can be connected to a single inverter model. The dc connecting leads of the module connect directly to the dc input of the inverter, and the 240 V ac outputs of the individual inverters are daisy-chained, with the maximum number of inverters being limited by the maximum load that can be handled by a circuit breaker, consistent with *NEC* requirements for the use of flexible cords. In addition, each inverter has its own built-in data-monitoring capability, so that a master system monitor can be used to observe the individual performance of each inverter as well as the overall system performance.

4.6.2 System Design

Because the inverters come separate from the modules, it is necessary to determine whether an inverter will operate under the output conditions of a particular module. The inverters still have a defined MPPT range and maximum operating voltages and powers. Because $V_m < 60$ V for most modules, microinverters must incorporate up converters rather than down converters or, perhaps, up-down converters. At the time of this writing, additional modules are being incorporated into the operating range of microinverters. And, of course, new modules are being introduced to the marketplace at a rapid rate, as well.

For the purposes of this example, it will be assumed that the installation will be in Albany, New York, on a structure that has a roof with which the architect took pride in incorporating six different slopes and orientations between east and west, not counting three more sections that face north or close to north. After carefully

Table 4.16 Summary of Module Placements on Dwelling Roof for Microinverter-Based PV System

Section #	Azimuth with South Ref = 180°	Slope	# Modules on This Section	Annual kWh (per PVWatts)
1	90°	45°	2	327
2	135°	45°	2	402
3	180°	60°	3	607
4	210°	45°	3	627
5	240°	60°	2	352
6	270°	45°	2	327
			Tot annual kWh	2642

reviewing module geometry and output power in order to maximize the number of modules that would fit on the roof, a 175 W module with $V_{OC} = 44.4$ V, $I_{SC} = 5.30$ A, $V_m = 35.8$ V, $I_m = 4.89$ A, $\Delta V_{OC}/\Delta T = -0.33\%/\text{K}$ and $\Delta V_m/\Delta T = -0.47\%/\text{K}$, and NOCT = 46°C has been selected. A suitable microinverter to match this module has $V_{DC}(\text{max}) = 54$ V, MPPT voltage range 25–40 V, module $I_{SC}(\text{max}) = 10$ A, and a maximum input current of 8 A. The rated inverter output voltage is 240 V and the rated output current is 750 mA, with maximum output power of 175 W. The maximum number of inverters per branch (i.e., connected in parallel to feed a single circuit breaker) is 16.

It will be assumed that the roof has room for 14 modules, to be located as indicated in Table 4.16.

Because the inverters are designed to be mounted to the module mounting rails, each module will have its inverter mounted essentially beneath the module. The connections are straightforward, because the two module leads connect directly to the inverter.

Inverters are interconnected as current sources in parallel. Each inverter has provisions for an ac cable at the ac input side and another ac cable at the ac output side. Thus, a single open-wire cable can be connected from inverter to inverter and then to a circuit breaker in the distribution panel. The sizing of wire and circuit breakers on the ac side of the modules are determined by the inverter maximum output characteristics, whether or not the inverter will ever achieve maximum rated output. Thus, the maximum current to a circuit breaker will be the sum of the output currents of the 14 inverters, or 14×0.75 A = 10.5 A. The wire and PUC circuit breaker must thus be rated at 125% of this value, or at least 13.12 A. Thus, a 2P15A circuit breaker is adequate, and presumably the wire supplied by the manufacturer will meet the system ampacity requirements as well as all UL and NEC requirements, because the cabling is custom-designed to fit the connectors on the inverters.

So the challenge here is to determine the expected annual and/or monthly and/or hourly output of the overall system. While PVWatts can handle the different azimuth angles and tilt angles, it is now necessary to take a closer look at the dc:ac conversion efficiency of the system, because module mismatch is no longer a factor in array performance. Referring back to Figure 4.2, and comparing the present system with a conventional system with source circuits and a single inverter, it becomes evident that

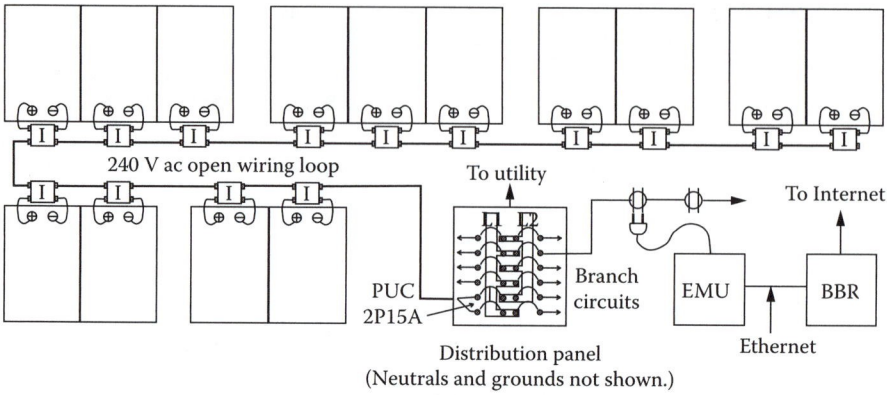

Figure 4.13 Schematic diagram of 2450 W microinverter-based PV system.

in addition to elimination of module mismatch, the dc wiring efficiency approaches 100% because of the short dc module leads, and if the module is guaranteed to have its output power within 3% of rated power under STC, then the dc:ac conversion factor increases to approximately 0.82 from its default value of 0.77. Using this default value in PVWatts, the annual performance of the individual modules can be calculated. The results are shown in Table 4.16, which shows an annual average kWh production of 1078 kWh/yr/kW. If this figure is compared with an array of conventional modules facing south with a latitude tilt, the annual performance can be expected to be 1161 kWh/yr/kW. Thus, the performance of the microinverter-based system on the assortment of roof sections is approximately 93% of the optimal performance of a single-inverter system.

4.6.3 Bells and Whistles (i.e., Monitoring Possibilities)

Another interesting feature of the microinverter-based system is the monitoring option. Each inverter has its own address, and communicates performance data via power line carrier connection to a communications gateway. The gateway enables performance monitoring of individual module/inverter systems and also can be connected via a broadband router to the Internet. Figure 4.13 shows the schematic diagram of the total system.

4.7 DESIGN OF A NOMINAL 21 KW SYSTEM THAT FEEDS A THREE-PHASE DISTRIBUTION PANEL

4.7.1 Introduction

Thus far, the systems that have been designed have been relatively small systems that connect into single-phase distribution systems. For the reader who has been wondering when an example of a larger system would be presented, the wait is over.

In general, any PV system between 10 and 500 kW is considered to be a medium-sized system. While systems at the lower end of this range may connect into a 208, 240, or 277 V (United States versions) single-phase supply, as system sizes increase, it becomes more likely that they will be connected into either 208, 240, or 480 V three-phase power distribution systems. These connections can be made with multiple inverters having single-phase outputs or with single, larger, inverters with three-phase outputs or with multiple inverters, each with three-phase outputs.

In this example, a nominal 21 kW system will be designed that uses three 7 kW single-phase inverters that feed a 120/208 V three-phase power distribution system. It will also be shown how these same components can be wired to feed a 277/480 V Y-connected three-phase power distribution system.

4.7.2 Array Configuration

The modules to be used are 305 W units with $V_{OC} = 64.2$ V, $I_{SC} = 5.96$ A, $V_m = 54.7$ V, $I_m = 5.58$ A, $\Delta V_{OC}/\Delta T = -0.27\%/°C$, and $\Delta V_m/\Delta T = -0.38\%/°C$. The installation is near Denver, Colorado, rack-mounted on a flat roof with a rack tilt of 45° to prevent snow from collecting on the modules. The range of module operation temperatures is between –30°C and +62°C. The 7 kW inverters have $V_{OC}(\max) = 600$ V, $250 < V_m < 480$ V, $I_{in}(\max) = 30$ A, internal GFDI device, four combining inputs, built-in dc and ac disconnect, and adjustable output voltages of 208, 240, or 277 V. The 208 and 240 V outputs can be connected line-to-line, but the 277 V output must be connected line-to-neutral. Jumpers can be adjusted to accommodate Y-connected or Δ-connected services at 208 and 240 V.

So, the first step is to configure an array that will hopefully deliver approximately 21 kW in a balanced fashion to the three inverters.

At the lowest operating temperature, V_{OC} increases to

$$V_{OC} = 64.2[1 - (25 - (-30))(-0.0027)] = 73.73 \text{ V}.$$

So $600 \div 73.73 = 8.13$. Rounding down results in 8 as the maximum number of series modules. At the highest temperature, V_m decreases to

$$V_m = 54.7[1 - (25 - 62)(-0.0038)] = 47.0 \text{ V}.$$

So $250 \div 47 = 5.32$. Rounding up results in 6 as the minimum number of series modules.

Table 4.17 shows possible module combinations that will result in a power close to 7000 W with no more than 4 source circuits. Note that there appear to be two options: four 6-module source circuits or three 8-module source circuits. Each case uses 24 modules. Because percent voltage drops are less in higher voltage arrays, it appears that using 3 source circuits of 8 modules each will be the best array configuration. Because there will be 3 inverters, this means the array will comprise 3 sets of 24 modules, or a total of 9 source circuits, which achieves the desired array balance of power with a total rated array power of 21,960 W.

Table 4.17 Array Power Based on Number of Source Circuits and Number of Modules in Source Circuits

Number of Source Circuits	Number of Modules in Source Circuits		
	6	7	8
1	1830 W	2135 W	2440 W
2	3660 W	4270 W	4880 W
3	5490 W	6405 W	**7320 W**
4	**7320 W**	8540 W	9760 W

4.7.3 System DC Wiring

Before wire lengths can be determined, it is necessary to determine the array layout. It will be assumed that the building faces south, with adequate east-west dimensions to allow three rows of 24 modules, adequately spaced so that no row shades any other row. Figure 4.14 shows the determination of row spacing for the array, showing necessary spacing between rows to ensure that the modules remain unshaded between 9 a.m. and 3 p.m. every day. Note that the modules are mounted in landscape position, and that the width of a module is 41.2 in.

Solving Equations (2.9) and (2.10) for Denver on December 21 shows that the altitude of the sun at 9 a.m. and 3 p.m. sun time is 14° and the azimuth is 42° east of south at 9 a.m. and 42° west of south at 3 p.m. Thus, the sun angles that define the module spacing are established. The assumption is that there are no other objects on the roof, such as fans, air conditioners, plumbing vents, drains, skylights, lightning rods, or fire hydrants that will either shade the array or require different module placement. In a real installation, all of these possibilities need to be accounted for. It is left as an exercise for the reader to determine that the dimensions indicated on Figure 4.14 are correct (Problem 4.9).

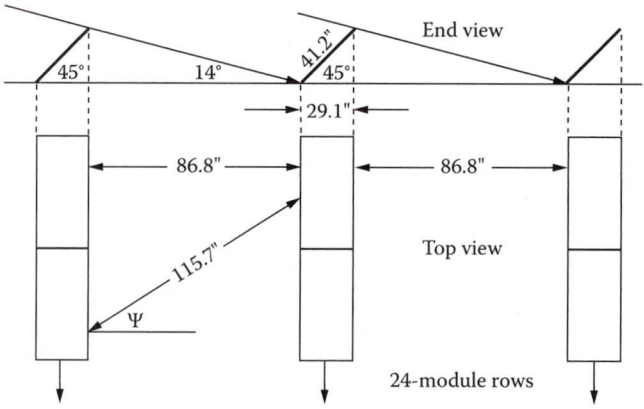

Figure 4.14 End view and top view of PV array row layout showing incident sunlight at 14° and array tilt at 45°.

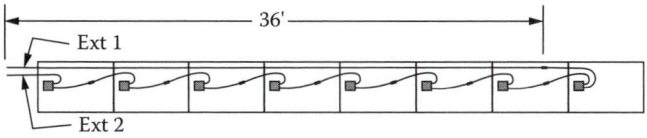

Figure 4.15 Determination of extension cable lengths from scale drawing of module row.

Next, assuming that a rooftop junction box is installed at the west end of each row of 24 modules, it is possible to estimate the length of open wiring between modules and junction box for each row. The long dimension of each module is 61.4 in., and it can be assumed that the spacing between modules is 0.4 in. Each module has a 3 ft positive lead and a 3 ft negative lead. Finally, the inverters will be located one floor down, a distance of 50 ft from the end of the center row and an additional 12 ft from the ends of the two outer rows.

Figure 4.15 shows a row of 8 modules drawn to scale, along with source circuit extension cables between module leads and the end of the row. The longest wire length will be from the eight modules at the east end of the row (right side). The next longest will be from the eight modules at the center of the row, and the shortest will be from the eight modules at the west end of the row.

Note that each section of 8 modules has a length of 41 ft 2 in., so that if 2 additional sections are located to the left of the section shown, an additional 82 ft 4 in. of extension cable will be needed to reach the end of the row of 24 modules. Allowing an additional 2 ft for entry into the junction box brings the additional extension cable length beyond the end of the first group of modules to 84 ft 4 in. Thus, the extension cable marked Ext 1 in Figure 4.15 will need to be 118 ft 4 in. long. Finally, because the other extension cable is attached to a lead from the left module in the group, the module lead itself will extend 2 ft toward the west, which is the equivalent of the additional 2 ft needed to reach the junction box. So the extension lead marked Ext 2 in Figure 4.15 needs only to be 82 ft 4 in. long. Be assured, however, that an installer will cut the leads to 119 ft and to 83 ft to be sure they are long enough.

So, for the easternmost group of 8 modules, the total open wiring length is 48 ft of module leads plus 83 ft extension for one lead and 119 ft extension for the other open lead, or a total of 250 ft of #10 USE-2 wire. This equates to a one-way distance of 125 ft.

For the center group of modules, the total open wire length will be 82 ft 4 in. less, because this group of modules is 41 ft 2 in. closer to the junction box. This equates to a one-way distance of 84 ft. For the west group of 8 modules, in addition to the 48 ft of module leads, only one extension wire will be needed, because this group ends at the junction box. The length of the extension cable will be 36 ft, so the total open wiring length will be 84 ft, which equates to a 42 ft one-way dimension.

Now that the equivalent one-way distances to the junction boxes have been determined, the overall one-way lengths of the source circuits to the inverters can be tabulated, along with the corresponding voltage drops if #10 wire size is used along the entire length. These results are shown in Table 4.18. In this table, source circuits

Table 4.18 Tabulation of Source Circuit Lengths and Voltage Drops for the 9 Source Circuits of the 21 kW System

Source Circuit #	Length to JB (ft)	Length JB to Inv (ft)	Total 1-way Length (ft)	I_m (A)	V_m (V)	% VD
1	125	62	187	5.58	437	0.58
2	84	62	146	5.58	437	0.45
3	42	62	104	5.58	437	0.32
4	125	50	175	5.58	437	0.54
5	84	50	134	5.58	437	0.42
6	42	50	92	5.58	437	0.29
7	125	62	187	5.58	437	0.58
8	84	62	146	5.58	437	0.45
9	42	62	104	5.58	437	0.32

1–3 are associated with the southernmost row of 24 modules, source circuits 4–6 are associated with the center row, and source circuits 7–9 are associated with the northernmost row.

Clearly, all voltage drops are well below 2%. It is left as an exercise for the reader to calculate the %VD values if #12 wire were run from junction box to inverter and if the only #10 wire were the module leads, with the remaining wiring #12 (Problem 4.10). Of course, it is still necessary to verify that the ampacity of #10 and #12 is adequate for the currents, temperatures, and conduit fills of this system.

To check on ampacity of the source circuit wiring, first note that 156% of I_{SC} = 9.30 A. Thus, the derated ampacity of the source circuit conductors must be at least 9.30 A. Assuming that each rooftop junction box contains terminal strips with 90°C rating, and that 6 current-carrying conductors, 2 for each source circuit, leave each junction box, and that the highest temperature of the junction box will be 50°C, the derating factors from Table 4.12 and Table 4.14 are 0.82 and 0.8. Hence, the ampacity of #10 will be $40 \times 0.82 \times 0.8 = 26.2$ A, and the ampacity of #12 will be $30 \times 0.82 \times 0.8 = 19.7$ A. So both wire sizes are adequate to handle 156% of I_{SC}. This completes the calculations relating to the system dc wiring, except for noting that the inverters should have fuses at the source circuit inputs that are no larger than recommended by the module manufacturer, which happens to be 15 A. Because the inverters have built-in ac and dc disconnects, this part of the design is also complete.

4.7.4 System AC Wiring

4.7.4.1 Wire and Overcurrent Protection Sizing

The ac wiring of the system will depend on the supply voltage to which the inverter is connected. It will first be assumed that the inverter will be connected to a 400 A, 120/208 V, Y-connected ac distribution panel. It will also be assumed that the outputs of the three system inverters will be combined in a separate distribution panel before the combined output is fed to the PUC. It will be assumed that the building has a spare space for a three-pole circuit breaker in the 400 A panel. But it will

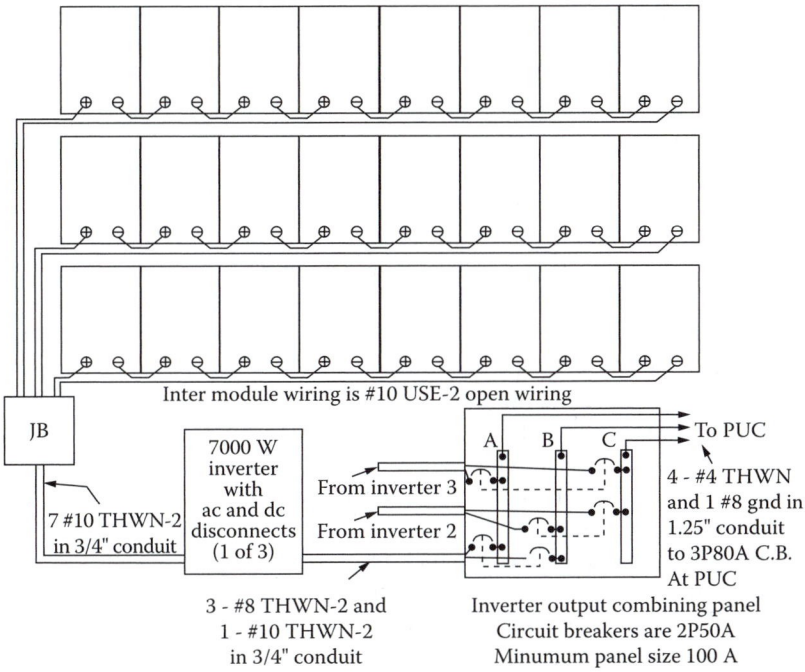

Figure 4.16 21 kW three-phase PV system showing 208 V, delta-connected, combining panel.

first be necessary to determine whether the combined outputs of the three inverters will exceed 20% of the 400 A panel busbar rating, or 80 A.

First, note that the rated output current of the inverter at 208V is 33.7 A, so 125% of 33.7 is 42 A. So a two-pole 50 A circuit breaker will be needed for connecting the inverter across 208 V, because this is the next standard size circuit breaker above 40 A. Figure 4.16 shows how the three inverter outputs can be connected to a single three-phase electrical panel to combine the outputs into a single three-phase output that will feed a single three-pole circuit breaker in the available 400 A distribution panel.

The output current from the combining panel will be $33.7\sqrt{3} = 58.4$ A. So, the circuit breaker connecting to the PUC must be at least $1.25 \times 58.4 = 72.96$ A. This means that an 80 A circuit breaker must be used. And, as luck would have it, 80 is exactly 20% of 400, so it can be installed in the 400 A panel as long as it is installed at the end of the busbar opposite the utility feed and labeled as a PV source that cannot be moved from that position, as required by *NEC* 690.64(B).

The previous calculations have also established the wire sizing for the ac side of the inverter. The inverter output must have an ampacity of at least 42 A, so #8, with an ampacity of 55 A, needs to be used. If the combining panel is close to the inverters, then voltage drop will not be a problem for this wire size. So, wiring from inverter to the ac combining panel will consist of 3 #8 (L_1, L_2, and neutral), and 1 #10 ground wire, as shown in Figure 4.16.

Next, the wire size from the ac combining panel to the PUC is determined to be #4, because it is rated at 95 A. In this case, because the PUC circuit breaker is a three-pole 80 A unit, the grounding conductor from ac combining panel to PUC must be increased in size to #8, as required by *NEC* 250.122. So the wiring between the ac combining panel to the PUC will consist of four #4 THWN-2 and one #8 THWN-2. A purist might argue that because the three sources are balanced, there will be no current on the neutral, so the neutral wire could be reduced in size. However, shading of one part of the array or a problem with an inverter or portion of the array could lead to an unbalance, which could result in some current on the neutral.

One should note that if the PUC happened to have room for three two-pole circuit breakers, then the ac combining panel would not be necessary. However, the current calculation used to determine the total current supplied to the 400 A busbars would be the same as the calculation that was used to determine the current out of the ac combining panel. The reader should note that by looking at the three two-pole 50 A circuit breakers in the ac combining panel, it would be easy to assume that because each phase has two 50 A circuit breakers connected to it, that the sum of the PV circuit breakers connected to the busbar would be 100 A. This would lead to the erroneous conclusion that the sum of the circuit breakers connected to the busbar is greater than 120% of the busbar rating. So the reader is cautioned to use all of her skills and knowledge of three-phase circuits, including, if necessary, a review of the chapter on three-phase ac circuits in a circuit analysis textbook. It should also be emphasized that although the preceding analysis was based on a balanced three-phase source, it is also valid for unbalanced sources.

4.7.4.2 Voltage Drop Calculations

Finally, assume that the ac combining panel is located 150 ft from the PUC. A check should be made on the voltage drop over this distance. But because this is a balanced three-phase circuit, the formula for voltage drop changes. Figure 4.17 shows a balanced, Y-connected, three-phase source and load, along with indications

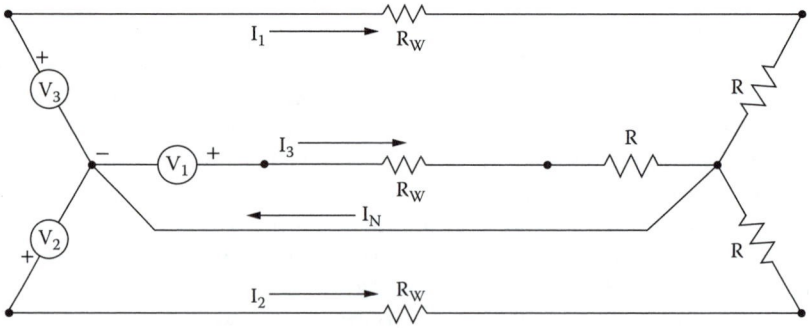

Figure 4.17 Determination of voltage drop in balanced three-phase system.

of all voltages and currents. It is assumed that the lengths of each phase and neutral conductor are the same.

The first observation is that if the magnitudes of I_1, I_2, and I_3 are the same, then $I_N = 0$, because the three phase currents are each 120° out of phase. Thus, the voltage drop in the wiring to and from each of the voltage sources is simply IR_W, because there will be no voltage drop in the neutral. Note that this is only half as much as would be the case if the same current were to flow in the neutral as in a phase conductor.

So in terms of the line-to-neutral voltage, V, the percentage voltage drop can be expressed as

$$\%VD = \frac{100IR}{V} = \frac{100I(\Omega/kft)L}{1000V} = \frac{I(\Omega/kft)L}{10V}. \tag{4.11a}$$

Now, because the line-to-line voltage, V_L, is $\sqrt{3}\ V_P = \sqrt{3}\ V$, Equation (4.11a) can be expressed in terms of the line-to-line voltage as

$$\%VD = \frac{100IR\sqrt{3}}{V_L} = \frac{100IR\sqrt{3}(\Omega/kft)L}{1000V_L} = \frac{I\sqrt{3}(\Omega/kft)L}{10V_L}. \tag{4.11b}$$

Thus, substituting $\Omega/kft = 0.308$ for #4 wire, $V_L = 208$ V, $I = 58.4$ A, and $L = 150$ ft into Equation (4.11.b) yields $\%VD = 2.25$. Although this is technically acceptable, because it is <3%, a more conservative approach would suggest using #3 wire with all the same circuit breaker sizes instead of #4. Substituting $\Omega/kft = 0.245$ gives $\%VD = 1.79$.

4.7.4.3 Connecting to a 277/480 V Y-Connected Source

Finally, suppose that the outputs of the three inverters are to be connected to a three-phase, 277/480 V source. In this case, the output current of each inverter would be 25.3 A, and 125% of this value is 31.58 A. So, it will require #8 wire and a single-pole, 40 A circuit breaker to connect to an ac combiner panel. Each inverter will connect to a different phase, such that this time the line currents from the ac combining panel to the PUC will be the inverter output currents, as the inverter outputs will now be Y-connected with line-to-neutral voltages of 277 V rather than delta-connected with line-to-line voltages of 208 V.

When line-to-line voltages are used, Equation (4.11b) is still valid for the Y-connected sources as long as the sources are balanced. So, checking voltage drop over the 150 ft distance using the new values for voltage, current, and Ω/kft gives $\%VD = 1.04$. So if the higher voltage can be used, then #8 wire size can be used rather than #3, at a significant cost savings and also an increase in efficiency because of the lower $\%VD$. Furthermore, the PUC circuit breaker can be a three-pole, 40 A unit, rather than a three-pole, 80 A unit.

4.8 DESIGN OF A NOMINAL 250 KW SYSTEM

4.8.1 Introduction

It is now time to design a larger system so that the concept of subarrays can be introduced. In fact, the last example used separate arrays for each inverter, so one could argue that each separate array should be classified as a subarray. But the last system could also be classified as three separate systems that were combined at the ac outputs of the inverters.

This example will use the modules of the last example, because they are intended for use in commercial installations. This does not necessarily mean that lower-power modules cannot be used in commercial installations, because megawatt systems have been designed around modules having power ratings less than 100 W. The main reason for using the modules of the last example is because it is already known that eight modules per source circuit will produce an output that falls within the input voltage range of many inverters.

Because this system is intended to be a 250 kW system, it will be designed around a 250 kW inverter. Again, this does not mean that a 250 kW system must be designed with a 250 kW inverter. There are also good arguments that a 250 kW system should be designed around ten 25 kW inverters or five 50 kW inverters. The argument is that if one inverter should fail, the remaining inverters would still be operational. So, the choice of how many inverters to use is the responsibility of the system designer. The array will be mounted on racks at ground level. Rack mounts will be further discussed in the next chapter.

There are a number of 250 kW inverters on the market, so the choice of inverter may depend on price, availability, weight, physical dimensions, reputation, or any number of other considerations. It will be assumed that the inverter to be used in this example is characterized by $V_{OC}(\max) = 600$ V, $320 < V_m < 600$ V, $I_{IN}(\max) = 814$ A, $V_{OUT} = 480$ V, and $I_{OUT}(\max) = 301$ A. The maximum output overcurrent protection is 361 A, so the closest standard fuse or circuit breaker rating to this value is 350 A. The unit can be configured with either 10 combining inputs rated at 160 A each, or 15 combining inputs rated at 100 A each. The unit has a CEC rated efficiency of 96%. It has built-in ac and dc disconnect switches and includes an isolation transformer at the output. The weight of the unit is 4500 lb.

4.8.2 Configuring the Array

The design goal is to configure an array that will deliver just over 250 kW to the inverter, because the inverter is rated for *output* power, and its dc-to-ac conversion efficiency is rated at 96%. So a reasonable array power to set as a goal is $250 \div 0.96 = 260$ kW. It will be assumed that the location of the installation will have the same ambient temperature conditions as in the last example, so the expected range of array output voltage for eight-module source circuits during system operation will be between 376 and 590 V, which is within the MPPT range of the inverter.

Noting that an eight-module source circuit will have a rated power of 2440 W, all that is now needed is to determine how many source circuits it will take to produce approximately 260 kW. To begin, divide 260 kW by 2440 to find the approximate number of source circuits. The result is 106.5574.

So, now what should be done with 106.5574? The answer is that when possible, it is preferred to be able to combine source circuits such that a balanced set of output currents can be obtained, consistent with the number of available combining inputs on the inverter. So a set of source circuit combiner boxes needs to be chosen that will hopefully each have the same output current. These combiner boxes will define the subarrays of the system that will feed the combining inputs of the inverter.

For example, if 100 source circuits were to be used, then the 100 source circuits could be split into 10 groups of 10 so that 10 source circuit combiner boxes could be used, each combining 10 source circuits. But 100 source circuits will only produce 244 kW, which is less than the desired array size. So it will be necessary to look for other combinations. If it is not possible to balance the array perfectly, then one group of combined source circuits may end up having fewer source circuits than the other groups. Before proceeding to narrow down the number of source circuits, it is first useful to look at the design current per source circuit, that is, 156% of I_{SC}, which is 9.30 A. So if 10 source circuits are combined, then the design output current will be 93.0 A, which is convenient for a 100 A fuse. If 15 source circuits are combined, then the design output current will be 139.5 A, and if 16 source circuits are combined, then the design output current will be 149 A, either of which works well with a 150 A fuse. So there are a large number of options for combining source circuits.

As it turns out, if it is decided to have 108 source circuits (263,520 W), this is 5.4% higher than the inverter rating. The 108 source circuits can be divided into 12 groups of 9 or 9 groups of 12, as long as the input current and voltage ranges match the inverter ranges. It has already been determined that the array output voltage matches the input voltage requirements of the inverter. The inverter manufacturer does not specify limits on short-circuit current, so the maximum input current will be $108 \times I_m = 602.6$ A < 814 A, which is well within the maximum input current limit of the inverter.

Using groups of 9 will require 12 source circuit combiner boxes, and use of groups of 12 will require 9 source circuit combiner boxes. So now it is time to go to the Web to explore what combiner boxes are available. A quick search reveals that a number of possibilities exist. One possibility is a combiner box that will combine 12 circuits, with an output bus rated at 192 A. Thus, if 9 combiner boxes are used, each combining 12 circuits, then a solution is at hand, because 156% of I_{SC} is $1.56 \times 5.96 \times 12 = 111.6$ A < 192 A. Note that the sizing of the combiner box busbar uses the same rules for sizing of wiring.

4.8.3 Source Circuit Voltage Drop Calculations

Before voltage drop calculations can be made, it will be necessary to determine the array layout. Because the array is ground mounted, there is an opportunity to make an effort to minimize the wiring cost from array to inverter. Figure 4.18 shows

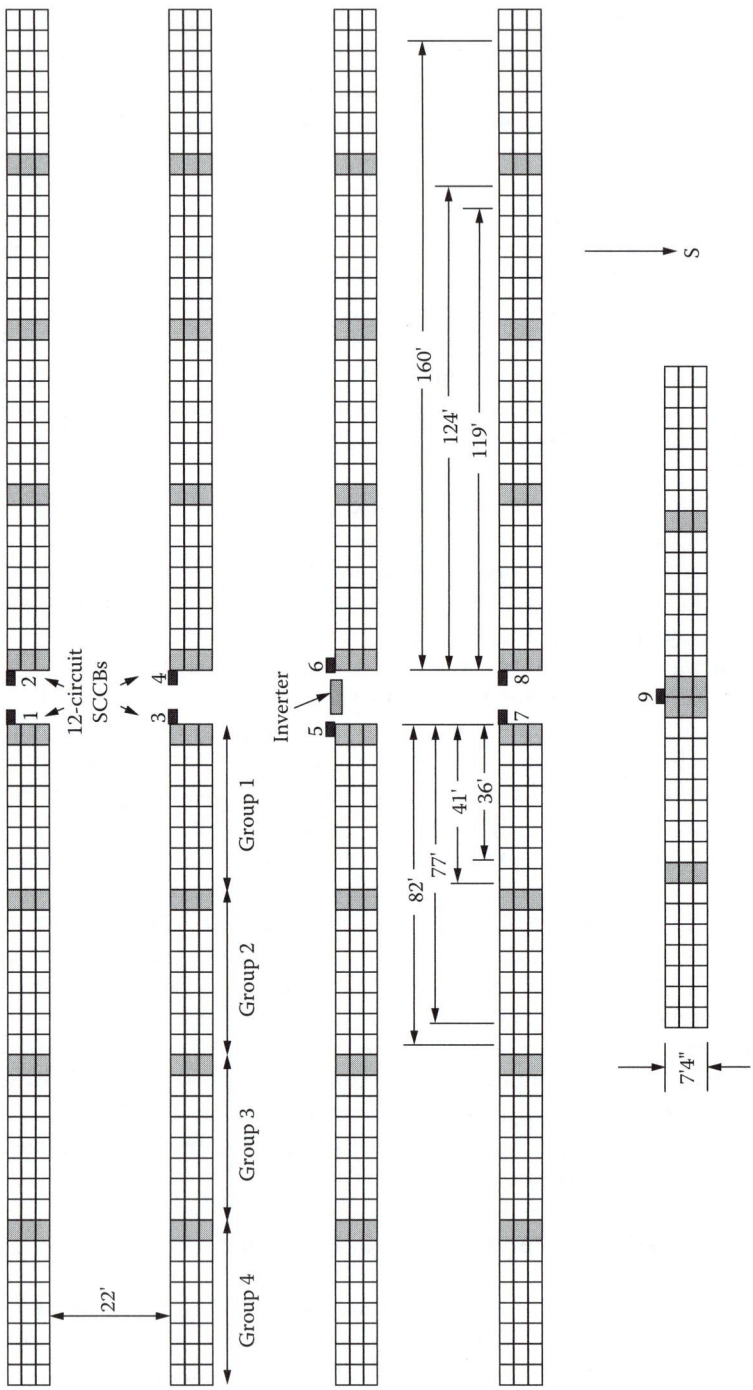

Figure 4.18 Array layout for 250 kW system showing locations of source circuit combiner boxes and inverter.

an attempt at minimizing wiring costs by locating the inverter at the center of the array. Locating the inverter at the center of the array, of course, presumes that the cost of connecting the inverter ac output to the utility will not cancel any savings on wiring on the dc side of the system.

The shortest wiring distances should result from the array taking the shape of a circle, but because generally rectangular areas are available, an attempt will be made to keep the array as "square" as possible. Because there are a total of 864 modules in the array, to determine the number of modules in each row, start with $\sqrt{864} = 29.39$ modules. Then, to keep the installation user-friendly, note that if a row contains 32 modules, this will be 4 source circuits. So, it will be convenient if the rows can consist of multiples of 8 modules.

The observation that there will be 27 rows suggests that grouping the modules 3 rows at a time will result in 9 rows with 96 modules per row. The groups of modules as shown in Figure 4.18 must have 22 ft between rows to avoid shading, assuming the array to be at the same latitude as the array of the last example, because the height above ground of the highest point is now three times as high as it was with a single row of modules in landscape orientation. A different latitude, of course, will require different row spacing. Because of the 22 ft spacing between rows, a better "square" is obtained by creating an "aisle" in the middle and then adding another 4 rows of 96 modules on the other side of the aisle. This leaves 96 modules for the last, southernmost, row, as shown in Figure 4.18.

Because the inverter is approximately 7 ft high, the row spacing also accommodates the shadow of the inverter if it is located as shown in Figure 4.18. Also, with 96 modules in each group, this conveniently allows for 12 source circuits to terminate at the combiner boxes as shown in the figure. Again, the extra array height and spacing makes it easier to mount and wire to the combiner boxes. It is then straightforward to run conduit from each combiner box underground to the inverter. All that is now needed is to calculate all the distances so wire can be sized to keep voltage drop under 2%.

Figure 4.18 shows how the source circuits can be grouped into four groups, working away from the center aisle. The symmetry of the array allows the designer to calculate the equivalent one-way distance from each group to the combiner box by adding up the total module lead length plus the lengths of the two extensions, as shown in Figure 4.18. The equivalent one-way distances, along with the percent voltage drop, based on the use of #10 wire, a source circuit current of 5.58 A and a source circuit voltage of 437.6 V, are tabulated in Table 4.19.

Table 4.19 Tabulation of Source Circuit Lengths and Voltage Drops for the Source Circuits of the 250 kW System

Group	Module Total Lead Length	Extension #1 Length	Extension #2 Length	Total Loop Length	Equivalent 1-Way Length	% VD
1	48	36	10	94	47	0.145
2	48	77	41	166	83	0.256
3	48	119	82	249	125	0.384
4	48	160	124	332	166	0.512

The next step is to determine the wire sizing between the source circuit combiner boxes and the inverter. Because 156% of the total I_{SC} of the 12 source circuits that are combined in each of the 9 combiner boxes is 111.6 A, the derated ampacity of the wiring must be at least this high. So, to meet the ampacity requirements of the PV output circuits, #2 THWN-2 copper wire is needed. Furthermore, the wire must be sized so the combined voltage drop of source circuits plus subarray output circuits (i.e., combiner box output currents) is less than 2%. Because the worst-case voltage drop in the source circuits is 0.512%, if the output circuit voltage drops are limited to 1.488%, the totals will not exceed 2%. So, to determine the voltage drops and select suitable wire sizing, it is now necessary to determine the distances between each of the 9 source circuit combiner boxes and the inverter.

It will be assumed that wiring from source circuit combiner boxes to inverter will be in conduit that is buried 18 in. below the surface, so that in addition to horizontal distances, vertical rises must also be included in the total distances. If the combiner boxes are mounted at a height of 4.5 ft above the ground, then an additional 6 ft is needed at the combiner box end, and if the inverter is mounted on a slab that is 12 in. above grade to keep water out, an additional 5 ft will be needed in order to account for the distance to the input terminals of the inverter.

Figure 4.18 shows the separation of rows to be 22 ft, and the width of the rows is close to 7 ft (see Figure 4.14), so the distances can now be tabulated between source circuit combiner boxes and inverter, if the northernmost two boxes are identified as 1 and 2, the next two are 3 and 4, etc. The resulting distances, allowing 5 extra feet for the distance between the end of each row and the center of the "aisle" between rows, are tabulated in Table 4.20.

Table 4.20 shows the percent voltage drop at STC and also at the expected maximum array temperature. To ensure that the voltage drop is always less than 2%, a prudent designer would use the high-temperature figure. As it turns out, for this case, the combined voltage drop is well below the 2% limit for either temperature because the wire runs are relatively short.

Table 4.20 Tabulation of Subarray Circuit Lengths, Voltage Drops, and Wire Sizes for the 9 PV Output Circuits

SCCB #	Dist to Inverter	Wire Size	%VD @ STC	%VD @ T_{max}
1	74	#2	0.349	0.406
2	74	#2	0.349	0.406
3	45	#2	0.212	0.247
4	45	#2	0.212	0.247
5	16	#2	0.075	0.088
6	16	#2	0.075	0.088
7	45	#2	0.212	0.247
8	45	#2	0.212	0.247
9	74	#2	0.349	0.406

The final check on the ampacity of the output circuits needs to account for any derating due to elevated ambient temperature. Because the array is ground mounted, there will not be an increase in ambient temperature that must be taken into account. Using the actual ambient temperature results in a derating of 0.96, so the ampacity of the #2 THWN-2 is reduced to 125 A, which is greater than 156% of I_{SC}. Finally, there will be only two current-carrying wires in each conduit between source circuit combiner box and inverter, so no conduit fill derating is needed.

4.8.4 DC Disconnects and Overcurrent Protection

The first location for overcurrent protection is the combiner boxes. The module manufacturer recommends fusing at no more than 15 A, so any fuse size between 156% of source circuit I_{SC} and 15 A is acceptable, because the ampacity of the source circuit wiring is higher than 15 A. The important point is that the fuses must be dc-rated with a voltage rating of 600 V. The next location for overcurrent protection is the inverter combining inputs. Because 156% of I_{SC} of the PV output circuits is 111.6 A, the next highest fuse is 125 A. So, the fuses at the inverter subarray combining inputs should be 125 A. An important point here is that the modules happen to be designed for use with a positive ground. Because fuses are prohibited in the grounded conductor, both the source circuit combiner boxes and the inverter combining inputs must be designed for grounding the positive inputs and fusing the negative inputs. It is essential that the positive conductors be grounded at a single point located internally at the inverter. Thus, no additional grounding should be made of the positive conductor. The grounding takes place at the ground fault detection and interruption (GFDI) circuitry of the inverter only.

The dc/array disconnect is a part of the inverter, so no additional disconnect is necessary. If maintenance is needed on any part of the array, the dc disconnect is shut down and then the section of the array that needs maintenance is disabled by removing the fuse to that section. The dc disconnect can then be turned on again, and the system can operate at a reduced power rating until the problem with the array section is repaired.

4.8.5 AC Wire Sizing, Disconnects, and Overcurrent Protection

The ac output of the inverter is 480 V @ 301 A, so wire sizing, disconnects, overcurrent protection, and the point of utility connection must take these values into account. Of the many options for the point of utility connection, the preferred options are to either transform the output up to distribution level voltage as soon as possible or to connect into a building panelboard with sufficient busbar rating to handle an additional 20% from the PV system circuit breaker. Because the rated PV ac output current is 301 A, the wiring, disconnects, and overcurrent protection must be rated at 125% of this value, or 375 A. But it also needs to be noted that the manufacturer specifies a maximum overcurrent protection of 361 A, so this will require a 350 A circuit breaker rather than a 400 A unit. Thus, 350 A must be no more than 20% of the busbar rating, which means the busbar rating must be at least $350 \div 0.2 = 1750$ A.

So, if the building closest to the array, hopefully owned by the owner of the array, has at least a 1750 A main service at 480 V, then the three-pole, 350 A circuit breaker from the inverter can be installed on the busbar on the end opposite the utility feed. As a possibly less expensive alternative, if there is room for additional lugs on the panelboard busbars, an external 400A disconnect fused at 350A can be used to connect to the busbar as long as the tap rules of NEC 240.21(B) are followed. Note that the *NEC* distinguishes branch circuit taps from feeder taps. Branch circuit taps are generally made with #10 or smaller wire that feeds loads less than 5000 W. Feeder taps are made with larger wire, so the feeder tap rule is appropriate for the PUC of this example.

So, what size wire should be used to achieve the 350 A ampacity? NEC Table 310.16 indicates that 350 kcmil THWN-2 will carry 350 A, so using 4 350 kcmil is one solution. Because of the large physical size of 350 kcmil wire, many electricians would prefer to use parallel conductors to carry this load instead. So an alternative wiring method would use two conduits, each containing 4 #2/0 conductors. Each conduit will contain conductors for each phase as well as the neutral, if required by the utility, and a ground wire, unless the wiring is in metallic conduit. The size of the grounding conductor is determined from *NEC* Table 250.122 to be #3 copper. Note that because NEC 310.4 only allows conductors of size 1/0 or larger to be paralleled, a #3 grounding conductor will need to be run in each conduit along with the 2/0 conductors. One exception to this rule is that if the neutral conductor is reduced in size, then parallel neutral (grounded) conductors may be run in sizes #2 or larger. In cases where the neutral is only used as a voltage reference but does not carry current, it may be reduced to the size of the grounding conductor, especially when the inverter itself does not have a terminal for a neutral.

Because the location of the PUC has not been specified, rather than make assumptions, the distance that the parallel 2/0 conductors can be run with a voltage drop less than 2% will be calculated instead. Using Equation (4.11b), solved for L, with 2% voltage drop, gives

$$L = \frac{10(\%VD)V_L}{\sqrt{3}I[\Omega/kft]} = \frac{10 \times 2 \times 480}{\sqrt{3} \times 301 \times .0484} = 380\,\text{ft},$$

where the resistivity of the parallel 2/0 conductors is shown to be half the resistivity of a single 2/0 conductor.

Thus, a distance of 380 ft is allowable with parallel 2/0 conductors connecting between the inverter output and the PUC. Note that although Equation (4.11) was derived for a Y-connected system, because Equation (4.11b) is expressed in terms of line-to-line voltage and line current, it is therefore valid for Δ-connected systems as well. This completes the design of the system. The reader is encouraged to imagine adaptations of the design to other situations, such as rooftops or as walkway or parking lot canopies. Also, this design can be considered to be a building block for a much larger system. The schematic is shown in Figure 4.19.

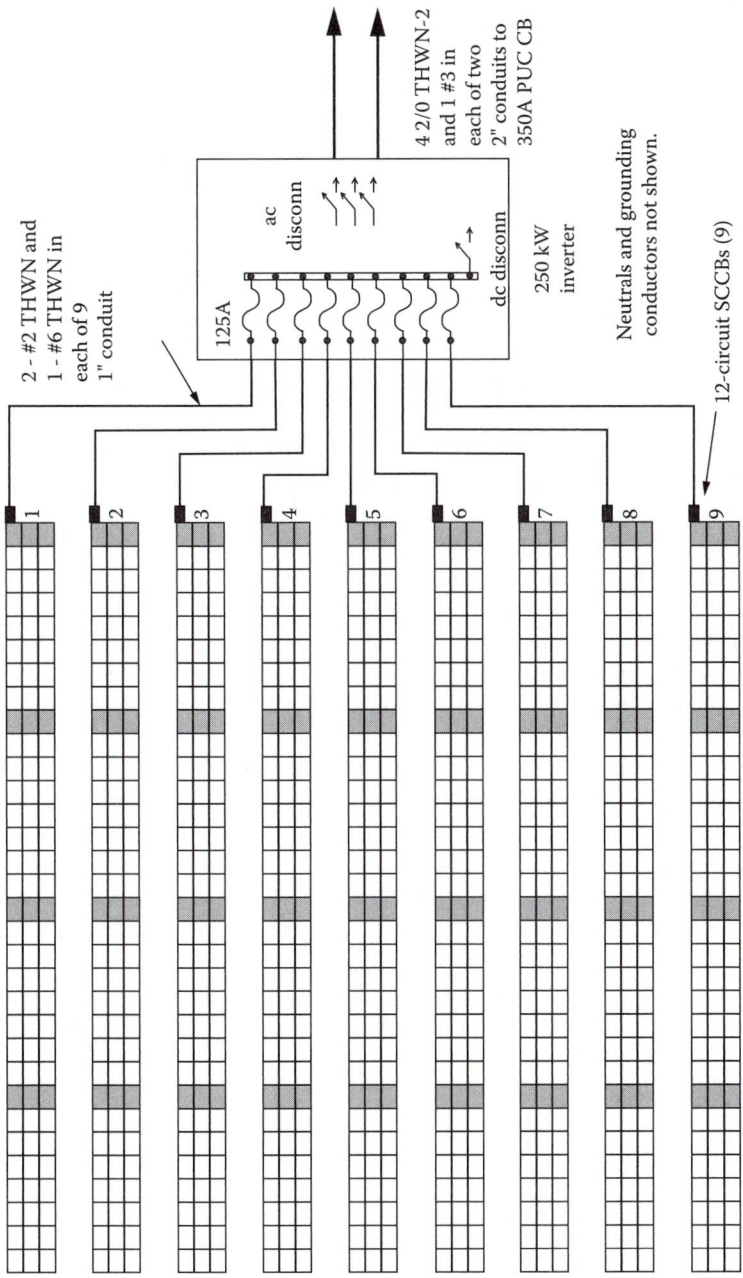

Figure 4.19 Schematic diagram of nominal 250 kW PV system.

4.9 SYSTEM PERFORMANCE MONITORING

It is not at all unusual for a PV system owner to wonder how well the system is functioning. One method is to compare this year's electric bills with last year's. Another way is to go out to the inverter each day and read what the display has to offer, if there is a display. But it is becoming more and more commonplace for system owners to want to be able to connect system performance data to the Internet so they can monitor the performance of their systems from wherever they may be. A system owner in Daytona Beach, Florida, compared "a PV system without monitoring as the equivalent of an automobile without a speedometer."

Monitoring systems can monitor a wide range of variables, such as source circuit current and voltage, inverter input power, inverter output power, inverter self-diagnostics, weather data, such as temperature, irradiance and irradiation, power purchased from the utility, and power sold to the utility. A wide range of communication protocols are available, including hard-wired data communication, radio frequency wireless data communication, and power line carrier data communication. Most inverters monitor input and output voltage, current, and power in order to accomplish functional tasks of the inverter, such as maximum power point tracking and self-diagnostics. So, the quantities of interest are digitized and made available to display devices, either on the inverter or remote to the inverter.

Web-based monitoring systems are becoming very popular. In general, PV system data and sometimes weather data are collected, digitized, and then transferred to a controller/display unit and then to a broadband router as shown in Figure 4.13. Some manufacturers provide their own communications equipment, others make data available at a port to be used with third-party hardware and software, and some do both.

As the number of performance-based systems increases, the need for utility-grade monitoring also increases. In some cases, third-party monitoring is required by the performance-based contract. As a result, equipment manufacturers are rapidly developing data-monitoring capability for their own products as well as providing for third-party monitoring of performance as well as fault diagnostics. As will be discussed in Chapter 8, performance-based contracts simply guarantee a certain level of energy production by the PV system. If the system does not meet the performance objectives, the system installer must pay the owner the difference. With this type of contract, the installer will want to be the first to know if there is a problem with a system.

Problems

4.1 Refer to Table 4.2 or to the *NEC* in order to:
   a. Determine the wire size needed to limit voltage drop to 3% for a 100 W, 24 V load at a distance of 75 ft from the voltage source.
   b. Using the wire size determined in part a, determine the actual voltage drop for the wiring.

4.2 An inverter has a balanced three-phase, 277/480 V output and is installed a distance *d* ft from the point of utility connection.

a. Derive an expression for the voltage drop between inverter and PUC in terms of d.
b. If the maximum inverter output current is 14.4 A and the inverter is located 200 ft from the PUC, determine a wire size that will have adequate ampacity if the wire has 90°C insulation and will have a voltage drop of less than 2% between inverter and PUC.

4.3 Show that $\alpha = \pi f_o / Q$ for a parallel RLC circuit.

4.4 Show that $f_o = \dfrac{\omega_d}{2\pi\sqrt{1-\frac{1}{4Q^2}}}$ for a parallel RLC circuit.

4.5 Design a resonant (RLC) load with $Q = 5$ that will dissipate 1000 W at 120 V (rms) at 60 Hz. Then use PSPICE or your favorite network analysis program to simulate the SFS and SVS algorithms by connecting a sinusoidal current source to the load and varying the frequency above and below 60 Hz and observing the total response (transient plus forced) as the excitation frequency moves farther away from 60 Hz.

4.6 Using module #1 as described in Section 4.4, and an inverter that has $V_{in}(\max) = 600$ V, $V_m(\min) = 250$ V, $I_{in}(\max) = 30$ A and is rated at 7000 W output power, with a maximum array power of 8750 W and provisions for combining 4 inputs, determine 3 configurations of the modules that will provide between 6500 and 8000 watts to the inverter. Assume the module operating temperature range will be between −20°C and +65°C.

4.7 For the 15-module array chosen in Section 4.4.3:
a. Determine the % voltage drop between rooftop junction box and inverter if #12 THWN-2 wire is used between the two points.
b. Verify that #12 THWN-2 has adequate ampacity if its operating temperature is 60°C.

4.8 For the module configuration in Figure 4.12:
a. Determine a suitable arrangement of source circuits. If necessary, eliminate one or more modules.
b. Select an inverter that will have input and output characteristics that will match your array of part a.
c. Complete the design of the rest of the system, including a schematic diagram of the entire system.

4.9 For the PV array of Section 4.7:
a. Verify the shadow length at 9 a.m. on December 21.
b. Account for the azimuth angle, and determine the minimum row spacing based on the shadow direction at 9 a.m.
c. Using the row spacing shown in Figure 4.14, determine the time interval during which the array will be unshaded on September 21.

4.10 For the PV array of Section 4.7:
a. Determine the %VD in each of the nine source circuits if #10 wire is used between modules and rooftop junction boxes and #12 wire is used between rooftop junction boxes and the inverters.
b. Determine the %VD in each of the nine source circuits if #12 wire is used everywhere except for the module leads.

Design Problems

4.11 Specify all the components for a nominal 1500 W ground-mounted, utility-interactive PV system. Assume the code in effect is the most recent version of the *NEC*. Use a utility connection point on the utility side of the meter.

4.12 Specify all the components and show the design for a nominal 5000 W residential rooftop-mounted, utility-interactive PV system, based on the most recent version of the *NEC*. The system will be located in Atlanta, Georgia, and will be

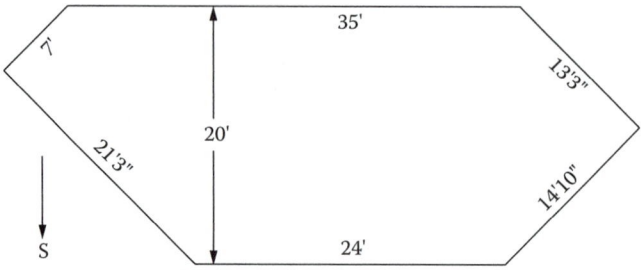

Figure P4.1

standoff-mounted on a roof with a 30° tilt. Connect the system on the customer side of the meter, specifying the allowable sizes of main circuit breaker and distribution panel busbars if the PUC is to be in the distribution panel. Use PVWatts to estimate the monthly and annual performance of the system.

4.13 For the roof shown in Figure P4.1, assume there are no obstructions in the area shown.

a. Make scale drawings that show how a set of modules could be mounted within roof zone 1 (in this case, more than 3 ft from the edge) to maximize the array power. Select whatever module you feel will best achieve the design goal. Ensure the modules can be configured to meet the input voltage requirements of an inverter that has $V_{OC}(\max) = 600$ V and $250 < V_m < 520$ V as the MPPT range. Specify the approximate power rating needed for the inverter.

b. Make scale drawings that show how a set of modules could be mounted on the entire available area of the roof, without extending beyond the roof. Select whatever module you feel will best achieve the design goal. Be sure the modules can be configured to meet the input voltage requirements of an inverter that has $V_{OC}(\max) = 600$ V and $250 < V_m < 520$ V as the MPPT range. Specify the approximate power rating needed for the inverter.

4.14 a. The south-facing portion of the roof shown in plan view in Figure P4.2 has a 5:12 slope. Select a module that has a rating of at least 175 W and lay out to scale an array configuration on the south roof. Make sure the array can be used with an inverter with the input characteristics indicated in Problem 4.12.

b. On the east-facing and west-facing portions of the roof, all of which also have 5:12 slopes, configure modules that use microinverters to fill as much of the roof as possible.

c. Assume your configurations of parts a and b are in Sacramento, California. Use PVWatts to estimate total PV production on a monthly and annual basis of the combined set of arrays.

4.15 Design a 36 kW commercial roof-mounted PV system that will feed a 120/208 V 3-phase 400 A distribution panel on the customer side of the meter. Comment on the bus capacity of the distribution panel and the allowable size of the main circuit breaker on the distribution panel feeder circuit.

4.16 A 120 kW utility-interactive system is to be installed as a parking lot canopy. It is to feed the grid with a 3-phase, 277/480 V balanced output. Design a system showing all necessary electrical components, including modules, inverters, wire sizes, and fuse/circuit breaker sizes and locations. Recommend a horizontal layout of modules that will provide shading for cars, and estimate how

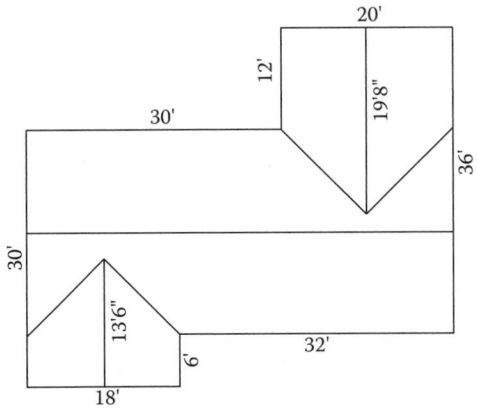

Figure P4.2

many parking spots will be covered. Comment on your choice of whether to use a single inverter or multiple inverters.

REFERENCES

1. IEEE 929-2000, IEEE Recommended Practice for Utility Interface of Residential and Intermediate Photovoltaic (PV) Systems, IEEE Standards Coordinating Committee 21, Photovoltaics, 2000.
2. IEEE 1547-2003, IEEE Standard for Connecting Distributed Resources to Electric Power Systems, IEEE Standards Coordinating Committee 21, Fuel Cells, Photovoltaics, Distributed Generation and Energy Storage, 2003.
3. *NFPA 70 National Electrical Code, 2008 ed.*, National Fire Protection Association, Quincy, MA, 2007.
4. UL 1741: 2001, Inverters, Converters and Controllers for Use in Independent Power Systems, Underwriters Laboratories, Inc., Northbrook, IL, 2005.
5. ANSI C84.1—1995, Electric Power Systems and Equipment—Voltage Ratings (60 Hertz), American National Standards Institute, New York, 1995.
6. Wohlgemuth, J., Testing for BIPV, *Proc. Photovoltaic Performance and Reliability Workshop*, Cocoa Beach, FL, November 1998, 113.
7. DeBlasio, R., Status of IEEE Standards Coordinating Committee (SCC 21) and IEC Standards Technical Committee 82 (TC82), *Proc. Photovoltaic Performance and Reliability Workshop*, Cocoa Beach, FL, November 1998, 236.
8. IEEE Standard 519–1992, *Recommended Practices and Requirements for Harmonic Control in Electrical Power Systems*, Institute of Electrical and Electronics Engineers, Inc., New York, 1992.
9. Zgonena, T., Safety Considerations for Static Grid-Tied Photovoltaic Inverters, Interconnecting Small PV Systems to Florida's Utility Grid, A Technical Workshop for Florida's Utilities, Cocoa, FL, October 22, 1998.

10. ASCE Standard 7-05, Minimum Design Loads for Buildings and Other Structures, American Society of Civil Engineers, Reston, VA, 2005.
11. Kern, G. A., Bonn, R. H., Ginn, J., and Gonzalez, S., Results of Sandia National Laboratories grid-tied inverter testing, *Proc. 2nd World Conf. and Exhibition on PV Solar Energy Conversion*, Vienna, Austria, July 1998.
12. Kern, G. A., New Techniques for Islanding Protection, Interconnecting Small PV Systems to Florida's Utility Grid, A Workshop for Florida Utilities, Cocoa, FL, October 22, 1998.
13. CFR 47, Federal Communications Commission, Rules and Regulations, Radio Frequency Devices, Part 15, 1994.
14. The PVWatts Solar Calculator, National Renewable Energy Laboratory, Golden, CO, 2008, http://www.pvwatt.com/.
15. Kern, G., SunSine™ 300: Manufacture of an AC Photovoltaic Module, Final Report, Phases I & II, 25 July 1995–30 June 1998, NREL/SR-520-26085, March 1999 (www.nrel.gov/docs/fy99osti/26085.pdf).
16. www.enphaseenergy.com for information on Enphase microinverter systems.

CHAPTER 5

Mechanical Considerations

5.1 INTRODUCTION

Because the primary function of a photovoltaic (PV) system is to convert sunlight to electricity, often the role and importance of the mechanical aspects of the system are overlooked. Most PV modules are designed to last 20 years or longer. It is important that the other components in the system, including mechanical components, have lifetimes equivalent to those for the PV modules. It is also important that the mechanical design requirements of the system be consistent with the performance requirements as well as with the operational requirements of the system. The mechanical design of PV systems cuts across a variety of disciplines, most notably civil and mechanical engineering and, to a lesser extent, materials science, aeronautical engineering, and architecture. More specifically, mechanical design involves:

- Determining the mechanical forces acting on the system
- Selecting, sizing, and configuring structural members to support these forces with an adequate margin of safety
- Selecting and configuring materials that will not degrade or deteriorate unacceptably over the life of the system
- Locating, orienting, and mounting the PV array so that it has adequate access to the sun's radiation, produces the required electrical output, and operates over acceptable PV cell temperature ranges
- Designing an array support structure that is aesthetically appropriate for the site and application and provides for ease of installation and maintenance

Each of these elements of the mechanical system will be discussed in more detail throughout this chapter.

5.2 IMPORTANT PROPERTIES OF MATERIALS

5.2.1 Introduction

Before discussing the mechanical design process, it is useful to review the properties of materials, especially the non-PV materials that are important components of PV systems. The properties of materials can be grouped into four general categories: electrical, mechanical, chemical, and thermal [1].

Electrical properties of semiconductor materials were introduced in Chapter 3 and are discussed in greater detail in Chapters 10 and 11.

The mechanical properties most important to PV systems are those associated with the strength of the structural members, including PV modules, in response to static and dynamic forces.

The most important chemical properties are those related to material degradation and deterioration due to corrosion and exposure to ultraviolet radiation. Also of interest are the rates of chemical degradation and the associated reduction in lifetimes of various materials, such as those used for weather sealing and insulation, that are caused by repeated exposure to high operating temperatures.

The thermal properties of most concern involve thermal expansion and contraction and the resulting thermal stresses.

5.2.2 Mechanical Properties

The PV system, in particular the PV array and its structural support members, is subjected to a variety of mechanical forces, both static and dynamic. These forces produce internal stresses and deformations. For common structural materials such as steel and aluminum, there are limits to these stresses and deformations that, if exceeded, may result in failure or irreparable damage.

Stress is defined as force per unit area. Uniform normal stress occurs when a force P is normal to and distributed uniformly over a cross-sectional area A. It can be calculated using the simple equation

$$S = \frac{force}{area} = \frac{P}{A}. \tag{5.1}$$

Why is the concept of stress important for the mechanical design of PV systems? Because it is the ability of a structural member to withstand stresses (i.e., forces per unit area) that determines its strength. It is the engineer's responsibility to first calculate or estimate the forces acting on the PV system and then to select and size the structural support members such that the maximum stresses experienced are well below allowable limits.

The most common static forces acting on PV arrays and supporting structures are due to the weight of the modules, mounting system and, in colder climates, snow and ice. These forces produce a combination of uniform normal, shear, and bending stresses that must not be overlooked in assessing the structural integrity of the array and all of its supporting hardware. Heavy accumulations of snow and ice can produce high stress levels. However, in the overwhelming majority of cases, these types of static forces are not large enough to exceed the stress limits of the array structure.

In addition to static forces, the PV array and its support structure will experience dynamic forces, most notably wind loads. Changes in wind direction and wind speed, including the rapid changes associated with gusting, complicate the job of the structural engineer. First, the changes in wind direction often result in a structural member experiencing alternating periods of tension and compression. Consequently, the structural members must be designed not only to withstand uniform tensile stress, but also not to buckle under compression. Second, the dynamic loads may give rise to a phenomenon known as fatigue failure. This occurs when the stresses in a structural member alternate between tension and compression over time and can occur at stress levels considerably below those that would produce a static stress failure. Fatigue first manifests itself as cracks on the surface of the structural member, often near a bend, angle, or other location where the stress may be concentrated. Following the initiation of these surface cracks, the fracture spreads inward over the cross section of the member until the load-carrying area of the member is reduced sufficiently to cause failure [1].

Figure 5.1 illustrates qualitatively the relationship between the fatigue stress limits, S_{max}, and the number of stress cycles experienced, N, for both ferrous (i.e., containing iron) and aluminum alloys. Note that ferrous materials, such as the steel alloys, have a well-defined fatigue (or endurance) limit. As long as the oscillating stresses experienced are below that limit, the number of cycles can increase indefinitely. A general rule of thumb for ferrous materials is that the fatigue limit is approximately half the tensile strength. This limit is also referred to as the reverse fatigue limit [1].

Fatigue is a bigger problem for aluminum and its alloys than for steel. No well-defined fatigue limit exists for aluminum. Because of this, the fatigue limit for an aluminum alloy is arbitrarily defined as a stress level for which the material can withstand a very large number of cycles (e.g., a million alternating tension and compression stress cycles).

What does this mean for the PV system design engineer? Obviously, problems with fatigue vary with geography. For areas prone to high, gusty winds, fatigue must be factored into the design. For coastal areas susceptible to hurricanes, buildings and

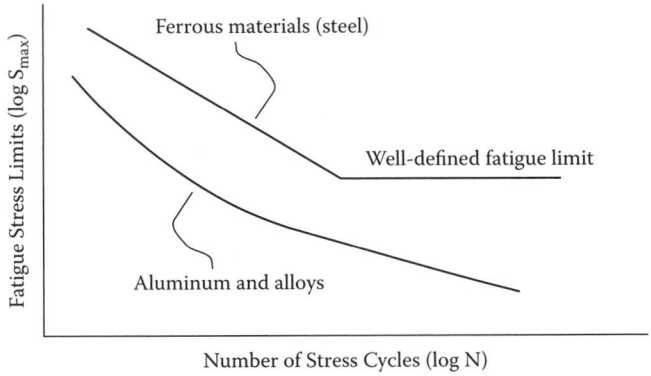

Figure 5.1 Fatigue stress versus cycles showing a well-defined fatigue limit for ferrous metals. (Adapted from Hanks, *Materials Engineering Science: An Introduction*, Harcourt, Brace & World, New York, 1970. With permission.)

other structures must meet stringent requirements resulting in more robust designs and higher strengths. Consequently, the stresses carried by these hurricane-tolerant buildings and structures typically will be well below the fatigue limits for the materials used.

In addition to stress, another important concept in discussing the strength of materials is strain, ε. Whereas stress is a measure of force intensity (i.e., force per unit area), strain is a measure of deformation per unit of length and can be defined by the equation

$$\varepsilon = \frac{elongation}{length} = \frac{\delta}{L} \tag{5.2}$$

where δ = elongation and L = length.

The cause-and-effect relationship between stress and strain should be intuitive to many. The intensity of the force, that is, stress, causes the structural member to deform, that is, strain. However, engineers need a more useful and quantitative relationship between stress and strain.

5.2.3 Stress and Strain

Figure 5.2 shows a bar of length L and cross-sectional area A. Force P acts uniformly over the cross-sectional area, putting the bar in tension. The experiment is simple: gradually increase the force P from small to larger values until the bar ruptures. During the experiment, the force and the elongation will be measured continuously. Remembering that stress is simply force per unit area and strain is elongation per unit length, stress can be plotted versus strain. The resulting graph from this experiment for a typical material such as steel is shown in Figure 5.3, which should be familiar to anyone who has studied strength of materials.

Note that the stress and resulting strain are directly proportional to each other up to a limit, appropriately named the proportional limit. If the force P is increased to produce stresses above the proportional limit, the stress–strain relationship is no longer linear. However, the bar may still be elastic to a slightly higher stress called the elastic limit. If the force P is removed at any point up to and including the elastic limit, the bar will return to its original dimensions. If the bar is stressed beyond the elastic limit, permanent deformation occurs. In addition to the elastic

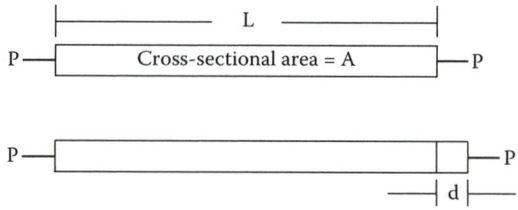

Figure 5.2 Simple pull test of a bar showing the resulting elongation.

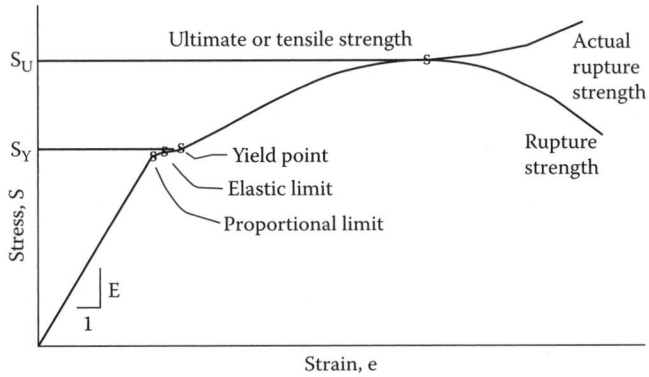

Figure 5.3 Stress–strain curve for steel.

limit, the yield point is defined as the point on the stress–strain curve corresponding to a specified permanent deformation (usually when the elongation per unit length equals 0.002). The yield point is used because it is easier to determine than the elastic limit for some materials. The stress corresponding to the yield point is defined as the yield strength.

Other key locations on the stress–strain curve include the maximum stress point on the graph, called the ultimate tensile strength, or, simply, the tensile strength, and two rupture strengths. The lower value for rupture strength results from using the original cross-sectional area A, while the higher value at the end of the dashed curve results from using the reduced area measured after rupture. Note that as material is stretched, its cross-sectional area decreases, resulting in higher stress for the same applied force.

In addition to the proportional limit, elastic limit, yield strength, ultimate strength, and rupture strength, another important parameter obtained from the stress–strain curve is the slope of the straight-line portion of the graph from the origin to the proportional limit. This slope is defined as the modulus of elasticity (also known as Young's modulus) and is usually represented by the letter E. It represents the ratio of stress to strain in the linear portion of the graph, which is where most structural materials are designed to operate. In equation form, it can be expressed as

$$E = \frac{stress}{strain} = \frac{S}{\varepsilon}. \tag{5.3}$$

A better name for E would be modulus of stiffness. The higher the value of E, the less the material deforms for a given stress. For a spring, the spring constant is analogous to the modulus of elasticity. The higher the spring constant, the harder you have to pull it to stretch it a given amount (i.e., the stiffer it is). Equation (5.3) is known as Hooke's law. For most materials, the modulus of elasticity is the same for both tension and compression.

Referring to Figure 5.3 once again, the area under the stress–strain curve has special physical significance in that it is proportional to the total energy required to rupture the bar. This energy is referred to as the toughness of the material.

For the simple case of one-dimensional, normal stress that has been discussed thus far, a useful formula for the elongation of the bar can be obtained by substituting for S Equation (5.1) and for ε Equation (5.2) into Equation (5.3), yielding:

$$\delta = \frac{PL}{AE}. \tag{5.4}$$

This equation shows that the amount of elongation depends on the force applied, P, the dimensions of the bar (length L and area A), and the bar's material, as represented by its stiffness E.

The normal stresses considered up to this point were one dimensional and were the result of normal forces acting uniformly over the cross-sectional area. In addition to normal forces, other forces act along or parallel to the area resisting the forces. These forces are called shear forces. Shear forces per unit area are called shear, or tangential, stresses. Figure 5.4 shows examples of structural members and fasteners subjected to forces that produce shear stresses.

In general, both normal and shear stresses are present in structural members. As with the modulus of elasticity for normal stresses, there is an analogous modulus of elasticity in shear, but it is more commonly referred to as the modulus of rigidity and is represented by G. Table 5.1 presents the modulus of elasticity, E, and the modulus of rigidity, G, for common structural materials.

The normal and shear stresses discussed up to this point are important and occur frequently in structural support members for PV systems. In computing these stresses, forces are assumed to be uniformly distributed over the cross-sectional

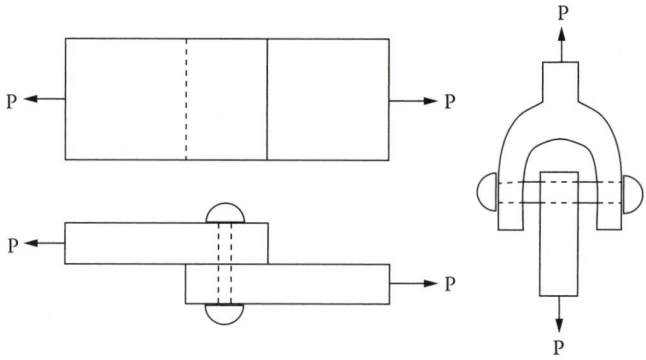

Figure 5.4 Two examples of shear loading.

Table 5.1 Moduli of Elasticity and Rigidity for Structural Materials

Metal	*E* (psi)	*G* (psi)
Stainless steel, 18–8	27.6×10^6	10.6×10^6
All other steels	$(28.6–30.0) \times 10^6$	$(11.0–11.9) \times 10^6$
Cast iron	$(13.5–21.0) \times 10^6$	$(5.2–8.2) \times 10^6$
Aluminum alloys, various	$(9.9–10.3) \times 10^6$	$(3.7–3.9) \times 10^6$
Titanium (99.0 Ti), annealed bar	$(15–16) \times 10^6$	6.5×10^6

Source: From Avallone and Baumeister, Eds., *Marks' Standard Handbook for Mechanical Engineers,* 10th ed., 1996. Reproduced with permission of the McGraw-Hill Companies.

areas resisting the forces. However, other forces may act on the structure and produce stresses that vary over the cross section. Examples include:

- Bending (or flexural) stresses. As an example, imagine a person standing on the end of a diving board. The top surface of the diving board would be in tension, the bottom in compression, and the normal stresses along the mid-horizontal plane would be zero.
- Torsional shear stresses. Imagine the torque applied to a drive shaft of an automobile. The shear stresses across the section of the shaft vary from zero at the center to a maximum at the outer radius.
- Combined stresses consisting of axial loading, bending, and torsion. All three of these types of loading may occur in various members of the PV array and support structure.

5.2.4 Strength of Materials

Referring again to Figure 5.3, the two most important values on the graph for most applications are the yield strength and the ultimate tensile strength. Although Figure 5.3 was representative of steel under tension, stress–strain curves for other materials and for different types of loading, such as compression and shear, are available [2,3]. With access to these values, how does the engineer use them to make sure the structure is sufficiently strong?

The first thing the engineer has to decide on is the allowable or working stress. This is the maximum safe stress that the material should carry. For PV systems, the allowable stress should be well below the yield strength and proportional limit. A common formula for computing the allowable stress is:

$$S_a = \frac{S_y}{N}, \tag{5.5}$$

where S_a is the allowable stress, S_y is the yield strength in psi, and N is a safety factor to compensate for any minor deviations of materials from the ideal.

For PV systems, N should be at least 2 to avoid accidental overloading. In areas where fatigue may be a problem due to high, gusty winds, a higher value should be used. Also, when fatigue is a concern, the structural engineer will often choose steel over aluminum for the support structure.

Table 5.2 Strength of Common Structural Materials

Metal	Ultimate Tensile Strength (psi)	Yield Strength (psi)
Cast iron	18,000–60,000	8,000–40,000
Wrought iron	45,000–55,000	25,000–35,000
Structural steel, ordinary	50,000–65,000	30,000–40,000
Stainless steel, 18–8	85,000–95,000	30,000–35,000
Aluminum, pure, rolled	13,000–24,000	5,000–21,000
Aluminum alloy, 17ST	56,000	34,000
Titanium 6–4 alloy, annealed	130,000	120,000

Source: From Avallone and Baumeister, Eds., *Marks' Standard Handbook for Mechanical Engineers,* 10th ed., 1996. Reproduced with permission of the McGraw-Hill Companies.

As an example, suppose the structural designer wants to use an aluminum alloy with yield strength $S_y = 34{,}000$ psi. Using an appropriate factor of safety of $N = 2$, the allowable stress, S_a, would be $S_y \div 2$, or 17,000 psi. This assumes fatigue is not a concern.

Table 5.2 shows the ultimate tensile strength and yield strength for selected metals. Note that titanium is more than twice as strong as steel.

5.2.5 Column Buckling

Often, structural members used to support a PV array have one dimension, their length, which is much larger than the other dimensions. Such members or columns, if long and slender enough, can fail due to buckling under compressive loading. These compressive forces may be considerably smaller than those that would cause failure due to crushing. Although there is no firm rule, generally a long, slender structural member is considered to be a column if its length is more than ten times as large as its least lateral dimension. For a column made of a material with a modulus of elasticity, E, a minimum moment of inertia (of its cross-sectional area), I, and length, L, there exists a critical compressive force, P, that will cause the column to buckle. This critical load is calculated using the formula

$$P = \frac{EI\pi^2}{L^2}. \tag{5.6}$$

This equation is known as Euler's formula [4].

5.2.6 Thermal Expansion and Contraction

Changes in temperature cause materials to expand or contract. For example, the linear deformation of a structural member of length L can be represented by

$$\delta_t = \alpha L \Delta T, \tag{5.7}$$

where α is the coefficient of linear expansion and ΔT is the change in temperature with units consistent with the units of α.

If the deformation is allowed to occur freely, no stresses will be induced in the structure. But if this deformation is restricted or constrained, internal thermal stresses result. The general procedure for handling thermal stresses is as follows [4]:

- Assume the structure is relieved of all applied forces and constraints.
- Allow the temperature deformations to occur freely and calculate the expansion or contraction using Equation (5.7).
- Show these deformations on a sketch.
- Determine the mechanical forces necessary to restore the structure to its constrained condition by solving Equation (5.4).
- Show these forces and deformations on the same sketch used previously.
- Use the equations of static equilibrium, thermal deformation, force deformation, and the definition of stress to compute all unknowns.

The computed stresses from using this procedure are the thermal stresses.

Example

Consider the case of a steel bolt in a bronze sleeve as shown in Figure 5.5. At 40°F, the sleeve fits snugly between the head of the bolt and the steel nut, but with no stress in any of the three parts. For the following dimensions and material properties, compute the thermal stresses in both the bolt and the sleeve when the temperature rises to 160°F:

D_1 = nominal diameter of bolt = inside diameter of sleeve = 1.00 in.
D_2 = outside diameter of sleeve = 1.50 in.
E_1 = modulus of elasticity for steel = 30 x 10^6 psi.
E_2 = modulus of elasticity for bronze = 10 x 10^6 psi.

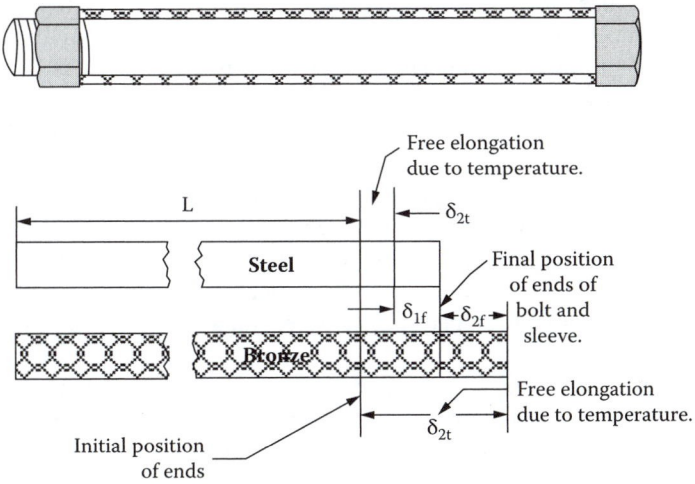

Figure 5.5 Elongations in steel and bronze due to a change in temperature.

α_1 = coefficient of thermal expansion for steel = 6.1 x 10^{-6} per °F.
α_2 = coefficient of thermal expansion for bronze = 10.1 x 10^{-6} per °F.

To solve this problem, note first that the elongation of the bronze sleeve would be greater than that for the steel bolt if both were unconstrained. For any length L, these elongations could be calculated using Equation (5.7). However, the bolt head and nut constrain the sleeve so that the elongations of both the sleeve and bolt must be equal. Hence, the sleeve is in compression and the bolt is in tension. Also, from the principle of action and reaction, the compressive force in the sleeve must be equal and opposite to the tensile force in the bolt.

From the sketch in Figure 5.5, the total elongations of the bolt and sleeve are, respectively,

$$\delta_1 = \delta_1\,(\text{thermal}) + \delta_1\,(\text{forces}) = \delta_{1t} + \delta_{1f} \qquad \text{(steel bolt)}$$

$$\delta_2 = \delta_2\,(\text{thermal}) - \delta_2\,(\text{forces}) = \delta_{2t} - \delta_{2f} \qquad \text{(bronze sleeve).}$$

Equating the forces, elongations, and applying Equations (5.4) and (5.7) yields

$$(\alpha_1)(L)(T_2 - T_1) + PL/[(A_1)(E_1)] = (\alpha_2)(L)(T_2 - T_1) - PL/[(A_2)(E_2)].$$

The length, L, is a common term and cancels. Solving for the force, P, gives $P = 3{,}330$ lb.

The normal stresses in the bolt are computed using Equation (5.1) and are

$$S_1 = 4240 \text{ psi (tension) and } S_2 = 3390 \text{ psi (compression).}$$

5.2.7 Chemical Corrosion and Ultraviolet Degradation

A PV system structure that can initially withstand all anticipated mechanical and thermal stresses with an adequate margin of safety might not be a good engineering design. The engineer must also consider the environment in which the system operates and how its component materials interact with the environment and with one another. Failure of metal components by corrosion is as common as failure due to mechanical stresses.

To design for corrosion resistance, knowledge of the various forms of corrosion is necessary.

Uniform attack is the most common form of corrosion. Chemical or electrochemical reactions proceed uniformly over the entire surface area. The metal becomes thinner and thinner and eventually fails. On a tonnage basis, uniform attack is the greatest cause of metal destruction, especially steel. However, it is not particularly difficult to control through proper material selection and appropriate use of protective coatings.

Galvanic corrosion occurs because of the potential difference that exists between two dissimilar metals in contact with a corrosive or conducting solution. Of the two dissimilar metals, the less resistant, or anodic, metal is corroded relative to the more resistant, or cathodic, metal. Table 5.3 presents an abbreviated version of the galvanic series for commercially available metals and alloys. The relative position in the galvanic series depends on the electrolytic environment and on the metal's surface

Table 5.3 Galvanic Series for Metal and Alloys

Most Noble Metal (Cathodic)	Platinum
	Gold
	Titanium
	Silver
	316 Stainless steel
	304 Stainless steel
	410 Stainless steel
	Nickel
	Monel
	Cupronickel
	Cu-Sn bronze
	Copper
	Cast iron
	Steel
	Aluminum
	Zinc
Least noble metal (anodic)	Magnesium

Source: Adapted from Dieter, *Engineering Design: A Materials and Processing Approach*, 2nd ed., 1991. Reproduced with permission of the McGraw-Hill companies.

chemistry. To minimize galvanic corrosion, the design engineer should use similar metals or metals close to each other in the galvanic series.

The use of a small anodic metal in contact with a cathodic metal of larger surface area should be avoided. If two metals far apart in the galvanic series must be used in near contact with each other, they should be electrically insulated from each other. Coating the anodic material may not protect it, because coatings are susceptible to pinholes, causing the coated surface to corrode rapidly in contact with the large cathodic area. If a galvanic couple is unavoidable, a third metal that is sacrificial to both of the other metals may be used. Note that zinc is sacrificial to both aluminum and steel and, consequently, is often used to protect these two most common structural materials.

Crevice corrosion is an intense form that frequently occurs at design details such as holes, gasket surfaces, lap joints, and crevices under bolts and rivet heads. Small quantities of stagnant liquid often form in these areas and allow this very destructive electrochemical process to occur. Unfortunately, stainless steel is especially susceptible to crevice corrosion.

Pitting is an extremely localized attack that produces holes in the metal. It is serious because it may lead to very premature failure of structural members. Pitting may require a relatively long initiation period but, once the process is started, it accelerates rapidly.

Intergranular corrosion is a localized attack along the grain boundaries of metals but not (or only slightly) over the grain faces. It is common in steel that has been heat-treated at high temperatures or heat-sensitized during welding. Intergranular corrosion due to welding is known as weld decay.

Selective leaching is the preferential removal of one or more of the alloying elements in a metal by the action of an electrolyte. The most common example of this phenomenon is the selective leaching of zinc from brass, leaving a spongy, weak matrix of copper—a process known as dezincification. Aluminum, iron, cobalt, and chromium are also susceptible to leaching. Where and whenever selective leaching occurs, the process leaves the alloy in a weakened, porous condition.

Erosion corrosion occurs when corrosive fluid flows over a metal surface and causes the gradual wearing away of the surface. Usually, the velocity of the fluid is high and mechanical wear and abrasion may be involved.

Stress corrosion cracking is caused by the combination of tensile stress and corrosion and leads to the cracking or embrittlement of a metal. The stress may result from applied forces or may be residual. Only specific combinations of alloys and the chemical environment produce stress corrosion cracking. Combinations include aluminum alloys and saltwater, some steel alloys and saltwater as well as mild steel and caustic soda. Preventing stress corrosion cracking involves selecting alloys that are not susceptible to cracking under expected operating conditions. If this is not possible, the structural members should be sized for low stress levels.

Dry corrosion involves reaction of the material directly with air. Virtually every metal and alloy reacts with the oxygen in air to form an oxide. This occurs in the absence of any liquid electrolyte and is an important form of corrosion in high-temperature applications.

In addition to corrosion, ultraviolet radiation can cause degradation and deterioration of some of the exposed materials used in PV systems. These include weather-sealing materials, conduit for wiring, wire insulation, some coatings and caulking, and other miscellaneous materials.

Ultraviolet radiation is that part of the electromagnetic spectrum having wavelengths between about 100 and 400 nm. The ultraviolet portion of the spectrum can be further subdivided into three regions: UV-A (315–400 nm), UV-B (280–315 nm), and UV-C (100–280 nm). UV-B and UV-C cause sunburn (erythema) and pigmentation (tanning). UV-B produces vitamin D_3. Long-term exposure to UV radiation results in loss of skin elasticity. Also, a link has been established between exposure to UV wavelengths below 320 nm and skin cancer. Because the human body is seriously affected by UV exposure, it should come as no surprise that other materials are also affected. In fact, ultraviolet radiation degrades both the optical properties and the physical properties of many materials. Consequently, the design engineer should ensure that all materials in the PV system that are exposed to sunlight, such as wire, conduit, and connectors, are resistant to UV degradation.

5.2.8 Properties of Steel

Iron is the base element of all steels. Commercially pure iron contains only about 0.01% carbon and is relatively soft and weak. The addition of carbon significantly strengthens iron. For example, the addition of 0.80% carbon can raise the tensile strength from about 40,000 psi to 110,000 psi [6]. Steel is an alloy of iron and carbon and usually contains small amounts of manganese and other elements. There are more than 3,500 grades of steel, of which 75% have been developed over the last 20 years. The many different grades of steel have many different properties: physical, chemical, and environmental. Carbon steel is steel that owes its distinctive properties to the carbon it contains. Alloy steel is steel that owes its distinctive properties to one or more elements other than carbon, or to the combination of these elements with carbon.

For any material, steel or otherwise, it is easy to measure its chemical composition and to determine its stress–strain curve using a pull test. The latter provides the yield strength, tensile strength, elongation, and reduction in area. However, the PV engineer must not forget to consider costs, corrosion resistance, and the possibility of structural fatigue as well.

Low-carbon steel is often used in the support structures for PV systems. Of the many low-carbon steel products, sheet and strip steels are becoming increasingly important and presently account for approximately 60% of the total steel production in the United States [2]. The relatively low cost of these steel products gives them a major advantage over more expensive aluminum, provided they can be protected adequately from corrosion.

Many different methods are used to protect steel from corrosion. In general, these protection or reduction methods include proper selection of materials, design, coatings, inhibitors, anodic protection, and cathodic protection [2]. One method of protecting steel structural members is hot-dip galvanizing. Hot-dip galvanizing is a process in which thoroughly cleaned steel or iron is immersed in molten zinc and withdrawn to provide a smooth, even coating that typically has a crystalline appearance. It is by far the most widely used method for protecting steel against corrosion [6].

Zinc is a less noble metal than steel (see Table 5.3) and is sacrificed to protect the steel as part of the corrosion process. How long the zinc protects the steel depends on the thickness of the zinc coating and the environment to which the structural member is exposed. The average minimum weight of zinc coating is 1.5 ounces per square foot of surface area when hot-dip galvanized according to the specifications of ASTM A123. For highly acidic industrial atmospheres, a 1.5 ounce per square foot coating may last less than 15 years. For a rural, clean atmosphere, the same coating may last 40 years.

The largest use of zinc is for protectively coating steel. Lead is sometimes added to the zinc to produce a surface pattern called *spangle*, which is often used on unpainted surfaces. The addition of aluminum to zinc improves its corrosion protection ability. *Galfan*, which contains 5% aluminum and 95% zinc, and *Galvalume*, which contains 55% aluminum and 45% zinc, are examples of zinc-aluminum coatings that effectively protect steel from corrosion [2].

Paints, lacquers, coal-tar or asphalt enamels, waxes, and varnishes are all used as organic coatings for corrosion protection. However, all paints are permeable to water and oxygen to some degree. They are also subject to mechanical damage and eventually break down.

5.2.9 Properties of Aluminum

Aluminum is the second most commonly used structural material next to steel. Some of the properties of aluminum that make it attractive for use in PV systems follow:

- It is very lightweight, with a density about one-third that of steel.
- It can be fabricated into many different shapes using a variety of different methods.
- It has a wide range of properties and is available at many different tensile strengths, depending on how it is alloyed.
- It has a high strength-to-weight ratio, thus making it attractive for many applications.
- It has good corrosion resistance and can be used in a wide variety of climates and weather conditions.
- It is available in a wide variety of finishes, making it attractive as an architectural metal.

Aluminum is one of the most fascinating of all the elements. It is very chemically active, with a strong affinity for oxygen. In powdered form, it is used as a fuel for solid rocket motors. How can it be that such a chemically active material is so commonly used in an unprotected form?

What makes aluminum unique is its special relationship with oxygen, which destroys other reactive metals such as sodium and seriously attacks less reactive metals such as iron. In the laboratory, most chemistry students have witnessed the spectacular reaction between sodium and the oxygen in water, in which the sodium is left unrecognizable and even more unusable. And everybody frequently encounters rusting steel. Neither of these phenomena occurs with aluminum. Rather, as soon as aluminum is produced, its surface reacts with the oxygen in the air and forms a very thin oxide layer over the entire surface. This oxide layer is hard, tenacious, and not only protects the aluminum from attack by other chemicals, but also from further oxidation. Because the oxide layer is transparent, it does not detract from the metal's appearance. Aluminum oxide must be removed from aluminum wire before connections are made.

Pure aluminum is relatively weak and only about one-third as stiff as steel. However, aluminum can be strengthened significantly by alloying. The most common alloying additions to aluminum are copper, manganese, silicon, magnesium, and zinc.

Aluminum and its alloys are divided into two major classes: wrought and cast. The wrought class is broad because aluminum can be formed by virtually every known fabrication process, including sheet and plate, foil, extrusions, bar and rod, wire, forgings and impacts, drawn or extruded tubing, and others. Cast alloys are poured molten into sand (sand casting) or into high-strength steel molds (i.e., permanent molds or die casting) and allowed to solidify into the desired shape. For PV systems, wrought aluminum is more commonly used than cast aluminum. The Aluminum Association has a designation system for wrought aluminum alloys that categorizes them by major alloying additions. Table 5.4 shows the various designations.

Table 5.4 Designation System for Wrought Aluminum Alloys

Alloy Series	Description or Major Alloying Element
1xxx	99.00% minimum aluminum
8xxx	Other element
2xxx	Copper
3xxx	Manganese
4xxx	Silicon
5xxx	Magnesium
6xxx	Magnesium and silicon
7xxx	Zinc
9xxx	Unused series

Source: Adapted from Kutz, *Mechanical Engineers' Handbook*, 2nd ed., 1998. Adapted with permission of John Wiley & Sons, Inc.

The first digit in the series classifies the alloy by alloy series, or principal alloying element. The second digit, if different from 0, denotes modification of the basic alloy. The third and fourth digits together identify the specific alloy within the series.

Alloys in the 6xxx alloy series, which use magnesium and silicon as alloying elements, possess a combination of properties, including corrosion resistance, that make them well suited for structures and architectural applications. They are also commonly used for marine applications, truck frames and bodies, bridge decks, automotive structures, and furniture. From the 6xxx series, 6061 and 6063 aluminum alloys are often used as structural members for PV arrays. 6061 aluminum contains 1.0% magnesium and 0.6% silicon. For the same tempering during fabrication, 6061 aluminum alloy has higher tensile strength than 6063 aluminum, which contains 0.7% magnesium and 0.4% silicon [3]. However, 6063 aluminum has better resistance to corrosion.

5.3 ESTABLISHING MECHANICAL SYSTEM REQUIREMENTS

5.3.1 Mechanical System Design Process

Mechanical system design is the *process of selecting, sizing, and configuring* a variety of structural and other components to meet predetermined design requirements. For most PV system designs, these components are usually available "off the shelf," although in some cases completely new components may need to be developed. The challenge is to bring together *components* of the appropriate materials, geometric shapes, and sizes to produce a *system* that best meets the design objectives.

The design requirements for the mechanical system consist of the following:

- Functional requirements
- Operational requirements
- Constraints
- Trade-offs

Developing specifications for each of the above four categories is not unique to PV system design. It is commonly used in the design of systems in general, including aircraft, satellites, and other products of modern society [7]. Determining the information to include in each of the four categories is an important responsibility assumed by the systems engineer at an early stage in the design process.

5.3.2 Functional Requirements

What are the primary functions of the mechanical parts of the PV system? First and foremost, the mechanical system must be capable of carrying all expected mechanical forces with an adequate margin of safety. In fact, the most important requirement for both the mechanical and electrical design of the PV system is the safety of individuals and the protection of structures and other property that could be damaged as the result of mechanical failure. The functional requirement related to structural considerations might take several forms, such as the *specification of factors of safety, or maximum allowable stresses, or limits on deformation.* This specification will affect the materials selected, as well as the geometry and size of the structural members.

For most designs, the PV array and support structure must withstand all forces and associated stresses to which they are subjected for 20 to 30 years or longer. Often, they must do so in harsh weather and chemical environments. These environments cause many materials to degrade, deteriorate, and eventually fail. Consequently, the *specification of the lifetimes of the structural members and other materials* that make up the PV system is an important functional requirement. This specification will affect not only the materials selected, but also the selection of protective coatings that may be necessary to safeguard these materials from corrosion and other forms of environmental degradation over the anticipated life of the system.

Some PV systems track the sun, using either single or double axis tracking. The function of the tracking subsystem is to provide increased exposure to the sun's radiation. Therefore, another type of functional requirement is the *specification of the required motion of the tracking system.* This specification is especially important for concentrating arrays and affects the design of the drive mechanism for the tracker.

5.3.3 Operational Requirements

Operational requirements involve the more important aspects of the design that affect the people who install, own, operate, maintain, repair, and/or service the PV system. An example of operational requirements would be *specification of the maximum labor hours required to install* a PV system and *specification that fuses or circuit breakers be easy to reach.* These types of specifications are becoming increasingly important as industry tries to reduce the costs of PV system installations. It has a major effect on the simplicity of the design and the ease of assembly of the system.

A specification related to accessibility, especially accessibility to individual modules and other critical components for inspection, maintenance, and repair, is another type of operational requirement. It affects the size and shape of selected mechanical components and how they are configured.

For some PV system designs, the specification of security requirements may be necessary. This type of specification is used to protect against vandalism, theft, and personal injury and affects the selection and configuration of components. For example, the operational requirement to protect PV modules from flying objects has led to the frequent use of shielding on the back of small arrays for area lighting systems. Security requirements have also led the engineer to designs that include buried or tamper-proof enclosures.

Another type of operational requirement is the specification of periodic inspection and maintenance. For example, the requirement for seasonally adjusting array tilt angles affects the selection and configuration of structural support components. Anticipated rain, snow, ice, dust, and the presence of birds, predisposed to target arrays with their droppings, are all examples of conditions that have affected the operational specifications for PV system designs.

5.3.4 Constraints

Invariably, the process of engineering design is constrained or limited in some way. The most common design constraint is *cost*. In the highly competitive renewable energy marketplace, it is important to keep costs to a minimum. Cost constraints have a major effect on system design. For example, aluminum may be the preferred structural material for the mechanical system because of its resistance to corrosion. But steel may be the material of choice because it is considerably less expensive, even after it has been hot-dip galvanized. Cost constraints are omnipresent and pervasive, and they challenge the ingenuity of the design engineer. Typically, the PV system must be designed and installed within a specified budget.

Another design constraint is time. Good design, in all of its creative aspects, takes precious time—and sufficient time is not always available to satisfy the inner needs of the design engineer. Typically, the PV system must be designed and installed according to a specified schedule.

From time to time, the design engineer will be confronted with other types of constraints. Limited access to the sun due to existing or imminent shading (e.g., due to proposed construction) is commonplace. Restrictive building codes, covenants, and zoning may prohibit the installation of a PV system or limit the design options. The same can be said for political, regulatory, and institutional issues. And finally, the effects of weather, location, and site layout often need to be considered before getting too far into the design process.

In summary, constraints are a reflection of the real world in which the engineer operates. They usually have a major effect on design decisions and, consequently, should be included in the specification of design requirements. A clear understanding of these constraints will assist the engineer in producing a better mechanical design.

5.3.5 Trade-Offs

Specifying the functional and operational requirements establishes the desired objectives for the mechanical system design. In a sense, it is a sophisticated wish list: the structure should be strong; the materials should last a long time; they should resist corrosion and other forms of degradation; and the array should have a certain orientation or possibly track the sun. In addition, the system should be easy to assemble in a short period of time; important components should be readily accessible to work on; it should be protected from vandals and thieves; and it should be easy to maintain. On the other hand, a listing of the constraints emerges: the costs are too high; the schedule cannot be met; the array must be moved to avoid shading; restrictive covenants may be violated; the grid interconnection requirements are unreasonable; and the weather is marginal for solar. In short, because of the constraints, the desired objectives cannot be met, and it is time to compromise.

When the desired objectives cannot be met, the following items are subject to compromise: quality or performance, cost, and schedule. There is an old saying: "Pick any two and trade down on the third." Assuming the cost and schedule constraints are rigid (they may not be), the engineer has to decide which trade-offs in quality and performance are acceptable. Examples of trade-offs to reduce costs and operate within budget include using steel rather than aluminum, using lower factors of safety, using lower-grade steel, using less expensive corrosion inhibitors, eliminating tracking, using less expensive and probably less skilled installers, and using less expensive security measures. Examples of trade-offs to meet schedule include substituting readily available structural members for preferred parts, eliminating certain design features, and using readily available labor.

5.4 DESIGN AND INSTALLATION GUIDELINES

5.4.1 Standards and Codes

Prior to the establishment of modern standards, structural members, nuts, bolts, screws, and so on were custom-designed and manufactured. For example, one manufacturer would produce ½ in. bolts with 9 threads per inch, while another would use 12 threads per inch. Some fasteners had left-handed threads and different thread profiles. In the early days of the automobile, mechanics would lay out fasteners in a row as they were disassembled to avoid mixing them during reassembly. A lack of standards leads to a lack of uniformity and precludes efficient interchangeability of parts. It is inefficient and costly.

A standard is a set of specifications for parts, materials, or processes intended to achieve uniformity, efficiency, and a specified quality. Another important purpose of standards is to limit the number of items in the specification and thereby provide a reasonable inventory of tooling, sizes, and varieties so that custom parts will not be required.

A code is a set of specifications for the analysis, design, manufacture, construction, or installation of something. The purpose of a code is to achieve a specified degree of safety, efficiency, and performance or quality. It should be noted that safety codes do not imply absolute safety, which is impossible to attain. Designing a structure to withstand 120 mph winds does not mean the designer thinks 140 mph winds are impossible. It just means the designer thinks they are highly improbable.

The following organizations are involved in developing standards and codes relevant to the mechanical design of photovoltaic systems:

- Aluminum Association (AA)
- American Institute of Steel Construction (AISC)
- American Iron and Steel Institute (AISI)
- American Society of Civil Engineers (ASCE)
- American Society of Metals (ASM)
- American Society of Mechanical Engineers (ASME)
- American Society of Testing and Materials (ASTM)
- Society of Automotive Engineers (SAE)

Building codes are design, construction, and installation guidelines. They are adopted and enforced by local jurisdictions to help ensure the safety of individuals and the protection of structures and other property in and around buildings. Building and electrical codes have value only if they are followed and enforced. The process begins prior to the installation of a PV system with an application for a building permit. After the installation, the building official inspects the system to determine compliance with the relevant codes. Unfortunately, many solar systems have been installed in the past without first obtaining a building permit. Most PV systems installed prior to the mid-1990s were stand-alone systems and probably would not meet present code requirements. However, with the renewed interest in PV systems for buildings as a result of financial incentive programs and government initiatives, most new systems comply with local codes. And utility-interactive PV systems are more likely to be inspected by building code officials than stand-alone systems.

5.4.2 Building Code Requirements

The most important standard affecting the mechanical design of PV systems is entitled *Minimum Design Loads for Buildings and Other Structures* [8]. It is a standard of the American Society of Civil Engineers and is periodically updated. It provides the minimum load-carrying requirements for buildings and other structures, and these requirements apply to PV systems. It is important to note that standards have meaning only when they are adopted by an enforcing jurisdiction and become part of their code. Many local jurisdictions reference parts of this ASCE standard in their building codes. However, local building codes are less uniform than local electrical codes, almost all of which use the *National Electrical Code®* (*NEC*).

The basic requirements of building codes are that the PV system, and any building or other structure to which it is attached, shall be designed and constructed to safely support any mechanical load or combination of loads to which they are subjected. In other words, the ASCE standard provides formulas for computing the mechanical loads from wind, snow, ice, and so on. The engineer computes these loads for the conditions expected at the site of the PV system. The stresses these forces produce must not exceed the appropriately specified allowable stresses for the structural materials being used.

In addition to establishing strength requirements, building codes often limit the deformations that can occur with physical structures such as PV systems. For example, there may be restrictions on the amount of deflection or lateral drift of a structure. Such a deflection might have an adverse effect on the use of the system or attached buildings or structures. Because vibration can lead to fatigue failure, the local building code may contain provisions that address the stiffness of the structure and the associated frequencies and amplitudes of structural vibration.

5.5 FORCES ACTING ON PV ARRAYS

5.5.1 Structural Loading Considerations

Because solar energy is a relatively dilute resource, PV modules and arrays are area intensive. These large-area solar electric devices are capable of transmitting a variety of forces to themselves, their support structures, buildings, and other frames or foundations to which they are mechanically attached. These forces include the dead weight of the array, the weight of array installation and maintenance people and their equipment, aerodynamic wind loading, seismic effects, and the forces due to rain, snow, and ice. Determining these forces under prescribed conditions can be an interesting and challenging mechanics problem.

Of the various types of forces mentioned earlier, aerodynamic wind loading presents the most concern. At most locations around the world, the force effects due to wind loading are much higher than the other forces acting on the structure. For example, both dead and live weight loads due to a PV array on a building are usually less than 5 pounds per square foot (psf). In contrast, the computed wind loads are typically between 24 and 55 psf, but sometimes greater than 100 psf, depending on location and the associated design wind speed for that location.

Because of the geographic variation in environmental conditions such as rain, snow, ice, wind speed, seismic activity, etc., local building codes tend to be much less uniform than electrical codes. Consequently, it is important for the PV systems engineer not only to become familiar with local building code requirements, but also to compute the minimum structural design loads that are relevant for a given location.

In addition to the aforementioned considerations, the design engineer must select and configure the array-mounting materials so that corrosion and ultraviolet degradation are minimized. Also, if the array is mounted on a building, the structural integrity of the building must not be degraded and building penetrations must be properly sealed so that the building remains watertight over the life of the PV array.

The ASCE standard [8] provides instructions and formulas for computing the following types of loads:

- Dead loads
- Live loads
- Soil and hydrostatic pressure and flood loads

- Wind loads
- Snow loads
- Rain loads
- Earthquake loads
- Ice loads—atmospheric icing
- Combinations of the preceding loads

5.5.2 Dead Loads

Dead loads consist of the weight of all the materials that are supported by the structural members, including the weights of the structural members themselves. As an example, consider a PV array mounted on the roof of a building. The modules are structurally tied together in panels using steel or aluminum structural members. The weight of the modules is carried by the structural members and transmitted to the roof structure of the building. The total dead load that the roof must carry is the combined weight of the modules, structural members used to form the panels, attachment hardware, and mounting brackets. These loads can be expressed in psf and assumed to act uniformly over the area covered by the array. Or the equations of static equilibrium can be used to determine the individual forces at each of the brackets that connect the panels to the roof. For PV systems, the dead loads are usually assumed to act uniformly over the supporting structure and are expressed in psf. Note that these are external loads—not internal stresses—even though similar units are used.

Dead loads for PV systems generally fall within the range of 2 to 5 psf and, even though they cannot be discounted, do not pose serious structural problems for the engineer.

5.5.3 Live Loads

Live loads associated with PV systems are those produced by individuals and their equipment and materials during installation, inspection, and maintenance. In designing structures to carry live loads, the engineer may treat them as uniformly distributed loads, that is, in psf, as concentrated loads (lb) or as a combination of uniformly distributed and concentrated loads. For PV arrays, live loads are usually assumed to be distributed uniformly and are small, on the order of 3 psf or less.

5.5.4 Wind Loads

The forces from the wind acting on PV arrays are aerodynamic forces. As such, their magnitudes depend partly on the properties of the atmosphere. Atmospheric properties include static pressure, temperature, viscosity, and density. Of these, density is one of the more important factors affecting the mechanical forces acting on PV systems. At any given location, density is usually assumed to be constant. However, density does vary with altitude—decreasing as altitude is increased. For example, the density in the mile-high city of Denver, Colorado, is about 86% of the atmospheric density for Florida cities, which are only slightly above sea level. The mass density of air for the standard atmosphere at sea level is 0.00256 lb-s^2/ft^4.

The air flowing over a PV array produces two types of forces: pressure forces normal to the surface and skin friction forces along the surface. Skin friction forces are simply surface shear forces resulting from the viscosity of the air as it contacts the surface.

Obtaining a better idea of the important relationships affecting the flow of air over structures requires some simplifying assumptions. First, assume the flow of air is inviscid, which implies that the viscosity of the air is zero. Next, assume that the flow of air is incompressible, which implies that the density of the air is constant. Finally, assume the flow of air is conservative, which implies that there are no other energy-dissipating processes. With these assumptions, the linear momentum equation for fluid flow reduces to [9]

$$\int \frac{1}{2} d(V^2) + \int g dz + \int \frac{dp}{\rho} = \text{constant}, \tag{5.8}$$

where V = wind speed, g = acceleration due to gravity, z = altitude or elevation, p = static pressure, and ρ = mass density.

Integrating each term yields

$$\frac{V^2}{2} + gz + \frac{p}{\rho} = \text{constant}. \tag{5.9}$$

Neglecting small changes in elevation in the vicinity of the PV system and rearranging terms yields

$$p + \frac{1}{2}\rho V^2 = \text{constant}. \tag{5.10}$$

Whereas p is the static pressure, the term $\frac{1}{2}\rho V^2$ is referred to as the dynamic pressure. This very important relationship between the two variables, static pressure and velocity, is known as Bernoulli's equation. It shows that an increase in velocity must be accompanied by a corresponding decrease in static pressure and vice versa.

To better understand the relationship between pressure and velocity, consider the case of uniform airflow over a cylinder as shown in Figure 5.6. Air flowing along the center streamline must slow down and stop as it rams into the cylinder at point A. At this point, the speed of the air is zero and point A is called a stagnation point. The velocity $V_A = 0$ and the pressure P_A is a maximum and substantially higher than the free stream pressure. Figure 5.6b shows an idealized pressure distribution on the surface of the cylinder. While the maximum pressure and minimum velocity occur at point A, the minimum pressure and the maximum velocity occur at point B. The pressure at point B is less than the free stream pressure P_∞ and represents the point of maximum uplift or suction. The pressure distribution shown in Figure 5.6b for the windward side of the cylinder is appropriate, but the pressure distribution shown for the leeward side is not. In the real world, friction and turbulence come into play

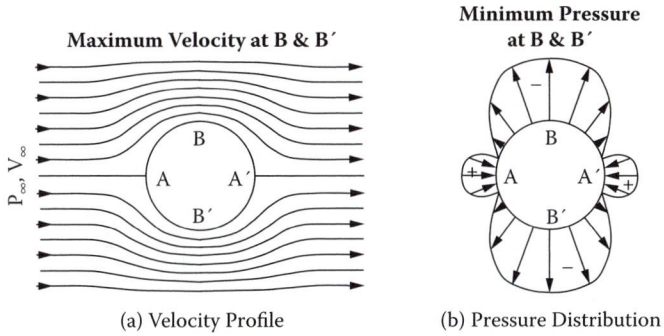

Figure 5.6 Idealized flow of air over a cylinder and the corresponding pressure distribution. (Based on Hurt, *Aerodynamics for Naval Aviators*, Navweps 00-80T-80, U.S. Navy, 1960.)

and cause the air to separate from the cylinder as shown in Figure 5.7. Note that this separation occurs at point C, leaving a trail of turbulent mixing downstream of the cylinder. Very little pressure recovery occurs on the right side of the cylinder, yielding a pressure distribution as shown in Figure 5.7b. This pressure distribution produces a resultant force, called drag, which acts to push the cylinder downstream. The cylinder is symmetric, but it is not streamlined. It experiences relatively high drag forces when immersed in flowing air.

Figure 5.8 shows another symmetrical body that has a streamlined shape. Streamlined bodies are shaped to reduce the separation of air, to maximize pressure recovery along their surfaces, and to produce relatively low drag forces. Note that for the same shape but different orientations to the flow, quite different pressure distributions are produced. Because they are streamlined, neither airfoil produces high drag forces. However, the airfoil tilted with respect to the oncoming air experiences

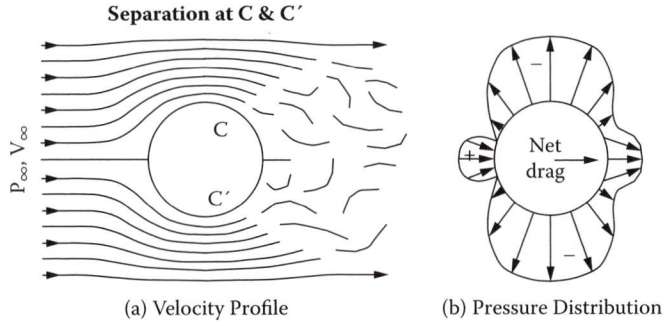

Figure 5.7 Actual air flow over a cylinder and pressure distribution, showing separation at C and C′. (Based on Hurt, *Aerodynamics for Naval Aviators*, Navweps 00-80T-80, U.S. Navy, 1960.)

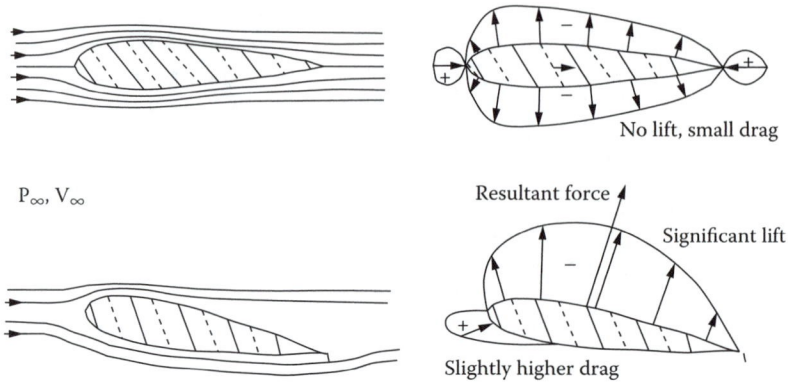

Figure 5.8 Airflow over symmetrical airfoils. The lower airfoil, which is at a positive angle of attack to the airflow, experiences significant lift. (Based on Hurt, *Aerodynamics for Naval Aviators*, Navweps 00-80T-80, U.S. Navy, 1960.)

a significant net upward force. This upward force, perpendicular to the direction of air flow, is called lift. The combination of lift and drag forces produces a resultant force on the airfoil.

To continue this line of thought, next consider two flat plates, similar to PV arrays, inserted into the stream of flowing air as shown in Figure 5.9. One of the flat plates has its area perpendicular to the direction of flow, while the other is cocked at an angle to the flow. The corresponding pressure distributions are shown to the right. The vertical flat plate produces an extremely high drag force and no lift. The other flat plate, however, produces a vertical force effect, in addition to fairly high drag. Actually, the vertical component of the resultant force can act either up (lift) or down (negative lift) depending on its angle of tilt.

How does all this relate to wind loading of PV arrays and support structures? Most PV arrays are essentially large flat plates that are tilted at various angles to

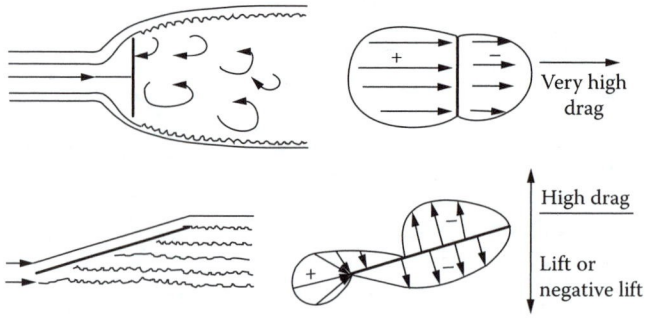

Figure 5.9 Flow over flat plates at different angles of attack to the airflow. The flat plate on the bottom may experience either an uplift or a downward force, depending on the angle of attack.

the direction of flow of the wind. Also, the buildings and structures to which they are attached and the terrain in their vicinity may significantly affect the flow around them. This flow often produces complex velocity and pressure distributions that, in turn, produce forces and stresses in the structural members. It should be obvious by now that the determination of wind loads on PV structures does not lend itself to theoretically derived solutions. Rather, aeronautical and civil engineers have together developed empirical formulas for estimating the structural loading resulting from exposure to various wind conditions.

At this point, it is useful to summarize the important variables and factors that affect the aerodynamic forces acting on PV arrays and structures. They include the following:

- Wind speed
- Effects of wind gusts
- Density of the air
- Orientation with respect to the wind direction
- Shape and surface area
- Elevation above the ground
- Effects of topographical and artificial features

The list is somewhat intimidating and shows that the best wind load analysis is not as accurate as some engineers would like it to be. Nonetheless, with "ballpark accuracy" and adequate factors of safety, good design is definitely within the engineer's grasp.

Figure 5.10 shows a view of a modular array field designed by Hughes Aircraft Company for Sandia National Laboratories [11]. Figure 5.11 shows the *net* pressure

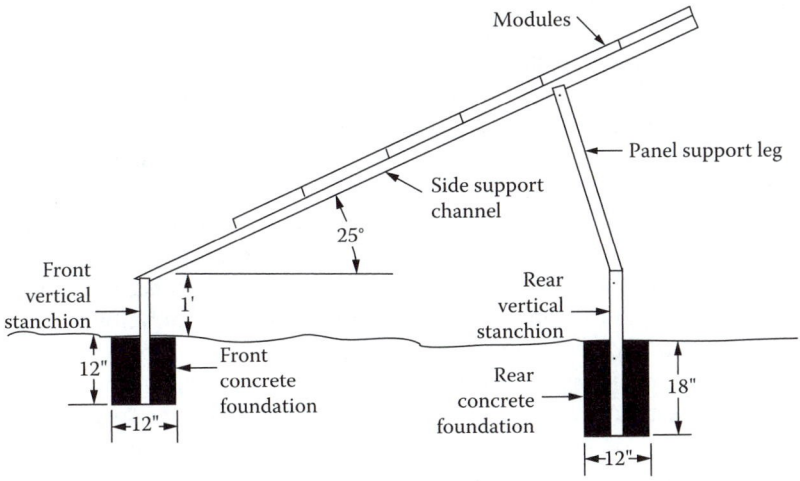

Figure 5.10 Sketch of the Hughes modular array field. This design was developed for Sandia National Laboratories. (Based on Modular Photovoltaic Array Field, prepared for Sandia National Laboratories, SAND83-7028, Hughes Aircraft Company, 1984.)

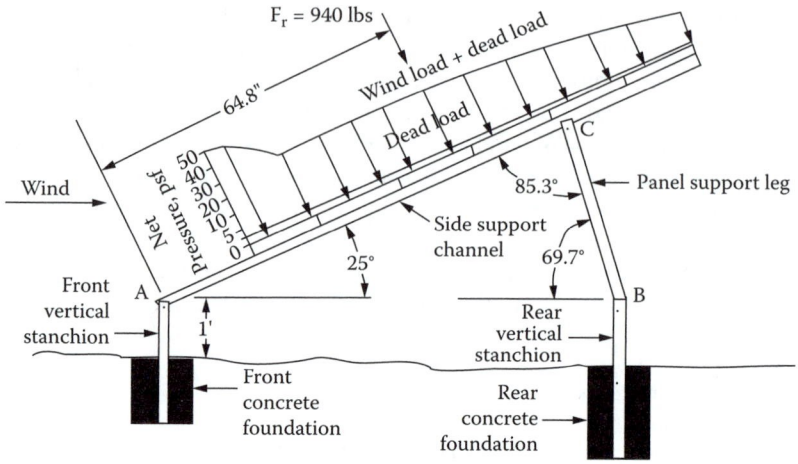

Figure 5.11 Net pressure on the array due to 100 mph winds toward the front of the array. Structural member BC is in compression.

distribution on the ground-mounted array and support structure for the combined wind and dead loads. The wind speed is 100 mph and is directed toward the front of the array. The resultant force acts down and puts structural member BC in compression. Hence, the structure must be analyzed for buckling under this loading. From the pressure distribution, based on wind tunnel testing, the forces and stresses in all the structural members can be computed.

Figure 5.12 shows the same array structure, but with the wind directed toward the back of the array. Note that the resultant force is trying to lift the array from its supports, and the structural support member BC is in tension.

Work done by Boeing showed that for single arrays, the wind forces were a minimum at array tilt angles of about 20° above the horizontal [12]. Surprisingly, maximum wind forces occur at array tilt angles of 10 to 15°, and again at 90°. At 10 to 15° tilt, the array acts as a reasonably efficient airfoil and large aerodynamic lift forces result. For tilt angles of 20° and higher, the air separates from the array—similar to the flow over the wing of an aircraft when it is going into a stall [13].

Ground clearance also affects wind loading. For example, it has been found that increasing the ground clearance from 2 to 4 ft increases the normal wind forces by 10% to 15% for an array with an 8 ft panel height [14].

For systems consisting of multiple rows of subarrays, wind tunnel tests have shown that the interior rows experience several times less wind loading than the exterior rows [15]. Consequently, windbreaks or wind fences are sometimes used to reduce the wind loading on the exterior rows.

Because of their large exposed surface areas, their elevation off the ground, and their orientation with respect to wind direction, PV arrays are often subject to exceptionally high mechanical forces. Not only do the modules in the array have to resist these forces, but also the attachments to roofs and/or other structures must be well secured.

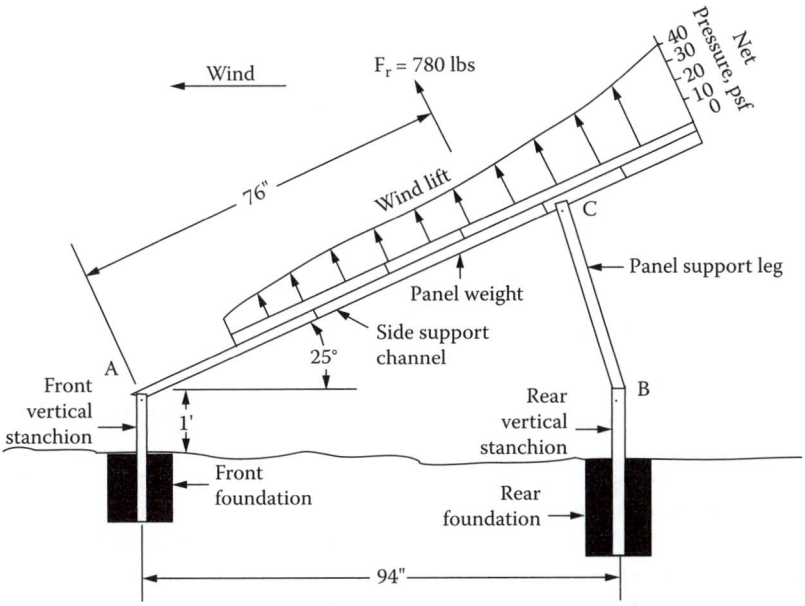

Figure 5.12 Net pressure distribution due to 100 mph winds toward the back of the array.

In designing PV arrays and structures to meet wind loads, the engineer must complete the following steps:

1. Establish the basic wind speed.
2. Determine the velocity pressure.
3. Determine the gust effect factor.
4. Determine the appropriate pressure or force coefficient.
5. Determine the wind loads on the array.
6. Determine the forces acting on critical members and attachment points.
7. Establish an appropriate factor of safety based on risk.
8. Select structural members and attachment hardware that can carry all loads with the prescribed factor of safety.

Fortunately, the American Society of Civil Engineers (ASCE) has developed standards and procedures for completing the foregoing steps [8]. For example, the ASCE has developed maps indicating the basic wind speeds throughout the United States (step 1); formulas, tables, and charts for computing the velocity pressure (step 2); gust effect factors for various downtown, urban, rural, and flatland sites (step 3); force coefficients for different shapes and structures (step 4); and equations for computing the wind loads (step 5). From these wind loads (forces per unit area) on the array, the equations of static equilibrium are used to compute the forces acting on the structural members (step 6). Knowing these forces, the engineer selects

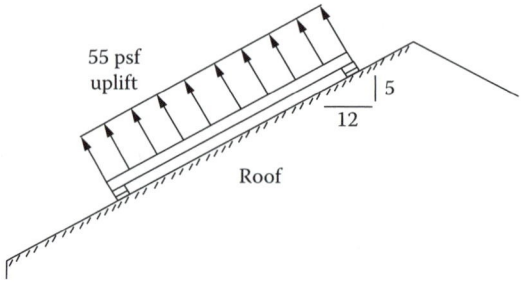

Figure 5.13 Standoff-mounted array with an uplifting pressure distribution.

the materials, structural shapes, and attachment hardware that meet all design requirements.

The ASCE procedures may be used to analyze the wind loads on standoff-mounted rooftop arrays, as shown in Figure 5.13. Wind directions from both the front and rear of the array-roof structure should be considered, resulting in both uplifting and downward pressure distributions. Uplifting pressure is usually of greater concern because of the tensile and withdrawal forces it exerts on supporting hardware.

In the ASCE standard, the highest basic wind speed in the United States is 150 mph. The velocity pressure, q, is computed using the equation

$$q = 0.00256\, K_z\, K_{zt}\, K_d\, V^2\, I, \tag{5.11}$$

where:

K_z = velocity pressure exposure coefficient at height z
K_{zt} = topographical factor
K_d = wind directionality factor
V = basic wind speed in mph
I = importance factor

Using the equations, tables, and charts in the standard yields

$K_z = 0.85$
$K_{zt} = 1.00$
$K_d = 0.85$
$I = 1.00$
$q = 42$ psf

The design wind pressure, p, is computed using:

$$p = q\, G\, C_f, \tag{5.12}$$

where:

G = gust effect factor = 0.85
C_f = force coefficient = 0.70 (for a normal gable or hip roof)
$p = 25$ psf

As an example, for an array having an area of 200 sq. ft., the total uplifting (resultant) force acting on the array would be 25 psf × 200 sq. ft. = 5000 lb. Knowing this resultant force, the design engineer can now determine the number of attachment points and the size of the mounting hardware necessary to safely carry this load.

Rather than computing the pressure distribution for each standoff array and roof combination, a more prudent approach is to assume a worst-case scenario for wind loading and design a mounting system capable of surviving the associated loading. Such a scenario might include maximum expected wind speeds (say 150 mph), array tilt angles that produce maximum uplifting pressure distributions (say 10° to 15°), topographical and shape effects that increase the pressure loading on the array and mounting system, and an adequate factor of safety. Using this approach and applying the methodology in the ASCE standard yields a reasonable upper limit on the design wind pressure of 55 psf.

Many manufacturers of PV modules supply structural members to support their arrays with an adequate factor of safety. In most cases, a registered professional engineer has performed a structural analysis based on a worst-case scenario such as that described earlier. However, such analyses are usually not site specific. Note that the array will be physically attached to and supported by a roof or some other type of structure at a specific location. It is important that the entire site-integrated system, including the attachment hardware and the roof or support structure to which the array assembly is attached, be capable of surviving the worst-case scenario of wind loading—not just the array and mounting members that come with the PV system package.

In summary, the largest mechanical forces acting on PV arrays and structures are due to high wind speeds. Suppliers of packaged PV systems should assume a worst-case wind loading of 55 psf (or higher) in designing their array mounting systems. Many have already done so. In addition, the entire integrated PV and building system must be able to survive the worst-case scenario of wind loading and comply with all site-specific building code requirements.

5.5.5 Snow Loads

Snow loads on PV arrays and on sloped roofs decrease as the tilt angles and slopes increase. There are three reasons for this. First, as the tilt angle increases, the component of the weight force parallel to the array surface increases relative to the component normal to the array surface, thus helping to shed the snow. Second, wind forces tend to shed the snow from the array. Third, melt water reduces the friction between the array surface and helps shed the snow. Most PV arrays have "slippery" glass surfaces that also aid the snow-shedding process. On the other hand, obstructions and rough surfaces hinder the shedding of snow and its weight. For example, some tile roofs contain built-in protrusions or rough surfaces that prevent the snow from sliding. The snow that accumulates on arrays mounted on such roofs may be obstructed from shedding [8].

An extreme case of snow loading occurs when unusually large amounts of snow accumulate on an array at a relatively small tilt angle. However, this is rare because shallow tilt angles are uncommon in colder, snow-prone parts of the world. In most cases, snow loads on PV arrays are less than 8 psf.

5.5.6 Other Loads

The PV array and support structure may experience mechanical forces due to other loads, such as rain, ice, hydrostatic pressure, and seismic activity. The latter is critical in earthquake-prone regions, and the ASCE has detailed procedures for designing for seismic loads.

Foundations, footings, or slabs that support PV arrays may experience uplift forces due to hydrostatic pressure in locations where groundwater is close to the surface. If expansive soils are present, the hydrostatic pressures can be quite large. In such cases, the expansive soils are sometimes excavated to a depth of at least 2 ft below ground level, followed by back filling with freely draining sands and gravel.

Regardless of the location, the PV engineer is responsible for ensuring that all of the relevant loads and load combinations have been properly accounted for in the mechanical system design.

5.6 ARRAY MOUNTING SYSTEM DESIGN

5.6.1 Introduction

Previous sections of this chapter discussed various properties of materials, the strength of materials, functional and operational requirements, standards and codes, and the mechanical forces acting on the array and support structure. Emphasis was placed on material selection and compatibility based on concerns for structural integrity, safety, and code compliance. This section focuses on those aspects of the array mounting system that affect costs, performance, maintenance, and appearance.

5.6.2 Objectives in Designing the Array Mounting System

In addition to meeting structural, code, and safety requirements, good design of the array mounting system should

Minimize installation costs
Enhance array performance
Provide reasonable accessibility for installation and maintenance
Make the system aesthetically appropriate for the site and application

5.6.2.1 Minimizing Installation Costs

Both good design and good installation practices can be employed to reduce installation costs. Studies in the 1990s by Sandia National Laboratories have shown that only about half the cost of PV systems is for modules. The other half is for nonmodule components and labor, with labor being the major contributor. For example, suppose a journeyman electrician installs a PV system. The electrical contractor may charge $75 per hour for the electrician's time. The value of reducing the number of hours for installation should be obvious. PV manufacturers and system suppliers can reduce installation costs by doing one or more of the following:

1. *Developing packaged kits:* For about the first two decades of applying PVs to terrestrial applications, system design was essentially customized for each installation. Since the mid-to-late 1990s, the trend has been to develop packaged kits that can easily be adapted to a wide variety of installations. Building applications of PV have helped to accelerate this trend. PV system suppliers now deliver standard models, which include modules, the structural support systems, attachment hardware, batteries, and inverters. Some include the module interconnect wiring, junction boxes, and other balance-of-system components. The development and refinement of packaged kits should help reduce the cost of PV system installations.
2. *Minimizing the total number of parts.* A second design approach to reducing installation costs is to minimize the total number of parts included in the design. As an example, some suppliers are moving toward the use of larger modules and fewer panels. The idea is to use less skilled personnel to assemble and prewire panels prior to shipment to the site. This reduces the time required by more highly skilled personnel to complete the installation at the site.
3. *Minimizing part variations.* A third design approach to reducing installation costs is to minimize part variations. This concept is easily applied to the size and type of structural members, fasteners, and attachment hardware. It often eliminates the need for special tooling.
4. *Designing parts to be multifunctional.* Designing systems such that some parts are multifunctional is a fourth approach to cost reduction. For example, a structural member may be used not only to carry the mechanical loads, but also to enhance the flow of air over the back surface of the array.
5. *Designing parts for multiple uses.* Structural members, brackets, attachment hardware, and fasteners should be designed for possible use with different kits, model numbers, and product lines.
6. *Designing for ease of assembly.* This not only reduces the time required to install systems but also often results in higher system reliability and reduced system maintenance.
7. *Avoiding separate fasteners.* One popular approach to reducing installation costs is to use slotted roof brackets that allow PV modules or panels to be simply dropped into the slots. It eliminates the need for on-site attachment of nuts and bolts to connect the panels to roof brackets. Another alternative approach is to use captive fasteners.
8. *Minimizing assembly direction.* This means that parts should be designed so that they fit or can be assembled only one way: the correct way. This is common practice in many industries.
9. *Maximizing compliance in assembly.* The installer should not have to struggle to assemble the panel framework, attach the modules to the panel, or attach the panels to the support structure. To maximize compliance in assembly, the designer should avoid tolerances that are too tight. Parts should fit together with a minimum of effort.
10. *Minimizing handling in assembly and installation.* This concept follows from a number of the other items, but also includes locating holes, slots, and other points of attachment to reduce handling.

In addition to the design practices suggested earlier [5], good installation practices can also be employed to reduce costs. The practice of preassembling and prewiring equipment off-site using less expensive personnel has already been mentioned. Another consideration for roof-mounted arrays is the number, type, and placement of roof attachment points. Often, this decision is made by the installer at the site rather than by the designer, and serious trade-offs must be considered. For example, minimizing the number of roof attachments saves time, but it also means that individual attachment points will be subjected to larger forces.

5.6.2.2 Building Integration Considerations

As the population grows and building density increases, space becomes more valuable, and rooftops become more attractive locations for PV arrays. The question arises: How can we most effectively integrate PV systems and buildings? For existing buildings, the most prudent approach is to use designs that minimize costs by incorporating the measures discussed earlier.

Building-integrated PV (BIPV) products are solar electric devices that replace conventional building materials. They include solar shingles, tiles, skylights, windows, overhangs, facades, and PV laminates bonded directly to metal roofing panels. They can be used on either new or existing buildings in a manner that is aesthetically pleasing and very appealing to buyers.

For broad acceptance by builders and homeowners, more BIPV products that are attractive and easily incorporated into building construction are needed. As these new BIPV products achieve attractive combinations of costs and durability, they should further stimulate a growing market for rooftop applications.

However, widespread BIPV applications may require more than just new BIPV products. They may also require buildings with new features that make them more receptive to PV systems. Examples include innovative roof, truss, and electrical system design. In addition, the integration of on-site generation and super-energy-efficient buildings offers the prospect of alleviating national energy problems—from both the supply and demand sides of the problem. The U.S. Department of Energy's Zero Energy Homes Program is a manifestation of the level of interest in this approach.

For new manufactured buildings, one interesting approach to building integration involves factory-installed PV systems. Approximately 30% of new housing stock in the United States is manufactured housing. To be effective, the manufacture of buildings in a factory requires careful planning, time-motion studies, and process optimization: areas of interest to industrial engineers. Not only are the buildings manufactured quickly, but also quality control becomes an ongoing part of the process. Opportunities exist to include PV systems in the building manufacturing process. The idea is to accelerate the installation process, reduce costs, and improve quality control. The complete PV system can be installed in the factory, part of it can be installed, or the manufactured building can be specially configured to more easily accept a PV system. The latter approach has been successfully demonstrated and involves installing the following items in the factory prior to shipping the building to the site: double roof trusses (to both increase load-carrying ability and allow for the use of larger mounting brackets); wiring conduit from the roof through the attic and walls to the crawl space under the building; a rain-protected opening and flashing (i.e., vent stack flashing) to accommodate wiring from the array into the conduit; and a prepared, structurally enhanced location for mounting the inverter. Once the building has been delivered and set up on the site, installation of the prewired array panels, inverter, and other balance-of-system components is achieved at a fraction of the time and cost of a custom installation [16].

5.6.2.3 Costs and Durability of Array-Roof Configurations

Life-cycle costing is used in economic analyses of renewable energy technologies because these technologies realize their value over time. They are usually characterized by high capital costs and low operating costs. For PV systems on buildings, the combined array-roof configuration is a major contributor to the life cycle cost. It is important that these configurations last 20 to 30 years. However, this is not always achieved without unexpected added cost. For example, asphalt shingles often have to be replaced about every 10 years in warm, humid climates. The life-cycle cost analysis should include the added cost of removing and reinstalling the PV array when reroofing.

The durability of some of the relatively new BIPV products is also a concern. For example, PV laminates bonded to either structural or architectural standing seam metal panels produce architecturally attractive arrays, and installation costs have been decreasing.

5.6.3 Enhancing Array Performance

The mechanical system can affect the array performance in several ways:

- Increasing the amount of incident solar radiation
- Avoiding shading
- Allowing the array to operate at lower cell temperatures
- Protecting the array from vandalism

5.6.3.1 Irradiance-Enhancement

Irradiance enhancement can be achieved by optimizing the array orientation, tracking the sun, and using concentrating collectors. The optimum array orientation depends on the type of system, application, and end user. For example, to maximize annual energy production, the optimum tilt angle is about 90% of the latitude of the site, and the optimum azimuth is true south for the northern hemisphere and true north for the southern hemisphere.

However, utility companies may be more interested in peak load reduction than total energy produced. Consequently, the optimum azimuth angle might point the array more closely to the west than to the south (in the Northern Hemisphere), because utility peak loads generally occur during summer afternoons.

Concentrating and tracking arrays work best in areas receiving a higher-than-average percentage of direct radiation, such as the desert southwest United

States, and may not be an attractive option in many locales. As discussed under the section on design requirements, the engineer must weigh the trade-offs.

5.6.3.2 Shading

Closely related to irradiance enhancement is the need to avoid shading. The performance characteristics of shaded arrays vary, but PV arrays usually are more seriously affected by shading than are solar thermal collectors. Consequently, shading should be avoided to the greatest extent possible.

5.6.3.3 Array-Cooling

PV modules, especially those using thick, crystalline cells, work better at lower cell operating temperatures. Consequently, the array mounting should be designed to allow air circulation along the back surface of the modules. Rack-mounted arrays typically operate at lower temperatures than other configurations. Experiments at Sandia National Laboratories show that the operating temperatures of direct-mounted arrays may be 18°C higher than the cell's nominal operating cell temperature (NOCT). For standoff-mounted arrays, a standoff height of 4 to 6 in., with no significant obstruction to airflow, will permit adequate passive cooling of the modules.

In general, the output power of thin film materials is degraded less by high operating cell temperatures.

5.6.3.4 Protection from Vandalism

Unfortunately, PV arrays are sometimes used as targets by vandals. Therefore, it may be necessary to include protection features. Back shields are often used for arrays on area lighting systems. Some of these vandal-guards do not permit adequate cooling of the array and can have a noticeable deleterious effect on performance. Back shields that allow more air circulation might be considered. For larger array fields, fencing, security lighting, motion sensors, and other protection measures are sometimes employed.

5.6.4 Roof-Mounted Arrays

PV array mounting can be categorized according to where the arrays are mounted, how they are supported, and whether they have a fixed or changing orientation. Arrays can be either roof-mounted or ground-mounted. Roof-mounted arrays typically use one of four different methods of support: rack, standoff, integrated, or direct. With the surge of interest in PV and building applications, better designs for each of these four roof-mounting approaches are being pursued.

5.6.4.1 Stand-Off Mounting

Stand-off arrays are mounted above and parallel to the roof surface, as illustrated in Figure 5.14. Stand-off mounts work well for buildings with sloping roofs. When

Figure 5.14 Standoff-mounted array using commercial array mount. The brackets that support the frame provide height to promote good air circulation under the array.

installing a standoff-mounted array, the PV panels or subarrays are often attached to the roof using point connections—usually at or near the corners of the panels or subarrays. At a minimum, the standoff height between the roof and the bottom of the module frame should be at least 3 in., but 4 to 6 in. is preferable. Lag screws that penetrate the roof rafters 2 to 3 in. may be used to fasten the mounting brackets to the roof. Spanner attachments, as shown in Figure 5.15, are stronger than lag bolts or lag screws and are commonly used in areas exposed to high winds or when no truss is located where the mounting bracket needs to be placed.

To promote passive cooling of standoff arrays, the engineer should consider the following designs:

- Designs that allow both lateral and vertical airflow along the back surface of the modules
- Designs that induce pressure differences between air inlet and exit regions
- Arrays with larger lateral dimensions than vertical dimensions (i.e., higher aspect ratios)

Figure 5.15 When trusses are not at locations where brackets are mounted, spanners can provide secure anchor points.

Figure 5.16 Example of rack mount on roof in 140 mph wind zone. Mounts are anchored to trusses below. (Photo courtesy of Vergona-Bowersox Electric, Inc.)

5.6.4.2 Rack Mounting

Rack-mounted arrays are above and tilted with respect to the roof, as illustrated in Figure 5.16. Rack mounts work well on flat roofs and roofs with a slope of 2 in 12 or less. They may be mechanically attached to the building structure or may employ ballast to resist wind and other mechanical loads. Rack mounts are usually subjected to higher structural loads, incur higher costs for mounting hardware, and are often less attractive than standoff mounts. However, for the same array area, the total energy output is often higher because of better orientation and lower average operating temperatures.

Ballasted rack assemblies offer the distinct advantages of simplicity and avoidance of roof penetrations, and should be considered if acceptable by local code jurisdictions. However, most rack mountings are firmly attached to the roof substructure. These are usually point connections, although several distributed attachment methods are sometimes used. Rack-mounted arrays typically run relatively cool compared with other mounts and they can reduce heat gain through roofs. Because the primary mechanism of rack array cooling is convection from the front and back surfaces of the modules, cooling can sometimes be enhanced by locating the rack in natural air channels and by reducing obstructions to airflow, such as screens, grates, and walls.

5.6.4.3 Integrated Mounting

For integrated arrays, also referred to as BIPVs or integral mounting, the array replaces conventional roofing or glazing materials as illustrated in Figure 5.17. Because integrated arrays replace conventional roofing and glazing materials, they become a significant architectural feature of a building and can be aesthetically very pleasing.

Figure 5.17 Solar tiles replace conventional roof tiles at a South Florida residence. Note that some tiles are located in roof zone 2, but will still withstand a 140 mph wind.

Dimensional tolerances may be tighter for some integrated arrays than for stand-off or rack mounts. If commercial curtain-wall glazing techniques are used, larger modules are preferable to smaller modules.

5.6.4.4 Direct Mounting

In direct mounting, the array is affixed directly to the roofing material or underlayment, with little or no airspace between the module and roof. Figure 5.18 illustrates a direct-mounted solar roofing tile.

Array operating temperatures for direct mounts are usually higher than for other mounting techniques. The use of direct mounting may become increasingly popular for new thin-film products that are not as sensitive to operating temperature. Some PV materials, such as solar shingles, fall into both the direct and BIPV mounting categories.

5.6.5 Ground-Mounted Arrays

Ground-mounted arrays are supported by racks, poles, or tracking stands. These arrays are secured to the ground to resist uplifting caused by wind loads. All ground-mounted arrays run relatively cool because good airflow is possible over both the

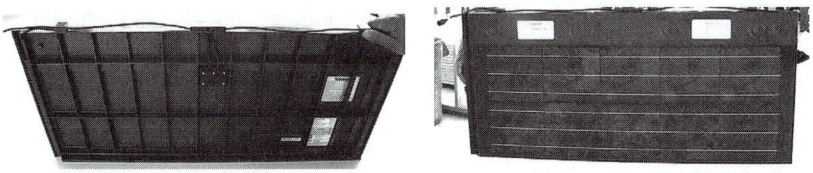

Figure 5.18 Front and back of solar roof tile. Note the 10 multicrystalline Si cells on the front view, which produce approximately 4 V @ P_{max}. It is one of a variety of new products that can replace conventional roofing materials.

Figure 5.19 Example of a rack mount at ground level. (Photo Courtesy of Solar Source, Inc.)

front and back surfaces of the modules. This cooling can be enhanced by minimizing obstructions to airflow such as shrubbery and fences.

5.6.5.1 Rack Mounting

Rack mounting is commonly used for mounting arrays on the ground. Simple structural hardware such as angles, channels, and metal tubing can be used for both small and large PV arrays. Most module manufacturers and equipment suppliers offer hardware for rack-mounting PV arrays on the ground. Figure 5.19 is an example of a rack-mounted array that has been designed for roof or ground mounting.

5.6.5.2 Pole Mounting

If the array consists of only a few modules, it can be mounted on a pole as shown in Figure 5.20. Depending on the number of modules and their height above the ground, the pole may need to be set in concrete to resist being overturned during windy conditions. Among others, PV-powered outdoor lighting is a good application for pole-mounted arrays.

5.6.5.3 Tracking-Stand Mounting

Because tracking arrays receive more sunlight than stationary arrays, their modules produce more energy. The value of a tracking array depends on the trade-off between

Figure 5.20 Pole-mounted arrays are often used for PV lighting systems.

additional energy produced and added cost and complexity. There are two types of trackers: active and passive. Active trackers use drive mechanisms, such as electric motors and gears, to point the array toward the sun. They may track about either one or two axes. The tracking motion may be controlled by a computer or by sun-seeking sensors.

Passive trackers normally track about only one axis and use a two-phase fluid, such as a refrigerant, that vaporizes and expands when it is heated by the sun. The expanding fluid causes the tracker to pivot toward the sun as the fluid's weight shifts from one side of the tracker to the other. An alternative design uses a hydraulic cylinder and linkage arrangement. For both tracker designs, sunshades are used to regulate the heating of the fluid and control the motion. A passive tracker is shown in Figure 5.21. Compared with a fixed-tilt array, these trackers show the greatest energy enhancement in the summer when the days are long and are less beneficial in the winter. Tracking arrays are often used for water pumping because they can produce high starting currents early in the day and because of the higher demand for water during the summer months.

5.6.6 Aesthetics

It is important for arrays on buildings to be aesthetically pleasing. They should be designed to blend into the building lines and colors. In designing PV arrays for building applications, the architect and engineer should consider the following suggestions:

- Mounting the array parallel to the roof (if appropriate)
- Using the roof dimensions to establish the array aspect ratio

Figure 5.21 Example of 2-axis tracker on pole mount.

- Avoiding harsh contrasts and patterns
- Making the mounting hardware as inconspicuous as possible
- Avoiding shading, even if appearance suffers somewhat

As mentioned previously, many of the new BIPV products add to the architectural attractiveness of buildings.

5.7 COMPUTING MECHANICAL LOADS AND STRESSES

5.7.1 Introduction

As indicated previously, the PV array is subjected to a wide variety of mechanical influences that affect the stresses and strains in the modules, structural support members, and attachment hardware. These include deadweight loads, live loads, wind loads, and snow and ice loads. In this section, we assume a pressure distribution on the array and discuss the methods used to calculate its effects on critical elements of the support structure.

5.7.2 Withdrawal Loads

A high percentage of standoff-mounted PV arrays are attached to the roof using lag screws or bolts. The lag screw typically passes through a hole in a mounting bracket, possibly a mounting pad, shingles, waterproof membrane, sheathing and, finally, into the truss, spanner, or other primary structural wood support member. The withdrawal strength of this and other attachment points is a function of the diameter of the lag screw, the length of thread embedded in the primary structural

Table 5.5 Allowable Withdrawal Loads for Lag Screws

Lumber	White Oak	Southern Yellow Pine	White Spruce	Douglas Fir
Specific gravity, G	0.71	0.58	0.45	0.41
Diameter, *D* (in.)	**Allowable Withdrawal Load, *p* (lb/in.)**			
1/4	381	281	192	167
5/16	450	332	227	198
3/8	516	381	260	226
7/16	579	428	292	254
1/2	640	473	323	281

Note: Values for *p* were computed using Equation (5.13) and include a safety factor of 4.5.

wood support, and the specific gravity of the wood. The allowable withdrawal load of lag screws and bolts per inch of embedded thread is given by [2]:

$$p = 1{,}800 D^{3/4} G^{3/2} \tag{5.13}$$

where:

p = allowable withdrawal load (lb./in.)
D = shank diameter of lag screw or bolt (in.)
G = specific gravity of oven-dry wood

Values for p computed using Equation (5.13) include a safety factor of 4.5. This relatively high value provides added protection just in case the actual wood used may have reduced strength due to variations in specific gravity, knots, grain irregularities, holes, and other defects [17].

The typical diameters of lag bolts or screws used to attach PV arrays are 1/4, 5/16, and 3/8 in., although they may be as large as 1/2 in. The specific gravities of various types of lumber can be found in the *Wood Engineering and Construction Handbook* [18]. Table 5.5 was developed using the preceding equation and the specific gravities for various types of lumber.

5.7.3 Tensile Stresses

In addition to withdrawal loads, uplifting pressure distributions produce tensile stresses in hardware used to secure the array to a roof or other support structure. This hardware could be lag screws, J-bolts, or threaded rods. The forces acting on the attachment hardware are computed using the equations of static equilibrium, and the tensile stresses are computed using Equation (5.1).

5.7.4 Buckling

In contrast to uplifting pressure distributions, forces distributed downward toward the plane of the array are usually not as serious a concern. However, an exception occurs when long, slender structural members are subjected to compressive forces that may result in structural instability known as buckling. In such cases, which may occur due to wind loads on rack-mounted arrays, the engineer must compute the critical buckling load using Equation (5.6), which can be expanded into the following form:

$$P = \frac{EI\pi^2}{L^2} = \frac{EAr^2\pi^2}{L^2} = \frac{EA\pi^2}{(L/r)^2} \tag{5.14}$$

where:

I = moment of inertia of the cross-sectional area = Ar^2
A = cross-sectional area of the column
r = radius of gyration of the cross-sectional area
L/r = slenderness ratio of the column

Note from the preceding equation that larger slenderness ratios and smaller cross-sectional areas yield lower critical buckling loads. If the length L of the structural member cannot be significantly changed, the engineer should consider selecting members with cross sections having larger radii of gyration. Conceptually, the radius of gyration is a measure of an equivalent distance from a reference axis at which the entire cross-sectional area may be assumed to be concentrated to produce the same moment of inertia. The type of cross section that yields higher radii of gyration is a hollow tube in which all the cross-sectional area is distant from the axis. Consequently, hollow metal tubing is often used for column support. Radii of gyration can be either calculated or obtained from tables for steel, aluminum, or other columns.

Example

Figure 5.22 shows a rack-mounted array supported by square 6063 T-6 aluminum tubing having cross-sectional dimensions of 1 × 1 × 1/8 in. The radius of gyration for this cross section is 0.361 in^2. The modulus of elasticity for the aluminum is 10×10^6 psi, and the cross-sectional area is 0.50 in^2. If the length of the rear structural support member is 8 ft (96 in.), compute the maximum force the member can withstand without buckling.

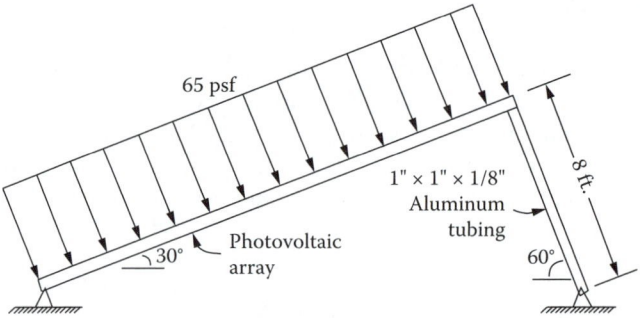

Figure 5.22 Rack-mounted array with rear support column in compression.

Substituting directly into Equation (5.14),

$$P = \frac{10^7(0.50)\pi^2}{(96/0.361)^2} = 698 \text{ lb.}$$

Thus, each rear support member should not be subjected to compressive forces that approach 698 lb.

5.8 STAND-OFF, ROOF MOUNT EXAMPLES

5.8.1 Introduction to ASCE-7 Wind Load Analysis Tabular Method

Figure 5.23 shows the PV array layout of Chapter 4, Section 4.5. Recall that this array is located in a hurricane-prone wind region, so the modules have been located in roof zone 1. Roof zones are identified in ASCE-7 [8] and are shown in Figure 5.24, along with the definition of the parameter, *a*, which defines the edges of the zones. For this example, it will be assumed that the roof has a 4:12 slope, the design wind velocity is 140 mph, and the roof, which is over an enclosed building, is located in a neighborhood characterized by Exposure B and has a mean roof height of 12 ft.

The exposure of a roof depends on the surface roughness of the land in the vicinity of the roof. According to ASCE-7, roughness B is characterized by areas with numerous obstructions the size of single-family dwellings or larger. Roughness C consists of flat, open country, grasslands, and all water surfaces in hurricane-prone regions. Roughness D has flat, unobstructed surfaces *outside* hurricane-prone regions.

Exposure B, then, has roughness B for a distance of 2600 ft upwind, unless the mean roof height is less than 30 ft, in which case roughness B prevails for at least 1500 ft upwind.

Exposure D has roughness D for a distance of 5000 ft upwind and distance of 600 ft downwind, even if the downwind roughness is B. Note, however, that Exposure D only applies to non-hurricane-prone regions because roughness D only applies to non-hurricane-prone regions.

Exposure C applies whenever exposure B or D do not apply. Thus, in a hurricane-prone region, any exposure different from Exposure B will necessarily be Exposure C.

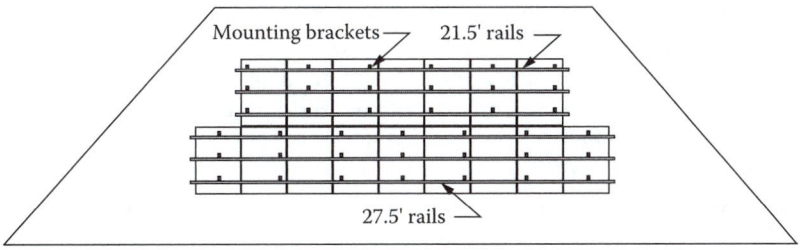

Figure 5.23 Mounting frame layout for high wind region.

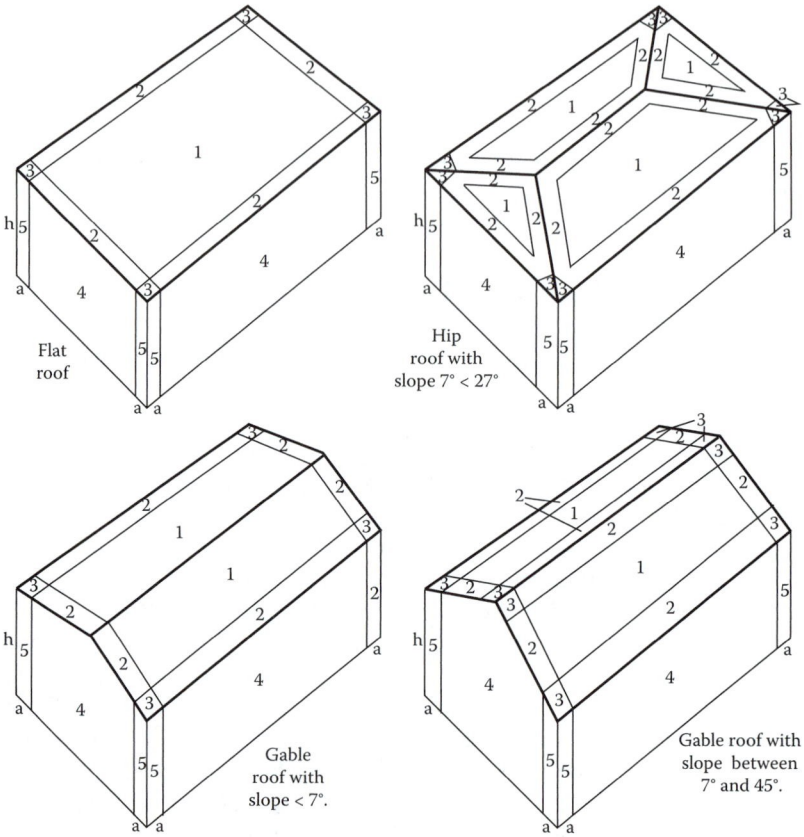

(Dimension "a" is 10% of least horizontal dimension or 0.4h, whichever is smaller, but not < 4% of least horizontal dimension or 3 ft)

Figure 5.24 Roof zones as defined by ASCE 7 for flat, hip, and gable end roofs. (Adapted from Figure 6.3, ASCE 7-05, with permission from American Society of Civil Engineers, Reston, VA. For educational purposes only. For design and permitting, refer to the current edition of ASCE 7.)

Chapter 6 of ASCE-7 presents three methods of determining wind load. One method involves table lookups and applies to a special class of "regular" buildings that are defined in the standard. Another method involves detailed calculations and can be applied to any structure. The third method involves wind tunnel testing and also can be applied to any structure. Because almost all enclosed buildings end up being classified as "regular," this example will assume the roof to be over a regular building. It will also use the table lookup method, because the method is very straightforward and also results in a conservative estimate of wind forces on the array. Because tables apply to a range of situations, the table entries must necessarily represent worst-case situations, so the designer can be comfortable that the table entries represent conservative estimates of wind forces on the array.

Table 5.6 Wind Pressures for Exposure Category B

Roof Slope	Roof Zone	EWA ft²[a]	Basic Wind Speed (Pressures in psf with + [Downward] and – [Upward]) 90 mph		100 mph		110 mph		120 mph	
0°–7°	1	10	5.9	–14.6	7.3	–18.0	8.9	–21.8	10.5	–25.9
	2	10	5.9	–24.4	7.3	–30.2	8.9	–36.5	10.5	–43.5
	3	10	5.9	–36.8	7.3	–45.4	8.9	–55.0	10.5	–65.4
7°–27°	1	10	8.4	–13.3	10.4	–16.5	12.5	–19.9	14.9	–23.7
	2	10	8.4	–23.2	10.4	–28.7	12.5	–34.7	14.9	–41.3
	3	10	8.4	–34.3	10.4	–42.4	12.5	–51.3	14.9	–61.0
27°–45°	1	10	13.3	–14.6	16.5	–18.0	19.9	–21.8	23.7	–25.9
	2	10	13.3	–17.0	16.5	–21.0	19.9	–25.5	23.7	–30.3
	3	10	13.3	–17.0	16.5	–21.0	19.9	–25.5	23.7	–30.3

Roof Slope	Roof Zone	EWA ft²[a]	Basic Wind Speed (Pressures in psf with + [Downward] and – [Upward]) 125 mph		130 mph		140 mph		150 mph	
0°–7°	1	10	11.4	–28.1	12.4	–30.4	14.3	–35.3	16.5	–40.5
	2	10	11.4	–47.2	12.4	–51.0	14.3	–59.2	16.5	–67.9
	3	10	11.4	–71.0	12.4	–76.8	14.3	–89.0	16.5	–102
7°–27°	1	10	16.2	–25.7	17.5	–27.8	20.3	–32.3	23.3	–37
	2	10	16.2	–44.8	17.5	–48.4	20.3	–56.2	23.3	–64.5
	3	10	16.2	–66.2	17.5	–71.6	20.3	–83.1	23.3	–95.4
27°–45°	1	10	25.7	–28.1	27.8	–30.4	32.3	–35.3	37.0	–40.5
	2	10	25.7	–32.9	27.8	–35.6	32.3	–41.2	37.0	–47.3
	3	10	25.7	–32.9	27.8	–35.6	32.3	–41.2	37.0	–47.3

Source: Excerpted from ASCE Standard, SEI/ASCE 7–05 *Minimum Design Loads for Buildings and Other Structures*, Figure 6.3. With premission of the American Society of Civil Engineers, Reston, VA, 2003.

[a] Effective wind area. Note that definition of EWA in ASCE-7 requires use of 10 ft² for all modules and arrays. The data in this table are for instructional use only. For design and permit documents, refer to the current edition of ASCE-7.

The basis of the table lookup of wind forces is Figure 6.3 of ASCE-7, part of which is shown as Table 5.6. The data of ASCE-7 apply to enclosed buildings, for components and cladding on walls and roofs in exposure category B, for roof heights less than 60 ft. The data are separated into three roof zones, with three ranges of roof slopes: 0°–7°, 7°–27°, and 27°–45°. Slopes and heights outside this range must be analyzed by either the computational method or the wind tunnel method. Pressures are tabulated for a wide range of wind speeds and are evaluated at mean roof heights.

5.8.2 Array Mount Design: High Wind Speed Case

The array mount will be based on a commercial mounting frame that has been designed to withstand wind speeds up to 125 mph in exposure category C. The mount uses two mounting rails under the modules for wind speeds less than 125 mph. Figure 5.23 shows the use of three mounting rails for the 140 mph wind region. Note that because wind pressures are proportional to the square of the wind speed, the wind pressure at 150 mph is $(150 \div 125)^2$ times higher than the 125 mph wind pressure, or 44% higher. But the third rail provides 50% more holding capacity, so the use of 3 rails on a 2-rail system extends the use of the mount to 150 mph winds. The specific forces will now be calculated for this array and the array mounting points.

Commercial array mounts are generally designed to have mounting feet located 4 ft apart along the rails. The proposed locations of the mounting feet for the rails of the example are shown in Figure 5.23. So the exercise here is to first determine the wind pressure on the array, then determine the total wind force on the array and, finally, determine the force that will be distributed among the mounting brackets.

The wind load can now be determined using the known information on roof slope, roof zone, and mean roof height. The pressures shown in Table 5.6 are valid for Exposure B for mean roof heights up to 30 ft. These tabulated pressures are based on an importance factor of 1.0 for the location of the roof. The 4:12 slope represents an angle of $\tan^{-1}(4/12) = 18.4°$, which falls within the 7°–27° range. Thus, because the mean roof height is less than 30 ft, the wind pressures at 140 mph in roof zone 1 are found from Table 5.6 to be 20.3/–32.3 psf. In this expression, the positive number represents the downward pressure on the array and the negative number represents the upward lift on the array. In other words, depending on wind direction, the pressure may vary between these two extremes.

Because there are two rows of modules, separate calculations of wind loading will be needed for each row. For the row of 7 modules, the total area is $35.2 \times 51.9 \times 7 \div 144 = 88.8$ ft^2. The total upward force on the array is $88.8 \times 32.3 = 2868$ lb. The maximum downward force on the array at 140 mph wind speed is $88.8 \times 20.3 = 1803$ lb.

Normally, a roof structure is designed to—and can easily—withstand the downward force, so the test is to ensure that the array can withstand the upward force without being lifted off the roof. This means that the 18 mounting brackets must be able to hold down the 2868 lb of upward pull. Although there will be some nonuniformity in the forces on each mounting bracket, it is acceptable practice to assume the load to be evenly distributed among the brackets. Thus, the force per bracket is simply $2868 \div 18 = 159$ lb/bracket. So the next check is to see how much thread must penetrate the roof truss to ensure adequate pull strength.

Most installers agree that for 2″ × 4″ or 2″ × 6″ trusses, a 5/16″ lag screw is a good choice for attaching a mounting bracket to the trusses. Thus, depending on the type of wood used in the trusses, Table 5.5 can be used to determine the length of lag screw needed to provide adequate pull strength to hold down the mounting bracket. In hurricane-prone regions, it is common practice to construct roof trusses of southern yellow pine, so it will be assumed that this will be the case in this example. Thus, using 332 lb/in. pull strength for that portion of the thread embedded in the truss, it is found that 0.480 in. of thread must penetrate the truss.

Of course, this does not mean the length of the lag screw only needs to be 0.5 in., because the screw must also pass through the thickness of the mounting bracket, the roof underlayment, the roof decking, and, perhaps, even some foam insulation, before it enters the truss. So the installer and designer need to verify the roof composition prior to selecting lag screws. Typically, the screw length will be 4 in., with 1.5–2 in. of thread embedded in the truss.

The bottom row of 9 modules in Figure 5.23 is held by 21 mounting brackets. Because the area of this section is 114.2 ft^2, the maximum upward force on each bracket is 175.6 lb. So for this row of modules, the lag screw threads must penetrate 0.529 in. into the trusses. Once again, if 1.5 in. of thread is in the truss, a comfortable safety factor is obtained.

5.8.3 Array Mount Design: Low Wind Speed Case

As a final example, consider the array layout of Chapter 4, Section 4.5, for the low wind speed case, as shown in Figure 5.25. It will be assumed that the roof measurements are the same, but the wind speed is now 90 mph. Two mounting rails will be used per row of modules, as shown in the figure. In this case, note that only 7 modules are completely in zone 1. Another 18 modules are partly in zone 1 and partly in zone 2. The remaining 6 modules bridge across all three zones. So what is done when modules cross zone boundaries?

It must be remembered that at the zone boundaries, the wind pressures undergo a gradual change, as opposed to undergoing an abrupt change. So if a very small fraction of a module that is predominantly in one zone encroaches on another zone, the additional wind pressure on the total area of the module is negligible. This would be the case for the two modules near the ends of the center row. For modules with major portions in each of two or three zones, however, the higher-pressure zones cannot be neglected.

One possibility is to determine the weighted average wind pressure on a module that bridges two zones. The more conservative possibility is to simply use the wind pressure in the higher-pressure zone as the pressure on the entire module. For example, the modules at the ends of the center row of modules are close to half in zone 1 and half in zone 2. So one might then use the average of the two pressures as the pressure on the modules. On the other hand, the top and bottom rows are substantially in zone 2, so it makes sense to simply use the zone 2 pressures for these modules.

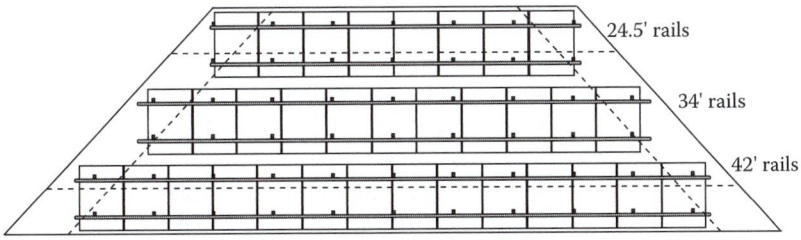

Figure 5.25 Lower wind speed region array mounting system.

From Table 5.6, the lifting components of the wind pressures on the modules are –13.3 psf for zone 1, –23.2 psf for zone 2, and –34.3 psf for zone 3. The total area of the top row of modules is 101.5 ft^2, the total area of the center row is 139.6 ft^2, and the total area of the lower row is 177.6 ft^2.

Using the conservative approach, the top row is broken down into 6 modules in zone 2 @ 23.2 psf and 2 modules in zone 3 @ 34.3 psf, which adds up to a total force on the 8 modules of $(6 \times 23.2 \times 12.7) + (2 \times 34.3 \times 12.7) = 2639$ lb. Distributing this total force among the 14 mounting brackets results in 188 lb/bracket.

The middle row has 9 modules in zone 1 @ 13.3 psf and 2 modules in zone 2 @ 23.2 psf, resulting in a total force on the 11 modules of $(9 \times 13.3 \times 12.7) + (2 \times 23.2 \times 12.7) = 2109$ lb, distributed among 18 mounting brackets. This equates to 117 lb/bracket. The bottom row has 12 modules in zone 2 @ 23.2 psf and 2 modules in zone 3 @ 34.3 psf, resulting in a total force of 4407 lb, or 200 lb/bracket. Because a rule of thumb in designing mounts is to be sure the pulling force per mounting bracket is less than 500 lb, it is evident that this array mount in this wind zone is adequate.

5.8.4 Exposure C, Exposure D, and Other Correction Factors

The simplified method of the previous section can be applied to other exposures and to heights between 30 and 60 ft by applying appropriate multipliers to the pressures presented in Figure 6.3 of ASCE 7. Table 5.7 indicates the multipliers to be used for various building heights and exposure categories.

If the building is near a hill (escarpment), additional multipliers apply. If the building is partially enclosed, or if the structure is open, such as a carport or walkway,

Table 5.7 Adjustment Factors for Building Height and Exposures

Mean Roof Height (ft)	Exposure		
	B	C	D
15	1.00	1.21	1.47
20	1.00	1.29	1.55
25	1.00	1.35	1.61
30	1.00	1.40	1.66
35	1.05	1.45	1.70
40	1.09	1.49	1.74
45	1.12	1.53	1.78
50	1.16	1.56	1.81
55	1.19	1.59	1.84
60	1.22	1.62	1.87

Source: Excerpted from ASCE Standard, SEI/ASCE 7–05 *Minimum Design Loads for Buildings and Other Structures*, Figure 6.3. With permission of the American Society of Civil Engineers, Reston, VA, 2003.

then the computational or wind tunnel approach must be used to determine the wind load. Design of array mounts for these situations is generally best left to the expertise and experience of a structural engineer.

Problems

5.1 Consider the following facts concerning a steel alloy and aluminum alloy: (a) steel is stronger, (b) steel is stiffer, and (c) aluminum stretches much more than steel. Sketch the stress–strain curves for both on the same stress–strain graph (see Figure 6.3 of ASCE-7). Show the relative positions of the yield strength, ultimate strength, and rupture strength. Also, indicate on the graphs how the moduli of elasticity compare for steel and aluminum.

5.2 For two bars having cross-sectional areas $A_1 = 0.25$ in.^{2} and $A_2 = 25$ in.^{2}, subjected to uniform axial forces $P_1 = 100$ lb and $P_2 = 1000$ lb, compute the uniform stresses in each bar.

5.3 During a simple pull test of a bar, the deformation per unit length was measured at 0.000167 in./in. The corresponding stress was 5000 psi. Further in the test, the elongation per unit length and corresponding stress were 0.000667 in./in. and 20,000 psi. If the proportional limit is 30,000 psi, what is the modulus of elasticity? Would this result be valid if the proportional limit was 18,000 psi? Explain.

5.4 A uniform bar has length = L and cross-sectional area = A. Rather than a force P acting on the end of the bar, the only force is the uniformly distributed weight of the bar. If the total weight of the bar is W, show that the total elongation is $\delta = WL/2AE$. If its weight per unit volume is w, also show that $\delta = wL^2/2E$.

5.5 A steel bar having a cross-sectional area of 0.5 in.^{2} and a length $L = 600$ ft is suspended vertically. The force P acting on the lower end is 5000 lb. If steel weighs 490 lb/ft^3 and $E = 30 \times 10^6$ psi, find the elongation of the bar. (Hint: Use the results of Problem 5.4.)

5.6 A PV array is to be mounted on the roof of a nursing home, which is adjacent to a hospital. It is located in the Topeka, Kansas, area and is subjected to high wind loading. List at least three possible functional requirements and three operational requirements. Also identify four constraints the designer must consider. Identify the trade-offs that might be weighed.

5.7 Why are building codes that affect the mechanical and structural design of PV systems much less uniform throughout the United States than electrical codes? Explain.

5.8 Give two examples of cases when the design loading for the PV array and structural supports is not dominated by wind loading.

5.9 What conditions must be satisfied before Bernoulli's equation can be used to relate the pressure and velocity between two points in the flow field? Does a PV array meet these conditions? Explain. Why is an understanding of the Bernoulli relationship important when considering wind loading on PV systems?

5.10 The design wind load calculation involves the product of the wind speed, a force coefficient, and an importance factor. How is the design wind speed determined? If the force coefficient (i.e., fudge factor) cannot be derived analytically, how might it be determined experimentally? The importance factor is related to the consequences of failure; the greater the consequences, the higher the factor. How would the importance factor for a PV system on a rural home compare with the application in Problem 5.6?

5.11 List four objectives in designing an array mounting system.

5.12 For a fixed-orientation array to be mounted on a south-facing roof, discuss the advantages and disadvantages of each of the four roof-mounting configurations.

5.13 Using Figure 5.5, assume the sleeve is made of aluminum for which $\alpha = 12.8 \times 10^{-6}$ per °F and $E = 10 \times 10^6$ psi. The initial temperature at which there are no stresses is 59°F, and the final temperature is 200°F. All other dimensions, properties, and information are the same as in the example for Figure 5.5. Compute the stresses in both the stainless steel bolt and the aluminum sleeve at 200°F.

5.14 Four lag screws of 5/16 in. diameter and 3.5 in. length are used to attach an 8 ft by 8 ft standoff array to an asphalt shingle roof. The total thread length is 2.5 in., and the combined thickness of the mounting bracket, pad, shingles, roof membrane, and plywood sheathing is 1.0 in. If the lag screw penetrates into the roof truss made of Southern yellow pine, what is the allowable withdrawal load on each lag screw? What is the maximum uplift loading in psf that the array can withstand?

5.15 A four-module PV array has an area of 40 sq. ft. and will be secured using six mounting brackets. Each bracket uses two 5/16 in. lag screws to secure the standoff-mounted array. The roof trusses are made of white spruce. The design wind load is 55 psf. What is the minimum thread penetration for each fastener?

5.16 A standoff-mounted array has a surface area of 50 sq. ft. and is secured to an asphalt shingle roof with four steel J-bolts of 3/8 in. diameter. If the allowable tensile stress for the steel is 40,000 psi, are the J-bolts of sufficient strength to carry a maximum uplifting wind load of 65 psf? What is the factor of safety?

5.17 An array of four PV modules, each with an area of 10 sq. ft., is to be mounted above and parallel to a standing seam metal roof using 6 L-brackets. The modules themselves can withstand a load of 75 psf. It is desired to support the 75 psf uplifting load on the full array with one lag screw in each bracket. If each lag screw has an allowable withdrawal resistance of 300 lb./in., what is the minimum thread length that must penetrate the roof truss?

5.18 Assume the ratio of direct to total irradiance is the same for two different locations, one at 30° N latitude and one at 60° N latitude. Which location would benefit the most from a two-axis tracking array (versus a fixed-orientation array)? Explain.

5.19 For more effective cooling of standoff-mounted PV arrays, would you recommend higher or lower aspect ratios? Explain.

5.20 A rack-mounted array similar to the one in Figure 5.22 is supported at the rear by two 4 ft columns of 6063 T-6 aluminum tubing. The modulus of elasticity of the aluminum is 10×10^6 psi. The cross-sectional area is 0.5 in^2 and the radius of gyration is 0.361 in. What is the moment of inertia of the cross-sectional area in units of in.^{4}? Compute the critical buckling load, P.

REFERENCES

1. Hanks, R. W., *Materials Science Engineering: An Introduction*, Harcourt, Brace & World, New York, 1970.
2. Avallone, E. A. and Baumeister, T., Eds., *Marks' Standard Handbook for Mechanical Engineers,* 10th ed., McGraw-Hill, New York, 1996.
3. Kutz, M., *Mechanical Engineers' Handbook*, 2nd ed., John Wiley & Sons, New York, 1998.
4. Singer, F. L., *Strength of Materials*, 2nd ed., Harper & Brothers Publishers, New York, 1962.
5. Dieter, G. E., *Engineering Design: A Materials and Processing Approach*, 2nd ed., McGraw-Hill, New York, 1991.
6. Clauser, H. R., Editor-in-Chief, *The Encyclopedia of Engineering Materials and Processes*, Reinhold Publishing, New York, 1963.

7. Larson, W. J. and Wertz, J. R., *Space Mission Analysis and Design*, 2nd ed., Microcosm, Inc. and Kluwer Academic Publishers, Torrance, CA, 1992.
8. *Minimum Design Loads for Buildings and Other Structures*, ASCE Standard, SEI/ASCE 7–05, American Society of Civil Engineers, Reston, VA, 2003.
9. Bertin, J. J. and Smith, M. L., *Aerodynamics for Engineers*, 2nd ed., Prentice Hall, Englewood Cliffs, NJ, 1989.
10. Hurt, H. H., *Aerodynamics for Naval Aviators*, Navweps 00-80T-80, U.S. Navy, 1960.
11. Modular Photovoltaic Array Field, prepared for Sandia National Laboratories, SAND83-7028, Hughes Aircraft Company, 1984.
12. Miller, R. D. and Zimmerman, D. K., Wind Loads on Flat Plate Photovoltaic Array Fields, Phase II Final Report, prepared for Jet Propulsion Laboratory, Contract No. NAS-7-100-954833, Boeing Engineering and Construction Company, Seattle, WA, 1979.
13. Marion, B. and Atmaram, G., *Preliminary Design of a 15–25 kWp Thin-Film Photovoltaic System*, prepared for the U.S. Department of Energy, Florida Solar Energy Center, Cape Canaveral, FL, November 13, 1987.
14. *Wind Design of Flat Panel Photovoltaic Array Structures*, prepared for Sandia National Laboratories, SAND79–7057, Bechtel National, Inc., June 1980.
15. Miller, R. D. and Zimmerman, D. K., Wind Loads on Flat Plate Photovoltaic Array Fields, Phase III and Phase IV Final Reports, prepared for Jet Propulsion Laboratory, Contract No. NAS–7–100–954833, Boeing Engineering and Construction Company, Seattle, WA, April 1981.
16. Ventre, G.G., *A Program Plan for Photovoltaic Buildings in Florida*, prepared for the Florida Energy Office/Department of Community Affairs and Sandia National Laboratories, Florida Solar Energy Center, Cocoa, FL, FSEC–PD–25–99, January 1999.
17. *Wood Handbook—Wood as an Engineering Material*, U.S. Department of Agriculture, Forest Service, Forest Products Laboratory, Madison, WI, 1999.
18. Faherty, K.F. and Williamson, T.G., *Wood Engineering and Construction Handbook*, 3rd ed., McGraw-Hill, New York, 1997.

SUGGESTED READING

Building Integrated Photovoltaic Project in the 21st Century Townhouses, Advanced Housing Technology Program, National Association of Home Builders Research Center, Inc., Upper Marlboro, MD, June 1997.

Epstein, S. G., *Aluminum and Its Alloys*, Aluminum Association, Inc., presented at the Annual Liberty Bell Corrosion Course, sponsored by the National Association of Corrosion Engineers, Washington, D.C., September 1978.

Healey, H., An Investigation of the Feasibility of a National Structural Certification Process, Contract Report for the Solar Rating and Certification Corporation, Healey and Associates, Merritt Island, FL, September 25, 1998.

Hollister, S. C., *The Engineering Interpretation of Weather Bureau Records for Wind Loading on Structures*, National Bureau of Standards, Washington, D.C., November 1970.

IEEE Recommended Practice for Qualification of Photovoltaic (PV) Modules, IEEE Standard 1262–1995, IEEE Standards Coordinating Committee 21 on Photovoltaics, Institute of Electrical and Electronics Engineers, Inc., New York, April 12, 1996.

Manual of Steel Construction, 8th ed., American Institute of Steel Construction, Chicago, 1980.

McCluney, W. R., *Introduction to Radiometry and Photometry*, Artech House, Inc., Norwood, MA, 1994.

Mehta, K. C. and Marshall, R. D., *Guide to the Use of the Wind Load Provisions of ASCE 7–95*, ASCE Press, Reston, VA, 1998.

Metal Roofing, *Roofer Magazine*, July 1997.

Orlowski, H., Corrosion protection for solar systems: methods of eliminating an old problem applied to solar components, *Heat/Piping/Air Conditioning*, July 1976.

Photovoltaic System Design, Course Manual, Florida Solar Energy Center, Cocoa, FL, FSEC-GP–31–86, April 1996.

Post, H. N., Low cost structures for photovoltaic arrays, *Proc. 14th IEEE Photovoltaic Specialists Conference*, San Diego, CA, January 7–10, 1980.

Post, H. N. and Noel, G. T., A comparative evaluation of new low-cost array field designs for grid-connected photovoltaic systems, *Proc. 19th Intersociety Energy Conversion Engineering Conference*, San Francisco, CA, August 19–24, 1984.

Roark, R. J., *Formulas for Stress and Strain*, McGraw-Hill, New York, 1954.

Russell, M. C. and Kern, E. C., *Stand-Off Building Block Systems for Roof-Mounted Photovoltaic Arrays*, Sandia Contract 58–8796, Massachusetts Institute of Technology, December 1985.

Shigley, J. E. and Mischke, C. R., *Mechanical Engineering Design*, 5th ed., McGraw-Hill, New York, 1989.

Sick, F. and Erge, T., Eds., *Photovoltaics in Buildings: A Design Handbook for Architects and Engineers*, International Energy Agency, Paris, 1996.

Smith, W. F., *Principles of Materials Science and Engineering*, 2nd ed., McGraw-Hill, New York, 1990.

Stefanakos, E. K. et al., Effect of row-to-row shading on the output of flat plate south facing photovoltaic arrays, *Solar Engineering–1986*, American Society of Mechanical Engineers, Anaheim, CA, April 13–16, 1986.

The Design of Residential Photovoltaic Systems, SAND87–1951, Photovoltaic System Design Assistance Center, Sandia National Laboratories, Albuquerque, NM, December 1988.

CHAPTER 6

Battery-Backup Grid-Connected PV Systems

6.1 INTRODUCTION

This chapter focuses on the design of grid-connected PV systems with battery backup. All one needs to do to discover an interest in these systems is to spend a week or two after a hurricane, an ice storm, or some other natural disaster without electric power. And, believe it or not, there are places in the world where the grid is not available on a 24 h per day basis. Those who have purchased fossil-fueled generators for backup power generally appreciate the idea of a quiet, reliable, pollution-free, non-fuel-consuming source of electricity. Rather than wait in long lines for gasoline for their generator, they can sit by the pool drinking cold beer. If they also have a solar water heater, they will have hot showers as well. In fact, they will probably be the most popular family on the block.

Battery-backup grid-connected PV system components still must comply with UL 1741 [1]. In the event of a utility disturbance or outage, the inverter must shut down for the allotted 5 min interval, monitor the grid, and reconnect again only if the grid has been stable in voltage and frequency for the 5 min. This means that the inverter must be a multipurpose device. It supplies power to the grid as a current source that continuously monitors the grid. It also acts as an alternate means of charging the system batteries in the event that the PV system for any reason has not kept them charged. Finally, it acts as a stand-alone voltage source to power designated standby loads in the occupancy while the utility is not available.

Recently, ac-coupled battery-backup PV systems have generated interest. For those who have installed non-battery-backup systems and then found they would have preferred a battery-backup system, it is possible to add a few extra components, including a battery-backup inverter and batteries, to achieve battery backup. There is also an argument that this type of system is inherently more efficient than the dc-coupled systems that have been popular because they immediately invert the PV array output to useful ac power, rather than having batteries as an intermediate stop. These systems are also adaptable to multiple ac sources, such as wind generators and low-head hydro systems.

Both types of systems will be discussed in this chapter. Figure 6.1 shows block diagrams of both types of system.

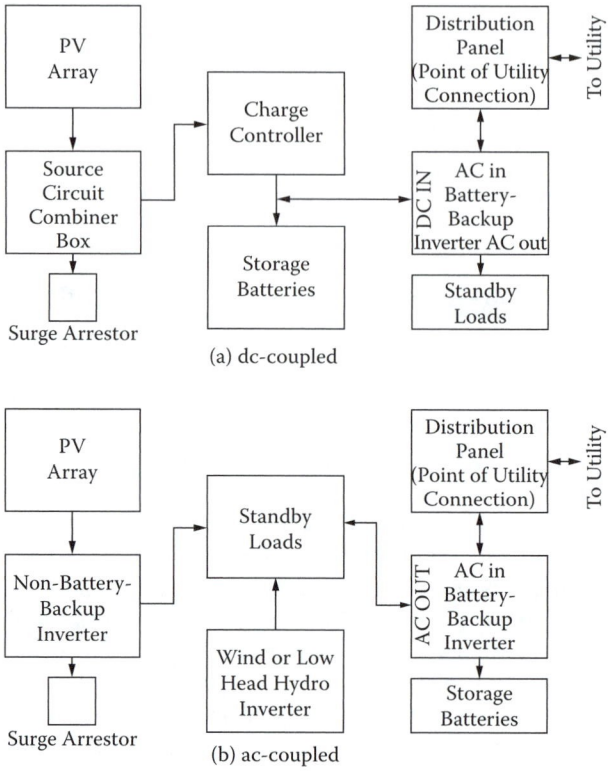

Figure 6.1 Block diagram of dc-coupled and ac-coupled battery-backup grid-connected PV systems. Grounding conductors and disconnects not shown.

The final step in system design is to understand how to program charge controllers and inverters and to understand monitoring options that may be available for the system. While newer versions of software for inverters and charge controllers may involve revised programming procedures, the designer should still understand how to determine the proper settings for charging parameters, "sell" voltages, and the relationships among these parameters when the system has more than one inverter or charge controller that contributes to either charging or discharging of the batteries.

6.2 BATTERY-BACKUP DESIGN BASICS

6.2.1 Introduction

Just as in the case of the non-battery-backup system, the selection of the right system for a given owner can involve a variety of selection criteria, such as the desired backup loads, the available roof space, the available space for batteries, or the budget. Often, those interested in battery-backup PV systems expect the systems to pay for themselves, even though no fossil standby generator ever paid for itself with the electricity it generated. The payback from a battery-backup system must be measured in terms of food not spoiled after a utility outage, the satisfaction of having a "green" source of standby power, or the comfort level of knowing that power will be available after a storm without having to start and fuel a fossil generator.

6.2.2 Load Determination

At some point in the design process, it will be necessary to determine the desired standby loads for the system. This is simply a tabulation of everything the system owner would like to have available on a standby basis. It is also a time of reckoning that probably the owner will need to forego the use of the 5-ton central air-conditioning unit, the 4500 W electric water heater, the 12 kW electric range, and the 5600 W electric clothes dryer. On the other hand, it will usually be found as a pleasant surprise that the refrigerator, TV, washing machine, microwave oven, computer, some lights, and some fans can be used. Sometimes even a small, mini-split air-conditioning unit can be incorporated into the standby load collection.

An important item to consider is whether there will be any 240 V loads on the standby list. While some battery-backup inverters deliver 120/240 V single-phase output, others deliver only 120 V unless they are connected in a "stacked" mode with a second inverter. The designer must also consider the total power requirements and the total energy requirements, as well as how the loads might be connected to a standby electrical panel to properly balance the loads. The example system designs to follow will demonstrate how to properly tabulate standby loads.

6.2.3 Inverter Sizing

The size of the inverter will be determined by the anticipated maximum instantaneous power requirements of the standby loads and whether there will be any 240 V loads. The selection of inverters will also be influenced by subtle items, such as whether the load served is a single-phase load on a 120/240 V single-phase supply, a 120 V or a 208 V single-phase load on a 3-phase supply or a 3-phase load on a 3-phase supply. These possibilities will be sorted out in the design examples that follow.

In addition, whether to use ac-coupling or dc-coupling of inverters will need to be decided. This decision may be based on economics, desired system performance, desired system configuration, or the available electric service. One thing is for certain—the inverter industry is growing so rapidly that additional options will no doubt be available either by the time this book goes to press or soon thereafter. So the most important part of this chapter will be to understand the design issues, so as new technology emerges, the designer will be able to understand how to evaluate the choices.

6.2.4 Battery Sizing

The sizing of the batteries will depend on the anticipated daily kWh consumption of the standby loads and on how many days of storage are desired. Once these two considerations have been established, there will likely be a trip back to the drawing board when it is discovered that either the choice of batteries is more costly than had been allowed for in the budget or there is not sufficient space for the batteries.

Another important consideration in battery selection is whether to use deep-discharge, flooded lead-acid batteries that will need maintenance and will require special venting or whether to use more expensive maintenance-free batteries. If maintenance-free batteries are chosen, whether to use gel or absorbent glass mat (AGM) battery types will need to be determined. With the advancement of hybrid vehicle and electric vehicle technologies, it is possible that battery technologies other than lead-acid may become the technologies of choice before this book is out of print.

The basic procedure for battery sizing involves the following:

1. Determining the standby load in kWh/day. This is the energy that must be supplied by the inverter *output*.
2. Allowing for a 2% energy loss in the wiring from inverter to loads.
3. Determining the energy required by the inverter to meet the standby load requirements. This takes into account the inverter efficiency by dividing the inverter output energy by the inverter efficiency to obtain the inverter *input* energy.
4. Determining the load on the batteries in kWh/day using

$$\text{kWh/day(batt)} = \text{kWh/day(loads)} \div \text{(inverter efficiency)} \div \text{(wiring efficiency)}.$$

5. Converting the load to Ah/day at the system battery voltage, using Ah/day = kWh/day(batt) × 1000 ÷ V_{batt}, where V_{batt} is the nominal battery voltage, that is, 12, 24, 48, etc.
6. Multiply the battery Ah load by 1.25. This factor accounts for the fact that it is desirable to ensure that the batteries do not supply more than 80% of their capacity before they are shut down for recharging.
7. If the batteries are located outside where they may be subjected to cooling, it may be necessary to use an additional multiplier to account for loss of capacity when the batteries are cold.
8. If additional days of storage are desired, multiply by the number of days desired. Fractional days are acceptable as multipliers.
9. Determine suitable batteries to connect in series and parallel as necessary to achieve the calculated storage capacity. Normally it is desirable to have no more than four parallel groups of batteries in order to ensure closely balanced currents in each series group of batteries.

There are a few other factors that are also important in designing grid-connected PV systems with batteries, such as rates of charging and discharging, because charging too fast or discharging too fast can result in more losses to the internal battery resistance. This can lead to the need for additional PV array size to overcome these losses. It may also lead to the need to increase the size of the battery bank to reduce losses from discharging too quickly. These considerations will be accounted for in the design examples that follow.

6.2.5 Sizing the Array

Once the standby loads have been determined, and the batteries have been selected, the array must be designed to provide sufficient daily energy to operate the standby loads when grid power has been lost. In other words, the array should be sized so the system can be operated as a stand-alone system. In this part of the exercise, it may be desirable to establish the month or months when it is most likely that the grid may fail because these will be the times when the array will need to produce the needed standby energy. So if PVWatts [2] is used to estimate system performance or if the NREL tables are used to determine peak sun hours, the estimated monthly system production can be checked against the anticipated monthly standby energy requirements.

Proper sizing of the array must account for the loss mechanisms described in Chapter 4 plus additional losses encountered in charging and discharging the batteries. It is common to estimate all loss mechanisms and then calculate a default overall loss percentage for the system for use in array sizing. For example, loss mechanisms will typically include the following:

1. Array temperature and mismatch losses, typically about 15% worst case.
2. System wiring losses, typically overall about 4% worst case.
3. Inverter losses, typically about 6% worst case, depending on whether ac or dc coupling is used.
4. Battery charging and discharging losses, typically assumed to be overall 10%.
5. Charge controller losses, typically in the 2%–4% range.

Using these loss factors, the worst-case dc:ac conversion efficiency can be estimated to be $0.85 \times 0.96 \times 0.94 \times 0.9 \times 0.98 = 0.68$, or 68%. On a cool, breezy, clear day, the conversion efficiency will probably be closer to 72%, and when the grid is operational, the conversion efficiency should approach the PVWatts default value of 77%. It should be noted that these performance figures are based on the use of an MPPT charge controller, so the array size can be computed in watts, as opposed to amperes. Also, noting the difference between worst-case and better-case performance, a reasonable choice for average system operation under standby conditions is 70%. This figure will be used in all the examples that follow as a default value. So, to determine the array size, simply divide the daily kWh needed for the standby loads by the peak sun hours for the month when the standby power is most likely to be needed, and then divide by 0.70 to account for losses. The designer can also vary the array size in PVWatts, using the 0.70 conversion factor, until the array size yields the desired daily kWh for the standby loads, or, if the appropriate loss factor is used, the array size can be calculated on the basis of the charge needed by the batteries.

There is one fine point, however. The array must be broken down into source circuits that have output voltages that fall within the input voltage and current limits of the system charge controller, preferably at a point close to the maximum operating efficiency of the charge controller.

At this point, the system will have been designed, except for balance of system (BOS) components. All that is left is to determine whether the cost of the system is within budget, whether there is room for the batteries and power-conditioning equipment, and whether there is room for the PV array. If any of these requirements cannot be met, then it is back to the drawing board. If they can be met, then the design can be completed by selecting all the BOS components in conjunction with *NEC* and ASCE-7 requirements, and any other local, regional, or national code requirements that may apply to the installation.

6.3 A SINGLE-INVERTER 120 V BATTERY-BACKUP SYSTEM BASED ON STANDBY LOADS

6.3.1 Determination of Standby Loads

For this example, it will be assumed that the owner would like to operate a shiny new Energy Star refrigerator, a microwave oven, a washing machine, a desktop computer, two lighting circuits, and two ceiling fans in the event of a power outage. The system will be located in Tampa, Florida, where the greatest likelihood of a hurricane-induced power outage is during the months of August through October.

Because the desired standby loads are known, the first step in the design is to construct a table that lists the loads, as shown in Table 6.1. The refrigerator is rated at 420 kWh/yr, which equates to 1.15 kWh/day. To determine the power consumption of the refrigerator, the best method is to find the model number and check the specifications. To estimate the power consumption of the refrigerator, it is possible to assume the compressor will be running 33% of the time and will be off 66% of the time. If this is the case, then the 1.15 kWh is consumed over an 8 h period. So dividing the 1.15 kWh by the 8 h results in 0.1438 kWh/h = 0.1438 kW = 144 W. Because the unit operates at 120 V, and because it has a motor, it can be assumed to operate at an 80% power factor. Thus P = IV(pf) can be solved to find a current of 1.5 A.

For the remainder of the loads, either the nameplate power and current, or estimates of power and current based on requirements of *NEC* Article 220, along with the estimated operating hours per day, are used to obtain the required table entries. For example, for the lighting circuits, local code administrators will generally require

Table 6.1 Summary of Standby Loads for 120 V DC-Connected Battery-Backup System

Load Description	Load Power (*W*)	Load Current (*A*)	Load (kWh/Day)
Refrigerator	144	1.5	1.15
Microwave oven	600	5.0	0.6
Washing machine	1500	12.5	0.4
Computer circuit	250	2.1	1.0
Lighting circuit 1	1200	10.0	1.2
Lighting circuit 2	1200	10.0	1.2
Ceiling fans	150	1.3	1.8
Total	5044	42.4	7.35

that the full possible load for the circuit be used for load determination purposes because most of the time the system will be operating under utility power as opposed to standby power.

6.3.2 Inverter Selection

The inverter must be capable of *passing through* adequate power to meet the load demands when the utility is operational and be capable of meeting reduced demand from standby loads when the utility is down. An important section in *NEC* is Section 690.10(A), which states that in a stand-alone PV system, the inverter does not need to be sized to meet the total load of the occupancy, as would be done in accordance with *NEC* Article 220 for utility power.

Battery-backup inverters have one power rating for when they are powering up the standby loads when the utility is down and another output current rating when the utility is operational. It is important to find each of these power ratings on the inverter specification sheet when the inverter is selected. For this example, one popular battery-backup inverter is rated at 3600 W backup power at 120 V, but will pass through 60 A on a continuous basis when the utility is available. Comparing the 60 A pass-through current with the total load current requirements of Table 6.1 shows that the load current requirements are easily met with this inverter. In fact, if a 60 A circuit breaker is used as the point of utility connection for the inverter output, for the purpose of determining load current, the load current must be considered to be continuous, which means that the circuit breaker must be rated at 125% of the load current. As a check on this requirement, 125% of 42.4 A = 53.0 A < 60 A, so the selected standby loads are acceptable. Furthermore, because it is very unlikely that all loads will be simultaneously operational, an additional safety factor is introduced.

If the utility is down, then the provisions of *NEC* 690.10(A) apply, meaning that although the maximum standby load on the inverter is 5044 W, the inverter will only supply 3600 W. So in this case, not all loads can be operated simultaneously while the grid is down. If the load on the inverter exceeds 3600 W, the inverter will shut down to protect itself and display an error message that it has been overloaded. When the overload is corrected, the inverter will restart. Normally, the system owner will have installed compact fluorescent lighting and will have incorporated other energy efficiency measures, so the loads may not be as large as allowed for in Table 6.1. Furthermore, the owner will know to use standby loads sparingly in order to be sure the system batteries do not discharge too far before the sun gets another chance to recharge them.

Now that the standby loads have been defined and the inverter has been selected, the batteries can be selected. Note that the selected inverter is designed for operation with a 48 V battery bank.

6.3.3 Battery Selection

The information in Table 6.1 is also needed for battery selection, following the procedure outlined in Section 6.2.4. Because the daily load requirements are 7.35 kWh, the energy that must be supplied *to* the inverter input is thus given by 7.35 ÷ 0.98 ÷

0.94 = 7.98 kWh, where 0.98 represents wiring losses between inverter and loads and 0.94 represents the inverter efficiency. Converting this figure to Ah that must be supplied by the batteries to meet the standby load requirements results in Ah = (7.98 × 1000) ÷ 48 = 166.2. Finally, accounting for 80% depth of discharge, divide again by 0.8 (equivalent to multiplying by 1.25), to obtain Ah(batt) = 207.8.

There is a subtle point that should be noted here that the reader has probably already observed. That is, the rating of a battery depends on how fast it is discharged, as was explained in Chapter 3. Fortunately, battery manufacturers rate their batteries at specified discharge rates. If 1 day of storage is desired, then the 24 h discharge rate should be used in determining the battery rating. If 2 days of storage are desired, then the 48 h discharge rate is used, etc. So before a battery can be selected, the system owner must decide on how many days of storage are needed. Admittedly, there are even finer points that might be argued in the battery selection process, such as how much battery discharging occurs during the day versus how much discharging occurs during the night. Obviously, all the nighttime loads must be supplied by the batteries, but during the day, all of the loads must be supplied by some of the energy from the PV array, while the remaining fraction of the PV array output will be used to make up the battery charge that was used the previous night. But, then, there are also those periods soon after sunrise and just before sunset where the standby loads are being satisfied partly by the PV array, and partly by the batteries. The ambitious designer may feel the need to document all the energy use habits of the PV system owner and then construct a minute-by-minute estimate of the power consumption of the loads, the power production of the PV array and the amount of battery charging taking place. And, by the way, be sure to take into account the randomness of cloud cover. In other words, if 1 day of storage is desired, use the 24 h discharge rate unless it is known for sure that all of the standby loads will only operate after sundown.

The decision of storage days may be influenced by space, because more storage days need more batteries, or by budget, because the cost of batteries is not trivial. In any case, unlike a stand-alone PV system, where the batteries are needed every day, assuming the loads are used every day, the batteries in a grid-connected system are hopefully rarely used for long periods of time. Thus, 1 day of storage is a common choice for battery-backup grid-connected systems, especially in hurricane-prone regions, where the sun generally shines brightly after the storm has passed, with no evidence of the storm except for debris all over the place. The idea is that if the sun shines every day following whatever caused the utility to fail, then the batteries will be recharged each day. If the sun does not shine enough for a full charge, then the owner needs to use the loads more sparingly. So, for this example, 1 day of storage will be used, so the 24 h discharge rate should be used for the batteries selected.

The final consideration before battery selection is whether the batteries will likely be cold when they are needed. If so, larger batteries or additional batteries will be needed for the system. In this example, because the system is in Florida and because the most likely time the batteries will be needed is the fall, no temperature correction will be needed.

So it is now time to look through the battery specification sheets. When doing so, the designer will note that nearly every manufacturer of PV batteries has a 6 V battery rated in the range of 200–250 Ah. So all that is needed is to decide whether to use flooded batteries or sealed batteries. If sealed batteries are chosen, it generally is preferable to select the AGM type because these batteries are designed to discharge somewhat faster than gel batteries. For more days of storage, then gel batteries are sometimes preferred. For this example, it will be assumed that the owner has shown a preference for maintenance-free batteries.

Thus, any maintenance-free deep-discharge battery with a 24 h discharge rating of 207.8 Ah or more is satisfactory. Coincidentally, one manufacturer makes a 208 Ah @ 6 V @ 24 h discharge rate sealed AGM unit, which comes close enough to the 207.8 Ah that is needed. If the battery is discharged in 8 h instead of in 24 h, according to the manufacturer's specifications, the capacity is reduced to 186 Ah. If the designer feels uncomfortable about this number, a somewhat larger battery can be selected.

Of course, because the batteries are for a 48 V system, 8 of the batteries will be needed for connection in series to obtain the necessary 48 V. Just as the current is the same for components in series, the Ah is also constant as batteries are added in series. The Wh, of course, increases with the addition of each battery, because the Wh is obtained by multiplying the Ah by the voltage.

6.3.4 Array Sizing

The array must be sized so that it will provide sufficient energy each day to meet the energy requirements of the standby loads. If the array is sized to fully charge the batteries, then it must supply enough energy to make up for the energy supplied by the batteries to the standby loads, inverter losses, and ac wiring losses, or 166.2 Ah. So the question is: How much energy must be supplied by the array so that 166.2 Ah can be removed from the batteries? This is where the battery charging and dc wiring efficiency enter the process. If it is assumed that only 90% of what goes into the batteries comes out of the batteries, then the array must supply $166.2 \div 0.9 = 185$ Ah per day to the batteries. Because the 185 Ah must be supplied at 48 V, it is convenient to convert this charge into energy by multiplying by the battery voltage. This gives 8880 Wh, or 8.88 kWh, produced by the array. So the next question is: What size array is needed to supply 8880 Wh/day?

To make this determination, the losses between array and batteries need to be taken into account. A summary of the losses includes 15% for array temperature and mismatch losses, 2% for wiring losses between array and batteries, and 2% losses for an MPPT charge controller. So, providing 8880 Wh to the batteries will require an array that is rated at $8880 \div 0.85 \div 0.98 \div 0.98 = 10{,}888$ Wh/day. The overall derating factor associated with this process is thus 0.82. The designer can now turn to PVWatts to determine the array size that will provide the necessary energy by using this derating factor along with a suitable tilt and iterating, noting that 10,888 Wh/day = 327 kWh/mo.

Using a south-facing roof with a slope of 18.43°, and starting with an assumed array size of 4 kW, one obtains monthly kWh production values of 526 for August, 461 for September, and 474 for October. Keep in mind that these numbers do not represent ac output from the system, but, rather, dc energy supplied to the batteries. If the 4 kW array were to be installed, there would be excess generation each of the months indicated, which, with the grid present, would either be sold to the grid or used by interruptible loads in the occupancy. But because the goal of this design exercise is to size the array to meet the standby load requirements during the most likely hurricane months, the array can be reduced in size from 4 kW to (327/461) × 4000 = 2837.31019523 W. A closer look at the PVWatts output data indicates that an array of this size will meet the minimum requirements for 8 months of the year. With this figure in mind, the designer can now proceed to selection of an appropriate module, along with appropriate charge controllers for maximizing array performance and protecting the batteries from overcharging.

6.3.5 Charge Controller and Module Selection

Before selecting modules and designing the array, the designer needs to know about the charge controller. Typical MPPT charge controllers are rated at either 60 A output or 80 A output. The 60 A output charge controllers normally specify a maximum array size of 3200 W and limit the maximum array open-circuit voltage to 150 V. For a 48 V battery system, these controllers generally operate at maximum efficiency when the array voltage is in the 70–80 V range. So at this point, an exercise similar to those used in sizing the arrays in Chapter 4 can be used.

As a check on charge controller output, when the array is operating at 2837 W, the *output* current will be approximately (2837 W) × (0.98) ÷ (53 V) = 52.5 A, which is within the limits of the controller. Because the controllers are rated for continuous output current, no multipliers are needed for the 52.5 A, other than for sizing the wire that carries the current, which will be done in the next section. So a 60 A MPPT charge controller can be used.

The careful reader might be wondering about the origin of the 53 V figure in the last calculation. This voltage was chosen because it is approximately what would normally be selected for the sell voltage of the inverter. This means that as soon as the battery voltage reaches 53 V when the grid is up, the inverter begins to use array current for producing electricity to sell to the utility. So the battery voltage is limited to 53 V. Other possible battery voltages will be discussed when the charge controller programming is discussed in Section 6.3.7.

At the time of this writing, only negative ground charge controllers are available, so it will be assumed that the modules used will have grounded negative conductors. At this point, it will be useful to generate a table that lists module characteristics for several different modules so that this and following design examples can refer back to the table. Table 6.2 shows performance characteristics for four different modules.

Table 6.2 Performance Characteristics for Four Different PV Modules

Module	Rated Power (*W*)	V_{oc} (*V*)	I_{sc} (*A*)	V_{mp} (*V*)	I_{mp} (*A*)	$\Delta V_{oc}/\Delta T$ (%/°C)	$\Delta V_{mp}/\Delta T$ (%/°C)
A	190	32.8	8.05	26.7	7.12	−0.35	−0.49
B	190	68.1	3.70	55.3	3.44	−0.25	−0.30
C	200	32.9	8.21	26.3	7.61	−0.374	−0.42
D	230	36.9	8.42	29.6	7.76	−0.33	−0.39

Table 6.3 Determination of Maximum Number of Series Modules per Source Circuit

Module	Rated V_{OC} (V)	V_{OC} (max) (V)	Rated V_{mp} (V)	V_{mp} (min) (V)	Max # Series Modules	Source Circuit Max V_{OC} (V)	Rated Source Circuit Power (W)
A	32.8	36.82	26.7	21.47	4	147.3	760
B	68.1	74.06	55.3	48.66	2	148.1	380
C	32.9	37.21	26.3	21.88	4	148.8	800
D	36.9	41.16	29.6	24.98	3	123.5	690

The design temperature range for the modules will be from −10°C to +65°C, so a table can be generated that summarizes the number of each module that can be connected in series to keep V_{OC} of the array below 150 V. This information is shown in Table 6.3.

In Table 6.3, the maximum values of V_{OC} per source circuit are calculated using Equation (4.8a), and the minimum values of V_{mp} per module are calculated using Equation (4.8b). Examining this table indicates that none of the four modules when connected for maximum charge controller efficiency can be configured into an array that will be rated at exactly 2837.31019523 W. A reasonable design decision would be to avoid array power ratings less than this value, but also to keep the array power below 3200 W so that an additional charge controller will not be needed. So it appears that reasonable choices for the array would be four source circuits of module A, eight source circuits of module B, or four source circuits of module C as indicated in Table 6.4. If the price per watt of all modules is the same and the warrantees are the same, then either A or B would represent good choices. Because module A only requires four source circuits, this means fewer wires and series circuit breakers, meaning a lower installation cost, so module A will be selected. The array will thus consist of 16 modules.

Once again, there is a subtle point. It is possible that one of either module A or module B will be more efficient than the other and will thus have a smaller module area. This could result in the need for less mounting hardware, fewer roof penetrations, and lower array mounting cost, so this is also a consideration in module selection. Because

Table 6.4 Determining Number of Source Circuits to Achieve Design Array Power

Module	Power/SC	# SC	Array Power
A	760	3	2280
		4	**3040**
B	380	7	2660
		8	**3040**
C	800	3	2400
		4	**3200**
D	690	**4**	**2760**
		5	3450

the module dimensions are not given, the designer is reminded to check module dimensions, include the cost of the mount in the overall system cost for both modules, and then select the module that results in the lowest overall installation cost, including wiring to the combiner box and mounting the modules to the roof.

6.3.6 BOS Selection and Completion of the Design

The remaining system components include the array mounting equipment, extension cables for array wiring, a rooftop junction box, a source circuit combiner box, a GFDI device, input and output disconnects for the inverter and charge controller, battery cables, battery enclosure, inverter input and output disconnects, inverter bypass assembly, circuit breakers and possibly a separate distribution panel for the standby loads, the equipment for connecting to the utility, and all conduit and wire for the system, including grounding conductors.

6.3.6.1 Array Mounting Equipment

Note that the occupancy is in an urban setting, so it is subject to wind load exposure B. Using the roof slope of 18.43°, a mean roof height of 15 ft, and the 120 mph design wind speed for Tampa, Florida [3], Table 5.6 (Chapter 5) shows a wind load of 14.9/–23.7 psf in roof zone 1. Thus, the modules can be mounted with two rails per row, as noted in Figure 6.2. The roof trusses are constructed of southern yellow pine, and 5/16 in. stainless steel lag screws will be used to attach the array mounting brackets to the roof.

The area of each module is shown to be 14.78 ft^2, so the total area of the top row of 7 modules is 103.4 ft^2 and the total area of the bottom row of 9 modules is 133.0 ft^2. The total upward force on the top row is thus $103.4 \times 23.7 = 2451$ lb, and the total upward force on the bottom row is 3151 lb.

With the mounting brackets spaced at 4 ft intervals, Figure 6.2 shows a total of 12 mounting brackets for the top row and 16 brackets for the bottom row. The force per bracket on the top row is thus $2451 \div 12 = 204$ lb, and the force per bracket on the bottom row is $3151 \div 16 = 197$ lb. Thus, from Table 5.5 (Chapter 5) the pull strength of a 5/16" lag screw is 332 lb/in., so for the top row of modules, the lag screws will need

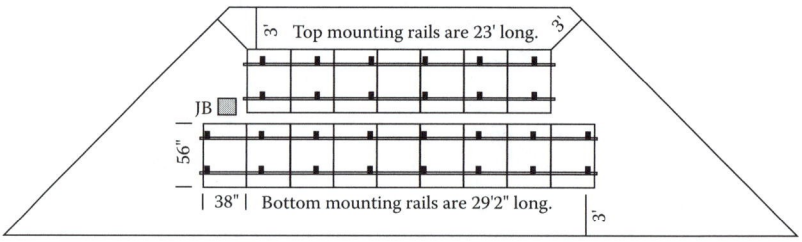

Figure 6.2 Module layout on roof showing mounting rails, mounting brackets, and junction box.

to penetrate 0.61" into the trusses, and for the bottom row of modules, the lag screws will need to penetrate 0.59" into the trusses. So, depending on the roof composition, lag screws in the 3.5" to 4" length range will probably be more than adequate. At this point, the designer need only check the manufacturer's installation instructions for the mounting rails to be sure that the mounting brackets are not mounted too far from the ends of the mounting rails. And, of course, it will be necessary to verify that either trusses are present close to the indicated locations or that the attic is accessible for any additional bracing that may be needed between trusses.

6.3.6.2 Rooftop Junction Box

The sizes of junction boxes are determined by the number and size of wires that enter the box as well as the volume of the devices that are installed in the box. Table 314.16(B) in the *National Electrical Code* [4] specifies the volume required in a junction box for conductors smaller than #6. For #10 wire, a volume of 2.5 in.^{3} is required for each conductor that terminates in the box. In this example, because there are four source circuits and a grounding conductor entering the box from the array, there will be a total of nine wires entering the junction box and nine wires leaving the junction box in conduit that will be run to the source circuit combiner box. Allowing 9 in.^{3} for an approved dc splicing mechanism, such as terminal blocks rated at a minimum of 90°C, plus 2.5 in.^{3} for each of the 18 wires entering the junction box results in a minimum required volume of 54 in.^{3} for the box. A popular junction box measures 6" × 6" × 4", providing adequate space for a neat wiring job in the box, assuming the array will be wired with #10 wire size. In fact, even if #8 wire size is used between junction box and source circuit combiner box, the junction box will provide adequate room because #8 requires 3 in.^{3} per conductor. For 9 conductors, this is 27 in.^{3} for #8 wire size versus 22.5 in.^{3} for #10 wire size.

6.3.6.3 Source Circuit Combiner Box and Surge Arrester

A number of manufacturers can supply source circuit combiner boxes. In particular, the manufacturers of the inverters generally also provide source circuit combiner boxes. In this case, a box capable of combining four source circuits will be needed. Because the array voltage will not exceed 150 V, 15 A dc circuit breakers can be used, per module manufacturer's specifications, in the combiner box as shown in Figure 6.3. Not shown in Figure 6.3 is the surge arrester, which fits into a 1/2" electrical knock-out hole in the combiner box, with leads that attach to the positive array output busbar, the negative array output busbar, and the ground busbar.

6.3.6.4 Wire and Circuit Breaker Sizing: dc Side

At this point, it is convenient to specify all the wire sizes for the system so the conduit sizes and disconnect sizes can be established. The most convenient way to do this is to construct a table to be sure no wires are missed. Table 6.5 shows one possible method of presenting the data and calculations.

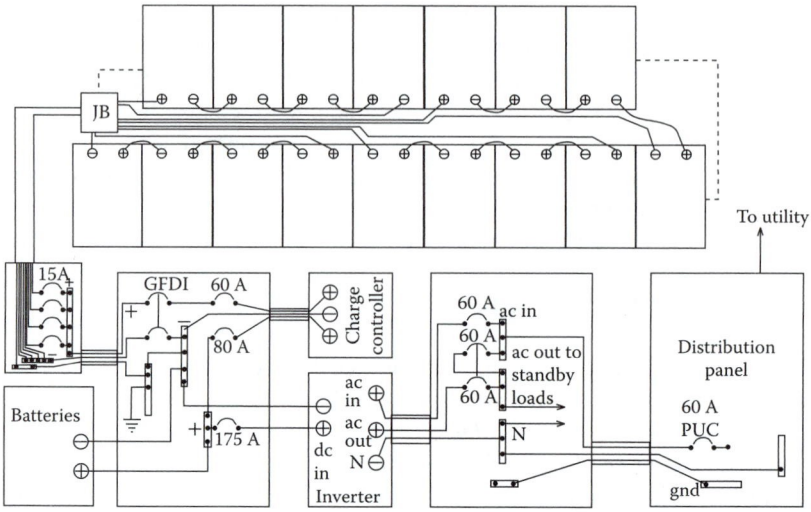

Figure 6.3 Schematic diagram of 3040 W battery-backup grid-connected PV system. Not all conduit and grounding conductors are shown.

Table 6.5 Determining Wire Sizes for the 3040 W Battery-Backup Grid-Connected PV System

Circuit	I_{sc} (A)	$1.56/_{sc}$ (A)	Wire Size	Wire 30°C Ampacity (A)	Temp Derate	Cond Fill Derate	Derated Ampacity (A)
PV source circuits	8.05	12.56	10	40	0.58	0.7	16.24
PV output circuit	32.2	50.23	8	55	1.0	1.0	55
	I	1.25I	Wire Size	Wire 30°C Ampacity (A)	Temp Derate	Cond Fill Derate	Derated Ampacity (A)
Ch cont output circuit	57.3	71.6	6	75	1.0	1.0	75
Inverter dc input I_{max}	85	106.3	2/0	195	1.0	1.0	195
Inverter ac input (A)	60	75	6	75	1.0	1.0	75
Inverter ac output (A)	60	75	6	75	1.0	1.0	75

Voltage Drop Calculations

Circuit	I_{op}	V_{op}	Wire Size	Max % VD	Wire Ω/kft	Max Wire L (ft)	Actual Wire L (ft)
PV source circuits	7.12	106.8	10	2	1.290	116	50
PV output circuit	28.5	106.8	8	2	0.8090	46	4
Ch cont output circuit	57.3	53	6	2	0.5100	18	3
Inverter ac input	60	120	6	2	0.5100	39	25
Inverter ac output	60	120	6	2	0.5100	39	3

The first calculations deal with the PV source circuits and PV output circuits. For these circuits, the ampacity of the wiring must be at least 156% of the short-circuit current values, as indicated in the table. Furthermore, because these wires are in a warm, ambient environment and are in conduit, derating factors must be applied to the listed 30°C ampacities, again as shown in the table. The derating factor for ambient temperature is listed very conservatively to show that even if the wire should reach a temperature of 65°C, it will still have adequate ampacity to carry the source circuit current. The derating factor for conduit fill is for eight current-carrying wires in a conduit.

It is assumed that the source circuit combiner box will be located near the power-conditioning equipment to make it easier to start up and troubleshoot the system. Thus, the PV output circuit will consist of a short run of wire inside the occupancy from the source circuit combiner box to a circuit breaker to part of the GFDI breaker and then to the charge controller, so no derating is necessary for these conductors. For circuit breakers, which also serve as disconnects, the PV source circuits will have the 15 A dc circuit breakers in the source circuit combiner box, and the PV output circuit will have a 60 A circuit breaker that also serves as the charge controller input disconnect and the array disconnect. This circuit breaker will be located in a dc component enclosure along with the charge controller output disconnect, the inverter input disconnect, and the GFDI device, as shown in Figure 6.3.

The next calculations deal with the charge controller output circuit and the inverter input and output circuits. First of all, it needs to be noted that the charge controller will begin to limit current if its input current exceeds 60 A, but the charge controller is rated at a continuous 60 A at its output. Thus, the wiring of the output circuit must be sized at 125% of this value, or 75 A. Furthermore, unless a continuous-duty circuit breaker is used as the charge controller output disconnect and overcurrent protection, the circuit breaker must also be rated at 125% of the charge controller output current, which is shown in the table to be 71.6 A. Because #6 wiring will be used, with a rating of 75 A, but because 75 A is not a standard circuit breaker rating, an 80 A circuit breaker can be used, according to *NEC* 240.4(B).

The inverter input is rated at 85 A dc, but on occasion, the inverter may need to supply a short surge current to start a motor or other load. It is thus commonplace to use 2/0 battery cables to the inverter that are protected with a 175 A circuit breaker. It is also advisable to keep the battery cables at lengths less than 10 ft to keep the voltage drop at an absolute minimum between batteries and inverter.

6.3.6.5 Wire and Circuit Breaker Sizing: ac Side

The inverter AC IN and AC OUT ports are both rated at 60 A @ 120 V. Note that ac power will only flow out of the AC OUT port, while ac power can flow either into the AC IN or out of the AC IN port. When the utility is down, the AC OUT port will supply a maximum of 30 A to the standby loads, while when the utility is up, the AC IN port will pass through 60 A to the AC OUT port. So, wiring of both ports must be sized for 60 A, which means #6, which is normally protected with a 60 A circuit breaker. Note that if the 60 A circuit breakers used are not rated for continuous duty, then the load at the AC OUT port must be limited to 80% of the 60 A, or 48 A.

The other component on the ac side that is mounted along with the circuit breakers that provide overcurrent protection and a disconnecting means for the inverter AC IN and AC OUT ports is the inverter bypass assembly, which is also shown in Figure 6.3. This combination of circuit breakers is installed such that it is not possible to have both on simultaneously, but it is possible to have one of them on at a time or both of them off. It is left as an exercise for the reader (Problem 6.1) to determine the positions of the circuit breakers that route power through the inverter and that bypass power from the AC IN to the AC OUT busbars in the ac component enclosure.

6.3.6.6 Wiring of Standby Loads

The standby loads indicated in Table 6.1 and implied by the two arrows in Figure 6.3 merit some discussion, especially if the installation is to be a retrofit in an existing occupancy.

There are essentially two types of 120 V circuits in an occupancy. The simplest is when a hot wire and a neutral wire, generally run as a piece of "romex" cable, run from an existing electric panel to a load or, perhaps to as many as ten 120 V outlets connected in parallel. Normally, these circuits are connected to single-pole 15 A or 20 A circuit breakers or sometimes to fuses. In these cases, the circuits can be transferred to either a separate distribution panel that has been wired for 120 V or to a circuit breaker in the ac component enclosure. That is, except for one detail. In the 2008 *National Electrical Code*, nearly every circuit in a dwelling unit must be protected by either an arc fault circuit interrupter or a ground fault circuit interrupter. The arc fault circuit interrupter is capable of detecting arcing at current levels below the circuit breaker trip rating and interrupting the circuit. The ground fault circuit interrupter is capable of detecting a difference of as little as 5 mA between the hot and neutral conductors of a circuit and interrupting the circuit if such a fault should occur.

The possibility is that a local inspector may interpret the installation of the PV system and subsequent transfer of circuits from the main panel to a standby panel to be an upgrade. In this case, even though the old circuit may have been protected by a standard circuit breaker, the moved circuit may be required to be fed by either an arc fault circuit breaker or by a ground fault circuit breaker. Before specifying the wiring or installing the wiring, it is advisable to discuss the situation with the local electrical inspector to determine whether the circuits will need to be upgraded.

The circuits that must be carefully checked, however, are the multiwire branch circuits. A multiwire branch circuit uses a common neutral as the return for wiring from opposite sides of a single-phase supply or for two or three lines that are connected to two or three phases of a three-phase system. These situations are illustrated in Figure 6.4. In the single-phase case, the neutral carries the *difference* between the two line currents, so the current on the neutral never exceeds the current on either line. For the three-phase case, the neutral carries the *sum* of the currents, but because the voltages in a three-phase system are 120° out of phase, if the currents on the three lines

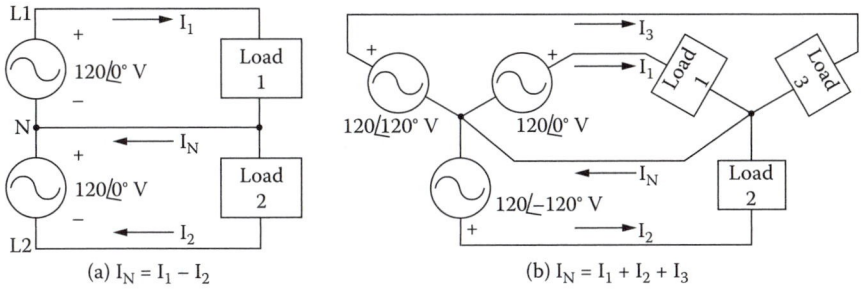

Figure 6.4 Illustration of multiwire branch circuits in single-phase and three-phase circuits.

are balanced, the neutral current is zero. If only two lines are used, and if the line currents are equal, then the neutral current will be equal in magnitude to the line current. If the line currents are unbalanced, the neutral current will be less than the larger of the two line currents (see Problem 6.2). Again, under worst-case unbalanced load situations, the current in the neutral does not exceed the current in any of the lines unless a significant third harmonic component is present (see Problem 6.3).

The problem with multiwire branch circuits is that they have been commonly used to feed bedroom circuits. A single three-conductor cable is run to a convenient location in or near two bedrooms, and then one line is fed to one bedroom and the other line is fed to the other bedroom. It is probably not too difficult to imagine one of the circuits going to the daughter's bedroom and the other going to the son's bedroom. So when the time comes to determine who gets power if the utility goes out, the answer is neither. The reason is that if both are fed from a single 120 V source, the neutral will return the *sum* of the two currents rather than the difference between the two. So, the only way that multiwire branch circuits can be connected to standby power is if the standby power is either 120/240 V single phase or 120/208 V 3-phase. Furthermore, the local inspector will need to agree that these circuits will not need to be upgraded. If the circuits need to be upgraded and if they are multiwire, then it may be necessary to run additional two-wire circuits to the desired locations.

In new construction, it is common to run standby circuits so, for example, a lighting circuit with five duplex outlets on it will have one outlet in each of five rooms so each room will have one standby power outlet. The bottom line is that owner, builder, architect, and PV designer need to carefully plan the locations of standby power outlets.

6.3.6.7 Equipment Grounding Conductor and Grounding Electrode Conductor Sizing

Equipment grounding conductors are sized in accordance with *NEC* Article 250.122. For this system, because most circuit breakers are 60 A or less, #10 equipment grounding conductors are acceptable. If the batteries are in a metal enclosure, then the grounding conductor for the battery enclosure will need to be sized according to the 175 A inverter disconnect, which means sizing to a circuit breaker between 100 and 200 A, which requires a #6 equipment grounding conductor.

Finally, if the PV power conditioning equipment is located more than 50 ft from the point of utility connection, it may be advisable to install a separate grounding electrode for the PV equipment. For example, if the PV equipment is in an outbuilding, then a separate grounding electrode is required. The grounding electrode conductor is then sized according to *NEC* Table 250.66, noting exceptions for when the grounding electrode is a rod or a plate.

This completes the design of the system. The schematic diagram, for those who have not looked at it yet, is shown in Figure 6.3. All that remains now is to program the inverter and charge controller once the system has been installed, and the installer may be depending on the designer to provide programming instructions for the system.

6.3.7 Programming the Inverter and Charge Controller

In battery-backup systems, both the inverter and charge controller need to be programmed to account for the types of batteries used in the system. While inverter and charge controllers normally come with default programs, and while the default programs of the charge controllers are generally close to the correct values, it is important for the designer to understand the programming capabilities of the inverter and charge controllers. Most importantly, however, the designer should be aware that the programming instructions that go into every small detail of programming possibilities for inverter and charge controller may be many pages long in the instruction manuals. The serious designer should make it a point to read and understand the entire installation and maintenance manuals for inverters and charge controllers in a system. The goal of this relatively short section is to stress the importance of understanding the purposes of the various programmable settings of the inverter and charge controller. The authors have already encountered a sufficient number of battery-backup system owners who read fervently the recommendations on various blogs and proceed to reset their inverter and charge controller programs without understanding what they are really doing. At some point, someone who understands the meanings of the settings will need to explain to them the reasons for the various programmed setpoints.

Inverters that are used with battery-backup grid-connected PV systems can also be used as stand-alone inverters or as inverters that are grid-connected only for charging batteries, but not for selling to the grid. So, the first item in programming the inverter is to establish the use of the inverter. If there is no PV system, then there is no need to sell to the grid. If there is no grid, then the system design will follow the procedure of Chapter 7. But when there is a PV system to charge the batteries, then the inverter needs to be programmed so that it will not be trying to charge the batteries when the PV system is trying to do so.

The battery-charging process was discussed in Chapter 3. Recall that, initially, the batteries are charged with the controller acting as a current source. If the current comes from the PV array, then presumably all the array current and, for that matter, maximum array power, will be used during the bulk charging stage. When the battery terminal voltage reaches the bulk voltage limit, which is programmable, then the charging process enters the absorption stage for a time that is also programmed. During the absorption stage, the charger acts as a constant voltage source at the bulk (or absorption, depending on whose instruction book is being followed) voltage level. After the absorption time has elapsed, the charger drops its voltage to the float voltage and continues to act as a constant voltage source at the float voltage as long as current is available from the charger.

The charging process can be achieved with either the inverter or the charge controller, so the idea is to program the inverter at a relatively low bulk charging limit, which will depend on whether flooded or maintenance-free batteries are being used. Because it is preferred to have the inverter sell electricity back to the grid, normally the sell voltage for the inverter, assuming a 48 V nominal system, will be set at approximately 52 V. The inverter bulk voltage is then set just below the sell voltage so that the inverter will only charge the batteries if their voltage drops below the sell level. The inverter float voltage will be set slightly lower than the inverter bulk voltage, and, if the inverter has a refloat voltage setting, this setting will be set somewhat lower than the float setting. The purpose of the refloat setting on the inverter is to have the inverter begin a float charge cycle if the battery voltage should drop to the refloat level for any reason.

In normal operation, with the inverter in the sell mode, as the PV system wakes up in the morning and begins to supply current to the charge controller, the charge controller will try to charge the batteries to the bulk charge level. But as soon as the batteries reach the sell voltage, the array current will be shifted from the batteries to the inverter for inversion to ac to power ac loads. So, under normal operation, the battery voltage will not reach the bulk voltage. Again, remember that the battery *terminal* voltage is not the same as the battery *cell* voltage. Under conditions of high array current, if any of the array current flows into the battery, then the battery terminal voltage will exceed the battery cell voltage. Thus, initially, the switch from charging batteries to selling through the inverter is not necessarily an instantaneous event. As charging current is drawn away from the batteries, the battery terminal voltage will tend to drop below the sell voltage, thus shutting off the sell function. But then the array current flows to the batteries and raises the terminal voltage above the sell level again. So, initially, the array current is divided between the inverter and the batteries until the battery *cell* voltage reaches the battery *sell* voltage.

The system designer should check with battery specification sheets for recommended setpoints for bulk and float voltages as well as absorption time. If the batteries are flooded batteries, then an additional equalize cycle should be run about once a month. During equalization, the batteries are purposely overcharged in order to cause the electrolyte to bubble as hydrogen and oxygen (and probably some acid) are given off. When the electrolyte bubbles, it tends to clean the electrodes and also to mix the electrolyte, which may have stratified with a higher acidity at the bottom and a lower acidity level at the top. But equalization is only used for flooded batteries. Maintenance-free batteries are not

Table 6.6 Typical Default Battery-Charging Setpoints for a 48 V Battery-Backup PV System

	Bulk Voltage	Float Voltage	Refloat Voltage	Sell Voltage	LBCO Voltage	LBCI Voltage
Charge Controller	57.6	54.4	—	—	—	—
Inverter	52	51.6	50	52.4	44	50–52

designed for gassing, and if pressure builds up from equalizing, the batteries can be damaged.

Recommended set points seem to change from manufacturer to manufacturer and from year to year. At the time of this writing, one charge controller manufacturer recommends one set of values as standard and another set of values for Australia [5].

Table 6.6 shows typical setpoints for inverter and charge controller for a 48 V battery-backup system. Two inverter setpoints not yet mentioned are the low battery cut out (LBCO) voltage and the low battery cut in (LBCI) voltage. The purpose of the LBCO is to disconnect the inverter from the batteries when the batteries reach approximately 20% of full charge, because further discharge could damage the batteries. Of course, as soon as the batteries disconnect from the inverter, the battery terminal voltage will rise above the LBCO voltage, because no further discharge current is flowing to lower the battery terminal voltage below the LBCO level. This would, of course, cause the inverter to "think" the battery voltage is not really as low as it thought it was. But inverters are smarter than that. They "know" perfectly well that if they turn on again, the battery voltage will drop below the LBCO again. So when the battery voltage reaches the LBCO and stays there for a programmed time, usually 15 min, the inverter shuts down and does not invert again until the battery voltage has reached the LBCI level. The inverter LBCI level is set sufficiently high that the batteries must be nearly fully charged before the inverter will begin to supply the standby loads again. The LBCO condition, of course, will only be realized if the grid is down, and the system is operating as a stand-alone system. If the grid is up, then as soon as the battery voltage drops to the refloat voltage, the inverter will recharge the batteries using grid power.

6.3.8 Fossil Fuel Generator Connection Options

Battery-backup grid-connected inverters can usually use a standby fossil fuel generator for backing up the sun on a cloudy day when the grid is down. The generator may be connected to a separate generator input on some inverters, or it may be connected in place of the grid by using a transfer switch on other inverters. The separate generator input is particularly convenient, because there is no need either for the owner to manually flip the transfer switch or to install a more expensive automatic transfer switch. But there are also subtle differences. For example, many generators do not have very stable frequency, so if the inverter is programmed to sell, it will be looking for a power source that meets the requirements of UL 1741 with tight frequency limits. Another subtle possibility is for the PV system to try to sell back to the utility, which, in this case, is just a small generator. This means the PV system will try to operate the fossil generator as a motor, which the prime mover may not "appreciate." So if a fossil generator is used as a replacement for the utility at the AC IN port of the inverter, the inverter should be programmed out of the sell mode so that it will not try to run the fossil generator backward, and it will not be sensitive to the frequency of the fossil generator. Of course, use of an inverter-based fossil generator with stable frequency should solve any frequency "doubts" the PV inverter may have.

Once the generator hookup has been established, the inverter can be programmed to start the generator at a predetermined battery voltage level and shut down the generator at a higher battery voltage level, allowing for the PV system to complete the charging process. The inverter can also be programmed to establish an operating time window for the generator so that it will not keep the neighbors awake at night (or during the day if they work nights). More generator options will be discussed in Chapter 7, where hybrid stand-alone PV systems will be discussed.

6.4 A 120/240 V BATTERY-BACKUP SYSTEM BASED ON AVAILABLE ROOF SPACE

6.4.1 Introduction

Because battery-backup grid-connected systems essentially use the utility grid to store/use any excess PV power generation, there is no reason to limit the size of an array to a size that will just keep a set of batteries charged or operate a particular set of standby loads. It may be the case that a larger array will be desired so the system can run certain loads during daytime hours if the grid is down but run a minimum of loads at night, thus reducing the need for a large battery bank. For this example, it will be assumed that a new house is being planned to make the greatest possible use of PV along with efficient overall design. It will be assumed that the house is designed to have enough useable south-facing roof space for mounting a PV system and a 4 ft by 10 ft solar domestic water heater. The house will be located in Burlington, Vermont, where the owner wants to have secure electrical power in the event of a winter power outage. The roof will have a 60° slope to discourage snow buildup and will face directly south. The distance between rooftop junction box and power conditioning equipment is 40 ft. The power-conditioning equipment is located next to the point of utility connection, so all ac wiring distances are less than 10 ft. Because roof mounting details have now been covered in several examples, this example will assume that the roof has been specifically designed to accommodate an array of thirty-six 200-W modules. Details of source circuit calculations will follow. Problem 6.4 offers the opportunity to design the mounting system for the array.

6.4.2 Module Selection and Source Circuit Design

The electrical characteristics of the selected PV module appear in Table 6.7. Because Burlington is not exactly a tropical winter paradise, the design must

Table 6.7 PV Module Electrical Characteristics

Rated Power (W)	V_{OC} (V)	I_{SC} (A)	V_{mp} (V)	I_{mp} (A)	$\Delta V_{OC}/\Delta T$ (%/°C)	$\Delta V_{mp}/\Delta T$ (%/°C)
200	32.9	8.21	26.3	7.61	−0.374	−0.42

incorporate a minimum array temperature of −30°F (−34.4°C) and a high temperature of 100°F (37.8°C) [6]. Using Equation (4.8a) yields, for one module,

$$V_{OC}(\max) = 32.9[1 - (-0.00374)(25 - (-34.4))] = 40.21 \text{ V}.$$

Because a source circuit $V_{OC}(\max) < 150$ V, this means no more than $150 \div 40.21 = 3.73$ modules can be connected in series. Thus, rounding down, the maximum number of modules in a source circuit is 3. For 3 modules, $V_{OC}(\max) = 120.63$ V. Next, the minimum source circuit V_{mp} on a hot day must be determined from Equation (4.8b), yielding

$$V_{mp}(\min) = 3 \times 26.3[1 - (-0.0042)(25 - (70.3)] = 63.89 \text{ V},$$

where the maximum cell temperature (70.3°C) is determined from the NOCT for the module, which is 46°C. Thus, at full sun ($G = 1.0$ kW/m^2), the cell temperature is found from

$$T_C = T_A + \frac{\text{NOCT} - 20}{0.8} G = 37.8 + 32.5 = 70.3°\text{C}.$$

Keeping in mind that this calculation is based on the hottest day on record, and that a more typical ambient temperature is 75°F in the summer, a more likely cell operating temperature will be 24 + 32.5 = 56.5°C, which equates to $V_{mp} = 3 \times 26.3[1 - (-0.0042)(25 - 56.5)] = 68.46$ V. These values of V_{mp} fall within acceptable limits for operation with an MPPT charge controller operating with a nominal 48 V output.

Thus, with 3 modules per source circuit, if 36 modules are used, then 12 source circuits will be required. Note that because the original plan was to use 36 modules, and 36 is divisible by 2, 3, 4, 6, 9, 12, and 18, the choice of 36 was not a very great risk. On the other hand, if it turned out that 5-module source circuits would have been preferable, then it would have been necessary to eliminate one module and have a 35-module array.

6.4.3 Source Circuit Combiner Box and Charge Controller Selection

Because the array size is 7200 W, with 12 source circuits, if an MPPT charge controller is used, the available current will be approximately $7200 \div 48 = 150$ A. This suggests that two 80 A MPPT charge controllers will be needed, because these charge controllers are rated for continuous duty. Because there are 12 source circuits,

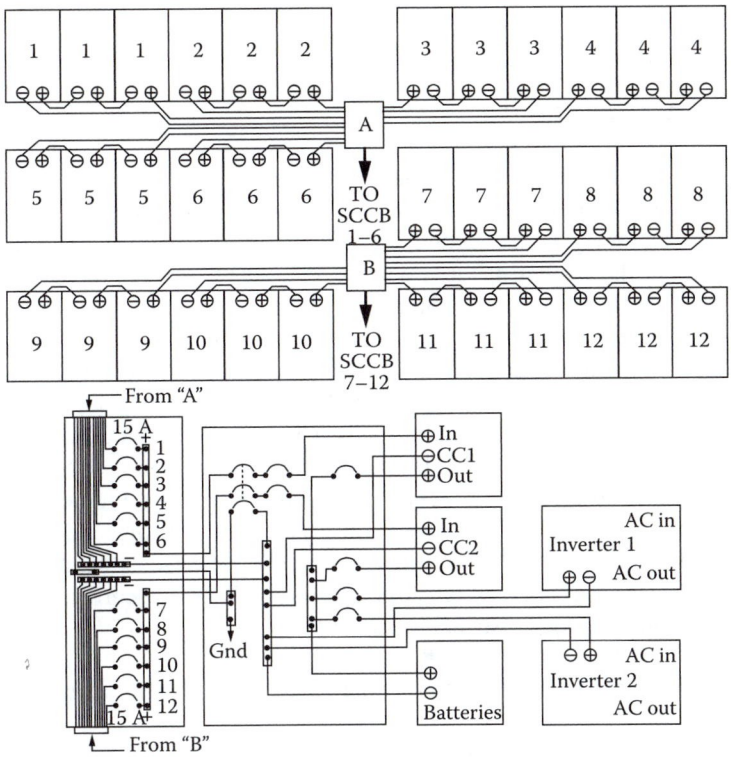

Figure 6.5a Wiring of dc portion of 7200 W battery-backup system.

the system can be nicely balanced by feeding each charge controller with 6 source circuits, as shown in Figure 6.5a.

This also means that a suitable source circuit combiner box or boxes must be found to feed the charge controller inputs. Because the maximum array output voltage is less than 125 V, it will be possible to use a combiner box that uses circuit breakers rated at a minimum of 125 V *dc*. The maximum circuit breaker current rating is determined by the PV module specifications, which, in this case, is 15 A. The minimum size circuit breaker would be 156% of I_{SC}, which is 12.8 A. The output busbars of the combiner box will need to be rated at a minimum of 156% of the combined short-circuit current of 6 source circuits, or 76.8 A.

It turns out that at least one such source circuit combiner box is manufactured that will accommodate all 12 source circuits and conveniently divide them into two separate output circuits, one for each charge controller. It is left as an exercise for the reader to find such a combiner box on the Web (Problem 6.5).

Finally, the designer should note that a maximum battery charging current of 150 A may occur. Normally, the charging current would be less, because the array will not normally be operating at full rating, and the charge controllers will likely have rated efficiencies of approximately 97%. Nevertheless, a substantial charging current is available from the array, and this should be noted when batteries are selected.

6.4.4 Inverter Selection

It makes sense to choose an inverter or inverters that will fully utilize the array output once the system batteries are charged. Several inverters are currently available that provide for grid connection and battery backup and will meet the design requirements of the system. It turns out that one manufacturer markets a 3600 W @120 V inverter that has all the features required by the design. Furthermore, this inverter can be combined in a "classic stacked" configuration with another identical inverter and programmed to provide an overall single-phase ac output of 120/240 V. So, when the system is operating independently of the utility grid, it will provide 7200 W of 120/240 V power to standby loads. The units are designed to pass through 60 A each when the grid is energized, so under these circumstances, a total of 14,400 W can be passed through to the standby loads as long as they are not continuous. However, if the standby loads were to run for 3 h continuously, then the pass-through load must be reduced to 80% of 14,400, or 11,520 W. Because the load must be derated, this means that a 2-pole, 60 A circuit breaker will be the correct size for connecting to the standby load distribution panel.

6.4.5 Determination of Standby Loads and Battery Selection

Once again, the standby loads are generally determined by discussing the options with the owner and generating a table that lists the kW and kWh associated with all the desired loads. In this case, because a 120/240 V source is available, the table is expanded to list which loads are connected to which side of the line, as shown in Table 6.8. Note also that in Table 6.8, loads are expressed in VA rather than in W, to

Table 6.8 Summary of Standby Loads for 120/240 V DC-Connected Battery-Backup System

Load Description	Load (*VA*)	Line 1 (*A*)	Line 2 (*A*)	Load (kWh/Day)
Refrigerator	144	1.5	—	1.2
Microwave oven	600	—	5.0	0.6
Washing machine	1500	12.5	—	0.4
Computer	250	—	2.1	1.0
Lighting circuit 1	1200	10.0	—	1.2
Lighting circuit 2	1200	—	10.0	1.2
Ceiling fans	150	1.3	—	2.4
Appliance circuit	1500	—	12.5	1.8
Mini-split AC	500	4.6	—	4.0
TV	120	—	1.0	1.0
Furnace fan (0.75 hp)	700	3.6	3.6	1.4
Total	7864	33.5	34.2	16.2

account for the presence of motor loads with lagging power factors in the refrigerator and the pool pump.

Next, notice that the total VA of all connected loads exceeds the total inverting capacity of the two inverters but is well within the pass-through capacity of the two inverters. Thus, the requirements of *NEC* 690.10(A) are satisfied because the inverters can supply the largest load connected and, when the utility is down, the inverters are not required to supply all the standby loads simultaneously. Also, notice that the load currents are balanced reasonably well if all loads are operating, and that one load, the pool pump, is connected to operate on 240 V. Finally, the total anticipated daily kWh requirements of all the standby loads is 16.2 kWh. It is thus important to verify that the batteries and array are capable of meeting the daily standby loads or that it is possible to reduce the standby loads on days of low sunshine levels.

To calculate the battery size, noting that the inverter efficiency is approximately 94% and the wiring efficiency is approximately 98%, the batteries must thus supply $16.2 \div 0.94 \div 0.98 = 17.6$ kWh/day to the inverters. If one day of storage is allowed, assuming an 80% depth of discharge, the batteries must be capable of storing 22.0 kWh, which, at 48 V, equates to 458 Ah.

The student is encouraged to search for batteries that will meet this storage requirement. One such battery will supply 256 Ah @ 12 V at a *C*/24 discharge rate. Thus eight of these batteries connected in two parallel sets of four in series, will deliver 512 Ah @ 48 V, which exceeds the system storage requirements. It might also be argued that because the most likely time the batteries will be needed will be winter and that in winter the batteries might be colder than in summer, the additional 54 Ah may provide a good safety factor in the event that utility power should be lost on a cold day.

As a final note, when the array delivers full power, the charging current will be 150 A, as determined in Section 6.4.3. This corresponds to a $512 \text{ Ah} \div 150 \text{ A} = C/3.4$ charging rate when maximum array current is flowing.

6.4.6 BOS Selection and Completion of Design

Once again, the remaining system components include the array mounting equipment, extension cables for array wiring, a rooftop junction box, a source circuit combiner box, a GFDI device, input and output disconnects for the inverter and charge controller, battery cables, battery enclosure, inverter input and output disconnects, inverter bypass assembly, circuit breakers and possibly a separate distribution panel for the standby loads, the equipment for connecting to the utility, and all conduit and wire for the system, including grounding conductors. The design of the array mount is the subject of Problem 6.4, so this section will focus on the remaining electrical design.

6.4.6.1 Rooftop Junction Box

The previous example used 4 source circuits. This example has 12 source circuits, so at least 25 wires (24 circuit conductors and a grounding conductor) must be run to the source-circuit combiner box. While it is possible to pull 25 wires in a single conduit, or, alternatively, to install the source circuit combiner box on the roof, the sizes of the conductors will first be determined and then suitable routing of conductors will be selected.

Because the distance between rooftop junction box and source circuit combiner box will be 40 ft if the source circuit combiner box is mounted near the charge controllers and inverters, the ampacity and voltage drop will be calculated for this configuration.

Noting first that 156% of I_{SC} = 12.81 A, and then derating for a wire temperature up to 60°C (0.71 derating factor) and a conduit fill of 24 (0.45 derating factor), #10 THWN-2 wires results in a derated ampacity of $40 \times 0.71 \times 0.45 = 12.78$ A. Because this is less than 156% of I_{SC}, pulling all the source circuit wires in a single conduit would mean that the derated ampacity is too small. So if a single conduit is used, #8 THWN-2 would be needed. This would require a 2 in. conduit for 24 conductors.

Before settling on the use of #8, it is useful to check voltage drops with #10 source circuit wiring. Using Equation (4.1), it is found that #10 wire can be run a distance of 73 ft with a voltage drop of 2%. Because the actual distance between the rooftop junction box and the source circuit combiner box is 40 ft, it makes sense to try using two conduits and two rooftop junction boxes. If this is done, then the number of current-carrying wires is reduced to 12 for each junction box and each conduit. This allows for the use of a derating factor of 0.50 rather than 0.45 for conduit fill, resulting in a derated ampacity of #10 THWN-2 of $40 \times 0.71 \times 0.50 = 14.2$ A, which is adequate to carry 156% of I_{SC}.

There are arguments for both options. Using #8 wire allows for a single conduit run, but requires a larger junction box (more than twice the volume of a single box using #10 wire), more expensive wire, and more effort to keep track of the wires. So, for this example, two rooftop junction boxes will be used with #10 THWN-2 source circuit conductors in each box and conduit. The junction boxes can thus be sized based on 2.5 in.^{3} per wire plus the volume of terminal blocks. With 26 wires entering each box, and 13 terminal blocks in each box, the box volume needed will be a minimum of $(26 \times 2.5) + 13 = 78$ in.^{3} Although a 6″ × 6″ × 4″ junction box has adequate volume, it is still necessary to verify that the box will hold 13 terminal blocks for wire junctions. This means knowing the actual inside dimensions of the box as well as the dimensions of the terminal blocks and terminal block mounting hardware. The remaining wire sizing for the system is relatively straightforward and is summarized in Table 6.9.

6.4.6.2 Source Circuit Combiner Box and Surge Arresters

In this case, a box capable of combining two groups of six source circuits will be needed. Because the array voltage will not exceed 150 V, 15 A dc circuit breakers can be used, per module manufacturer's specifications, in the combiner box as shown in Figure 6.5a. In this case, because there are two sets of source circuits and two charge controllers, each will need a separate surge arrester. If the reader has tried homework Problem 6.5 and found a single source circuit combiner box, then that would be the box of choice for this example. If the reader has not found such a box, it will be necessary for the reader to believe that at least one such box exists.

Table 6.9 Determining Wire Sizes for the 7200 W Battery-Backup Grid-Connected PV System

Circuit	I_{SC} (A)	1.56/$_{SC}$ (A)	Wire Size	Wire 30°C Ampacity (A)	Temp Derate	Cond Fill Derate	Derated Ampacity (A)
PV source circuits	8.21	12.81	10	40	0.71	0.50	14.20
PV output circuit	49.3	61.6	6	75	1.0	1.0	75

Charge Controller and Inverter Calculations

	I	1.25I	Wire Size	Wire 30°C Ampacity (A)	Temp Derate	Cond Fill Derate	Derated Ampacity (A)
Ch cont output circuit	65.4	81.75[a]	3	110	1.0	1.0	110
Inverter dc input I_{max}	85	106.3	2/0	195	1.0	1.0	195
Inverter ac input (A)	60	75	6	75	1.0	1.0	75
Inverter ac output (A)	60	75	6	75	1.0	1.0	75

Voltage Drop Calculations

Circuit	I_{op}	V_{op}	Wire Size	Max % VD	Wire Ω/kft	Max Wire L (ft)	Actual Wire L (ft)
PV source circuits	7.61	78.9	10	2	1.290	73	40
PV output circuit	49.3	78.9	6	2	0.5100	13	4
Ch cont output ckt	65.4	53	3	2	0.2540	34	3
Inverter ac input	60	120	6	2	0.5100	39	6
Inverter ac output	60	120	6	2	0.5100	39	5

[a] Charge controller output limited to 80 A continuous.

6.4.6.3 Wire and Circuit Breaker Sizing: dc Side

The remaining wire and circuit breaker sizing on the dc side of the system is shown in Table 6.9 and Figure 6.5a. The only differences are that each inverter is wired with 2/0 cable at the input, and each inverter requires a separate 175 A circuit breaker as its input disconnect. The 175 A circuit breakers are supplied with a predrilled busbar that is attached to the circuit breaker battery connection sides and attached to a 4/0 cable that is routed to the batteries, as shown in Figure 6.5a. An alternative to the 4/0 cables is to run parallel 2/0 cables from the circuit breaker input busbar back to the batteries. The outputs of the charge controllers are also connected, via 100 A circuit breakers, to the input positive busbar.

The inverter input is rated at 85 A dc, but on occasion, the inverter may need to supply a short surge current to start a motor or other load. It is thus commonplace to use 2/0 battery cables to the inverter that are protected with a 175A circuit breaker. It is also advisable to keep the battery cables at lengths less than 10 ft to keep the voltage drop at an absolute minimum between batteries and inverter. If large surge currents are anticipated, parallel 2/0 cables will experience lower voltage drop during the surges. Otherwise, 4/0 cables are rated in excess of 125% of the combined inverter maximum ratings of 170 A.

6.4.6.4 Wire and Circuit Breaker Sizing: ac Side

The AC IN and AC OUT ports of each inverter are both rated at 60 A @ 120 V. In this case, the inverters will be programmed so that if the utility goes down, they will operate 180° out of phase. So, one inverter will act as Line 1 of a 120/240 V single-phase supply system, and the other inverter will follow instructions, via the system controller, to operate as Line 2. Thus, because the inverters are connected in tandem, as opposed to parallel, line currents are still limited to 60 A, but now the total available voltage is 240 V.

When the utility is operational, the AC IN port of one of the inverters will be connected to utility Line 1, and the AC IN port of the other inverter will be connected to utility Line 2. Because these phases are passed through each inverter to the AC OUT port, the AC OUT ports of each inverter supply power in the form of a 120/240 V single-phase supply to the AC OUT L1 and L2 busbars, because this is the utility configuration. The two outputs have a common neutral at the AC IN and AC OUT ports. In any case, because the AC IN and the AC OUT ports each must carry 60 A, #6 wire and a 60 A circuit breaker will be needed for each of the ports for each inverter.

An inverter bypass assembly is also included to bypass utility power directly to the AC OUT Line 1 and Line 2 busbars of the AC component enclosure. The inverter bypass assembly operates identically to the simpler assembly of the 120 V system, but requires four circuit breaker poles to achieve the bypass function, as shown in Figure 6.5b.

Figure 6.5b also shows the point of utility connection in the main distribution panel. In this case, the main distribution panel has busbars rated at 225 A and is fed by a 200 A main circuit breaker. As a check on the largest allowable circuit breaker at the point of utility connection, note that 120% of 225 is 270 A. Because the sum of all circuit breaker ratings of circuit breakers feeding power to the panel must not exceed 120% of the busbar rating and because the sum of the PV PUC circuit breaker and the main circuit breaker ratings is 260 A, the PUC is compliant with *NEC* 690.64(B)(2). Furthermore, note that the PUC circuit breaker has been located on the busbar at the opposite end of the utility feed, as required by *NEC* 690.64(B)(7).

6.4.6.5 Wiring of Standby Loads

With a 120/240 V standby system, the only consideration with regard to multiwire branch circuits is that if one phase of a multiwire branch circuit is selected for standby operation, then the other phase must also be selected for standby operation. The reason is that the *NEC* requires that for multiwire branch circuits the circuit breakers that supply these circuits must emanate from the same distribution panel and must be supplied by either a two-pole circuit breaker or by two adjacent circuit breakers that are supplied with a handle tie.

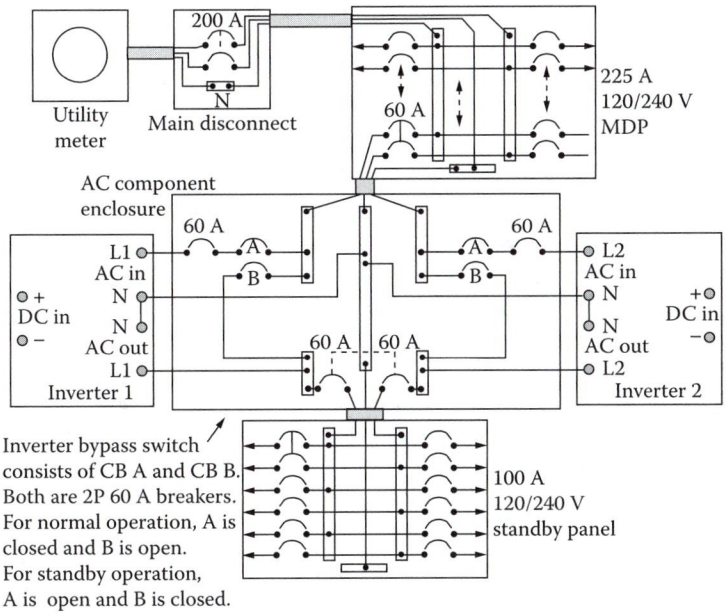

Figure 6.5b Wiring of ac portion of 7200 W battery-backup system.

The standby loads should be wired in the standby panel as shown in Table 6.8 in order to properly balance the loads on the two phases of the panel. Again, be sure to check with the local code authority to determine whether any of the standby circuits will need to be protected with either ground fault circuit interrupters or with arc fault circuit interrupters.

6.4.6.6 Equipment Grounding Conductor and Grounding Electrode Conductor Sizing

Once again, equipment grounding conductors are sized in accordance with *NEC* 250.122. For this system, because most circuit breakers are 60 A or less, #10 equipment grounding conductors are acceptable. If the batteries are in a metal enclosure, then the grounding conductor for the battery enclosure will need to be sized according to the 175 A inverter disconnect, which means sizing to a circuit breaker between 100 and 200 A, which requires a #6 equipment grounding conductor. The same criteria of the previous example apply to whether an additional grounding electrode conductor and grounding electrode should be required.

When the wiring design is complete, the designer will need to provide recommendations for programming setpoints for the installer to check. The same considerations used in the previous example also apply to this example, with the additional requirement that the two inverters be programmed in the "stacked" configuration to produce 120/240 V when operating in the backup mode.

6.5 AN 18 KW BATTERY-BACKUP SYSTEM USING INVERTERS IN TANDEM

6.5.1 Introduction

As commercial enterprises experience prolonged power outages, the interest in larger battery-backup systems is becoming more common. Only a few years ago, most battery-backup systems were less than 11,000 W. It is now possible to find fully UL 1741-compliant battery-backup inverters in the 100 kW range [7]. This example will show the design of an 18 kW battery-backup system that supplies 120/240 V backup power.

Fitting 18 kW on a residential installation would require quite a substantial roof. But it is not at all unusual for a commercial installation to have a roof that will accommodate an 18 kW system. This design will be for a commercial building with a flat roof in Lexington, Kentucky, located at 38° N latitude. The ambient temperature range for the design, based on historic Lexington weather data [8], has been chosen to be from –10°F (–23.3°C) to +100°F (37.8°C).

The roof is a membrane roof applied over a layer of insulation that is attached to a corrugated steel structure supported by metal trusses that will support an average of 20 psf. In fact, the maximum daily snowfall rarely exceeds 9 in. [9]. Assuming a "wet" snowfall, would imply a density of approximately 20% [10], or 9.4 lb/ft^2 for a 9 in. snowfall. The building is located in exposure category B with a design wind load of 90 mph. The roof height is 15 ft, and the roof has a 3 ft parapet around the roof perimeter.

After a serious ice storm that left them without utility electrical power for 10 days, the owners have decided to invest in a battery-backup PV system to provide standby power for their furnace blower motors and igniters as well as some lighting and commercial refrigeration equipment, as shown in Table 6.10. The owner already has a 15 kW fossil fuel backup generator, and the equipment of Table 6.10 is already connected to the backup generator via a manual transfer switch. The building is served by a 400 A 120/240 V single-phase system. The plan is to configure the standby system so the generator will only operate if the PV system does not meet the standby load requirements. Figure 6.6 shows the roof, including air-conditioning units and vent fans.

6.5.2 Inverter and Charge Controller Selection

6.5.2.1 Inverter Selection

Because it has already been determined that it is desirable to install an 18 kW PV system, this time the inverters will be selected first, along with accompanying charge controllers. The inverting for this system will be accomplished by a parallel combination of three 6 kW inverters that operate with a 48 V nominal dc input from batteries. Each inverter has a 120/240 V single-phase output. The three inverters can be connected in parallel such that a system controller will ensure that the inverters all provide outputs that are in phase in the event that utility power is lost. When the

Table 6.10 Summary of Standby Loads for 18 kW 120/240 V DC-Connected Small Commercial Battery-Backup System

Load Description	Load (*VA*)	Line 1 (*A*)	Line 2 (*A*)	Load (kWh/Day)
Refrigerator # 1	144	1.5		1.2
Refrigerator #2	144		1.5	1.2
Furnace blower #1	700	3.6	3.6	6.0
Furnace blower #2	700	3.6	3.6	6.0
Furnace blower #3	700	3.6	3.6	6.0
Lighting circuit 1	1200	10		2.5
Lighting circuit 2	1200		10	2.5
Lighting circuit 3	1200	10		2.5
Lighting circuit 4	1200		10	2.5
Computer circuit 1	300	2.5		2.4
Computer circuit 2	300		2.5	2.4
Computer circuit 3	120	1.0		1.0
Computer circuit 4	120		1.0	1.0
Coffee maker	1000	8.3		1.0
Microwave oven	900		7.5	1.0
Total	9,928	44.1	43.3	39.2

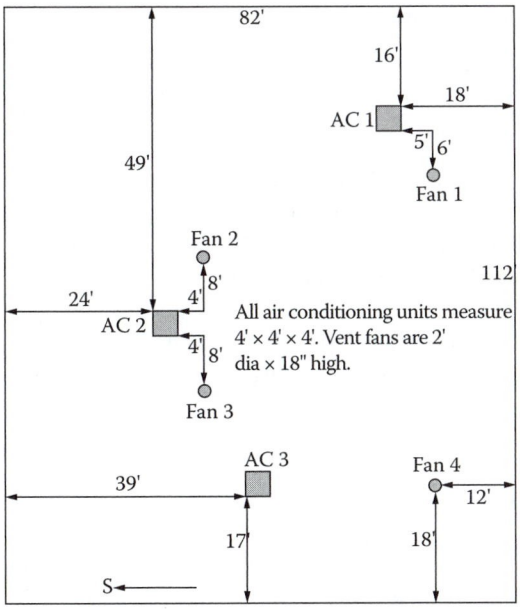

Figure 6.6 Commercial flat roof showing air conditioners and vent fans.

utility is operational, it synchronizes the inverter outputs. It should be noted that the installation manual for this particular inverter is 146 pages long. This illustrates the importance of developing an understanding of the capabilities of this type of system with the knowledge that when designing a system, the designer will need to be familiar with the contents of the installation manual.

Each inverter has the following basic specifications. Additional specifications will be mentioned as needed:

Output power	6,000 W continuous or 12,000 W for 10 s
Maximum surge output current	105 A @ 120 V or 40 A @ 240 V
AC input connections	Grid and generator
AC input breaker	2P 60A
CEC weighted efficiency	92.5%
AC output voltage	120/240 V
DC input current at rated power	130 A
Nominal DC input voltage	50.4 V
DC input voltage range	44–64 V

The inverters come with optional conduit enclosures and an ac/dc power distribution panel that houses all the system disconnects. The overall integrated power-conditioning system also includes MPPT charge controllers, a system control panel, and automatic generator start. The system control panel and automatic generator start will be discussed as the system design progresses.

6.5.2.2 Charge Controllers

In order to determine the number of charge controllers that will be needed in this system, it will first be necessary to look at the specifications of the charge controllers that are available. It makes sense to select charge controllers manufactured by the inverter manufacturer because they can then be installed to communicate with the inverters.

In this case, at the time of this writing, only a 60 A charge controller is available from this manufacturer. Because 60 A MPPT charge controllers should normally be limited to 3200 W of array power for a 48 V system, and because this system will have a nominal 18 kW array, the system will require six charge controllers as long as the array size does not exceed 19,200 W and as long as the array output power can be distributed evenly among the charge controllers. As a final note, the charge controllers have internal ground fault detection and interruption, so no additional GFDI devices will be needed.

6.5.3 Module Selection and Array Layout

The roof of the building is typical of many flat-roofed buildings. It has an abundance of other equipment on it, all of which needs to be checked for shading of the PV array. This book could not possibly be complete without presenting such an example. So the first step in the design is to determine how to locate the PV modules, which means doing a shading analysis on the existing rooftop-mounted equipment. As mentioned in Section 2.5.2.2, there are a number of good shading-analysis computer programs, but in this case, use will be made of Chapter 2 equations that describe the position of the sun in order to prove that the equations can be useful.

Usually it is adequate to look at the position of the sun on the first days of each of the four seasons over approximately an 8 h period centered on solar noon. More than likely, the exposure period will need to be reduced to 6 h on December 21, because otherwise shadows will be quite long. To optimize system performance, it is desirable to be sure that the array will receive at least 6 h of unshaded sunlight every day of the year, neglecting cloud cover. This analysis can be done quite readily with the help of a spreadsheet that is equipped with the formulas for solar altitude and azimuth as functions of latitude, solar declination, and time of day. In other words, it will be necessary to tabulate values of ϕ, δ, and ω in order to compute values of α and ψ using Equations (2.9) and (2.10) in Chapter 2. Once α and ψ are known, the length and direction of shadows can be determined in order to determine both the spacing between module rows and the clearances between modules and rooftop equipment. But before the spacing between module rows can be determined, a module must be selected.

6.5.3.1 Module Selection

In this case, the building owner's cousin is dating the sister of a distributor who can supply a particular multicrystalline silicon module at an attractive cost. The modules are cosmetically blemished, but carry the full UL listing and full manufacturer's warranty. The local building official has approved their use. The modules are 210 W units that have $V_{OC} = 23.1$ V, $V_{mp} = 18.7$ V, $I_{SC} = 12.2$ A, and $I_{mp} = 11.23$ A. The temperature coefficients are $\Delta V_{OC}/\Delta T = -0.32\%/°C$ and $\Delta V_{mp}/\Delta T = -0.43\%/°C$, and the module NOCT = 44.8°C. The modules measure 65.0" × 37.5".

Thus, at the lowest ambient temperature, V_{OC} will increase to 23.1[1 − (25 − (−23.3))(−0.0032)] = 26.7 V. At the highest temperature and highest irradiance level, the module will operate at a cell temperature that is 31°C above ambient, or 68.8°C (see Problem 6.7). This will reduce V_{mp} to 18.7[1 − (25 − 68.8)(−0.0043)] = 15.18 V.

Noting that the array V_{OC} must not exceed 150 V so that the charge controllers will not be damaged, the maximum number of modules in a source circuit is limited to 150 ÷ 26.7 = 5.62, which rounds down to 5. Thus, the maximum array V_{OC} will be 133.5 V. At high temperatures and high insolation levels, the array V_{mp} will decrease to 75.9 V. At an array temperature of 25°C (i.e., STC), V_{mp} will be 93.5 V.

Because the rated power of a 5-module source circuit is 1050 W, it will take 18,000 ÷ 1050 = 17.14 source circuits to obtain 18,000 W. If 18 source circuits are used, then the array power will be 18,900 W. If 17 source circuits are used, the array power will be 17,850 W. So how many source circuits should be used? As far as the design is concerned, either will work. With 18 source circuits, 6 charge controllers will be used, each of which will control the output of 3 source circuits. With 17 source circuits, 5 charge controllers will control 3 source circuits each, and one will control 2 source circuits. Because no additional charge controllers or inverters will be needed if the additional source circuit is included, the overall cost-effectiveness of the system is improved if 18 source circuits are used. Because the owner has already demonstrated an appreciation of cost-effectiveness, the system will be designed with 18 source circuits, such that a total of 90 modules will be used in the system.

6.5.3.2 Array Layout

In this case, the owner has expressed great concern over roof penetrations and has shown a strong preference for somehow mounting the array without penetrations. Because this owner is not the first to express such concerns, racking manufacturers have developed a collection of ballasted array mounts that use concrete blocks in pans to hold down the array rather than bolts through the roof. So, it is simply a matter of selecting a suitable array mount and determining the row spacing.

Because the latitude of Lexington is 38°, it would be nice to have a tilt of at least this much, if possible, to maximize winter solar harvest. So, the search will be for an array mount that will allow a tilt of approximately 45° without roof penetrations, and will have a dead load low enough that the roof will not collapse under the combined weight of the array mount, array, and a good snowfall. Because the building is in a 90 mph design wind speed zone, which is a typical wind speed for most of the United States, it can be expected that if a commercial array mount can be found that will allow a nonpenetrating, 45° tilt, it would be engineered to withstand a 90 mph wind speed. It will also be assumed that if the wind were to blow at a speed of 90 mph, it would blow most of the snow off the roof, so maximum wind load and maximum snow load will not occur simultaneously. In fact, the likelihood of a 90 mph wind in the winter is extremely small.

Figure 6.7 shows how the row spacing can be determined. The procedure is to solve Equations (2.9) and (2.10) in Chapter 2 on an hour-by-hour basis over the

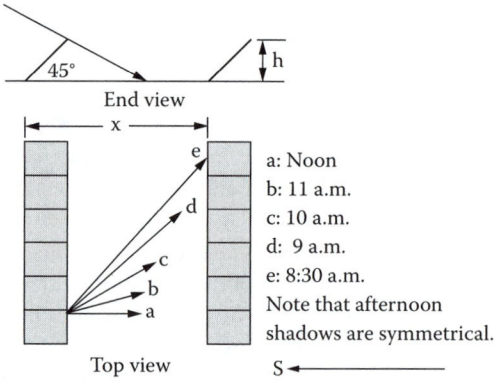

Figure 6.7 December 21 shadow analysis to determine row spacing.

period during which it is desired to have the array unshaded. This is probably most easily done by developing a spreadsheet to do the repetitive calculations. By solving for altitude and azimuth angles during the period over which the modules are desired to be unshaded, it is possible to determine the length and direction of shadows during this period. Then, either a little trigonometry or a little playing with a scaled drawing can produce the desired shadows, and, subsequently, the necessary spacing of modules from one another and from objects on the roof, such as fans, air conditioners, and parapets. Normally the spacing will be determined by the shadows on the first day of winter, but because the sun follows a limited arc from southeast to southwest in the winter, there may be obstructions to the east, northeast, west, or northwest that might cause shading during either the swing season or during summer months. So these possibilities should be checked. It is left as an exercise for the reader to calculate the necessary distance between rows if it is desired to have 7 h of unshaded operation on December 21 (Problem 6.8).

Figure 6.8 shows the analysis of shading of the three air conditioners and the parapet walls. Each air conditioner is a cube, 4 ft on a side. The parapet walls are 3 ft high. The dotted line in the figure shows the extent of shadows caused by the parapet wall on December 21 at 8:30 a.m. and 3:30 p.m. sun time. On all other days, the shadows are shorter, and during the 7 h interval between 8:30 a.m. and 3:30 a.m., the shadows are also shorter.

The air conditioner shade patterns show the extent of shading on the first day of winter, the first day of fall, and the first day of summer. So the modules must be placed outside these shadow zones. Finally, Figure 6.8 shows one possible configuration of the modules to avoid shading as much as possible. Note that the array layout has been designed to be as installer-friendly as possible. Because source circuits consist of groups of five modules, the modules are grouped in multiples of five. The ten small dark rectangles indicate reasonable locations for junction boxes to transition from open wiring to wiring in conduit. The idea is to bring six source circuits down in each of three conduits to each of three source circuit combiner boxes that will feed two charge controllers each. To avoid roof penetrations, the conduit to the source circuit combiner boxes will pass through the parapet, down along the outside of the building, and into the equipment room.

For ballasted array mounts, the manufacturer will normally specify the number of ballast blocks needed for a given module orientation and a given module tilt. In this case, the modules are mounted in portrait at a tilt of 45°. The manufacturer's specifications call for 72 blocks per 5-module group, each of which weighs 14.5 lb [11]. Thus, for an array of 90 modules, where each module weighs 38 lb and the frame weighs 45 lb per module, the total weight of all array components is 26,262 lb. Because the area of the roof is 9184 ft^2, this amounts to an additional average dead load on the roof of 2.86 lb/ft^2, which is well within the allowable load on the roof.

Finally, one can note that the per-module weight of a ballasted system is normally calculated to be 150% of any uplift from wind forces. Furthermore, the downward wind forces will not exceed the upward forces, so any additional downward wind force that must be added to the dead load will not exceed the dead load itself;

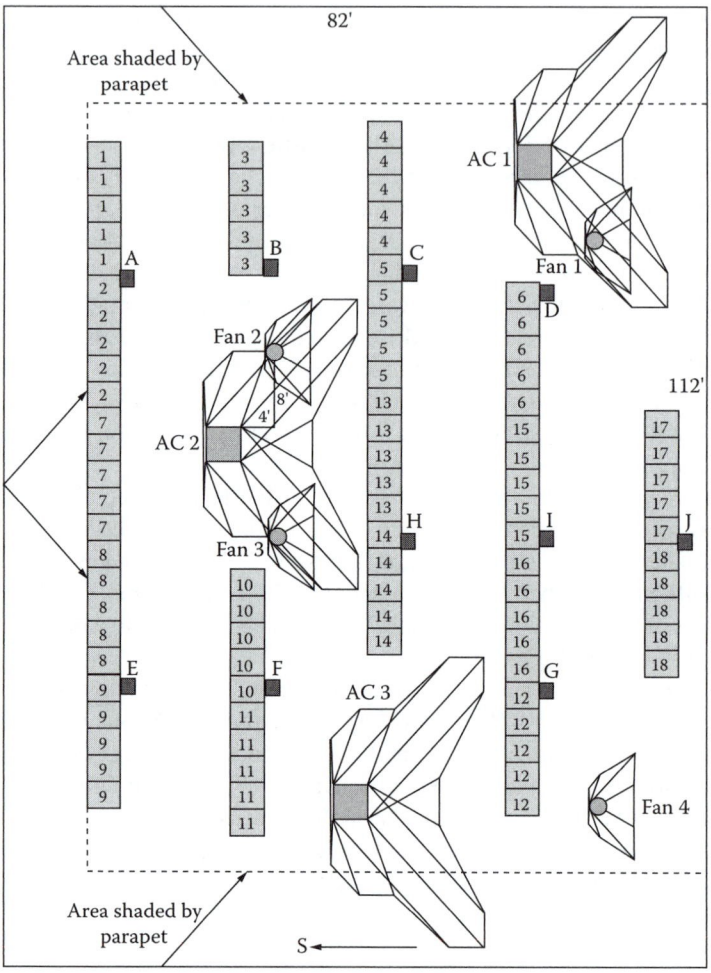

Figure 6.8 Array layout showing equipment and parapet shading patterns.

so it is safe to add another 2.86 lb/ft^2 as a wind load to obtain a total load on the roof of 5.72 lb/ft^2 as a worst-case load on the roof by the PV system. Thus, the maximum load from wind, snow, and the PV system will not exceed 15.1 psf, so the roof is adequately designed to withstand all the forces simultaneously.

6.5.3.3 Array Performance

Now that the array has been defined, PVWatts [2] can be used to estimate the monthly and annual system energy production. Using an array size of 18,700 W, with a 45° tilt in Lexington, Kentucky, PVWatts predicts an annual energy production of 22,021 kWh, with a winter monthly average of 1,437 kWh/mo, a fall average of

1,767 kWh/mo, a spring average of 2,052 kWh/mo, and a summer average of 2,084 kWh/mo. Thus, the average daily kWh production is 47.9 in winter, 58.9 in fall, 68.4 in spring, and 69.4 in summer. Comparing these figures with the desired standby loads shown in Table 6.10 shows that for each season, the predicted system energy production exceeds the energy needed to power the standby loads. When the utility is up, this energy will be supplied to other building loads, and possibly some will be sold back to the utility.

6.5.4 Battery and BOS Selection

The battery size is determined by examining the energy requirements of the standby loads as shown in Table 6.10. Assuming a wiring efficiency between inverter and loads of 97%, an inverter efficiency of 92%, and a 90% battery charging/discharging efficiency, this means that the amount of energy that must be delivered to the batteries to meet the load requirements is $39.2 \div 0.97 \div 0.92 \div 0.90 = 48.8$ kWh/day.

Because the owner is serious about keeping the business open, but also because the owner has a fossil generator for additional backup power, only one day of battery storage will be included in the design. The batteries are sized according to the energy that must be delivered by the batteries, which is $0.9 \times 48.8 = 43.9$ kWh. Of course, in order to deliver 43.9 kWh, the batteries will dissipate some energy in the process of discharging, so it is a good idea to allow an extra 5% for battery losses. Allowing for an 80% depth of discharge and 5% battery discharge loss, the batteries must be capable of storing 57.8 kWh @ 48 V at a 24 h discharge rate. This means the Ah rating of the batteries must be $57{,}800 \div 48 = 1{,}204$ Ah.

Because good design practice is to design the battery bank with fewer than 4 parallel sets of batteries, a search can be made for batteries that are either rated near 1204 Ah or 602 Ah or 401 Ah or 301 Ah. A decision also needs to be made as to whether the batteries should be flooded or sealed.

Flooded or sealed, it turns out that a typical L-16 size battery will be rated at approximately 400 Ah @ 6 V @ 24 h discharge rate. So 24 of these batteries, connected in 3 series sets of 8 batteries each, will store the necessary energy. It will be up to the owner to check the prices and then decide whether to purchase flooded or sealed units.

6.5.5 Wire Sizing

As in previous examples, Table 6.11 summarizes the calculation of wire ampacities. Note that for this system, it will be assumed that the power-conditioning equipment, including the batteries, will be located in a room in the center of the north side of the building. The cabling between batteries and equipment will be less than 10 ft. Source circuit combiner boxes will be located near the power-conditioning equipment. Lengths of source circuit conductors include the equivalent two-way length of open wiring plus the length of conductors from junction box to source circuit combiner box. Three separate conduit runs will carry the source circuit outputs to three source circuit combiner boxes, each of which will feed two of the six

Table 6.11 Determining Wire Sizes for the 18 kW Battery-Backup Grid-Connected System

Circuit	I_{sc} (A)	$1.56/_{sc}$ (A)	Wire Size	Wire 30°C Ampacity (A)	Temp Derate	Cond Fill Derate	Derated Ampacity (A)
PV source circuits	12.2	19.03	#8	55	0.71	0.50	19.5
PV output circuit	36.6	57.1	#6	75	1.0	1.0	75

Charge Controller And Inverter Calculations

	I	$1.25I$	Wire Size	Wire 30°C Ampacity (A)	Temp Derate	Cond Fill Derate	Derated Ampacity (A)
Ch cont output circuit[a]	57.2	71.5	#4	95	1	1	95
Inverter dc input, I_{max}	130	162.5	#4/0	260	1	1	260
Inverter ac input (A)[b]	50	62.5	#6	75	1	1	75
Inverter ac output (A)[b]	50	62.5	#6	75	1	1	75

[a] Charge controller output limited to 60 A continuous.
[b] Surge current ratings. Continuous ratings are 25 A.

charge controllers. Thus, each conduit will have 12 current-carrying conductors and a ground wire. In order to determine voltage drop, it will be necessary to determine the length of the wires.

Starting with the modules, each module has #10 positive and negative leads, each of which is 42 in. long. Thus, for a 5-module source circuit, the total length of module leads will be 420 in., or 35 ft. Note that this is equivalent to a one-way distance of 17.5 ft for two side-by-side conductors. This will be the distance, *d*, used in the voltage drop formula.

Next, note that most of the junction boxes shown in Figure 6.8 are located at the ends of source circuits. Thus, one of the source circuit leads will connect directly into the junction box, and one additional extension cable will be needed for connecting between the junction box and the last module of the source circuit. The length of this single wire will need to be approximately 13 ft (see Problem 6.9), which is equivalent to a one-way distance of 6.5 ft for 2 wires. So the total equivalent one-way length of open wire for source circuits with junction boxes at one end will be approximately 24 ft. The only exception will be source circuit 7, where open wiring will need to extend to junction box E. So this adds another 16 ft to the one-way length, resulting in a total of 40 ft for the open wiring of source circuit 7.

The next step is to determine the percentage voltage drop between module and rooftop junction box. For all but source circuit 7, using Equation (4.1) gives %VD = 0.70. For source circuit 7, %VD = 1.16. Thus, for the remainder of each source circuit, no more than 1.3% VD can be allowed for 17 circuits, and 0.84% is all that can be allowed for source circuit 7. So, to determine the necessary wire sizes between rooftop junction boxes and source circuit combiner boxes, it is now necessary to determine the distance between junction boxes and combiner boxes for each source circuit.

Noting that at each junction box there will be a 1 ft vertical rise for each conduit entering the box, the conduit lengths between box and parapet can be determined. For example, a conduit with wiring from source circuits 1 and 2 will run from junction box A to junction box B. Then another conduit will run from B to C with source circuits 1–3. Another conduit will then run from C to D containing conductors from source circuits 1–5. Finally, a conduit will run from D to the wall and then across and down to the source circuit combiner boxes with source circuits 1–6.

On a scaled drawing, one can measure the distances between junction boxes to determine the distances between junction box and wall, and then the remaining distance to the combiner boxes can be added. Alternatively, if the scaled drawing is on a computer, it may be possible to use the drawing program to determine the conduit length.

Similar conduit runs will be installed between E, F, and G and the wall and between H, I, and J and the wall. The E-F-G run will contain source circuits 7–12, and the H-I-J run will contain source circuits 13–18. The resulting distances between junction box and combiner box are listed in Table 6.12, along with the wire sizes needed between junction box and combiner box for each source circuit to meet the voltage drop constraint. Once again, it is left as an exercise for the reader (Problem 6.11) to calculate the actual total voltage drops for each source circuit to verify that the wire sizes listed are adequate.

An important observation in Table 6.12 is that, while #10 conductors will meet the voltage drop requirements for source circuits 15–18, it will be necessary to use #8 anyway to meet the ampacity requirements of the wire, as noted in Table 6.11 for conduit filled with 12 current-carrying conductors.

The inverter manufacturer recommends using 4/0 battery cables with 250 A circuit breakers for each inverter in order to handle surge currents. Under normal operation, the battery cable current will not exceed 130 A.

Table 6.11 also shows the wire sizes for the ac side of the inverters. Up to the busbar side of the inverter terminals, #6 is used, protected by 60 A circuit breakers. The outputs of all the inverter inputs are combined at the busbars so that the wires

Table 6.12 Summary of Source Circuit Lengths between Junction Box and Combiner Box Showing Wire Sizes Needed to Meet Voltage Drop Constraints

Source Circuit	**1**	**2**	**3**	**4**	**5**	**6**	**7**	**8**	**9**
Length	102	102	82	67	67	55	103	103	103
Wire	#6	#6	#6	#8	#8	#8	#4	#6	#6
Source Circuit	**10**	**11**	**12**	**13**	**14**	**15**	**16**	**17**	**18**
Length	85	85	50	55	55	37	37	18	18
Wire	#6	#6	#8	#8	#8	#10	#10	#10	#10

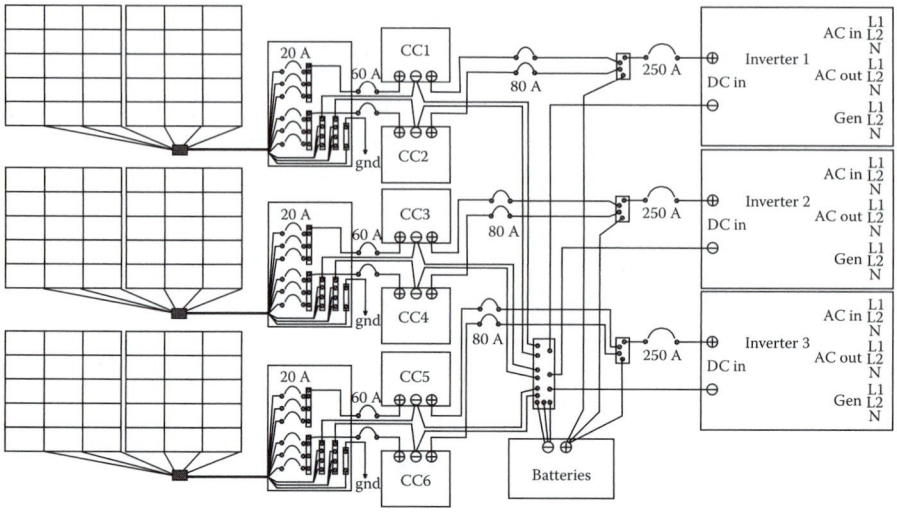

Figure 6.9a DC wiring of 18 kW battery-backup system.

to the utility interconnect point, the generator, and the standby loads will be larger than #6. These wire sizes will be determined in the next section.

6.5.6 Final Design

Figure 6.9 shows the final system design. Note that the dc schematic of Figure 6.9a shows the array to be broken down into six subarrays of 15 modules each, such that 3150 W of array is connected to each charge controller array input. Source circuits are protected from backfeed by the 20 A circuit breakers in the source circuit combiner boxes, each of which combines source circuits for two charge controllers. The 60 A circuit breakers at the inputs of the charge controllers serve as overcurrent protection as well as array disconnects. The 80 A circuit breakers at the outputs of the charge controllers serve as disconnects for the charge controller outputs. The inverter inputs can be connected as shown with three busbars to provide connections to the battery for an inverter and two charge controllers each, or by a single large busbar with connections to all inverter inputs, three parallel 4/0 or equivalent to the batteries, and #4 wiring from the charge controller outputs. Note also the three parallel 4/0 conductors from battery negative to the negative busbar. The busbars must be rated to carry 125% of the full load inverter input currents, depending on the number of inverters connected to each busbar.

It should also be noted that the combined outputs of the six charge controllers, at 60 A each continuous, is 360 A. This is a high charging current for the batteries. Because the batteries have a total capacity of 1200 Ah @ 48 V, a charging current of 360 A represents a $C/3.33$ charging rate. Of course, this is when the charge

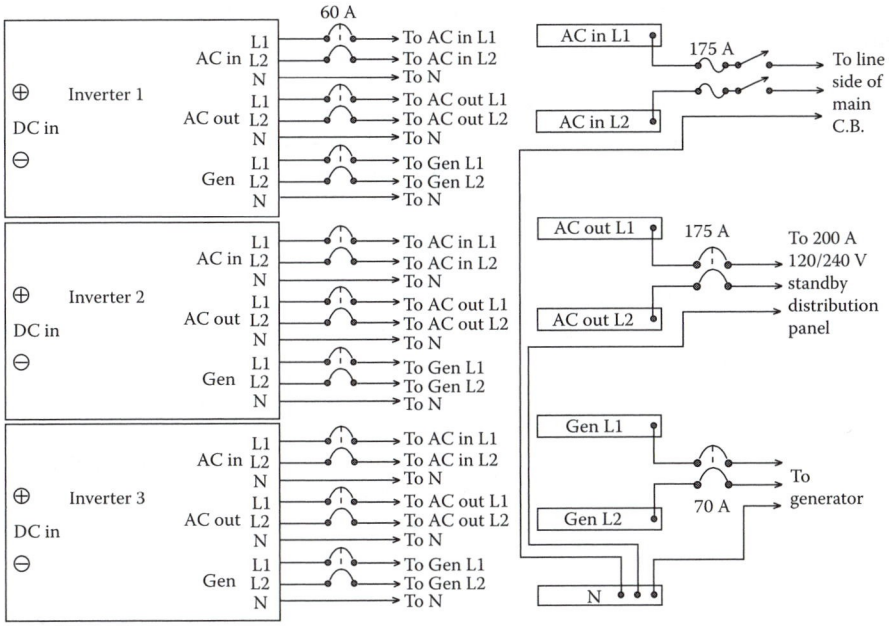

Figure 6.9b AC wiring of 18 kW battery-backup system. Grounding conductors and control wiring not shown.

controllers are operating at full power, which will likely not be the case, at least on a continuous basis. The charge controller manuals should be consulted for appropriate settings of absorption time for this charging rate.

The ac sides of the inverters are wired as shown in Figure 6.9b. In this figure, the author had to decide whether to present a confusing mess of wires crossing wires, or to simply indicate the appropriate busbar to which each inverter ac-side wire should be attached. The figure shows which choice was the winner. The bottom line is that all of the ac ports of the three inverters are paralleled, so the busbars simply provide the points for achieving the parallel connections. Thus, the input/output currents can be triple the ratings of single inverters.

The generator circuit breaker is only 70 A because the generator is only rated at 15 kW. Note that 15 kW ÷ 240 V = 62.5 A. It is good practice to avoid overloading mechanical generators because overloading can cause the prime mover to stall. Hence, the circuit breaker is sized at just under 115% of this value, or 70 A. The inverters can be programmed to limit the generator current to 90% of its power rating so that it will operate at maximum efficiency when it is running. When the generator runs is another programming task. The inverters can be programmed to start the generator when the battery voltage reaches a predetermined value, such as 48 V, and to stop the generator when the batteries reach 52 V. The system controller also incorporates a real-time clock so that the generator can be prevented from starting after a certain hour in the evening, or before a certain hour in the afternoon. Note that the generator is only expected to be used on cloudy days or at night when there is a utility outage. This means that monthly maintenance/starting of the generator will be necessary to ensure that the generator is available when it is needed.

Because the AC IN and AC OUT busbar outputs are protected by 175 A circuit breakers, this means that 2/0 copper will be needed, because the distance between inverter output and either the utility connection or the standby distribution panel is relatively short.

For this system, the manufacturer provides a system control panel and an automatic generator start accessory [10]. The system control panel enables programming of all the programmable system components, including inverters, charge controllers, and the automatic generator start accessory. The automatic generator start has three built-in relays to support 13 preset generator configurations. It monitors the battery voltage as well as the inverter charge mode. It also monitors the system output power and can start the generator if the system output power needs cannot be met by the inverter. It also can be set up to automatically exercise the generator so the generator field will not become demagnetized or the generator gasoline will not turn into gelatin.

Note that the system output must be connected to the line side of the 400 A building main disconnect, because 175 A exceeds 120% of the busbar rating of a 400 A electrical panel. The only way this 175 A of PV current could be fed to a 400 A electrical panel would be if the panel main circuit breaker were no larger than 300 A because 120% of 400 A is 480 A and $300 + 175 < 480$. If the system AC IN is connected to the line side of the main breaker, then the AIC (asymmetrical interrupting capacity) rating of the 175 A overcurrent devices at the point of utility connection must be no smaller than the AIC rating of the main breaker. The AIC rating of a circuit breaker indicates the maximum short-circuit current that can flow through the circuit breaker and still cause the circuit breaker to trip.

The critical reader may also have noticed the absence of any inverter bypass circuit breakers in this system. There are several reasons for this. One is that the component enclosure only has room for nine 2-pole ac circuit breakers. The other reason is that the likelihood of all three inverters failing simultaneously is remote. Thus, if one inverter should fail, it can be isolated and repaired while the other two inverters continue to operate.

Finally, note that the schematic diagrams do not show all the grounding conductors and do not show any surge arresters. Good design practice would involve incorporating surge arresters at the source circuit combiner boxes at the outputs to each charge controller. The grounding conductors between various system components are selected in accordance with *NEC* Table 250.122. For example, because the wiring from busbars to the PUC is protected by a 175 A circuit breaker, the grounding conductor must be no smaller than #6 copper. On the array side of the charge controllers, where the conductors are protected by 60 A circuit breakers, #10 grounding conductors will suffice. If the batteries are in a metallic case, then the case will need to be grounded with a #1/0, because the total circuit breaker rating for battery cables is 750 A.

6.6 AC-COUPLED BATTERY-BACKUP SYSTEMS

6.6.1 Introduction

While the dc-coupled battery-backup systems of the previous examples have been popular, there are situations where ac-coupled battery-backup systems may be preferred. One reason might be that a system owner has installed a straight grid-connected system without batteries and has then experienced a lengthy power outage. Regrettably, the system owner then had an idle PV system until the utility came back on. Another possibility is that the designer or owner has noticed that dc-coupled battery-backup inverters tend to have efficiencies in the 92%–94% range, while straight grid-connect inverters have efficiencies in the 95%–97% range. Furthermore, the dc-coupled systems encounter additional losses to the batteries, even though they remain essentially fully charged. If the charge controller outputs are pulsed, to some extent the pulse enters the battery and then discharges to the inverter while the pulse is off, causing charge/discharge battery losses to the battery internal resistance. If the array output is converted to ac first, then feeds the standby loads, and then passes through to the grid, the process can be more efficient.

Another possibility is having more than one renewable source in the system, such as wind or low-head hydro. Then it will likely be desirable to incorporate all of these as sources of standby power if the grid should fail. Because all of these sources must comply with UL 1741 for grid connection, the easiest way to incorporate them all is via ac coupling. Another reason for ac-coupled systems is to take advantage of high-voltage source circuits.

These systems are not without disadvantages. One disadvantage is higher first cost that will presumably be justified by any or all of the preceding reasons. Another disadvantage is the possible power limitations of the battery-backup inverter, as well as the energy storage limitations of the batteries. These systems tend to be best suited when the standby loads are coincident with the power available from the PV array and the nighttime standby loads are minimal.

The trick is to keep the straight grid-connected inverters operating when the utility grid is down by introducing a battery-backup source that will look like the utility to the other inverters and will pass excess power production back to the utility when the standby loads are satisfied. This section will not attempt a complete system design, because all the component systems have already been designed. Rather, the focus will be on the basic ac-coupling concept. The array, load, and battery calculation procedures are the same for these systems as they were for previous systems. It should be emphasized that before attempting to design ac-coupled systems, the designer should be very familiar with the installation manuals and any associated application notes available from the inverter manufacturers [12].

6.6.2 A 120/240 V Battery-Backup Inverter with 240 V Straight Grid-Connected Inverter

The ac coupling problem can be straightforward if a 120/240 V battery-backup inverter is used in conjunction with a 240 V straight grid-connected inverter. The system takes advantage of the well-kept secret that the AC OUT port of most battery-backup inverters can allow power flow in either direction. Normally these ports are considered as supply-only ports. Ideally, the battery-backup inverter will have a contactor or electronic switch that will connect directly from AC OUT to AC IN ports when the grid is up. This way, when the grid is up, the grid functions as the synchronization signal for the straight grid-connected inverters. If the grid goes down, then the contactor opens and the AC OUT port functions as an ac voltage source that has utility voltage characteristics, namely, stable amplitude and stable frequency of a good sine wave, while the AC IN port is disconnected from the utility except for monitoring capability. There are limits to this system, because the grid can only supply as much power as the battery-backup inverter will pass through, which is typically limited to 60 A for each battery-backup inverter used. The other limits are the battery capacity and the inverter output power rating when the grid is down, which may be less than the pass-through capacity. During hours of sunshine, depending on the power ratings of the array and the straight grid-connected inverters, the maximum loads that can be served are essentially dictated by the array power and the dc:ac conversion efficiency. But at night, the loads must be restricted to whatever the batteries can supply unless another source is present, such as wind or hydro.

This system is shown in Figure 6.10, which illustrates the internal switching arrangement of the battery-backup inverter along with a necessary control link between the two inverters. The diagram shows that it is also possible to incorporate an external fossil fuel generator as well. No grounding conductors are shown in the diagram.

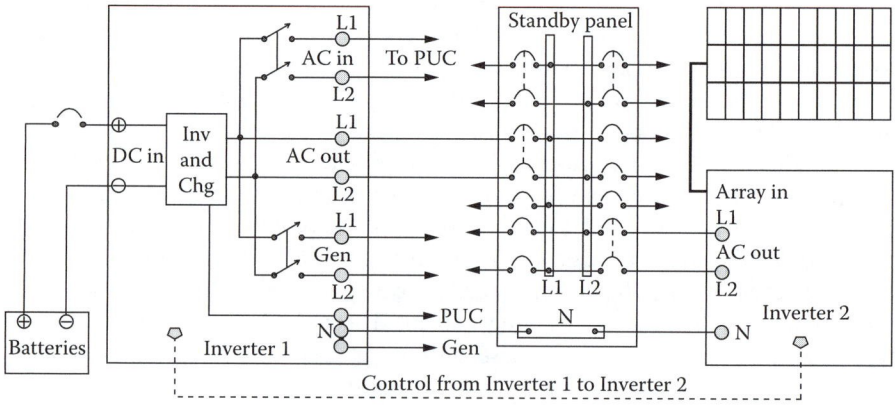

Figure 6.10 Basic ac-coupled system with 120/240 V battery-backup inverter and 240 V straight grid-connected inverter.

The first thing to notice is that the PV array is connected in the usual fashion to Inverter 2, as though Inverter 2 were operating in a straight grid-connected configuration. No PV is connected to the input of the battery-backup inverter. If it were to be connected, then the system would involve dc coupling of the array to the inverter via batteries, and the ac coupling advantage would be lost.

Next, note the internal switching of Inverter 1. The AC OUT port is always connected directly to the ac output of the invert/charge electronics of the inverter. The AC IN port will be connected to the AC OUT port if the utility stability requirements of UL 1741 are met. If the utility power is lost, then the switch to the AC IN port is opened automatically by the inverter, and power to the AC OUT port is supplied by the inverter, using battery backup if no power is available from the PV array. The generator input port switch is left open unless the inverter is programmed to accept power from a generator.

The system has several possible operating modes. Normal daytime operation with the grid available involves the grid being directly connected to the ac port of Inverter 2 via Inverter 1 and the standby panel. This enables the operation of Inverter 2. If the PV array produces sufficient power to operate the standby loads with leftover power, then the leftover power flows back to the grid through Inverter 1 as long as the batteries are fully charged. Note that this means the batteries are charged by rectifying power from the ac side of Inverter 1, the origin of which may be either the grid or the output of Inverter 2. It also means that Inverter 1 will have the charger settings that would normally be assigned to a charge controller if a charge controller were present. But because the array is on the ac side of this inverter, no external charge controller is needed.

Another possible operating mode is when the grid is present, but the sun is not. Thus, the straight grid-connected inverter is not capable of supplying any power to the standby loads. This power must be supplied by pass-through of the AC IN grid connection to the AC OUT port. This means that the pass-through current rating of Inverter 1 must be adequate to supply the nighttime standby loads. If the load on the AC OUT port of Inverter 1 is too large, the inverter will shut down unless the system has a generator that is programmed to assist the battery-backup inverter in such cases.

Another possible operating mode is when the grid is not present, but the sun is present. As long as the battery voltage is high enough to operate Inverter 1, Inverter 2 will interpret Inverter 1 as the grid and will deliver power as soon as power is available from the PV array. If the sun produces adequate power via Inverter 2 to operate the standby loads, any excess power now cannot be sold back to the utility, but can only be used for charging the batteries. But unless a mechanism is present to prevent overcharge of the batteries, it is possible to overcharge and possibly damage the batteries. This is why the control cable is present between Inverter 1 and Inverter 2. If the batteries are fully charged and power is still flowing to Inverter 1 from Inverter 2, then Inverter 1 tells Inverter 2 to shut down to prevent overcharging the batteries. But, then, what happens to the standby power if Inverter 2 is shut down? It is supplied by the batteries via Inverter 1 until the battery voltage has dropped to a level such that Inverter 1 will allow Inverter 2 to start up again.

Finally, it is possible that both the grid and the sun may be down. In this case, the batteries need to supply the standby loads via Inverter 1. If the standby loads are too large, it is possible that the batteries will discharge to the low-voltage cut-out point and the standby loads will go without power. This also presents a control dilemma for the inverters, because if Inverter 1 is shut down, then Inverter 2 will think the utility is down and will remain off. The result is no power available to recharge the batteries. So this challenge must also be met by communication between the two inverters. If an attempt is made to construct such a system, then it is important for the designer to determine whether the inverters are capable of overcoming this situation. If not, it means back to the drawing board.

6.6.3 A 120 V Battery-Backup Inverter with a 240 V Straight Grid-Connected Inverter

Figure 6.11 shows an ac-coupled system with a 240 V output straight grid- connected inverter and a 120 V output battery-backup inverter. When the outputs of two inverters with different output voltage are combined, a transformer must be used to match the voltages. The system shown uses a full sized transformer with isolated primary and secondary windings. Although this is not the common method of coupling the two inverters, it illustrates how the single-voltage primary input can be transformed into a dual-voltage secondary output by using a center-tapped secondary winding. Note that the neutral of the primary side is carried through as the neutral of the secondary side of the transformer. this maintains the ground reference from primary to secondary, while enabling the secondary to provide either 120 V or 240 V as may be required by the loads and the straight grid-connected inverter. Note that although 240 V is available on the secondary side, each of the two ungrounded conductors has a voltage of 120 V to ground, Figure 6.12 shows how the dual-winding transformer

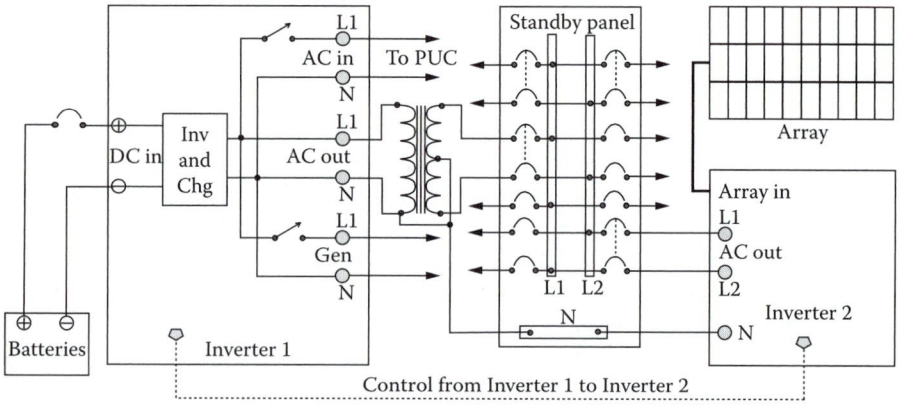

Figure 6.11 Basic ac-coupled system with 120 V battery-backup inverter and 240 V straight grid-connected inverter.

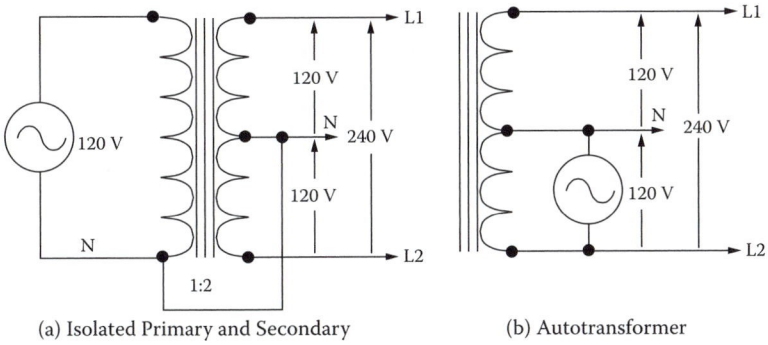

Figure 6.12 Comparison of transformers showing connection of neutral associated with autotransformer.

can be replaced by an auto-transformer as long as the center tap of the autotransformer is used as the grounded conductor.

Just as in the previous example, the battery-backup inverter must be programmable to sell power to the utility when power is being generated by the PV array, the standby loads have been satisfied, and the batteries are fully charged. It must also be programmable to shut down battery operation if the battery voltage is too low and to limit charging if the battery voltage becomes too high if the grid is disconnected. The inverter would be expected to sell to the grid when the grid is available and the batteries are fully charged. It is possible that auxiliary relay contacts on the battery-backup inverter will need to be used to achieve some of these features if they are not directly incorporated into the inverter. The possible operating conditions of this system are the same as the possible operating conditions of the previous example.

6.6.4 A 120/208 V 3-Phase AC-Coupled System

In larger buildings, it is common for the service to be at 277/480 V 3-phase. This voltage is then transformed down to 120/208 V to serve lower voltage loads, while the medium voltage is used for 277 V lighting or 480 V motors. The 120/208 V supply is sometimes delivered as a single-phase 3-wire supply to a single-phase distribution panel and sometimes is fed as a 3-phase, 4-wire supply to a 3-phase distribution panel. For the single-phase supply, the phases remain 120° out of phase rather than 180° out of phase, which accounts for the 120 V line-to-neutral voltage and the 208 V line-to-line voltage. Although essentially all straight grid-connected inverters will operate at either 208 or 240 V (and sometimes 277 V and even 480 V), only a few battery-backup inverters can be controlled to operate 120° out of phase if the utility fails. At the time of this writing, most can only be programmed to operate in phase or 180° out of phase to produce a 120/240 V supply.

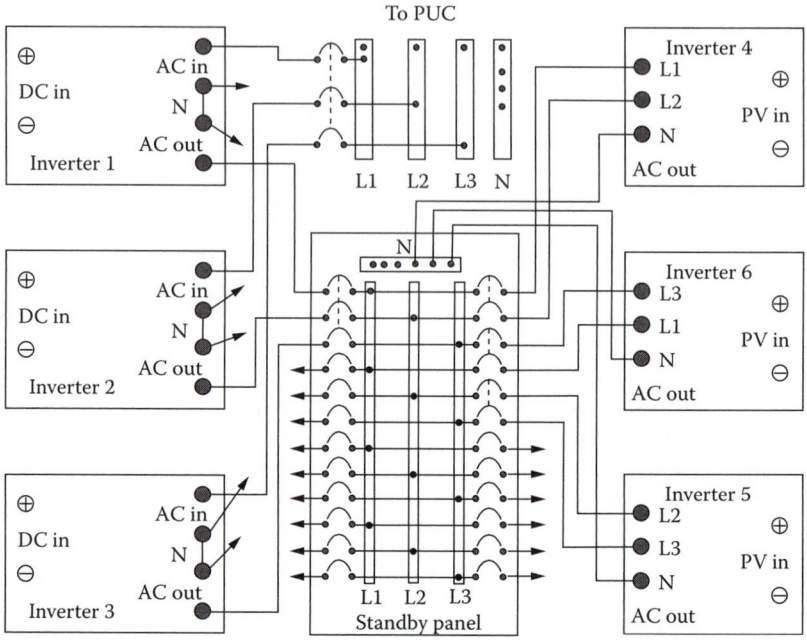

Figure 6.13 Three-phase battery-backup ac-connected system. Grounding conductors and control wiring not shown.

Thus, the primary requirement for a 3-phase system is to have battery-backup inverters that can produce a 3-phase supply that has the same phase sequence as the utility line so that any 3-phase motors on the system will not change rotation direction. Thus, the battery-backup inverters need to remember the phase sequence in the event of utility power loss. Figure 6.13 shows a balanced system that uses three single-phase 208 V straight grid-connect inverters and three 120 V battery-backup inverters. Once again, the design constraints and possible operation modes are the same as those of previous examples.

With 3-phase supplies, however, several other interesting options exist. The reader may be familiar with the open delta utility transformer configuration in which a proportionately larger 120/240 V center-tapped transformer is used to supply a predominance of 120/240 V single-phase loads. If this transformer has utility phases A and B connected to its primary, then a second smaller transformer with a 240 V secondary winding is connected between either phases B and C or between C and A, such that its output will be 120° out of phase with the larger transformer. This transformer configuration is shown in Figure 6.14a. Note that if the voltage drop from A to B is $240\angle 0°$, then the voltage drop from B to C will be $240\angle -120°$, and the voltage drop from C to A will be $240\angle 120°$, for a positive phase sequence. The phasor diagram of these voltages is shown in Figure 6.14b. Thus, a three-phase set of voltages can be produced with two transformers. This suggests that two inverters with the

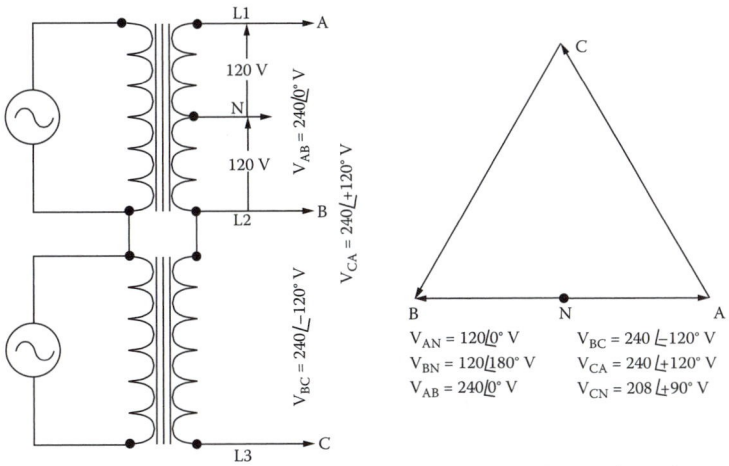

Figure 6.14 Open delta three-phase system.

proper phase relationships between their outputs could also produce a three-phase output. This should be true for either the battery-backup inverters or the straight grid-connected inverters as long as they are properly synchronized. Once again, before attempting to design any such system, it is important to check installation manuals and application notes for any idiosyncrasies in the operation of the inverters. It is also important to verify whether the three-phase supply is a balanced 120/208 V system or a 120/240 V delta system with a hi leg (phase C in Figure 6.14b).

6.7 BATTERY CONNECTIONS

An important, but often overlooked, component of good PV system design and installation is the proper connection of batteries to ensure a balanced current flow in all batteries in the system. If connecting wires did not have resistance, the manner in which batteries are connected would be relatively unimportant. But wire does have resistance, and therefore one needs to consider this resistance when hooking up batteries. In fact, even the terminal lugs have resistance, but this resistance is more difficult to characterize because it will depend on the specific lug type and on how tightly the lug is connected to the wire and to the battery. Connecting lug resistance will also increase over time if any corrosion should occur at the lug. In the examples to follow, the connecting lug resistance will be assumed to be incorporated into the Thevenin equivalent (internal) resistance of the batteries.

Figure 6.15 shows three possible ways to connect eight batteries in a series-parallel configuration. If the batteries are 12 V batteries, the system will produce 24 V. Note that options 1 and 2 show battery-to-battery parallel connecting wires to be

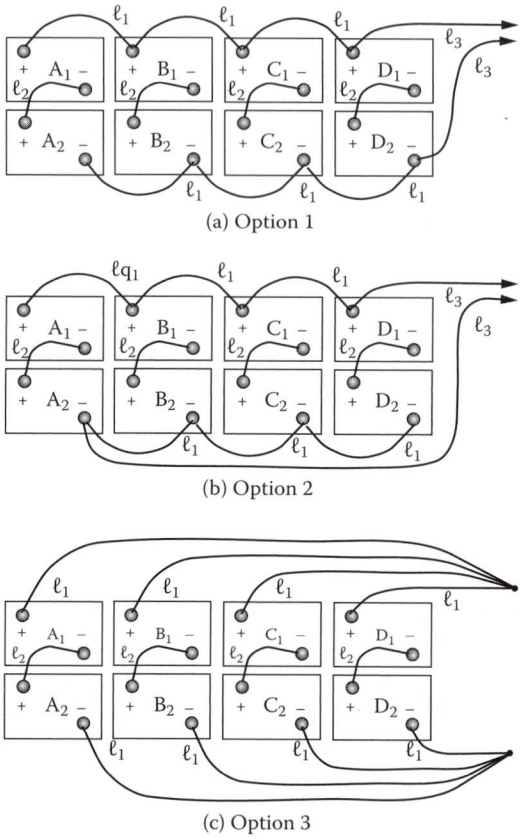

Figure 6.15 Three possible battery hook-up configurations.

of equal length, and thus of equal resistance. The series-connecting wires are also of equal length. If 2/0 copper cables are used, and if $l_1 = l_2 = 1$ ft, then the cable resistance will be 0.0000967 Ω for each of these 1 ft lengths of cable. Typically, the distance from battery to inverter input is about 6 ft. Thus, if $l_3 = 6$ ft, the resistance will be 0.00058 Ω.

It is interesting to calculate the currents that will flow under charging and discharging conditions in the series battery strings for connection options 1 and 2 if all batteries in the system are identical. For example, consider lead-acid batteries for which the open-circuit battery voltages are all 12.60 V and the Thevenin equivalent resistances of the batteries are 0.01 Ω each. According to Table 3.1 in Chapter 3, this would mean the batteries are about 25% discharged. Note that these parameters suggest a possible short-circuit current of 1260 A.

Consider first the charging situation. Figure 6.16 shows the equivalent circuit for option 1. Note that if the current source negative lead is connected to point A

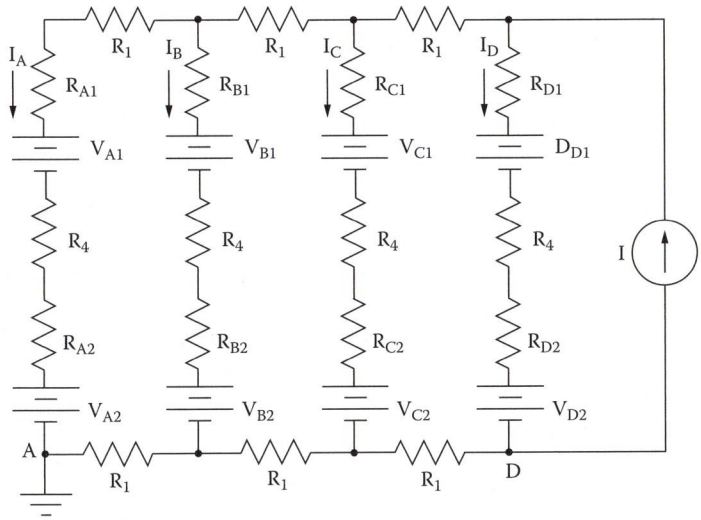

Figure 6.16 Equivalent circuit for battery connection Option 1.

rather than to point D, then the circuit will be equivalent to option 2. Setting $I = 60$ A as a nominal value for either charge or discharge and enlisting the assistance of a convenient network analysis program yields results for charging and discharging for options 1 and 2 as shown in Table 6.13.

It is thus evident that option 2 should be the preferred option for several reasons. First of all, the currents are more closely balanced for all series strings of batteries. Furthermore, the charging currents are equal to the discharging currents, so even though the A and the D batteries are cycled somewhat deeper than the B and the C batteries, the starting and ending points of a full cycle of charge and discharge are the same for all the batteries. Furthermore, when the batteries are in a state of higher discharge, the cell voltage decreases and the Thevenin equivalent resistance increases such that the rate of discharge of the batteries tends to be self-regulated. In other words, as the batteries become more discharged than a parallel set, the batteries at higher charge supply more current to the load than the batteries at lower charge. Under charging conditions, the batteries at lower charge levels should tend to charge faster, depending on their internal resistance.

Table 6.13 Comparison of Charging and Discharging Currents for Options 1 and 2

Situation	I_A (A)	I_B (A)	I_C (A)	I_D (A)
Option 1 charge	14.65	14.79	15.07	15.50
Option 1 discharge	14.65	14.79	15.07	15.50
Option 2 charge	15.07	14.93	14.93	15.07
Option 2 discharge	15.07	14.93	14.93	15.07

The problems with the option 1 connection are also somewhat mitigated by batteries at a lower state of charge delivering less current. But there is still a greater difference in charging and discharging currents for option 1 than for option 2. The option 1 connection causes the D batteries to cycle between a higher level of charge and a lower level of discharge than the A, B, or C batteries. Over many cycles, this may cause the D batteries to degrade sooner than the other batteries. The battery life versus average daily discharge curve of Figure 3.11 (Chapter 3) illustrates this point.

Option 3 presents a somewhat different approach to equalizing battery currents. For this option, smaller wire, such as #6, is used to connect to each series battery string. The idea is to terminate the #6 ends at terminal blocks near the inverter and then use very short lengths of larger wire between the terminal blocks and the inverter. If all the lengths of wire are the same on a round-trip basis, then each string of batteries will experience the same voltage drop in the battery cabling, and currents will be exactly balanced under charge or discharge conditions. The higher resistance of the #6 wire tends to produce a current-limiting effect. So, if for some reason one series set tends to discharge or charge at a higher rate than another series set, the discharge or charge current will be limited by the resistance of the connecting wires, thus creating a balancing effect.

The disadvantage of option 3 is that each individual battery string will require a separate fuse or circuit breaker, similar to the source circuit fuses in a PV array, but much larger. So in this case, there would be four battery disconnects instead of the single disconnect needed for options 1 and 2. However, it is possible that the higher resistance of the connecting wires of option 3 will limit the battery short-circuit current sufficiently that fuses or circuit breakers of a lower interrupting capacity (AIC), and thus lower cost, can be used. Recall that the interrupting capacity of an overcurrent device is a measure of the ability of the device to interrupt the circuit under short-circuit conditions, where there is a possibility that arcing may occur across open switch contacts. As a result, the overall cost of option 3 may still be attractive. However, *NEC* 230.71(A) limits the number of switches allowed to disconnect a circuit to no more than six. An interesting approach to this problem is to use 2- or 3-pole circuit breakers, because each device has a single lever for operating the device and hence will count as a single disconnect even though it will disconnect two or three strings of batteries with a single flip of the switch handle.

In summary, the key to optimizing battery performance is proper cabling. The use of equal lengths of cables is essential. Problems 6.13 and 6.14 offer the reader an opportunity to explore the effect of unequal cable lengths and battery parameters on battery system charging and discharging. Option 2 is clearly better than option 1 for battery wiring, and option 3 is also potentially attractive.

Furthermore, new batteries tend to have lower internal resistance than older ones of the same type, resulting in greater unbalance of currents in parallel strings of new batteries. This generally means that if one battery is replaced, all should be replaced, probably with the exception of an early failure due to a battery defect. In any case, if not all batteries are replaced, the condition of all batteries should be carefully checked to ensure that currents are well balanced. If this is not the case, future premature failures may continue to occur. A clamp-on dc ammeter is most useful for this kind of analysis.

Problems

6.1 Explain the operation of the inverter bypass switch in the ac component enclosure of the example in Figure 6.3.

6.2 Assume that a 3-wire branch circuit is derived from a 120/208 V distribution panel where the voltage on one phase is $120\angle 0°$ and the voltage on the other phase is $120\angle -120°$.

a. Prove that, for resistive loads on either side of the circuit, the current in the neutral will not exceed the largest phase current.

b. Prove that when the loads on each phase are equal, the magnitude of the neutral current will equal the magnitude of the current in either phase.

6.3 Show that in the case of a 3-phase, 4-wire set of branch circuits, that

a. The current in the neutral will not exceed the maximum phase current if only the fundamental component of the current is present on each branch circuit.

b. If a balanced, nonlinear load across all three phases causes a third harmonic current component to flow in each phase conductor, the neutral current can exceed the individual line current.

6.4 Design a mounting system for the array of Section 6.4 if the 36 modules are arranged in rows of 9, 12, and 15 modules. The dimensions of the modules are 37″ × 56″. You may assume a wind speed of 90 mph and exposure category B with all modules located in either roof zone 1 or roof zone 2. The mean roof height is 20 ft. and the slope of the roof is 23°.

a. Specify the rail lengths and the numbers and locations of mounting brackets.

b. Calculate the total lifting force on each section of array and the force on each mounting bracket.

c. Determine the necessary penetration depth of the threads of a 5/16″ lag screw if the trusses are made of Douglas Fir.

6.5 Find a suitable source circuit combiner box that meets the requirements of Section 6.4.3. List the specifications of the unit that prove that the unit is suitable for the application.

6.6 Determine the size of schedule 40 PVC conduit that will be needed to accommodate 13 #10 THWN-2 conductors between rooftop junction box and source circuit combiner box. Then, assuming that the source circuit wiring will be routed through an attic to the power-conditioning equipment and array disconnect inside the structure, determine the size of electrometallic tubing (EMT) that will be required to accommodate the 13 wires. Finally, if a metallic pull box is a part of the raceway between rooftop junction box and source circuit combiner box, recommend a suitable box size if the wires entering the box pass through the box without splicing. You may want to refer to *NEC* 314.16(B)(1).

6.7 Verify that the modules of the array of Section 6.5 will have maximum operating temperatures of 68.8°C when ambient temperatures reach their maximum design values and the irradiance level is 1000 W/m^2.

6.8 For the 18 kW array of Section 6.5, calculate the necessary distance between rows if it is desired to have 7 h of unshaded operation on December 21.

6.9 Verify the lengths of the open-wire extension cables that will be needed for each of the source circuits shown in Figure 6.8.

6.10 Calculate the overall voltage drops for each of the 18 source circuits of the 18 kW array of Section 6.5, using the lengths shown in Table 6.12.

6.11 Search the Web for a battery-backup inverter that can be configured to produce a 3-phase output.

6.12 Prove that the voltage between phase C and neutral as shown in Figure 6.14b will be $208\angle 90°$.

6.13 Use a network analysis program to explore the effect of unequal open circuit battery voltages that might be expected for charge levels of 50% and 75%. Assume, for example, that the "A" batteries of Figure 6.15 are charged to 50% and the rest of the batteries are charged to 75%. Then solve for the battery string currents in options 1, 2 and 3 under charge and discharge conditions. Cable lengths remain unchanged. Use charge/discharge currents of 60 A.

6.14 Use a network analysis program to explore the following effects of unequal battery cable lengths for battery connection options 1, 2 and 3 of Figure 6.15. Solve using charging/discharging currents of 60 A.

a. In option 1, change l_2 as follows: From A_1 to A_2, 1 ft; from B_1 to B_2, 2 ft; from C_1 to C_2, 3 ft; from D_1 to D_2, 4 ft.

b. In option 2, use the replacement values for l_2 that were specified in part a.

c. In option 3, assume the l_1 cables are #6 Cu and the distance from the charging/discharging point to the A_1 and A_2 batteries is 12 ft; the distance from the charging/discharging point to the B_1 and B_2 batteries is 10 ft; the distance from the charging/discharging point to the C_1 and C_2 batteries is 8 ft and the distance from the charging/discharging point to the D_1 and D_2 batteries is 6 ft.

Design Problems

6.15 Design a nominal 12.0 kW, customer-owned, commercial rooftop utility-interactive PV system with a 120/240 V output and 30 kWh of battery backup. Specify all components and connections.

6.16 Using PV roofing tiles that are characterized by V_{OC} = , I_{SC} = , V_{mp} = , I_{mp} = , $\Delta V_{OC}/\Delta T$ = and $\Delta V_{OC}/\Delta T$ = ,

a. Determine the maximum number of tiles that can be installed in series with $V_{OC} < 150$ V for the series combination if the lowest annual temperature will not fall below –20°C.

b. Determine the minimum number of tiles that can be installed in series and still have $V_{mp} > 65$ V for the string if the highest anticipated module (cell) temperature will be 70°C.

c. Determine the range of power available from a single source circuit at STC, and then determine how many tiles (modules) will be needed to achieve an array power of approximately 3 kW @ STC. Choose source circuits to achieve a power as close to 3 kW as possible without letting the values of V_{OC} or V_{mp} fall out of range for the system charge controllers.

d. Select a utility-interactive inverter, charge controllers, and batteries to provide a minimum of 12 kWh of storage with an 80% depth of discharge of the batteries. The inverter must supply a maximum load of 3 kW.

e. Complete the system electrical design, and draw a schematic showing all wire sizes, disconnects, and overcurrent protection.

f. If the system is located in Birmingham, Alabama, estimate the monthly and annual system performance if the roof has a tilt of 30°, facing directly south. You might want to consult.

6.17 Repeat Problem 6.16, but this time look up the performance characteristics of a laminate module that can be applied to a metal seam roof, and use this module. Determine the amount of roof space and a preferred roof geometry to best accommodate this type of module.

6.18 Design a 5 kW STC, ac-coupled, grid-connected, battery-backup PV system that will supply power to a maximum load of 4 kW @ 120/240 V ac. The system should have 16 kWh of battery backup @ 80% depth of discharge. Use a battery-backup inverter that has a 120/240 V AC OUT port. Be sure to specify all the control functions needed to avoid overcharging or overdischarging of the batteries.

6.19 Design a battery-backup, grid-connected PV system for use in Nashville, Tennessee, that will provide adequate power to supply a collection of 120 and 240 V ac loads that are expected to consume 15 kWh/day during summer months. The maximum power to be consumed at any one time will be 4 kW. The system can be ground-mounted at whatever angle the designer specifies, on a mounting system to be specified by the designer. Show the system electrical schematic diagram, and specify all system components, including the array mount. Upon completion of the design, run PVWatts and show the expected monthly and annual system performance.

REFERENCES

1. UL 1741, Inverters, Converters and Controllers for Use in Independent Power Systems, Underwriters Laboratories, Inc., Northbrook, IL, 2005.
2. The PVWatts Solar Calculator, National Renewable Energy Laboratory, Golden, CO, 2008, http://www.pvwatt.com/.
3. ASCE Standard 7-05, Minimum Design Loads for Buildings and Other Structures, American Society of Civil Engineers, Reston, VA, 2005.
4. NFPA 70 National Electrical Code, 2008 Ed., National Fire Protection Association, Quincy, MA, 2008.
5. *FLEXmax™60, FLEXmax™80 Maximum Power Point Tracking Charge Controller User's Manual, Installation and Programming,* Outback Power Systems, Inc, Arlington, WA, 2008. www.outbackpower.com/pdf/manuals/flexmax.pdf.
6. NOAA, National Weather Service Forecast Office, Burlington, VT http://www.erh.noaa.gov/btv/climo/BTV/extremes/extremetemps.shtml.
7. PowerGate 100 kW Hybrid Solar PV Inverter, Satcon Technology Corporation, 2009 http://www.satcon.com/pv_inverters/100kwhybrid.html.
8. NOAA, Lexington, KY climate information. http://www.crh.noaa.gov/images/lmk/lexington_cli_pdf.
9. NOAA, Information on snowfall extremes for Lexington, KY. http://www.crh.noaa.gov/lmk/?n=lexsnowextremes.
10. CSAC Snow and Avalanche Glossary http://www.avalanche-center.org/Education/glossary/density.php.
11. Ballasted Roof Mounts—Non-penetrating Mounting Systems, Direct Power & Water Corporation, Albuquerque, NM, http://power-fab.com/pricing/ballasted200803.pdf.
12. Application Note—AC Coupling of Xantrex Inverters: Forming mini-grids with XW and GT products www.xantrex.com.

CHAPTER 7

Stand-Alone PV Systems

7.1 INTRODUCTION

There are many similarities between stand-alone PV systems and grid-connected battery-backup systems. As previously discussed, when a grid-connected battery-backup system is disconnected from the grid, it becomes a stand-alone system until grid power is restored.

To some extent, stand-alone systems have more flexibility, because if an inverter is used, it does not need to comply with UL-1741. Stand-alone systems may have the option of being designed for dc only, for a mix of dc and ac loads, or for ac loads only. Some stand-alone systems will have battery storage, and some will not. Grid-connected battery-backup systems are normally limited to ac loads only.

On the other hand, design examples in Chapters 4 through 6 have shown that the tilt of the array for a grid-connected system is generally not critical for annual performance, as long as it is within about 15° of latitude. But stand-alone systems may well need to be designed for seasonal performance, meaning more critical consideration of array tilt angle because no grid is available for backup. In fact, in the far north or the far south, it may be necessary to supplement the energy production of the PV system with another form of generation, such as fossil or wind.

In the grid-connected battery-backup examples, only one day of battery storage was normally included in the design. In stand-alone systems, it is common to have several days of backup, again, because the PV system is the only source of electrical power year-round.

Furthermore, monthly electrical loads may differ. This means a more critical consideration of the loads that must be served by the PV system, often on a monthly basis. For example, for solar street lighting, there is generally abundant sun in the summer to charge batteries, but the light is on less because the sun is up longer. The converse is true for winter. Thus, the array will probably need to be tilted for optimal winter performance in the hope that the winter tilt will provide adequate battery charging for summer use.

This chapter will introduce the design process for several complete self-contained PV systems. The first system will be a simple dc fan powered by a PV module, followed by a dc water pumping system that incorporates a linear current booster. A parking lot lighting system, a cathodic protection system, and a highway sign will also be designed as dc systems. Finally, a collection of ac systems will be designed, including a critical-need ac refrigeration system, a remote cabin, and a hybrid permanent residence that will operate off the grid. The hybrid residence will be designed with an emphasis on the process of selecting an appropriate mix of PV and propane electrical generation.

In any design, the first task is normally to determine the load. Often, there are several choices. For example, a choice might be made between an inexpensive incandescent lamp and a more expensive fluorescent lamp. In almost all such cases, the choice of the more efficient load is more cost-effective. In fact, the most efficient load may not necessarily be the most expensive one.

The load voltage, current, and power are needed for proper sizing of fuses, wires, batteries, and other system components. In some cases, not all of these quantities are given. For example, the specifications for a refrigerator may only give the voltage and kWh per month, or Ah per day. In these cases, it may be necessary to either calculate or estimate the other parameters.

Once the load has been determined, the amount of battery backup needs to be determined. Some systems will not need batteries, some will have minimal storage, and some will require sufficient battery storage to meet critical performance requirements in which the system must operate more than 99% of the time. It will also be the responsibility of the engineer to determine whether to use flooded lead-acid batteries or to use maintenance-free batteries at nearly twice the initial cost.

After battery selection, the size of the PV array must be determined. Then the electronic components of the system, such as charge controllers and inverters, are selected. Finally, the balance of system (BOS) components are selected, including the array mounts, the wiring, switches, fuses, battery compartments, lightning protection and, perhaps, monitoring instrumentation.

While computer programs may be available for assistance in system design, as in previous chapters, emphasis in this chapter will be placed on sound, common-sense reasoning and the use of good engineering judgment. In order to keep the design examples as realistic as possible, many of the system components described in the examples of this chapter have been found on an assortment of Web sites. It is important for the reader to note that the use of a particular component or Web site in an example does not constitute an endorsement of the component or the company. Vendors and component availability change on a daily basis, so when a real system is designed, the designer should carry out an extensive search for the components that best meet the design goals. Hopefully, the procedures for component selection outlined in this chapter will prove to be useful when the opportunity to design a real system presents itself.

7.2 THE SIMPLEST CONFIGURATION: MODULE AND FAN

Figure 7.1 shows the simplest of PV systems, a fan motor connected to a PV module. The figure also shows the superimposed performance (*I-V*) characteristics of the fan and the module. The operation is simple: as the sun shines brighter, the

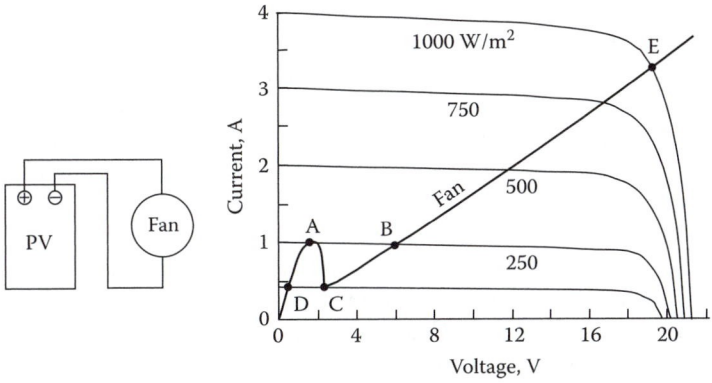

(a) Connection (b) Operating Characteristics of Fan and PV

Figure 7.1 A simple PV-powered fan, showing performance characteristics for fan and PV module.

fan turns faster. If the designer has no concern for the exact quantity of air moved, the design becomes nearly trivial. However, if the amount of air moved must meet a code requirement or other constraint, then it will be necessary to consider the design in more detail.

The system operation point is determined by the intersection of the performance characteristics. Note that as the sun shines brighter, making more PV current and voltage available, the fan consumes more power. It is reasonable to assume that as the fan consumes more power, it will move more air.

Perhaps the second observation the reader will make regarding the performance characteristics intersections in Figure 7.1 is that the module is not operating anywhere near maximum power at low light levels. If a module having higher short-circuit current is used, the fan will remain at nearly constant speed over a wider range of light levels, but at high illumination levels, the module is delivering only a fraction of its maximum power capability. Hence, the designer must decide how much air movement is needed at various irradiance levels, and choose the module accordingly. So, even in this relatively simple design example, the designer must use discretion. Larger modules will cost more, but will deliver more air at lower irradiance levels.

Figure 7.1b also shows the hysteresis effect encountered in starting the fan. Under stalled rotor conditions, the fan motor does not produce a back emf, and thus the fan will draw stalled rotor current until sufficient armature current is present to overcome the starting torque. The irradiance level at point A on the curve is just adequate to provide this current, and the operating point then jumps to point B. As the irradiance level continues to increase, the operating point moves toward point E. When irradiance levels decrease, fan performance follows the fan characteristic to point C, after which the fan stalls and the operating point jumps to point D, eventually approaching the origin as darkness falls.

Another question for the designer to ask is whether it would be better to use a different fan to meet the design requirements. The obvious answer is "maybe." And that is what makes the design of PV systems so much fun. It should be clear from Figure 7.1b that regardless of the choice of fan or module, there will be a significant power mismatch over a relatively wide range of irradiance. Thus, no matter what the choice, there will be some portion of the fan or PV characteristic where maximum power will not be transferred to the fan. If it is desired to optimize fan power for all illumination levels, a maximum power point tracker will need to be incorporated into the design.

The tracker can be particularly useful at irradiance levels between the start and stop irradiance levels of Figure 7.1b, where it will enable the fan to start at a lower irradiance level and to stop at a lower irradiance level, with greater airflow at irradiance levels between these points. Here the interesting part of the trade-off is whether including a maximum power point tracker with a smaller module will cost less than using a larger module to obtain comparable system performance.

For example, the PV curve upon which points A and B lie has an approximate power available at the maximum power point of 18.13762 W. But at the intersection of the fan curve and the PV curve, the fan is using only 5.78623 W, which means that an additional 12.35139 W is available, but not used. So the fan is only using 31.9% of the available power. In the ideal case, $P \propto v^3$, where v is the air velocity moved by the fan and P is the power consumed by the fan. So as the power triples, the air velocity increases by a factor of 1.44. Thus, if the simple fan system does not move enough air at low sunlight levels, then an MPPT can make a difference in fan performance.

7.3 A PV-POWERED WATER-PUMPING SYSTEM

7.3.1 Introduction

One of the most common stand-alone PV applications is water pumping, especially when the water to be pumped is a long distance from a utility grid. Water-pumping applications do not normally require battery backup unless the water source will not produce an adequate supply of water to meet the pumping needs during the period of peak sun. Another reason for using batteries is that the water can be pumped over a longer time at a slower rate with a smaller, less costly pump. Under these circumstances, it is common practice to charge a battery so the pump can run for an extended period. When the water supply can meet the pumping capacity of the system, then it may be more cost-effective to pump all the water the pump is capable of delivering and store any excess in a storage tank. In effect, the storage of water replaces the storage of electricity in batteries. It still represents conversion of kinetic to potential energy. In fact, it is conceivable that the pumped water could be used during dark periods to turn a generator to generate electricity while the water is being delivered to its final use. The sacrifice is loss of pressure at the final delivery point.

When designing a water-pumping system, it is necessary to determine a number of parameters in order to properly size the system components. First of all, the daily water needs must be determined. Second, the source must be characterized in terms of available water and vertical distance over which the water must be pumped. Once these factors are known, along with the number of hours per day available for pumping, the pumping rate can be determined. The pumping rate along with the pumping height equates to the pumping power, once again the product of a pressure quantity and a flow quantity. The pumping power can then be converted to horsepower so that the size of the pump motor can be determined. It should be noted that this approach is somewhat simplified, because a pump motor does not produce constant horsepower as the flow and pressure are varied. Normally, depending on the exact type of pump, higher volumes at lower pressures involve higher horsepower than higher pressures and lower volumes from the same pump.

Once the size of the pump motor is known, the ampere-hour requirements of the motor can be determined, and, finally, the size of the PV array needed to provide the ampere-hours can be determined. As in the fan example, inclusion of a linear current booster (LCB) or MPPT extends the useful pumping time of the pump motor and enables the use of a smaller motor and a smaller array that is utilized more efficiently.

7.3.2 Selection of System Components

To quantify the pumping problem, it is useful to note that a gallon of water weighs 8.35 lb and that 1 horsepower = 550 ft-lb/s = 746 W, assuming 100% conversion efficiency. This means that pumping a gallon of water to a height of one foot involves 8.35 ft-lb of work. In the MKS system, converting gallons to liters and feet to meters gives the result that pumping 1 L of water to a height of 1 m requires 7.23 ft-lb = 9.83 J. If the pumping time is given in hours, and it is desired to determine the pump horsepower, then the pump horsepower can be determined from

$$\mathrm{HP} = (4.22 \times 10^{-6}) \frac{(\mathrm{GPD})(h)}{(\mathrm{PT})(\mathrm{PTF})(\eta)}, \tag{7.1a}$$

where GPD is the gallons per day to be pumped, PT is the pumping time, PTF is the pumping time factor, h is the effective height, and η is the wire-to-water efficiency of the pump-motor combination. In MKS, the horsepower is given by

$$\mathrm{HP} = (3.66 \times 10^{-6}) \frac{(\mathrm{LPD})(h)}{(\mathrm{PT})(\mathrm{PTF})(\eta)}, \tag{7.1b}$$

where now LPD is the pumping requirement in liters per day, and h is the effective pumping height in meters.

The effective height is the sum of the distance from the top of the water supply to the delivery point, including piping friction losses, which, in a properly designed system, will be limited to about 5% of the total effective height. The pumping time will normally be the same as the peak sun hours, and the pumping time factor is a modifier to account for the use of either batteries, an LCB, or a tracking array mount. The product of PT and PTF then represents an effective pumping time.

If batteries are used, then the PTF is simply the ratio of the actual time the pump operates each day to the peak sun hours. Then, expressing PT as the peak sun hours and multiplying by this PTF gives the actual operating hours per day. If an LCB or MPPT is used, so that the pump performance curve more closely matches the PV performance curve maximum power point, then more water will be pumped during low sun hours than would otherwise have been pumped. Use of an LCB or MPPT in the system normally increases the daily volume pumped by an additional 20%. Hence, a reasonable default value for PTF when an LCB or MPPT is used is 1.2. If the pump is connected directly to the PV array, then the PTF is 1.0. If batteries and an MPPT are both used, then PTF × PT is still the actual daily pump operating time, but the MPPT allows for a smaller array to be used for battery charging.

The wire-to-water efficiency, η, will be specified by the pump manufacturer. For fractional horsepower pumps, it is typically about 25%, while larger pumps will be more efficient.

Piping friction loss is determined by the type and diameter pipe used, just as voltage drop is determined by the size and material of the wire used, although the relationship between pressure and flow for a water pipe tends to be somewhat more nonlinear than the *I-V* relationship for a wire. However, at relatively low flow rates, the flow versus pressure curve for a piping system can be approximated by a linear relationship. Figure 7.2 shows pressure versus flow curves for several sizes of piping.

Although it is reasonably straightforward to select the horsepower for a pump, selecting the pumping system that will perform at maximum efficiency is somewhat more involved. The reason is that some pumps are designed to deliver higher pressure than others. The pumps that can deliver higher pressure are needed for lifting water to greater heights. Figure 7.3 shows performance curves for two pumps of equal horsepower, one of which is a high-head (pressure) pump and the other of which is a medium-head pump.

Note that the medium-head pump will deliver more volume than the high-head pump at low pressure, but the high-head pump will overcome a greater pumping

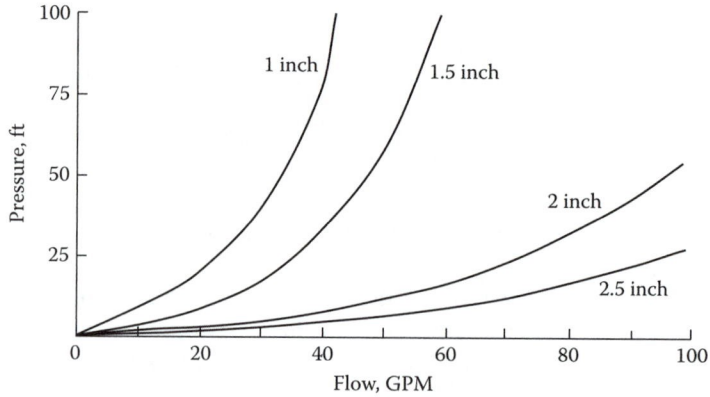

Figure 7.2 Pressure versus flow curves for equal lengths of piping of different diameters.

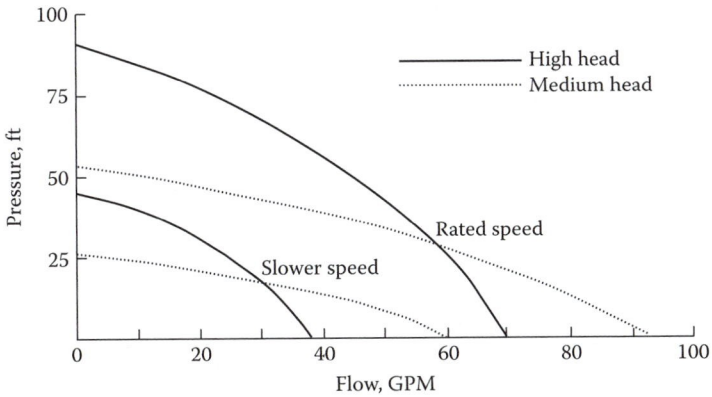

Figure 7.3 High-head and medium-head pump performance characteristics at two different speeds.

height. Note also that the performance of a pump depends on the speed at which the pump is operated. If the pump speed decreases, both pressure and flow capacity decrease. It is thus important to select a pump that will be able to overcome the lift requirement under low-sun conditions.

7.3.3 Design Approach for Simple Pumping System

If pumping data are available from a pump manufacturer, then all the designer needs to know is the daily amount of water needed and the overall pumping height. The daily water needs can be converted to gallons or liters per minute over the time the pump will operate, and a suitable pump can be selected from manufacturers' tables.

Using 2000 gal per day, 200 ft pumping height, and 6 h of sunlight, means that 2000 gal must be pumped in 6 h if no batteries are used. This means $2000 \div 6 \div 60 = 5.56$ gal per minute must be pumped a height of $200 \times 1.05 = 210$ ft to account for piping losses. Table 7.1 tabulates the gallons per minute (GPM) delivered at specific pumping heights at specific current and voltage levels for one model of dc submersible pump. Because 5.56 GPM and 210 ft are not on the table, it is necessary to interpolate to determine the appropriate pump voltage to deliver the required GPM. The PV power is 125% of the product of the pump current and pump voltage for this sizing algorithm.

7.3.3.1 Determination of Pump Operating Current and Voltage

It would be nice if a linear interpolation could be used. Because hydraulic power is proportional to the product of lift and GPM, one might expect electrical input power would be proportional to hydraulic power. A look at the figures for pumping lifts of 150, 175, and 200 ft shows this to be approximately the case, because the ratio

Table 7.1 Pumping Characteristics of a Typical DC Submersible Pump

Lift (ft)	GPM	Pump Current	Pump Voltage	PV Watts
150	6.4	6.00	90	675
150	12	8.95	120	1340
175	6.2	5.56	90	625
175	13.7	8.82	120	1320
200	7.6	6.64	105	875
200	11.0	8.42	120	1260
250	6.4	7.76	120	1164

Source: Based on data from *1997–98 Design Guide and Catalog*, Alternative Energy Engineering, Redway, CA.

of electrical powers is approximately equal to the ratio of GPM for these values of lift. In fact, a convenient normalization involves determining W/GPM at each lift. At 200 ft the result is 115 W/GPM, and at 250 ft the result is 182 W/GPM. Linear interpolation between 200 and 250 ft then yields a W/GPM ratio at 210 ft to be

$$\frac{\text{W}}{\text{GPM}} = \frac{210-200}{250-200}(182-115)+115 = 128.$$

Because the pumping requirement is 5.56 GPM, the PV power required is 128 × 5.56 = 712 W.

Next, the voltage necessary to produce 712 W must be determined. One might expect the ratios of the powers to be proportional to the square of the ratios of the voltages. Testing this hypothesis gives 1260 ÷ 875 = 1.44 whereas $(120 \div 105)^2 = 1.31$, suggesting perhaps a different power relationship. If it is true that

$$\frac{P_2}{P_1} = \left(\frac{V_2}{V_1}\right)^n, \tag{7.2}$$

then one can solve for n, using 1260 and 875 for P_2 and P_1 and 120 and 105 for V_2 and V_1. The result (see Problem 7.4) is $n = 2.73$. Now, $n = 2.73$ is used in estimating the operating voltage associated with the 712 W of PV power needed to pump the 5.56 GPM through the 210 ft lift. The voltage, V_2, can be determined by using the values $P_2 = 712$ W, $P_1 = 875$ W, and $V_1 = 105$ V in Equation (7.2), with $n = 2.73$. The result (see Problem 7.4) is $V_2 = 97.4$ V. Hence, if $V_2 = 97.4$ V, the PV current must be 712 ÷ 97.4 = 7.31 A.

7.3.3.2 Module Selection

The challenge now is to find a module combination that will come as close as possible to producing 7.31 A at 97.4 V. But because most of the time the array will

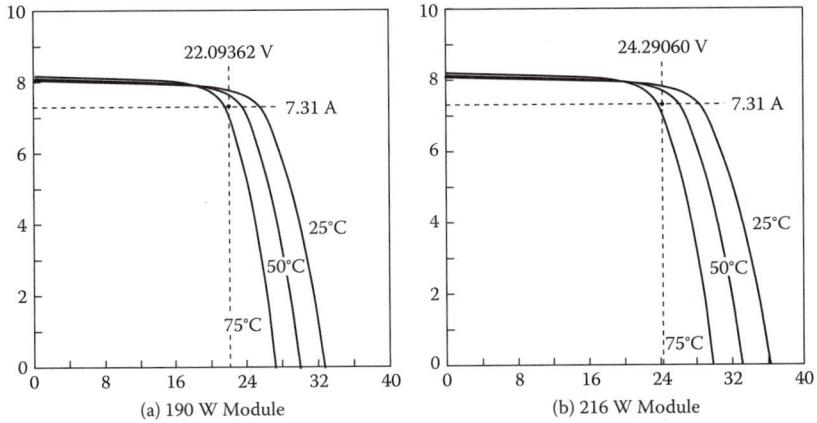

Figure 7.4 Comparison of two possible modules for pump operation.

not be operating at maximum rated array power, either because of module temperature or soiling or mismatch, pump manufacturers generally recommend oversizing the PV array by 25%. Furthermore, pump manufacturers also recommend using an LCB with the pump. In fact, often the pump is sold with an LCB. So, the final system should probably include an LCB for better performance at lower light levels and to better match the array characteristic to the pump characteristic.

Figure 7.4 shows two sets of module characteristics. Assuming that the cost/watt for each module is the same, all that remains is to determine which module will come closest to meeting the system needs. What needs to be done is to attempt to achieve $I = 7.31$ A and $V = 97.4$ V with some combination of modules. Because each module is capable of producing more than the necessary current, the idea is to choose the necessary current on the module characteristic for the assumed operating temperature of the module and note the resulting module voltage that will produce the needed current. Noting that each module has nominal operating cell temperature (NOCT) $\cong$ 46°C, and if the temperature on a particularly warm pumping day can be expected to reach 37°C, then the highest module temperature can be estimated to reach 69.5°C (see Problem 7.5). Because the water is for cattle to drink, and because they will drink more water on a hot day, it is especially important to deliver the water on this hot day. Thus, the designer can choose to interpolate between the 50°C and the 75°C curves, or simply use the 75°C curve to remain on the conservative side. Figure 7.4 shows that at 69.5°C and $I = 7.31$ A, module A will operate at 22.09362 V, and module B will operate at 24.29060 V.

So, the number of module A needed to reach 97.4 V will be $97.4 \div 22.09362 = 4.40851$, and that of module B will be $97.4 \div 24.29060 = 4.00978$. Thus, in order to achieve sufficient voltage, it will be necessary to use five of module A, whereas four of module B will just produce the needed voltage.

It is interesting to ask about the operating point of module A if five are used. In this case, it can be expected that the pump will operate at a higher voltage and current. Perhaps the easiest way to determine the system operating voltage and current at full sun when five of module A are used is to create a composite *I-V* characteristic for the five modules and to superimpose the *I-V* characteristic for the pump from manufacturer's data. This is left as an exercise for the reader (Problem 7.6).

So what about the final design? Using the manufacturer's recommendation that the STC array power rating be 25% higher than the necessary pump power would require a rated array power of 890 W, which can be achieved with 4.68 of the 190 W modules (module A) or 4.12 of the 216 W module B. Clearly, four of module A will not suffice, but five will. On the other hand, four of module B with an LCB that increases pumping output by 20% should comfortably achieve the design pumping goal. Hence, it becomes a matter of whether the price of four modules and an LCB will cost less than five modules without an LCB. If each of the modules is priced at \$3.75/W, then the cost of five A modules will be \$3,562.50, and the price of four B modules will be \$3,240.00. So if the LCB can be purchased for less than \$322.50, then the combination of module B and LCB may be the best option. However, one should keep in mind that an additional component may result in an additional increment to the installation cost, so this should also be accounted for.

As a final note, however, the cost of this 1 HP pump is over \$1800, whereas a 1/4 HP pump that will deliver 2.15 GPM while consuming 186 W can be purchased for \$495. This suggests the entire system, including the battery storage needed for the slower pumping rate, should probably be evaluated when a serious design exercise is in progress.

Other manufacturers provide tables for their pumps that list lift, GPM, and power consumed for a fixed pump voltage. These tables are very convenient and result in simple interpolations and reliable results, especially if pump voltage is maintained nearly constant with batteries.

7.4 A PV-POWERED PARKING LOT LIGHTING SYSTEM

7.4.1 Determination of the Lighting Load

It should be evident that if a PV system is to power a parking lot lighting system, a battery storage system will be needed, because if the sun is shining, the lights are not needed, and vice versa. The first step is to determine the lighting load, followed with battery selection and, finally, the number and type of PV modules to use. In order to determine the lighting load in watts, it is first necessary to determine the amount of light needed and the area over which the light is needed. Hence, the design begins with the determination of the necessary illumination level.

While illumination levels could be measured in W/m^2, in the United States, illumination levels are most commonly measured in foot candles. A foot-candle is the amount of light received at a distance of 1 foot from a standard candle. A standard candle is a candle that emits a total amount of light equal to 4π lumens. The lumen is thus the basic quantity of light in the foot-candle system of measurement of light intensity. It compares to the coulomb in the electrostatic realm.

Table 7.2 Approximate Luminous Efficacy for Several Light Sources

Source	Luminous Efficacy (l/W)	Lamp Lifetime (h)
25 W incandescent	8.6	2500
100 W incandescent	17.1	750
100 W long-life incandescent	16.0	1125
50 W quartz incandescent	19.0	2000
T-8 fluorescent	75–100	12,000–24,000+
Compact fluorescent	27–80	6,000–10,000
Metal halide	80–115	10,000–20,000
High-pressure sodium	90–140	10,000–24,000+
3.6 W LED array	~130	100,000+

Sources: Based on data from *9200 Lamp Catalog*, 22nd ed., GE Lighting, General Electric Co., 1995; *Lamp Specification and Application Guide*, Philips Lighting Co., Somerset, NJ, 1999; www.cetsolar.com for information on LED arrays.

If a closed surface surrounds the standard candle, then all of the 4π lumens of light must ultimately pass through the surface. If the light from the source is emitted uniformly in all directions, and if a sphere of radius 1 ft is centered on the light source, then the light will be uniformly distributed over the surface of the sphere with a density of (4π lumens) ÷ (4π ft^2). This light intensity of 1 lumen/ft^2 defines the foot-candle (f-c).

The Illumination Engineering Society publishes guidelines for illumination levels for various spaces [2]. For example, parking lot lighting should normally be lighted to an average illumination level of approximately 1 f-c, depending on the degree of security desired. A desk for normal work is generally adequately lighted with 50 f-c. Direct sunlight provides about 10,000 f-c [3].

The luminous efficacy of a source is a measure of the efficiency with which the source transforms electrical energy to light energy. It is measured in lumens per watt. Table 7.2 shows the luminous efficacies for several light sources.

Determination of the wattage of light needed to accomplish a specific lighting task, then, will depend on the required illumination level and the area to be lighted. It also depends on the luminous efficacy of the source. Other factors include whether the available light can be directed only to the area where the light is needed and whether some of the light will be absorbed by walls or other absorbers, such as the light fixture itself, before it reaches the surface to be illuminated. Dust on the fixture and lamp also absorbs useful light output.

In addition to the intensity of light, the color temperature of light is sometimes also a factor to be considered. The color temperature of light refers to the equivalent spectral content of radiation from a blackbody at a particular temperature. The color temperatures with which the reader is most familiar are the 5800 K temperature of the sun, which produces its characteristic white color, and the 2700 K temperature of a tungsten incandescent light filament, which is more toward orange. Not all light sources can be characterized by a color temperature, because the concept is based on blackbody radiators. Sources with discrete spectral components, such as lasers or gas discharge lamps, can be assigned equivalent color temperatures to indicate the temperature to which the spectrum of the source is most closely matched, but the color temperature is not a precise measure of the color of the source. For example, xenon produces a very white flash, which, on photographic film, appears to be close to the color of daylight. Although the output spectrum of a xenon lamp differs from the AM 1.5 solar spectrum, xenon lamps are commonly used in solar simulators with appropriate correction factors.

Consideration of color temperature can be an important factor in the choice of light sources. For example, low-pressure sodium has a very high luminous efficacy, but the light consists primarily of the sodium d_2 lines. When low-pressure sodium sources are used, anything not yellow in color will not be accurately perceived, because, if a source does not contain a particular color, then that color cannot be reflected back to the eye to be perceived as such. For PV applications, generally the most popular and efficient sources are fluorescent, metal halide, and high-pressure sodium. Occasionally, incandescent sources are used for special-purpose applications, while LED sources are now becoming increasingly popular as replacements for fluorescent and incandescent sources.

In any case, once a light source and a suitable fixture for the source are chosen, for an outdoor lighting system, the available light from the fixture, expressed in lumens, can be obtained from the formula

$$\text{Lumens} = (\text{FC}) \times (\text{area}) \div (\text{CU}) \div (\text{MF}) \div (\text{RCR}), \tag{7.3}$$

where FC is the desired illumination level in foot-candles, area is the area to be illuminated, measured in ft^2, CU is the *coefficient of utilization* of the fixture, MF is the *maintenance factor* of the fixture, and RCR is the *room cavity ratio*. The coefficient of utilization of a fixture is a measure of the fraction of light available from the lamp that is directed to the surface to be illuminated. The maintenance factor of the fixture provides a means of estimating the amount of light from a fixture that can be lost as the fixture and lamp get dirty, and the room cavity ratio is a measure of the amount of light from the fixture that will be absorbed by the room and its contents. A good fixture will have a CU of 0.8 or more and an MF of 0.9 or more. Of course, the MF is dependent on the environment in which the fixture operates and the maintenance interval for the fixture. The RCR depends on room size, wall color, floor color, and room contents. A room full of dark colors will appear darker because it *is* darker. Outdoors, the RCR is generally assumed to be 1.0. Indoors, the RCR is generally <1.

7.4.2 Parking Lot Lighting Design

Figure 7.5 shows a parking lot for which it is desired to provide an average illumination of 2 FC. The figure also shows suitable locations for fixtures and an approximate lighting pattern that would be obtained with a well-designed sharp

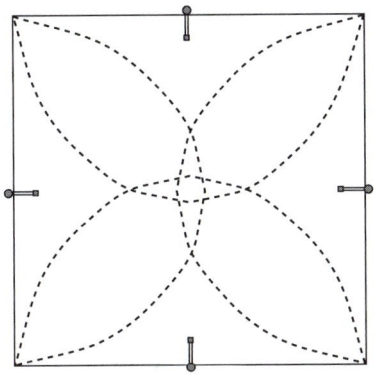

Figure 7.5 Parking lot showing locations of fixtures and approximate coverage patterns for each fixture.

cut-off fixture. The fixture has a CU of 0.8 and an MF of 0.9. The sharp cut-off feature of the fixture helps to prevent spillage of light onto adjacent property, so it will be assumed that any loss of light to adjacent property will be accounted for by the CU. The lot measures 160 × 160 ft, so the total area of the lot is 25,600 ft^2. With the spacing of the fixtures as shown, the distance between fixtures is 113 ft. For parking lots, the spacing of lighting poles is typically set at 4 pole heights. Thus, in this case, the pole height should be approximately 28 ft.

7.4.2.1 Determination of Lamp Wattage and Daily Load Presented by the Fixture

First, the total lumens needed to illuminate the parking lot must be determined, and then a suitable lamp can be selected. The total lumens can easily be found from Equation (7.3) to be $2 \times 160 \times 160 \div 0.8 \div 0.9 = 71{,}111$. If four fixtures are to be used, as shown in Figure 7.5, then each lamp will need to supply 17,778 lumens. Next, assuming that a metal halide lamp is chosen, Table 7.2 can be used to determine the lamp wattage required by each fixture. Using the average luminous efficacy of metal halide of approximately 100 lumens/W indicates that the lamp size should be 178 W. Coincidentally, 175 W metal halide lamps are available, so this is a reasonable choice of lamp.

However, the lamp wattage is not necessarily the power consumed by the fixture. Because metal halide lamps are gas discharge lamps, they must operate with a ballast. The ballast is used for starting the lamp and for stabilizing the operating lamp current. The power loss in the ballast is approximately 10% of the lamp wattage, so it is advisable to use 200 W as the power consumption of the fixture for design purposes.

If this parking lot is in Miami, Florida, and if the lights are to be on from sunset to sunrise, then the maximum amount of time the fixture will need to operate will be 13 h during the month of December if the lights come on 15 min after sunset and shut off 15 min before sunrise. During the month of June, the fixture will only need to operate for 9.8 h [7]. Thus, the winter worst-case daily Wh = 200 × 13 = 2600 Wh, and the summer energy consumption is 200 × 9.8 = 1960 Wh.

7.4.2.2 Determination of Battery Storage Requirements

The value of stand-alone outdoor lighting systems is enhanced if each fixture can operate independently of others. This saves the effort of running wires from one fixture to another. So all that is needed is to design one system and then build three more identical to it. Because Miami does not normally have many consecutive totally cloudy days, especially in the winter, 2 days of storage will be used. For a 24 V system, this means the winter daily *connected* Ah for the load will be 2600 ÷ 24 = 108 Ah, which needs to be supplied by the batteries. Allowing for 2% wiring losses and an 80% depth of discharge, the batteries will need to have a 24 h discharge rate capacity of 108 × 2 ÷ 0.98 ÷ 0.8 = 275 Ah @ 24 V. Note that the 24 h discharge rate is used because over the 2-day period, the batteries will be discharging for approximately 12 h each day. Total discharge will thus take approximately 24 h. One way to achieve this is to use four 6 V @ 275 Ah batteries in series, if such a battery can be found.

7.4.2.3 Determination of Array Size

The array size can be determined by first calculating the energy that must be supplied to the batteries by the array. Because the connected load on the batteries is 2600 Wh, and because losses of 10% can be expected for battery charging/discharging and an additional 2% for wiring losses, a total of 2600 ÷ 0.98 ÷ 0.9 = 2948 Wh must be supplied to the batteries by the array each day.

Assuming the array will be connected to the batteries through an MPPT charge controller and assuming the combination of charge controller and wiring from array to charge controller have an efficiency of 96%, the array must generate 2,948 ÷ 0.96 = 3071 Wh each day. Allowing for 12% array losses due to elevated temperatures, soiling, and mismatch, the array must be sized to generate 3071 ÷ 0.88 = 3490 Wh/day at STC.

At this point, it would be possible to go to PVWatts to size the array, but this time the sizing will be done using the table for Miami in Appendix A. The array will be tilted at latitude + 15° to optimize winter performance, so that in December, the average daily psh will be 4.9. Thus, the array rating in watts is found from 3,490 ÷ 4.9 = 712 W.

One way to come very close to this array power is to use four 180 W modules in series. As done in previous examples, the module open-circuit voltages must be checked for the maximum low-temperature V_{OC} to ensure the maximum charge controller input voltage is not exceeded.

As a final check on the array sizing, one should look at the array output during other months to be sure that it will be adequate to meet the lighting loads for these months. Of course, during other months, the lighting loads will be less, because all other months have more daylight hours than December. The loads will be

Table 7.3 Comparison of Daily Loads (Wh) and Energy Availability (Wh) for Lighting System

	Jan	Feb	Mar	Apr	May	Jun
Load	2560	2440	2300	2140	2020	1960
Avail	2683	2897	3005	3058	2683	2414
	Jul	**Aug**	**Sep**	**Oct**	**Nov**	**Dec**
Load	1980	2100	2260	2400	2540	2600
Avail	2575	2683	2629	2736	2629	2629

proportional to the nighttime hours, and the supply will be proportional to the psh. So they can be tabulated relatively quickly, as shown in Table 7.3. Once again, the nighttime hours used for determining the loads in Table 7.3 are determined to be ½ h less than the number of hours of between sunset and sunrise, as determined from Naval Observatory data [7] for each month. The available energy is determined by noting that the selected array will deliver 2629 Wh and then multiplying by the ratio of psh for the month in question to the psh for December.

7.4.2.4 Final System Schematic

At this point, all major components have been selected except for the charge controller. Some attention needs to be paid to this component to be sure it will be adequate. The controller output current will be determined by dividing the rated array power by the battery voltage and multiplying by the controller efficiency. The result is 29.1 A, so a charge controller with a rated continuous output current of 30 A @ 24 V needs to be found.

Then the wire size from array to charge controller must be determined. The method here is the same as it has been in all previous design examples: calculate all deratings due to temperature, and then check voltage drops. The same method is used for the wiring to the dc fixture that will operate at 24 V, if such a fixture can be found. The alternative is to select an inverter and operate the fixture at either 120 or 240 V, because these voltages are more common for these fixtures. However, if an inverter is used, then the inverter efficiency will need to be accounted for in determining the battery size and the array size.

If a 24 V fixture can be found, it will require $200 \div 24 = 8.33$ A to operate. Thus, one might quickly conclude that #14 wire is adequate. However, to be prudent, a voltage drop check should be made. Because the distance between batteries and fixture will be approximately 25 ft, the voltage drop with #14 wire is found to be 5.3%. To reduce the voltage drop to less than 2%, it will be necessary to use #8 wire, with a resulting 1.32% voltage drop. Use of #10 results in %VD = 2.1. What this illustrates is that at low voltages, wire sizes need to be larger than might be expected to limit voltage drop to acceptable levels.

If the modules are operated at STC, then $V_{mp} = 25.9$ V and $I_{mp} = 6.95$ A for one possible module choice. Thus, for four modules in series, V_{mp} increases to 103.6 V. For a distance of 25 ft between array and charge controller, with a 2% limit on

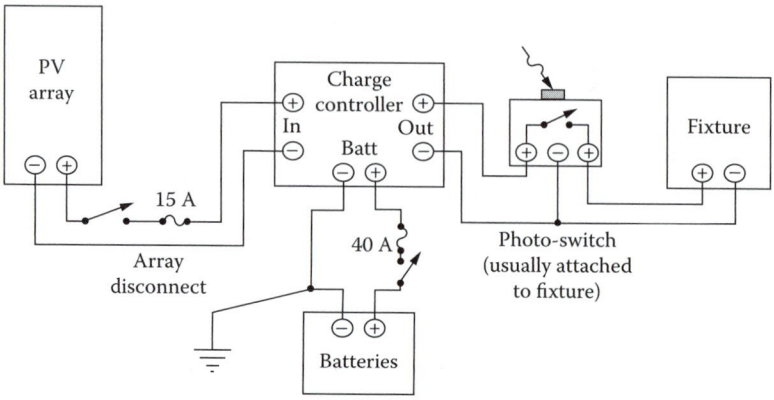

Figure 7.6 Schematic diagram of dc parking lot lighting system.

voltage drop, the wire must have Ω/kft < 5.96. The ampacity of the wire must be at least 156% of I_{SC} of the modules, or 1.56 × 7.78 = 12.13 A. A temperature derating will not be needed for this wire run because it will not be run over a roof space. Thus, in this case, #14 will be acceptable.

Next, appropriately sized fuses and switches need to be selected for the input and output of the charge controller and also for the inverter, if an inverter is used. Figure 7.6 shows a schematic diagram of a dc system. Note that it includes a photocontrol to turn on the light when it is dark outside. Finally, it needs to be noted that the charge controller (or charge controllers) must also be capable of shutting off the light if the battery voltage drops too low. This may be possible with a single charge controller, or it may require two charge controllers, one to provide maximum power point tracking (MPPT) and to prevent overcharging and one to prevent overdischarging of the batteries. In this circuit, the 15 A fuse is provided at the charge controller input, because the 40 A battery fuse will not protect the PV array from excessive backfeed, the only possible origin of which might be the batteries. As a practical matter, both the 15 A fuse and accompanying switch and the 40 A fuse and accompanying switch can be replaced by 15 and 40 A circuit breakers, because dc circuit breakers with voltage ratings less than 150 V are relatively inexpensive. The circuit breakers then serve dual roles as overcurrent protection and disconnect.

Because the array voltage is in excess of 50 V, it is required to be grounded, which is also shown in the schematic. All metal parts, including the mounting pole if it is metal, will also be connected to this same ground, making sure that the grounded conductor is connected to ground at only one point.

7.4.2.5 Structural Comment

Although the optical and electrical design requirements have been met for this system, it is important to note that it will also be necessary to design a sufficiently robust 28 ft pole to hold all the system components. Because the design wind speed for Miami is 146 mph, this may be a nontrivial exercise. The actual design of such a structure is beyond the scope of this text, and, as such, should be done by a structural engineer.

In the event that the structure should prove to be too expensive, the electrical engineer will be confronted with a redesign of the system. This process may involve exploring the use of more fixtures on shorter poles, different light sources with higher luminous efficacies (LEDs?), shorter lamp operating times, or a compromise on the light level. The point here is that the engineer should not be discouraged if the first try does not work. Often, the development of an acceptable solution to a problem will involve a number of redesigns.

7.5 A CATHODIC PROTECTION SYSTEM

7.5.1 Introduction

Material can be electroplated onto another material by immersing the two materials in a suitable electrolyte and applying a voltage between an anode composed of the desired plating substance and a cathode consisting of the material to which the material is to be plated. The result is transfer of material from the anode to the cathode.

When a metal is buried in the ground, it is highly likely that it will become a part of an electroplating system resulting from galvanic action between two dissimilar metals. If the metal assumes a higher potential than its surroundings, that is, becomes an anode, then metal will be removed as a result of ion loss from the metal. However, if the metal is deliberately connected as the cathode of a system, then electrons will flow from the voltage source negative terminal to the metal. The positive terminal of the voltage source is connected to a buried anode material, so electrons flow from the anode material to the positive terminal of the voltage source. Removal of electrons from the anode material creates positive ions that can enter the electrolyte (i.e., the ground) and flow toward the cathode. The process is shown in Figure 7.7.

The U.S. government requires that any underground storage of toxic materials or petrochemicals have cathodic protection. Cathodic protection involves using the material to be protected, usually steel, as a cathode, while an anode (or anodes), typically of graphite, is buried nearby. In addition to protection of toxic waste containers, billions of dollars of infrastructure are built with steel-reinforced concrete, much of which is under water. Cathodic protection of the steel in the concrete can prolong the life of buildings, bridges, and other important infrastructure components.

To prevent ion loss from the cathode, different current densities are required for different materials, ranging generally from a fraction of a mA/ft^2 to several mA/ft^2. The total current needed to protect a cathode is thus the product of the necessary current density and the surface area of the cathode. The voltage needed to supply this current is thus determined by the product of the current and the resistance from anode to cathode. In this example, the methods for determining current density,

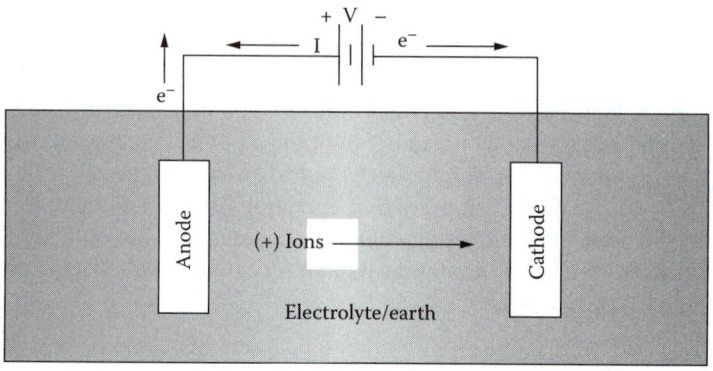

Figure 7.7 Schematic diagram of flow of current and charge carriers in an electrolyte system.

resistance, and voltage are explored, along with determining necessary battery storage and array size.

If all soils were identical, cathodic protection system design would be trivial. Fortunately for the corrosion engineer, the earth has been blessed with a wide range of soil types, often with diversity in a small area. This relatively wide range of soil conditions, sometimes varying with time in the same location, makes the job of protection of critical systems sufficiently challenging to warrant the high fees of the corrosion engineer. Considering the fines and/or lost revenues that can result from a spill of petrochemicals or toxics, design of systems to protect these systems cannot be left to amateurs. While this example will not convert the reader into a professional cathodic protection designer, it will at least convey some background needed for cathodic protection design.

7.5.2 System Design

The first step in the design of a cathodic protection system is to determine the system current needs. Suppose the item to be protected is an uncoated steel tank in sandy soil. Suppose the tank has an exposed surface area of 100 ft^2. The current density required for use in this environment for exposed steel is 1 mA/ft^2, so the total current needed will be 100 mA. If more current is generated, the cathode will remain protected, but the anode will be sacrificed at a higher rate.

The next step is to choose an anode. A typical anode will carry a maximum current of 2 A, so the choice of anodes will presumably not be significantly affected by current. The other consideration in choosing an anode is the resistance between the anode and the cathode. This resistance depends on the soil resistivity and on the size of the anode. Because the anode is generally cylindrical, the current from the anode travels more or less radially outward. The reader may recall from an electromagnetics course that a cylindrical geometry with an infinitely long charged cylinder produces an electric field that varies inversely

Table 7.4 Anode Resistance to Ground in Standard 1000 Ω-cm Soil

Anode Diameter (in.)	Anode Length (ft)				
	4	5	6	7	8
3	5.0 Ω	4.3 Ω	3.7 Ω	3.3 Ω	3.0 Ω
4	4.7 Ω	4.0 Ω	3.5 Ω	3.1 Ω	2.8 Ω
6	4.1 Ω	3.5 Ω	3.1 Ω	2.8 Ω	2.5 Ω
8	3.7 Ω	3.2 Ω	2.9 Ω	2.6 Ω	2.3 Ω
10	3.5 Ω	3.0 Ω	2.7 Ω	2.4 Ω	2.2 Ω

Source: Based on data from *Stand-Alone Photovoltaic Systems: A Handbook of Recommended Design Practices*, Sandia National Laboratories, Albuquerque, NM, 1995.

with the distance from the cylinder. This results in a logarithmic variation in voltage and a nice, nonlinear relationship between the length and diameter of the anode and the resistance from anode to ground. Fortunately, the resistance to ground for anodes of different diameter and length in a uniform soil having a resistivity (ρ) of 1000 Ω-cm has been tabulated. The resistance to ground for different soil resistivity is then in proportion to the resistance at standard conditions. The resistance to ground for the anode is used as the resistance between anode and cathode.

In this case, suppose a 3 in. diameter, 5-ft-long anode is chosen. Table 7.4 shows the resistance to ground for such an anode in 1000 Ω-cm soil to be 4.3 Ω. But the resistivity of sandy soil is closer to 25,000 Ω-cm, so the resistance to ground of this anode at the tank location will be approximately 25 times higher, or $25 \times 4.3 = 107.5$ Ω.

The required voltage is thus the product of the current and resistance, which in this case is $0.1 \times 107.5 = 10.75$ V, which can readily be supplied by a standard nominal 12 V module (i.e., $V_{OC} \approx 20$ V). Assuming the current is needed on a 24 h/day basis, a 12 V storage battery will be needed.

Using a battery charging efficiency factor of 0.9 and a wire efficiency factor of 0.98, the daily corrected system load is determined to be $(0.1\text{ A}) \times (24\text{ h}) \div 0.9 \div 0.98 = 2.72$ Ah.

The next step is to determine the size of the battery needed, along with the current rating of the PV array. Assuming deep-discharge batteries with 5 days of storage time and an allowable discharge of 80%, the battery needs will be $(2.72\text{ Ah}) \times (5\text{ days}) \div 0.8 = 17$ Ah. This is about the size of the battery in a small uninterruptible power supply for computer backup power.

Finally, if the minimum daily psh for the location of the tank is 4 h, using a module derating of 90%, the module current is found to be $(2.72\text{ Ah}) \div (4\text{ h}) \div 0.9 = 0.76$ A.

The module can thus be a 10 W, 12 V unit, and the battery can be a small battery. However, if the maximum sun is significantly more than 4 h, then the battery may overcharge, so a charge controller will be needed.

One might expect an alternative to a charge controller may be to use a larger battery. However, with an excess of 2 or more Ah/day from the module, in 30 days' time, the battery will accumulate an additional 60 Ah. Only if the battery loses charge over time with no load or if it is significantly oversized will the battery be safe from overcharge. Specifically, *NEC* 690.72(A) requires a charge controller whenever the PV array provides more than 3% of the battery rating in 1 h at maximum array current.

Suppose the combination of protection current and anode resistance to ground had resulted in the need for more than 12 V. Several means of solving this problem are available. One is simply to use a higher voltage module. Another is to use a larger anode or to use anodes in parallel. One might expect an anode with twice the surface area to have half the resistance to ground, but due to the nonlinearities of the system, this is not the case, as shown in Table 7.4.

The parallel anode solution results in a lower resistance to ground, but, perhaps not surprisingly, the resistance to ground of two identical anodes is not simply half the resistance of a single anode. The cylindrical geometry again adds a nonlinear twist to the problem, resulting in the resistance to ground of two anodes depending on the separation of the anodes. Table 7.5 shows multiple anode adjustment factors, assuming all anodes are identical and that the soil is uniform in composition. The factors in the table are multipliers for the single anode resistance to ground. Thus, for example, if the resistance to ground for a single anode is 100 Ω, then the resistance to ground for three anodes spaced 15 ft apart will be $100 \times 0.418 = 41.8\ \Omega$. This reduces the voltage required to 41.8% of that required for a single anode. It is thus a matter of calculating the life-cycle cost (coming up in Chapter 8) of the three-anode system versus the single-anode system. Keeping in mind that each anode now will only carry 1/3 of the system current, the anodes should last three times longer.

A somewhat more elegant solution to the problem would involve an electronic constant current source that would provide the required current regardless of soil conditions or PV output. However, wet soil requires more current than dry soil, so the current source would need to be compensated for soil resistivity, rendering the current source design somewhat more challenging.

Table 7.5 Multiple Anode Adjusting Factors

Number of Anodes	Separation of Anodes				
	5′	10′	15′	20′	25′
1	1.000	1.000	1.000	1.000	1.000
2	0.652	0.576	0.551	0.538	0.530
3	0.586	0.460	0.418	0.397	0.384
4	0.520	0.385	0.340	0.318	0.304
5	0.466	0.333	0.289	0.267	0.253
6	0.423	0.295	0.252	0.231	0.218
7	0.387	0.265	0.224	0.204	0.192
8	0.361	0.243	0.204	0.184	0.172
9	0.332	0.222	0.185	0.166	0.155
10	0.311	0.205	0.170	0.153	0.142

Source: Based on data from *Stand-Alone Photovoltaic Systems: A Handbook of Recommended Design Practices*, Sandia National Laboratories, Albuquerque, NM, 1995.

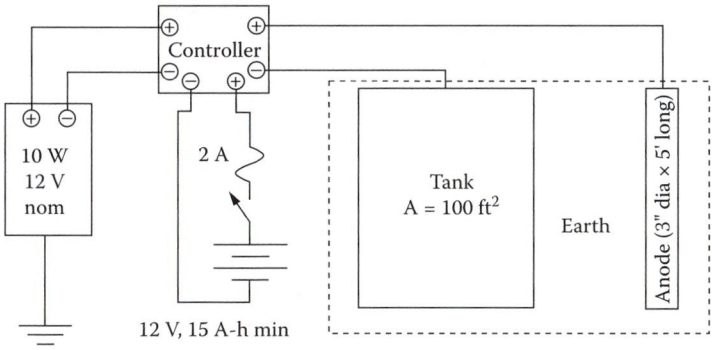

Figure 7.8 Cathodic protection system schematic diagram.

Whenever possible, it is advisable to make soil resistance measurements so that empirical data can be used to size and locate system components properly. This eliminates many of the assumptions made and provides for greater confidence in the performance of the system. Figure 7.8 shows the final system.

7.6 A PORTABLE HIGHWAY ADVISORY SIGN

7.6.1 Introduction

Once upon a time, illuminated highway signs either had to be connected to the power grid or else had to be self-contained with their own portable fossil-fueled generator. Furthermore, their messages were often hard-wired, so the sign could only convey a single message. In this example, rather than determining the number of PV modules and number of batteries for a given load, the energy available for a load and the corresponding average load power will be determined. This approach is taken for several reasons, one of which is that it has not been used yet and the other being that sometimes a PV system may be limited by size or cost.

Anything to be transported on a roadway should normally be 8 ft or less in width. Assuming a sign of 8 ft width and assuming that PV modules will be mounted horizontally on top of the sign because the orientation of the sign will be random, suggests that 4 modules, each 2 × 5 ft, can be used conveniently. The sign will be mounted on a trailer of standard size along with the battery pack and BOS. While it may be possible to mount additional modules elsewhere on the trailer, it is assumed that to avoid module shading and minimize damage from vandalism, only placement of the modules on top of the sign is practical.

Because each month has different peak sun hours, each month will have different available average power for the sign. Thus, during some months, it may be possible that the sign will not be used 24 h per day or it may be possible to convey longer or brighter messages during some months. For this example, it will be assumed that the sign will be used in the vicinity of Atlanta, Georgia, where freeway construction

Table 7.6 Summary of Available Average Monthly Irradiance for Atlanta, Georgia (peak sun h/day)

	Jan	Feb	Mar	Apr	May	Jun
Horizontal	2.6	3.4	4.5	5.7	6.2	6.4
	Jul	**Aug**	**Sep**	**Oct**	**Nov**	**Dec**
Horizontal	6.2	5.7	4.8	4.1	2.9	2.4

has been in progress for the past 50 years and may continue into the foreseeable future.

7.6.2 Determination of Available Average Power

If modules with 18% efficiency are used, then they will be able to generate approximately 16.7 W/ft^2 under full sun irradiance. The four modules will thus have a maximum power output of $40 \times 16.7 = 668$ W, provided that they are kept clean. Assuming, however, that they will commonly be used in dusty construction sites, it is probably best to assume an 80% degradation factor due to the dust and elevated cell temperature. The maximum output is thus reduced to 534 W. This output, of course, is only achieved if the sun is directly overhead. Irradiation data are available for Atlanta, Georgia, in Appendix A and are repeated in Table 7.6 for a horizontal fixed array.

Next, the amount of energy available for charging batteries can be determined. Assuming that an MPPT charge controller with an efficiency of 96% is used, and that wire losses are 2%, $0.96 \times 0.98 \times 534 = 502$ W is available at the battery terminals; using a nominal system voltage of 12 V leaves $502 \div 12 = 41.8$ A of effective charging current from the charge controller. Thus, during the worst month, the average daily charge delivered to the batteries will be $41.8 \times 2.4 = 100.4$ Ah.

If 100.4 Ah is delivered to the batteries, and if the charge/discharge efficiency of the batteries is 90%, then 90% of the charge (energy) that entered the batteries will be recoverable for use by the load. Thus, on average, approximately 90.4 Ah will be available from the batteries each day to operate the sign during the worst month for sunlight.

If 5 days of storage is desired, the batteries should be sized on the basis of 100.4 Ah/day, with an allowable depth of discharge of 80%, which results in the need for $502 \div 0.8 = 627.5$ Ah of batteries. There are a number of ways to obtain this battery capacity, one of which is a set of four 6 V batteries, each of which is rated at 315 Ah at a $C/120$ discharge rate. Alternatively, a set of two 12 V batteries, each of which is rated at 315 Ah at $C/120$, will also be adequate. In fact, a 12 V, type AGM, maintenance-free battery is available with a 305 Ah rating at $C/120$, which should be adequate for this design.

Finally, the average daily power available to the sign can be computed. The assumption is that the batteries are fully charged when sign operation is begun. Under these circumstances, the batteries will undergo a net discharge on some days and a net charging on other days, depending on whether the irradiation is above the

Table 7.7 Average Daily Power Available to the Sign for Each Month of the Year

	Jan	Feb	Mar	Apr	May	Jun
Peak sun (h)	2.6	3.4	4.5	5.7	6.2	6.4
Avg power (W)	49	64	85	107	117	120
	Jul	**Aug**	**Sep**	**Oct**	**Nov**	**Dec**
Peak sun (h)	6.2	5.7	4.8	4.1	2.9	2.4
Avg power (W)	117	107	90	77	55	45

monthly average or below the monthly average for the particular day. If the average daily energy used by the sign is 10% less than the average daily energy provided to the batteries, then the net discharge of the batteries is close to zero. So all that is needed is a tabulation of the average daily energy in Wh available *from* the batteries for each month of the year. For each day, this energy is found from

$$\text{Wh} = 0.9 \times V_B \times I_{CH}(\text{max}) \times (\text{psh}), \tag{7.4}$$

where V_B is the nominal battery voltage and I_{CH}(max) is the charge controller output current at peak sun after accounting for array, charge controller, and wiring losses. This value has already been calculated to be 41.8 A for this example. The average power available over a 24 h period is simply the daily Wh divided by 24 h. Table 7.7 tabulates the daily average power available for the 12 months of the year.

With a microcontroller in the system, it is straightforward to program the unit to inform the user of the average power that will be used to implement any particular program. For that matter, the system can even be programmed to give a warning to the programmer if the average daily power is exceeded by the proposed announcement. If the sign is not programmed to use maximum available power, then the controller needs to have the capability to disconnect the PV array from the batteries.

7.7 A CRITICAL-NEED REFRIGERATION SYSTEM

7.7.1 Introduction

It is desired to provide electric power for 10 ft^3 of refrigeration for storage of medication. The medication is costly and difficult to obtain, so the system must have 99% availability. The system will be located in a village in Angola where villagers will be able to keep a careful watch on system performance so that any problems will become immediately evident. This design will be based on an ac refrigerator that has one of the highest energy efficiency ratings. The refrigerator is a 10.12 ft^3 unit that is rated at 171 kWh/yr, which equates to 0.47 kWh/day [9]. It will be powered by an inverter that has a 94% efficiency rating that produces a sine wave output that is acceptable as a source of power for the refrigerator. The inverter will have a 12 V input and will have low-voltage cut-out and cut-in capability.

7.7.2 Load Determination

To convert 0.47 kWh/day to Ah per day, multiply by 1000 to get 470 Wh per day. Then assume the refrigerator is well insulated such that the compressor only needs to run for 6 h/day. This means the compressor will require 470 ÷ 6 = 78 W for operation. Note that this estimate is for the purpose of determining the output power rating of the inverter. So, a 300–500 W range for inverter power should easily cover any starting surge of the compressor. In fact, if the inverter size is much larger, the low power efficiency of the inverter may drop below the 94% level.

The next step is to determine the daily load on the batteries. Starting with the refrigerator load and then compensating for inverter and wiring losses results in a load on the batteries of (470 ÷ 0.94) ÷ 0.98 = 510 Wh/day. This will be defined as the *connected load.*

7.7.3 Battery Sizing

The batteries must be sized to power the connected load over the duration of time determined as the critical load storage period. For a single day, the amount of energy that must be transferred to the batteries will be the *connected* load adjusted for battery charge/discharge losses, wiring losses and, possibly, losses due to lower operating temperature or rapid charging rates. In equation form, this is

$$Ah = \left(\frac{Ah}{day}\right)\left(\frac{days}{D_T D_{ch}(disch)}\right), \tag{7.5}$$

where Ah/day represents the *corrected load* on the batteries, *days* represents the number of days of autonomy, D_T is the temperature derating factor, D_{ch} is the charge/discharge derating factor, and *disch* is the depth of discharge expressed as a fraction. In Chapter 3, Section 3.5.2, it was noted that lead-acid battery capacity decreases at lower temperatures and higher discharge rates. Higher charge rates also can result in greater losses.

The capacity of a battery decreases as temperature decreases. For lead-acid batteries, the capacity reduction can be approximated by the following empirical relationship, for battery temperatures between 20°F and 80°F.

$$D_T = \frac{C}{C_o} = 0.00575T + 0.54, \tag{7.6}$$

where C is the battery capacity at temperature T (in °F) and C_o is the rated battery capacity at 80°F [8].

If the anticipated load will exceed the specified discharge rate for more than 10 min, an additional correction factor should be applied to the total corrected Ah. For example, if the battery is rated at a discharge current of 20 A, and the actual discharge rate will be 30 A, then D_{ch} = 20/30 = 0.67. This factor reflects the fact that there are greater I^2R losses to the internal battery resistance as the current increases, thus decreasing the energy available from the batteries to power the load.

If the battery charge rate exceeds the rated charge rate, an alternate calculation of battery capacity should be performed and compared with Equation (7.5). Lead-acid batteries are typically designed for charging at a $C/10$ rate of charging. Hence, if 10 h times the PV charging rate in amperes exceeds the battery capacity in Ah calculated in Equation (7.5), it means the PV array is capable of fully charging the batteries faster than the $C/10$ rate. If this condition should occur, the battery capacity should be determined by the product of the rated charging time and the available charging current, rather than by Equation (7.5). The capacity calculated in this manner should result in a value lower than the $C/10$ rated capacity because greater battery internal I^2R losses occur at the higher charging rate.

For the present design, assuming a combined loss of 12% for battery charge/discharge and wiring losses and neglecting other factors means that the *corrected* load will be 510 ÷ 0.88 = 580 Wh/day. Next, allowing for 80% depth of discharge, the amount of battery storage for 1 day of operation must be 580 ÷ 0.8 = 725 Wh. Dividing by the nominal battery voltage (12 V) converts the 725 Wh to 60.41 Ah.

At this point, it will be necessary to determine the number of storage days needed for critical system availability. Using the minimum daily average psh for Luanda, as obtained from Appendix A, in Equation (3.11a) yields $D_{crit} = (0.2976 \times 3.93^2) - (4.7262 \times 3.93) + 24 = 10$ days. Thus, a total of 604 Ah @ 12 V of storage at a $C/240$ rate will be needed. Because most battery manufacturers only specify discharge rates up to $C/120$, this is the discharge rate to use when selecting the batteries.

Additional factors to consider when selecting batteries include the additional connections required when using larger numbers of batteries and the corresponding additional time required for installation and maintenance. Furthermore, the greater the number of connections, the greater the possibility of failure of a connection; the cost of battery cables is not negligible. On the other hand, with a smaller number of batteries, if one should fail, the system is impacted more significantly than in the case of a larger number of batteries. The weight of the battery and the difficulty of obtaining the battery may also be considered. A local supplier with the ability to provide timely delivery and warranty service is generally more desirable.

One battery that nicely meets these requirements is a 12 V sealed unit that has a rating of 305 Ah @ $C/120$. So, two of these in parallel will provide 610 Ah of storage, which nicely meets the design criterion as long as it is available at an acceptable cost.

7.7.4 Array Sizing

7.7.4.1 Fixed Array

The array must be sized to meet the daily *corrected* load requirements during the month with the lowest daily psh. In this case, the worst-case psh has already been determined to be 3.93 Ah for a fixed array with a latitude + 15° tilt. However, note that for a tilt of latitude, the worst-case psh is 3.71 and for a tilt of latitude − 15°, the worst-case psh is 3.36 psh. So one might be wondering how 3.93 is the worst case when there are two other cases that are worse.

Consider the fact that the larger the worst-case psh, the smaller the number of storage days for critical system availability. Because the designer has the choice of array tilt, it is logical to determine the worst-case psh for each array tilt and then to use the *best* worst-case result. This will result in the lowest number of storage days and, therefore, the least cost for batteries. The system is then designed with the tilt that produces the *best* worst-case array performance, which, in this case, is latitude + 15°. Furthermore, the latitude + 15° tilt has the least annual variation in peak sun hours. This will result in a minimum of wasted PV energy production.

When specifying the array tilt, it is important to recognize that the latitude of Luanda is 8.82° south of the equator, or −8.82°. Thus, latitude + 15° means a tilt of +6.18°, or facing north with a tilt of 6.18°. This is because during the "winter" in Angola, the sun is north of the equator, while during the winter in the Northern Hemisphere, the sun is south of the equator, meaning the declination, $\delta < 0$. So Angola gets the least amount of sun during the Northern Hemisphere summer season and the most sun during the Northern Hemisphere winter season.

Because the array tilt and the *corrected* load are now known, all that is left is to determine the needed size of the array. If an MPPT charge controller is used, then the array size can be calculated in watts. If a non-MPPT charge controller is used, then the array current needs to be calculated at a reasonable array operating voltage. The reason for this is that with a non-MPPT charge controller, the controller output *current* is the same as the controller input *current*, whereas with an MPPT charge controller, the controller output *power* is the same as the controller input *power* except for minimal loss in the controller. In this example, the array size will be determined for both types of charge controller and then the effect of a two-axis tracking mount on array and battery sizing will be considered.

For the MPPT charge controller, the array must deliver the *corrected* load to the batteries. Thus, after array losses, losses in wiring between array and charge controller, and charge controller losses, 580 Wh must still be left for the batteries. From previous examples, the total of these three loss mechanisms has been found to be approximately 15%; thus, in order to deliver 580 Wh to the batteries, the array must be able to produce 580 ÷ 0.85 = 682 Wh/day.

Dividing by the worst-case psh for the year results in an array size of 682 ÷ 3.93 = 174 W. Thus, a single module rated at 174 W or more will deliver the necessary energy to the batteries, and there is an abundance of such modules available in the marketplace.

For the non-MPPT charge controller case, the *corrected* load must be expressed in Ah. Because a 12 V battery system will be used, dividing the *corrected* load in Wh/day by 12 results in 48.3 Ah/day needed for the batteries. Note that this figure differs from the battery Ah because the batteries must be capable of storing an additional 25% so that they will not discharge fully. Again, using 3.93 psh, the daily Ah can be converted to required array current. Dividing 48.3 by 3.93 gives 12.3 A of array current. So, in this case, a module or modules must be found that will deliver 12.3 A as opposed to 174 W. The 12.3 A must be at a voltage that is capable of charging the batteries when the array is operating at its highest temperature. This means the array V_{mp} must be at least 14 V at the highest array temperature. This condition is automatically achieved by the use of a nominal 12 V module, which has 36 cells and typically has $V_{mp} \approx 17$ V at STC. It should also be noted that the module will operate at just above the battery-charging voltage, because a small voltage is lost across the switching devices in the charge controller. It should also be noted that the selection of the module should also allow for up to 10% degradation in module output to account for the possibility of the module getting dusty or soiled in any other manner as well as possibly having an output power below the advertised value. So, a conservative selection of module will make this allowance and require a total operating current of $12.3 \div 0.9 = 13.7$ A. A check on available 12 V nominal modules reveals a module that has $V_{OC} = 21.8$ V, $I_{SC} = 7.99$ A, $V_{mp} = 17.2$ V, and $I_{mp} = 7.15$ A. Note that this is a 123 W module, so two of them will have a power rating of 246 W, which is 72 W higher than the rating needed with an MPPT controller. So the check to be made is to compare cost of the additional module power with the difference in cost between the conventional charge controller and the MPPT charge controller.

7.7.4.2 Tracking Array Mount

For a two-axis tracking array mount, the minimum peak sun hours is 5.14 for July (Appendix A). Note that for the two-axis tracker, there is no need to look for the best worst-case psh, because the two-axis tracker tracks both altitude and azimuth of the sun. Thus, with the new value of minimum psh, a new calculation for storage days can be made with Equation (3.11a). The result is $D_{crit} = (0.2976 \times 5.14^2) - (4.7262 \times 5.14) + 24 = 7.57$ days. There is no need to round this number up or down, because it is simply used as a multiplier in determining needed battery capacity. In this case, as the capacity for 1 day has already been calculated to be 60.41 Ah, the capacity for 7.57 days will need to be 457 Ah, again at the $C/120$ discharge rate. So the tracking array saves almost 150 Ah of battery capacity, which means lower battery cost.

For the MPPT charge controller, the array power needed with the two-axis tracker is (682 Wh) ÷ (5.14 h) = 133 W.

For the standard charge controller, the array current needed with the two-axis tracker is (48.3 Ah) ÷ (5.14 h) = 9.4 A. Allowing for possible degradation due to dust and possible module STC rating deviation, the conservative designer will choose an array capable of producing $9.4 \div 0.9 = 10.4$ A. Again, two modules in parallel can be found to generate this amount of current, but this time it is also possible to find a single 200 W module that has $I_{mp} = 11.05$ A and $V_{mp} = 16.7$ V.

So the tracking mount will save on both battery size and module size. Again, the test is whether the additional cost of a tracking mount compared to the cost of a fixed mount is less than the savings in array and battery cost.

7.7.4.3 Special Cases

In some cases, especially where low-light or high-temperature performance is desired, the designer must examine the module output curves under a range of light

Table 7.8 Summary of Wire Sizes and Fuse Sizes for Refrigerator System

Wire Location	VD A	One-Way Length (ft)	System Volts (V)	Max Allow (R, Ω)	Wire (Ω/kft)	Wire Size	Fuse Size (A)
Array to controller	4.95	40	35	0.0716	0.8950	#8	10
Controller to batteries	13.9	6	12	0.00863	0.719	#6	25
Controller to inverter	6.9	8	12	0.0174	1.0875	#8	10
Inverter to refrigerator	0.65	15	120	1.846	61.5[a]	#14	15

[a] Wire size must be based on ampacity rather than voltage drop.

intensity and a range of temperature to be sure the module will have adequate output under anticipated adverse operating conditions.

7.7.5 BOS Component Selection

At low voltages, careful attention must be paid to voltage drop in the wiring. It is generally useful to lay out the system and then tabulate the lengths of the various wire runs along with the allowable voltage drops. This enables a simple calculation of the required resistance per 1000 ft of wire to keep the wire voltage drop within allowable limits. Once the resistance per 1000 ft is known, the proper wire size can be selected. As in previous examples, the correct wire size is the larger of either the size needed to carry the rated current or the size needed to meet system voltage drop constraints.

Table 7.8 summarizes the wire sizing for a system with fixed array tilt and an MPPT charge controller. A word of explanation may be useful for some of the table entries. In the first column, current used to calculate voltage drop, the array to controller current shown is I_{mp}, which is generally used for voltage drop calculations. It is recognized that $1.56I_{SC}$ is required for sizing based on ampacity, so this check will be made after the voltage drop calculation for wire size is done. Note that all but the inverter to refrigerator wire sizing ends up being determined on the basis of voltage drop as opposed to ampacity.

The controller to battery current is 95% of the quotient of the 175 W array power and the battery voltage. The 95% figure compensates for array to controller wiring loss plus controller loss. The controller to inverter current is the inverter input power divided by the inverter input voltage. The inverter input power is the power to the refrigerator divided by the wiring efficiency (close to 100%) and the inverter efficiency.

The next two columns are self-explanatory. The allowable resistance is the wire resistance that would produce a 1% voltage drop, and the Ω/kft column is the allowable resistance divided by the round-trip kft length of the wire run. Wire sizes are based on the allowable Ω/kft value except for the wire to the refrigerator, which is set at #14 because #14 is the smallest wire size that is allowed to be concealed in a 120 V system.

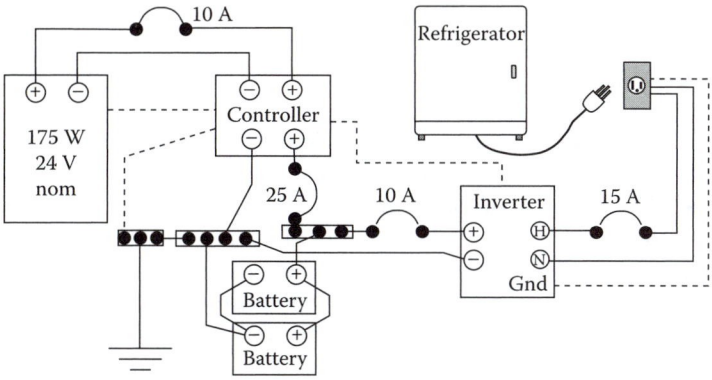

Figure 7.9 Refrigerator system schematic diagram.

7.7.6 Overall System Design

At this point, all system components have been selected for the system. It is now possible to develop a system schematic diagram. Figure 7.9 shows the schematic diagram of the system that uses a fixed array and an MPPT charge controller. Because the maximum system voltage is below 50 V, grounding of the wiring is optional for systems wired in accordance with the *NEC*. However, if the system is left ungrounded, then disconnect switches must open all ungrounded conductors. It is thus generally better to establish a single point where the *grounded* conductor is connected to the *grounding* conductor, because all metallic enclosures must be grounded anyway.

The grounded conductor carries current and is normally chosen as the negative conductor of the PV system. The grounding conductors do not carry current, and are used for the purpose of connecting all exposed metal system parts to ground to protect against shock hazards. As long as the grounded and grounding systems are connected at only one point, current will not flow in the grounding conductors under conditions of normal operation. Grounding conductors are shown along with the single point of connection between grounded and grounding conductors. Sizing of grounding conductors will be considered in later design examples.

The single grounding point is usually a terminal block with multiple set screw terminations for connecting grounded and grounding conductors. In systems that require a ground fault detection and interruption device (GFDI), the grounded conductors will generally terminate at one terminal block, and the grounding conductors will terminate at a separate terminal block. The two blocks are then connected to each other through the trip mechanism of the GFDI device. Any current flowing through the GFDI device would indicate that more than one point of the grounded system is connected to the grounding system.

7.8 A PV-POWERED MOUNTAIN CABIN

7.8.1 Introduction

Figure 7.10 shows the floor plan of the cabin, located west of Denver, Colorado, to be outfitted with PV-powered electrical loads. The cabin will have a kitchen light, a dining room light, a light in each bedroom, a bathroom light, a living room light, a motion sensor light outside each door, a ceiling fan in each bedroom, an entertainment center, a deep-well-water pump, and an Energy Star® 16 ft^3 refrigerator/freezer. The PV array will be located approximately 10 ft from the back of the cabin as shown. The 200-ft-deep well will be located 30 ft from the cabin in the same relative direction. The pump will be a submersible unit, so the total distance to the pump will be 230 ft. All loads will operate at 120 V ac, so an inverter will be needed to power the loads. The inverter input voltage will be determined after the inverter power requirements have been determined.

The system loads, with the exception of the refrigerator, will operate 3 days per week to accommodate weekend use over long weekends. The refrigerator will be left on continuously in order to prevent inadvertent odor buildup in the event it should get too warm inside the unit. At least, this is the stated reason. It is suspected that the real reason is to have cold beer ready as soon as someone arrives at the cabin.

Because the *NEC* requires that a building be wired as though it were to be connected to the grid, the cabin will be wired for all 120 V loads. The PV array will be rack-mounted at ground level, so the feasibility of seasonal adjustment of array tilt will be explored as a means of optimizing system performance. Also,

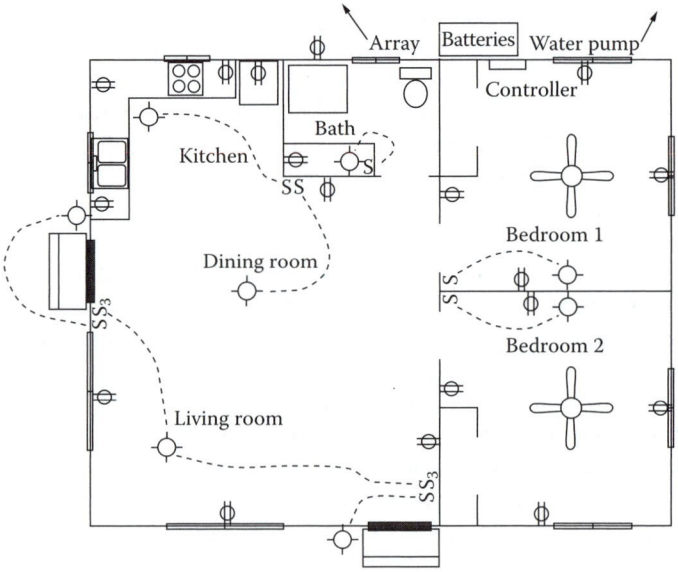

Figure 7.10 Floor plan of mountain cabin.

maintenance-free, valve-regulated lead-acid batteries will be used as a part of keeping system maintenance to an absolute minimum.

7.8.2 Load Determination

At this point, it should be abundantly clear that the only reasonable choice for the refrigerator and other loads will be high-efficiency units. The refrigerator that will be used has separate freezer and refrigerator compartments and is rated at 306 kWh/yr, which equates to 0.84 kWh/day. Assuming 6 h of compressor operation per day equates to a power consumption of 140 W.

For the water pump, it is first necessary to determine the water needs for the cabin. This determination can be quite a wild guess unless the uses of the water are reasonably well defined. Assuming the only means of heating water is a propane stove, it is reasonable to assume that showers will not be too lengthy. However, if a solar water heater is added later, shower length may increase. A reasonable estimate for water usage might be 40 gal per day per person, with anticipated occupancy of 4 persons. This amounts to a total water requirement of 160 gal per day. If more people use the cabin, it may be necessary to do some water rationing. In the worst case, where drinking is the only use made of the water, 160 people would still have a gallon each per day for drinking over the 3-day weekend. It is assumed that less water is used in winter, spring, and fall than the 160 gal per day of summer use.

The next step is to go to the manufacturers' catalogs or to the Internet. For example, [11] lists a number of different water pumps. For deep wells, normally a submersible pump is the best choice. The first check to make on a pump to be used in a system with battery storage is whether it can pump the needed water in either a day or a week. If the needed water can be pumped in a day, then only a 160 gal storage tank will be needed. If it takes all week to pump the water for the weekend, then a 480 gal storage tank will be needed. The pumping power, will depend on the flow rate and the lift distance. If the top of the storage tank is 10 ft above ground to provide pressure for the system, and, allowing for 5% piping losses, the equivalent lift for the pump will be $210 \times 1.05 = 220.5$ ft.

In Section 7.2 it was observed that it is generally less costly to use a small pump that pumps over a longer period of time than a larger pump that pumps the needed water in a short time. Of the many pumps on the market, the ac pumps tend to have higher horsepower than the dc pumps. One pump, however, that will run on either dc or ac, will pump 1.15 GPM at a 225 ft head, using 115 W. This is not exactly a cheap pump, but it has a reputation for performance and efficiency. The 1.15 GPM pumping rate amounts to about 31.5 ft-lb/s, or 0.057 HP. This means the wire-to-water pumping efficiency is approximately 37%, which is indeed quite efficient for this size pump. At 1.15 GPM, the time to pump 160 gal is 139 min, or 2 h 19 min. So a 160 gal storage tank should be adequate. The energy required by the pump for pumping this amount of water is $(115\ \text{W}) \times (2.317\ \text{h}) = 266$ Wh/day.

The remaining loads are listed in Table 7.9 as corrected for inverter and wiring losses. Thus, the actual loads are divided by 0.92 to determine the load on the

Table 7.9
Summary of Monthly Variation in Weekly Ah Loads for Mountain Cabin

Load Description	P Watts	Days/Week	Nov-Feb		Mar	
			h/Week	Ah/Week	h/Week	Ah/week
Kit light	32	3	12	8	10.5	7
BR 1 light	17	3	6	2.1	4.5	1.6
BR 2 light	17	3	3	1.1	3	1.1
LR light	17	3	15	5.3	12	4.3
Outdoor lights	17	3	1.5	0.5	1.5	0.5
DR light	32	3	12	8	9	6
Bath light	17	3	6	2.1	6	2.1
Refrigerator	152	7	35	110.8	38.5	121.9
Water pump	125	7	5.5	14.3	6.0	15.6
BR 1 fan	32	3	0	0	0	0
BR 2 fan	32	3	0	0	0	0
Receptacles	2500	3	2	104.2	1.75	91.1
TOTAL	2990			256		251

Load Description	Apr, Oct		May, Sep		Jun, Jul, Aug	
	h/Week	Ah/Week	h/Week	Ah/Week	h/Week	Ah/Week
Kit light	10.5	7	9	6	7.5	5
BR 1 light	4.5	1.6	3	1.1	3	1.1
BR 2 light	3	1.1	3	1.1	3	1.1
LR light	9	3.2	6	2.1	6	2.1
Front door light	1.5	0.5	1.5	0.5	1.5	0.5
DR light	7.5	5	6	4	4.5	3
Bath light	5	1.8	4	1.4	3	1.1
Refrigerator	38.5	121.9	42	133	45.5	144.1
Water pump	6.5	16.9	7	18.2	8	20.8
BR 1 fan	3	2	8	5.3	24	16
BR 2 fan	3	2	8	5.3	24	16
Receptacles	1.5	78.1	1.25	65.1	1	52.1
TOTAL		241		243		263

batteries. Note that in this example, the monthly variations in loads are accounted for. For example, the ceiling fans will be used more in the summer, and the lights will be used more in the winter. The monthly variations in loads are important to use when determining the array tilt.

Note that the 120 V branch circuit loads are computed at the values required by *NEC* 690.10, even though the total power and energy used will be limited by the array size, the inverter size, and the battery size. The idea is that one day when the grid is everywhere, some future owner may elect to connect to the grid. For the moment, however, *NEC* 690.10(A), (B), and (C) will apply while the system is a stand-alone system. In other words, the inverter will not be required to simultaneously power all loads, and the power distribution panel will need to have a label indicating that the panel is a single 120 V supply.

Per *NEC*, each duplex outlet, except in kitchen or bathroom, must be rated at 180 VA, so the total load assigned to the duplex outlets in the two bedrooms and the living/dining room area is 2160 W. The load assigned to the kitchen circuit is 1500 W as is the load assigned to the bathroom (hair dryers). Even though all the lighting outlets will have no more than a 20 W compact fluorescent lamp, they still must be rated at 90 W each. But to calculate the load on the inverter and the batteries, the expected off-grid loads on these circuits are used in Table 7.9, with the understanding that the circuits are wired adequately to handle the loads as required by *NEC* 210 and 220.

For each load, the average Ah/week if the load operates for n hours per day and d days per week is determined by

$$\frac{Ah}{week} = \frac{Pnd}{\eta V}, \tag{7.7}$$

where P is the load power, η is the conversion efficiency of the inverter, and V is the 48 V system dc voltage. The 48 V system voltage at this point is a guess, because a specific inverter has not yet been selected. In fact, the total power in column 1 suggests that an inverter rated at approximately 3000 W will be suitable for the cabin, and it is likely that a 3000 W inverter will operate at 48 V. Note that the power assigned to the receptacles is unusually high, but the hourly use of the receptacles is estimated to be comparatively low to correspond to a few minutes of microwave or hair dryer time and longer times for other, relatively low-power plug-in devices, such as a TV or a computer or a small reading lamp.

The entries in Table 7.9 are based on 3 days per week of usage for all loads except the refrigerator, which remains on for 7 days per week and the water pump, which will pump as needed to keep the storage tank full. The pump is controlled by a float switch in the storage tank.

It is important to note that the average Ah/week represents the weekly average, taking into account that the water pump and the refrigerator may run 7 days per week, but the other loads are only operational for 3 days per week while the cabin is occupied. The array and batteries will thus be selected to meet the average weekly system corrected loads.

The next step is to compute the system corrected loads. Using a 98% wire efficiency factor and a 90% battery efficiency factor, the corrected Ah loads for each month are obtained by dividing the load Ah/week by 0.98×0.9. The corrected loads are shown in Table 7.10.

7.8.3 Battery Selection

Because the batteries need to hold sufficient charge to power the cabin loads for a week, the batteries will not discharge in less than 10 h. In addition, because the

Table 7.10 Corrected Weekly Ah Loads for Cabin after Accounting for Wire and Battery Losses

Nov, Dec, Jan, Feb	Mar	Apr, Oct	May, Sep	Jun, Jul, Aug
290	284	273	275	298

batteries will take a week to charge fully, the charge rate will also not be excessive. Thus, the charge and discharge derating factors will both be unity.

The factors that affect battery selection in this case are the monthly variation in average daily energy use and the monthly temperature variation of the batteries. In winter months the batteries will be colder, even though they will be in an insulated enclosure, so temperature correction factors are needed. Table 7.11 shows the calculation of required battery capacity on a monthly basis. It is assumed that a deep-discharge unit will be chosen with a maximum depth of discharge of 80%. Once the total battery capacity is determined, a battery can be selected that has a capacity close to an integral divisor of the total capacity. Because the batteries are allowed 3 days to discharge, the rating at $C/72$ can be used for sizing. In this case, a 244 Ah, 12 V, 8-year lifetime sealed lead-acid battery is a reasonable choice, as indicated by the ratio of total capacity to battery capacity. The system will require eight of these units configured in two parallel sets of four units in series. The last row in Table 7.11 indicates the percentage of the battery capacity used each month.

7.8.4 Array Sizing and Tilt

Array sizing involves determining the optimum system design current by computing the design current for each month of the year at each of three tilt angles, then choosing the tilt angle that yields the lowest design current. Because the corrected load is a weekly load rather than a daily load but the peak sun hours are listed as daily peak sun hours, it is necessary to divide the corrected load by the peak sun hours and then by 7. This is equivalent to dividing by the *weekly* peak sun hours.

Table 7.11 Determination of System Battery Capacity Requirements

Month	Jan	Feb	Mar	Apr	May	Jun
Ah/week	290	290	284	273	275	298
Temp derate	0.80	0.80	0.85	0.90	0.95	1.00
Total cap req	453	453	418	379	362	373
% cap used	93	93	86	78	74	76
Month	**Jul**	**Aug**	**Sep**	**Oct**	**Nov**	**Dec**
Ah/day	298	298	275	273	290	290
Temp derate	1.00	1.00	0.95	0.90	0.85	0.80
Total cap req	373	373	362	379	426	453
% cap used	76	76	74	78	87	93

Table 7.12 Determination of Optimum Design Current and Array Tilt Angle

Month	Corr Load Ah/wk	Latitude – 15°		Latitude		Latitude + 15°	
		psh/day	A	psh/day	A	psh/day	A
Jan	290	4.32	9.6	5.07	8.2	5.51	7.5
Feb	290	4.94	8.4	5.54	7.5	5.81	7.1
Mar	284	6.42	6.3	6.8	6.0	6.8	6.0
Apr	273	6.69	5.8	6.65	5.9	6.24	6.3
May	275	7.07	5.6	6.69	5.9	5.97	6.6
Jun	298	7.22	5.9	6.67	6.4	5.78	7.4
Jul	298	7.32	5.8	6.84	6.2	6.01	7.1
Aug	298	6.84	6.2	6.66	6.4	6.13	6.9
Sep	275	6.78	5.8	7.02	5.6	6.85	5.7
Oct	273	5.92	6.6	6.53	6.0	6.75	5.8
Nov	290	4.37	9.5	5.05	8.2	5.43	7.6
Dec	290	4.05	**10.2**	4.81	**8.6**	5.28	**7.8**
Design current for tilt			10.2		8.6		7.8
Optimum design current					7.8		

Note: Boldface data indicates optimal design choice.

Table 7.12 shows the results for the cabin. Note that for the specified loads, the optimum tilt angle is latitude + 15°, and the optimum design current is 7.8 A. Peak sun hours (psh) are for Denver, Colorado (Appendix A).

The data in Table 7.12 can be rearranged to show the optimal array tilt angle for each month and the corresponding array current, as shown in Table 7.13. Because the corresponding design current represents the minimum allowable current for that month, the maximum of these values over the 12-month period must be chosen to provide adequate system design current for the worst month. In this case, by adjusting the tilt three times per year, the system can be optimized to obtain a design current of 7.8 A, which is the same as the result for the fixed tilt array. If the loads were different, this may have resulted in a smaller number. The difference is that in the summer, more excess energy will be produced if the array tilt is adjusted.

The next step is to account for module degradation by dividing the optimal design current by the estimated degradation factor. In this case, 90% will again be used as a default value. The array will be located where reasonable maintenance will

Table 7.13 Optimization of Array by Seasonal Tilt Adjustment (+15 = lat + 15°, etc.)

Mo	Jan	Feb	Mar	Apr	May	Jun
Tilt	+15	+15	+15	–15	–15	–15
A	7.5	7.1	6.0	5.8	5.6	5.9
Mo	**Jul**	**Aug**	**Sep**	**Oct**	**Nov**	**Dec**
Tilt	–15	–15	lat	+15	+15	+15
A	5.8	6.2	5.6	5.8	7.6	7.8

be possible so that it will remain relatively clean. This yields derated design currents of 8.7 A for each mounting system. This will be the design current if a conventional charge controller is used.

If an MPPT charge controller is used, then the design current can be multiplied by the battery-charging voltage and divided by the controller efficiency and the wiring efficiency to determine the array power needed to produce the design current. In this case, using a battery-charging voltage of 54 V, the result is $P_{\text{array}} = 7.8 \times 54 \div 0.97 \div 0.98 = 443$ W. In this case, because the array will produce excess power in the summer, even at elevated temperature operation, using a factor of 10% for losses will be adequate for sizing the array for winter operation. So for adequate annual performance, an array power of $443 \div 0.9 = 492$ W @ STC should be more than adequate. The decision now is whether or not to use the MPPT charge controller. A look at available modules can shed some light on this question.

For the conventional charge controller, an array current of 8.7 A @ 54 V minimum is needed. To be sure that the array is capable of generating the 8.7 A @ 54 V, it is a good idea to be sure that the STC array voltage will be at least 12% higher, as in the summer, at elevated temperatures, the array voltage may decrease by about 12%.

So one begins the search for modules that will produce either 8.7 A or 4.35 A. One module that comes close to this criterion has $I_{mp} = 4.46$ A and $V_{mp} = 34.8$ V. Thus, four modules in a parallel combination of 2 series groups will have $I_{mp} = 8.92$ A and $V_{mp} = 69.6$ V. The total array power at STC will thus be 620 W.

For the MPPT charge controller, an array power of 492 W is needed at an STC voltage higher than 60 V. It turns out that three 165 W modules have a power of 495 W. One particular 165 W module has $V_{OC} = 44.5$ V and $V_{mp} = 35$ V. This means three modules in series will have $V_{OC} = 133.5$ V. This is close to the allowable 150 V at the input of the charge controller, so it will be necessary to check the value of V_{OC} at low temperatures, which may go as low as −20°F (−29°C). For this module, $\Delta V_{OC}/\Delta T = -0.29\%/\text{K}$, so at −29°C, $V_{OC} = 133.5[1 + (54 \times .0029)] = 154.4$ V, so this module is not a good choice for the array, despite the good power match. Of course, the series combination could be limited to two modules, but then a total of four modules would be needed to obtain a minimum power level of 492 W.

Another possible module is a 130 W unit that has $V_{mp} = 17.6$ V, $I_{mp} = 7.39$ A, $I_{SC} = 8.02$ A, and $V_{OC} = 21.9$ V. If four of these modules are connected in series, the result is $V_{mp} = 70.4$ V and $I_{mp} = 7.39$ A, with a total power of 520 W @ STC. For this module, it will be necessary to check on V_{mp} at elevated temperatures, because V_{mp} is relatively low for the series combination. At 60°C, V_{mp} drops to approximately 15.5 V, according to the published temperature dependence curves for the module. So four modules in series will still have $V_{mp} = 62$ V @ 60°C, which is acceptable. Thus, the use of an MPPT charge controller saves 100 W of array power. If the module cost is \$4/W, this could easily make up the difference in cost between the conventional and the MPPT charge controller.

Before proceeding to controller selection, it is interesting to look at the amount of excess Ah produced by the system on a month-by-month basis, because the

Table 7.14 Average Weekly Excess Ah Produced by the Selected Array for the Cabin

Mo	Jan	Feb	Mar	Apr	May	Jun
Ah	31	48	112	90	73	39
Mo	**Jul**	**Aug**	**Sep**	**Oct**	**Nov**	**Dec**
Ah	52	59	124	120	26	18

array has been selected to produce adequate energy during the worst month. For the 520 W array and the MPPT charge controller, the four modules will produce a battery-charging current of $0.9 \times 520 \times 0.96 \div 54 = 8.32$ A. When multiplied by seven times the average daily peak sun hours at a latitude tilt, the average weekly Ah available from the array can be determined for each month. If the required average weekly Ah is subtracted from the available average weekly Ah, the result is the average weekly excess energy produced on a monthly basis. Table 7.14 shows this result. Of course, this is somewhat of a simplification, because during the cooler months, the array can be expected to generate higher power and, thus, a current >8.32 A.

The next logical challenge for the designer is to figure out what to do with all this extra electricity. Comparison with the Ah requirements of the cabin loads shows that the spring and fall excess is approximately equal to the daily refrigerator Ah. As it is unlikely that the cabin owner will want to bring in another refrigerator for spring and fall use only, perhaps it is more likely that the fans might be run longer. As each fan will run for 8 h on 5.3 Ah of electricity, each fan could be operated continuously for 3 days on 31.8 Ah during the summer months. Doing so, however, is contingent on having sufficient battery storage to capture the additional Ah, which otherwise would have been diverted from the batteries by the controller at the point of full charge.

Another possible use for the excess electricity is to heat some water. In this case, the controller would need to divert the array output to a water-heating element when the batteries reach full charge. If this is done, it is interesting to calculate the amount of heating that will take place.

Because it takes 1 Btu to raise the temperature of 1 lb of water by 1°F, and because 1 Wh is equivalent to 3.413 Btu, one need simply convert the Ah to Wh. For example, in April, the excess 90 Ah at a system voltage of 48 is $90 \times 48 = 4{,}320$ Wh. As a gallon of water weighs 8.35 lb, it takes 8.35 Btu, or 2.45 Wh, to raise the temperature of a gallon of water by 1°F. Hence, the leftover 4,320 Wh per week will add 14,744 Btu to a tank of water. For a 30 gal tank, this amounts to about a 59°F temperature increase per week.

Another possibility would be to pump extra water. It is left as an exercise for the reader to determine the additional water that could be pumped with the leftover electricity (Problem 7.22).

It is up to the engineer to decide which use, if any, will make economic sense for this extra electricity. The determination will ultimately be dependent on specific preferences of the cabin owner. Who knows, because it is turning out to be such a nice cabin, maybe the owner will want to spend 4 days per week at the cabin during the months with the greatest excess production.

7.8.5 Controller and Inverter Selection

7.8.5.1 Charge Controller

The charge controller needs to be selected on the basis of its maximum continuous output current. To make this estimate, no array degradation is assumed, and the wire and controller efficiencies are assumed to be 100%, to ensure that a conservative estimate is made. Then, another correction is made for cloud focusing by multiplying the array rating by 125%. Finally, the batteries are assumed to be charging at their nominal voltage. So, the calculation is simply $I_{out} = 1.25P_{array}/48 = 13.5$ A. So any MPPT charge controller with an output current rated at 13.5 A or more will be more than adequate for the system as long as it can handle the array size and has a 48 V output. Keep in mind that if a controller is rated at 30 A output, it can control any current up to this value. At the time of this writing, at least four such units are available, and it is likely that more will enter the marketplace by the time this book is in press.

Note that this controller will not be used to control the discharge of the batteries. It will do three things: (1) prevent the batteries from overcharging, (2) maximize the energy delivered by the array, and (3) act as a diversion controller and send excess array current to whatever diversion load may have been selected by the owner.

7.8.5.2 Inverter

Although a modified sine inverter may meet the waveform requirements of all the loads, because the load on the inverter may go as high as 2990 W, an inverter that produces a utility-grade sine wave will be selected. The selected inverter will need to (a) meet the power requirements of the loads and (b) monitor battery voltage and provide for low battery cut-out and low battery cut-in operation. Additional bells and whistles may also come in the inverter package, but (a) and (b) along with the highest conversion efficiency possible are the key selection criteria. For example, some units may be available in 120/240 V output models. Many will have battery-charging capability, so that an external ac generator can be connected to charge batteries if the sun does not shine enough. If the inverter has battery-charging capability, then it may have features such as current limiting, automatic generator start, multiple-stage charging, etc. Once again, there are enough inverters presently available that will meet the design specifications that the reader is encouraged to do a bit of surfing on the Web to become familiar with some of the options. For the balance of system calculations, it will be assumed that the inverter selected will be rated at 3000 W, with a nominal 48 V dc input and an efficiency of 94%.

Table 7.15 Summary of Component Maximum Currents for Cabin

Array to Controller (A)	Controller to Battery (A)	Battery to Inverter (A)	Inverter to Panel (A)
8.02 (I_{SC})	13.5	72.5	25

7.8.6 BOS Component Selection

7.8.6.1 Wire, Circuit Breaker, and Switch Selection

The nice thing about an ac cabin is that no extraordinary effort is required for sizing the wiring to most of the loads, as long as they are no more than 100 ft from the distribution panel. In this case, everything is close, with the exception of the well pump, for which a voltage drop calculation will be needed. Table 7.15 summarizes maximum system component currents, and Table 7.16 summarizes one possible selection of circuits along with wire sizing and sizing of fuses. For the dc circuits, ampacity and voltage drop calculations will be needed, but the procedure is the same as in all previous examples. For example, array wiring will need to be sized at a minimum of 156% of I_{SC}. The inverter input current can be determined by dividing the inverter output power rating by the minimum inverter input voltage and the inverter efficiency. This gives the battery-to-inverter current of Table 7.15. The reason the battery-to-inverter wiring is nearly twice the size based on the inverter input current is that most inverters can produce surge currents that are double the normal maximum operating current. Thus, the larger wire and larger circuit breaker ensure

Table 7.16 Summary of Circuits, Wiring, and Overcurrent Protection for the Cabin

Wire Location	Max A	*NEC* A[a]	Length (ft)	Max Ω/kft	Wire Size	C.B. Size (A)
Array to controller	8.02	12.5	30	3.17	#12	15
Controller to batteries	13.5	21.0	6	5.92	#10	25
Batteries to inverter	72.5	90.6	6	1.10	#2/0	175
Inverter to panel	25	31.3	5	9.60	#8	40
Refrigerator	16	20	<100	1.93	#12	20
Water pump	1.0	1.25	230	5.21	#14	15
Kitchen receptacles	16	20	<100	1.93	#12	20
LR receptacles	12	15	<100	3.07	#14	15
BR 1 receptacles	12	15	<100	3.07	#14	15
BR 2 receptacles	12	15	<100	3.07	#14	15
Bath and outside receptacles	12	15	<100	3.07	#14	15
Lights and fans	12	15	<100	3.07	#14	15

[a] Required for ampacity calculations. All wire sizes based on ampacity requirements.

that the inverter will be able to generate the surge current that may be needed to start motors or other loads that encounter large surges when starting. The inverter output current is simply the quotient of the inverter rated power and the inverter ac output voltage (120 V). The rated power of the inverter is used, because the wiring between inverter and distribution panel must be capable of carrying 125% of the rated inverter output current.

In Table 7.16, the NEC currents incorporate the appropriate multipliers, such as 156% for array and charge controller currents and 125% for the remaining currents. It turns out that all the calculations in this table lead to wire sizing based on necessary ampacity. Note that the cabin lights and outlets are wired with #14 wire as would be done in a grid-connected residence. The kitchen circuits are 20 A circuits that are wired with #12. The 2008 *NEC* requires arc-fault circuit breakers on all the 15 A circuits, except the well pump and ground-fault circuit interrupters on kitchen, bathroom, and outside receptacles.

Switches should have the same current rating as the fuses or circuit breakers along with adequate voltage rating for the purpose. For lighting, the switches sell for less than $1 at your favorite home improvement store. Fans come with fan controllers rated for the purpose. The pump will need to be controlled with a level switch in the storage tank and possibly with a time switch to limit pumping to certain hours. These switches need to be adequately rated to carry the pump motor current, which is a comparatively small amount.

7.8.6.2 Other Items

Other items that need to be included are array mount, distribution panels, lightning arresters, ground rod, battery container, battery cables, and miscellaneous connectors and junction boxes. Most of these components, except perhaps those related to the batteries, will be used in the electrical system regardless of the source of electricity.

Once wire length and size are known, wire cost can be calculated. For the cabin, a distribution panel with ac circuit breakers rather than fuses is used to supply individual circuits to the various loads, with some doubling up of loads on a single circuit as indicated in Table 7.16. Lightning arresters are important for protection of the total system and are normally located at the controller.

The array mount may take many shapes and forms, from commercially available units to homemade wooden frames. For the cabin, a commercial pole mount has been selected.

Figure 7.11 shows the cabin system electrical schematic diagram. Equipment grounding conductors are shown as dashed lines. On the dc side, #10 conductors are used because the array grounding conductors will be exposed to the weather and thus will be bare wire. Also, the 25 A circuit breaker at the charge controller output requires a #10 grounding conductor for the charge controller. The remaining equipment-grounding conductors will also be #10. If the batteries are in a metal container, the metal container will need to be grounded with #6 because of the 175 A circuit breaker. If the battery enclosure is nonconductive, then no ground is needed to the enclosure.

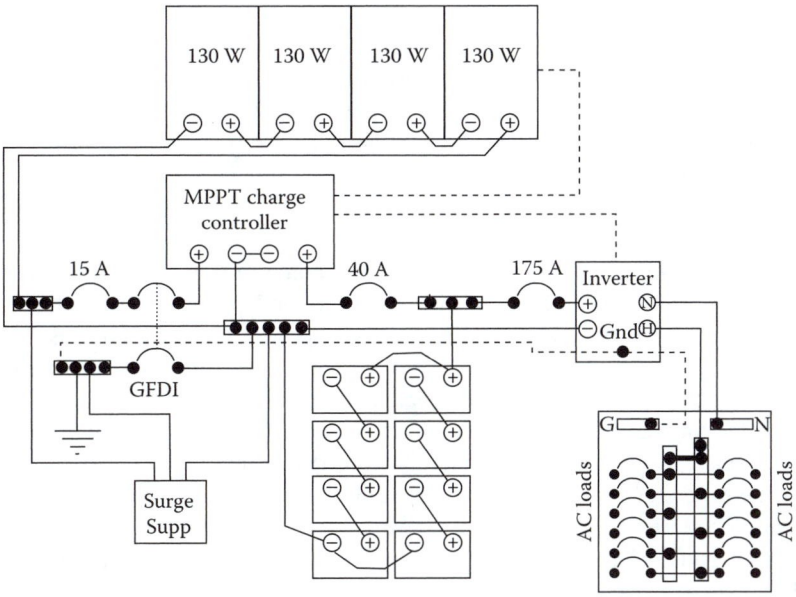

Figure 7.11 Cabin system based on MPPT charge controller and array tilted at latitude + 15°.

7.9 A HYBRID-POWERED, OFF-GRID RESIDENCE

7.9.1 Introduction

This design example investigates the method of choosing alternate generation capacity to supplement the output of the PV array when there is a large discrepancy between month-to-month system needs versus month-to-month PV generation capacity. If installation of a PV array to meet minimum sun availability results in significant excess generation for a number of months, then much of the PV output is wasted. In such cases, it often makes better economic sense to use a generator to supplement the PV output during the months of low PV output and size the PV to meet most of the needs during months of higher peak sun.

It was observed in Chapter 3 that there is a significant cost increase between sizing a PV system to provide 95% of system electrical needs versus providing 99% of system needs. Hence, use of a generator for increasing system availability from general to critical may also be cost-effective. In this chapter, the design will be completed with several array and generator options. In Chapter 8, a life-cycle cost analysis will be performed on the options.

In the present example, it is desired to power a residence near Bismarck, North Dakota, but away from the utility grid, with a combination of PV and generator. The residence will be occupied 7 days per week all year long. All loads will be either 120 V or 240 V ac as in typical residences, except that energy efficiency has been taken into account in the selection of the loads and the design of the dwelling. For example, a high-efficiency refrigerator has been selected, and all lighting will be fluorescent. The dwelling is designed with fans and is highly insulated with triple-pane glass in the windows. Much of the heating will be by passive solar, but propane will be used for supplemental heating, solar water heating backup, and cooking. Because propane will be on-site, it will also be used for the generator. The goal of the design is to arrive at a combination of PV and propane generator energy production that will result in the lowest life-cycle cost.

Because PV and generator output capability come in steps, this design process will explore several combinations of PV and generator. Up to this point, little discussion of generator properties has taken place, so a part of the design discussion will deal with the selection of generators. The design process is somewhat modified from the previous examples, because the propane generator also needs to be sized and incorporated into the system. This means a more complicated system block diagram, resulting in a somewhat more challenging control problem.

As the generator is available for PV system backup, it may seem unnecessary to incorporate batteries into the system. However, for generators to operate efficiently, they need to run at close to 90% of their output capacity. Operation of a generator at a small fraction of capacity will result in significant decrease in efficiency, as was noted in Chapter 3. Hence, batteries are used so that the generator can charge them at a rate close to its capacity. Furthermore, at the time of this writing, the authors are unaware of any stand-alone inverters that will operate without batteries. Sometimes, a generator can be used to "fool" a straight grid-connected inverter into believing that the generator is the utility, but if the inverter then tries to "sell" electricity to the grid, it ends up trying to sell it to the generator, thus turning the generator into a motor. When this is done, the "motor" tries to run the prime mover, rather than vice versa. Prime movers are not the least bit "happy" with such a role reversal. They will generally change speed, and, hence, frequency, resulting in shutting down the inverter per UL 1741 requirements.

Because charging batteries too quickly tends to result in an inefficient charging process, the generator/battery system should be sized so that the generator will take at least 5 h to charge the batteries [8]. The battery selection criteria previously discussed appear to present a more conservative requirement of at least 10 h to fully charge the batteries. Actually, these criteria are not necessarily inconsistent, because the batteries will not normally be charged from full discharge to full charge. Normally, the generator will charge the batteries from about 20% to about 70%. Charging the batteries from 20% to 70% in 5 h requires a charging rate of $C/10$.

The bottom line, then, for batteries, is to provide a few days of storage so the charging rate will not be excessive. More storage will normally result in somewhat lower use of the generator, because the generator will not necessarily need to back up the PV array in the event of cloudy weather for a few days. Longer storage times may be desirable in areas where summers have periods of sunny days followed by periods of cloudy days. In general, fewer batteries will be used in a hybrid system, because the propane generator will supplement the sun. Choice of the number of days of autonomy for the system, however, becomes more dependent on other factors, such as how long it may take to implement emergency repairs on the generator.

After loads and batteries are selected, the mix of PV and propane generation is determined. Then the array size and the generator size are calculated, followed by calculation of generator fuel use and maintenance costs. Then controllers, inverters, battery chargers, fuses, wires, and other BOS components are selected.

7.9.2 Summary of Loads

Table 7.17 summarizes the estimated loads for the residence by season. All loads are 120 V ac loads, and the input of the inverter will be 48 V dc. The loads are tabulated in Ah at 48 V dc. A conversion efficiency of 92% is assumed for the inverter and wiring from inverter to load, so the connected loads include this figure and thus represent the energy that must be supplied by the batteries to the inverter to operate the loads. The corrected load figures in Table 7.17 represent the connected loads

Table 7.17 Estimated Seasonal Loads for Hybrid Residence

		Dec, Jan, Feb		Mar, Apr, May Sep, Oct, Nov		Jun, Jul, Aug	
Load Description	**P, W**	**h/Day**	**Ah/Day[a]**	**h/Day**	**Ah/Day[a]**	**h/Day**	**Ah/Day[a]**
Kitchen lights	80	4	7.25	3.5	6.34	2	3.62
Dining room lights	20	2	0.91	1.5	0.68	1	0.45
Living room lights	40	1	0.91	1	0.91	1	0.91
Family room lights	40	4	3.62	4	3.62	3	2.72
BR 1 lights	20	1	0.45	1	0.45	1	0.45
BR 2 lights	20	1	0.45	1	0.45	1	0.45
BR 3 lights	20	1	0.45	1	0.45	1	0.45
Bath 1 lights	20	1	0.45	1	0.45	1	0.45
Bath 2 lights	20	1	0.45	1	0.45	1	0.45
Refrigerator	150	7	23.8	7	23.8	7.5	25.5
Microwave oven	600	0.5	6.79	0.5	6.79	0.5	6.79
TV	80	4	7.25	4	7.25	4	7.25
Stereo	100	2	4.53	2	4.53	2	4.53
Water Pump	200	1	4.53	1	4.53	1	4.53
BR 1 fan	50	0	0.00	0	0.00	8	9.06
BR 2 fan	50	0	0.00	0	0.00	8	9.06
BR 3 fan	50	0	0.00	0	0.00	8	9.06
Family room fan	50	0	0.00	0	0.00	8	9.06
Furnace fan	500	8	90.6	6	67.9	0	0.00
Washer	600	0.5	6.79	0.5	6.79	0.5	6.79
Vacuum and small appliances	2400	0.5	27.2	0.5	27.2	0.5	27.2
TOTAL	5110		186		163		129
CORRECTED TOTAL			207		181		143

[a] @ 48 V dc.

divided by the battery charge/discharge efficiency and are therefore the Ah needed to be supplied to the batteries by either the PV array, the generator, or both.

It is assumed that the household will have common electric small appliances such as toasters, blenders, vacuum cleaners, hair dryers, etc., and that no more than 2400 W of small appliances will be used at any one time. It is also assumed that the fan motor of the propane furnace and the ceiling fans will not run simultaneously. These assumptions may affect sizing of the generator and the wiring, depending on whether any of the load will need to be supplied by the generator while it is running to charge batteries. In this particular design example, the assumption is that the generator is used only for battery charging. Wiring, fusing, and switching between batteries and inverter and between inverter and distribution panel, however, are based on the rating of the inverter as required by the *National Electrical Code.*

7.9.3 Battery Selection

The batteries will be reasonably well protected from the cold North Dakota winter, but will experience somewhat lower winter temperatures. The same derating schedule will be used for the North Dakota installation as was used for the Colorado cabin battery installation.

The design goal for the system will be to maximize PV energy production without unreasonably high system life-cycle cost or unreasonable waste of PV generated electricity. Noncritical battery backup is considered satisfactory for this system, with a choice of 3 days of storage to allow the sun 3 days to reappear before calling in the generator as a replacement for the sun. This period also provides time for emergency repair of the generator or other system components. If a generator is warranted (and hopefully this will be the case because this design example calls for one to be used), it will be selected to charge the batteries in 10 h or more so there will not need to be any compensation for quick charging of the batteries. Also, with 3 days of storage, it is highly unlikely that quick discharge of the batteries will occur. Finally, in this case, Equation (3.11b), Chapter 3, was not used as the means of determining the number of days of storage, even though the problem stated noncritical storage. The presence of the generator adds another source of energy that was not used in determining Equation (3.11b). But, just for the sake of using Equation (3.11b), it is interesting to compare the result of using Equation (3.11b) with the decision to provide 3 days of battery backup. The result is $D_{non} = 0.1071(T_{min})^2 - 1.869T_{min} + 9.4286 = 4.54$ days. One might argue that the use of the generator as backup saves 1.54 days of battery backup.

On this basis, battery capacity is shown in Table 7.18 with temperature compensation taken into account, with an assumed depth of discharge of 80%, based on the corrected loads of Table 7.17. Table 7.18 also shows the number of batteries required if a 12 V sealed lead-acid battery with a capacity of 295 Ah @ $C/72$ is selected for the system. Note that the indicated number of batteries is four times the number of necessary batteries in parallel to produce the required total Ah capacity, because the system voltage will be 48 V dc. Note also that the selection of this battery will provide more than 3 days of storage for much of the year. However, at 80% capacity

Table 7.18 Battery Storage Requirements for Hybrid Residence

Month	Dec, Jan, Feb	Mar, Nov	Apr, Oct	May, Sep	Jun, Jul, Aug
Corr Ah/day	207	181	181	181	143
Temp derate	0.8	0.85	0.9	0.95	1
Total 3-day cap req	970	799	754	714	536
# Batt	12	12	12	12	12
Avail cap	885	885	885	885	885
# Days storage	2.74	3.32	3.52	3.72	4.95

rating during the winter, they will provide 91% of the capacity required for 3 days of storage. Thus, during winter months, the batteries will provide 2.7 days of storage, while in the summer they will provide nearly 5 days of storage.

7.9.4 Array Design

Sizing of the array for a hybrid system is generally an iterative process. The first step is to size the array for a system with no generator and then to gradually reduce the number of modules in the array while simultaneously computing the percentage of the annual energy needs provided by the PV array. Once the collection of designs is complete, the economic advantages of each system can be calculated. The designs will be completed in this chapter, and an economic analysis will be made in Chapter 8.

A somewhat different approach to array sizing will be taken in this example. Because December has the lowest psh for the year, regardless of array tilt, and because December also has the highest load, PVWatts [12] will be used to determine an array size and tilt that will meet the December load requirements. The output of this array will then be compared with system load requirements for the remaining months of the year. Assuming MPPT charge control, the array daily energy output can be estimated from the battery input energy requirement. Because the December corrected daily Ah requirement is 207, if the battery-charging voltage is 54 V, then the charging energy is $54 \times 207 = 11{,}178$ Wh, or 11.2 kWh. Allowing for approximately 15% combined loss due to wiring loss, charge controller loss, and array loss means that a conversion factor of 0.85 can be used in the PVWatts calculator to determine the array power that will produce 11.2 kWh/day for *battery charging*. This is still dc power, even though it shows as ac power in the PVWatts table. Table 7.19 shows the annual results for a tilt of 70° (latitude + 23°) and array size (3.6 kW) that will meet December requirements. The table also shows the excess energy production of the system for each month of the year. Note that the excess energy production corresponds to *corrected* load energy requirements, as opposed to actual array energy production.

Now that the array size has been determined, a module needs to be selected for which the source circuit $V_{OC} < 150$ V when the temperature is −40°C (−40°F) and a sufficient number of source circuits can be incorporated to reach the required 3.6 kW array power.

Table 7.19 Determination of Array Size and Tilt for Hybrid Residence

Month	Corr Load (Ah)	Corr Load (kWh)	Array Rated (kW)	Daily Avg Array (kWh)	Excess Daily (kWh)
Jan	207	11.2	3.6	13.4	2.2
Feb	207	11.2	3.6	15	3.8
Mar	181	9.8	3.6	16.3	6.5
Apr	181	9.8	3.6	14.3	4.5
May	181	9.8	3.6	13.1	3.4
Jun	143	7.7	3.6	12.5	4.8
Jul	143	7.7	3.6	13	5.3
Aug	143	7.7	3.6	13.7	6.0
Sep	181	9.8	3.6	13.7	4.0
Oct	181	9.8	3.6	12.9	3.1
Nov	181	9.8	3.6	10.3	0.5
Dec	207	11.2	3.6	11.4	0.2

One might imagine that it may be a good idea to look at available 200 W modules, because this would require 18 modules. If the modules can be connected in 3-module source circuits, then the system will require 6 source circuits. If the modules can be connected in 6-module source circuits, then the system will require 3 source circuits. Alternatively, one could look for 180 W modules, such that it would take 20 modules to achieve 3.6 kW. Then the source circuits might consist of either 4 or 5 modules, depending on the source circuit V_{OC} limits. But, then, there is also the case of the 150 W modules, for which a total of 24 modules will be needed. These modules might be connected in either 3-module, 4-module, or 6-module source circuits. It is unlikely that 8- or 12-module source circuits will have $V_{OC} < 150$ V. And, of course, there might be the 120 W modules, etc. So with so many options, where does one start?

At this point, it is useful to remember that a propane generator will also be incorporated into the system. Thus, it would be helpful to have a larger number of source circuits rather than a smaller number, if it is desired to explore the effect on life-cycle system cost as PV generation is replaced with propane generation. This way, as each source circuit is removed, an additional increment of propane generation will need to be added.

Checking the possible options again suggests that the 150 W module option may meet this requirement better than the other module sizes, especially if either 3- or 4-module source circuits will be possible.

One module rated at 150 W has $V_{OC} = 40.26$ V, $I_{SC} = 5.14$ A, $V_{mp} = 31.98$ V, and $I_{mp} = 4.80$ A, with NOCT = 48°C and $\Delta V_{OC}/\Delta T = -155$ mV/K. Thus, at −40°C, $V_{OC} = 40.26 + (65 \times 0.155) = 50.34$ V, which is too high if 3 are connected in series. Thus, source circuits must be limited to 2 modules in series, meaning 12 source circuits, which will increase wiring costs. So, back to the drawing board.

This time, a 200 W module with $V_{OC} = 32.9$ V, $I_{SC} = 8.21$ A, $V_{mp} = 26.3$ V, and $I_{mp} = 7.61$ A, with $\Delta V_{OC}/\Delta T = -123$ mV/K will be tried. For a single module at −40°C, $V_{OC} =$

Table 7.20 Monthly Excess kWh Capability of PV Array for Six Array Sizes

		Monthly Excess Array Energy Production (Negative ⇒ Deficiency)					
Month	**Corr Load kWh/mo**	**18 Mod 3600 W**	**15 Mod 3000 W**	**12 Mod 2400 W**	**9 Mod 1800 W**	**6 Mod 1200 W**	**3 Mod 600 W**
Jan	347	69	0	−69	−139	−208	−277
Feb	313	107	37	−33	−103	−173	−243
Mar	303	203	119	34	−50	−134	−219
Apr	293	135	63	−8	−79	−151	−222
May	303	104	36	−32	−99	−167	−235
Jun	232	144	82	19	−44	−106	−169
Jul	239	164	96	29	−38	−105	−172
Aug	239	187	116	45	−26	−97	−168
Sep	293	119	50	−19	−87	−156	−225
Oct	303	97	30	−36	−103	−170	−236
Nov	293	16	−36	−87	−139	−190	−242
Dec	347	5	−53	−112	−171	−229	−288
Ann PV %	3505	100	97.5	88.9	69.3	46.2	23.1

$32.9 + (65 \times 0.123) = 40.9$ V. Thus, source circuits can consist of 3 modules in series, or 600 W @ STC. So 6 source circuits will be needed to reach the 3.6 kW level.

The next step is to tabulate the available energy for battery charging from 1–6 source circuits for each month of the year, as shown in Table 7.20. In this table, the excess PV is calculated on a monthly basis, but the annual percentage of the total load requirements met by the PV array counts the months with positive excess as just meeting the needs for the month, because the rest of the energy produced is either wasted or used for unanticipated purposes. But the months with negative excess are months where the generator will need to make up the difference, so the amount the generator will need to supply is subtracted from the total annual requirement, divided by the total annual requirement, and then converted to a percentage. This gives the percentage of annual energy needs supplied by the PV system. The remaining percentage must be supplied by the generator. Table 7.20 indicates that the 15-module array, which supplies 5/6 as much energy as the 18-module array, needs a boost from the generator during November and December, but supplies all the needed energy during the rest of the year. Each time a source circuit is removed, another 1/6 of the 18-module array output is removed. By the time the array is reduced to 3 source circuits (1800 W), it needs supplementary energy every month of the year, but the PV system still provides nearly 70% of the annual energy needs.

The next step is to tabulate the amount of energy that must be supplied by the generator on a monthly basis under the six different PV array scenarios. This listing is shown in Table 7.21, and is simply the PV deficits from Table 7.20.

At this point, it is possible to determine a size for the generator and then to determine the generator monthly operating hours.

Table 7.21 Required Monthly kWh to Batteries from Generator for Six PV Array Sizes

		Array Size					
Month	**kWh Needed**	**18 Mod**	**15 Mod**	**12 Mod**	**9 Mod**	**6 Mod**	**3 Mod**
Jan	347	0	0	69	139	208	277
Feb	313	0	0	33	103	173	243
Mar	303	0	0	0	50	134	219
Apr	293	0	0	8	79	151	222
May	303	0	0	32	99	167	235
Jun	232	0	0	0	44	106	169
Jul	239	0	0	0	38	105	172
Aug	239	0	0	0	26	97	168
Sep	293	0	0	19	87	156	225
Oct	303	0	0	36	103	170	236
Nov	293	0	36	87	139	190	242
Dec	347	0	53	112	171	229	288
TOTAL	3505	0	89	396	1078	1886	2696
Ann gen %		0	2.5	11.1	30.7	53.8	76.9

7.9.5 Generator Selection

The generator should be sized to charge the batteries at approximately *C*/10 as indicated previously. At this rate of charging, the generator should be operating at approximately 80% to 90% of its rated output in order to ensure maximum generator efficiency. Hence, it is first necessary to determine what *C*/10 means for this system.

The first question when analyzing Table 7.18 is, which capacity should be used in determining *C*/10? With temperature derating applied, the batteries have a capacity of 708 Ah in the winter and 885 Ah in the summer. Because most generator operation will be in the winter, as indicated by Table 7.21, the winter capacity is the appropriate choice. Hence, 70.8 A for 10 h will produce 708 Ah of charge.

The generator will be rated in watts. Thus, it is necessary to determine the charging power, rather than the Ah requirements of the batteries. This is obtained by incorporating the battery-charging voltage. For a *C*/10 charging rate, the battery charger output power will thus be the product of the charging voltage and the charging current, or $48 \times 1.2 \times 70.8 = 4078$ W, because the charging voltage is generally about 120% of the nominal battery voltage. If the electrical conversion efficiency of the battery charger is 90%, this would require an input power of $4078 \div 0.9 = 4531$ W to provide 70.8 A of charging current. Because the generator is to be used only for battery charging, a 5000 W generator would be a good choice, as it will run at 91% of its rated output when charging at a *C*/10 rate.

If it is desired to have the generator run part of the load while charging the batteries, or to charge at a faster rate, it may be reasonable to select the next larger size generator. However, the 5000 W unit operating at 4531 W will charge the batteries from 20% to 70% in 5 h. This is normally an acceptable charge rate, especially because it will take 1.7 days to use this amount of charge during winter months. Charging the batteries to only 70% allows for the PV array to top off the charge if it can. If not, and the batteries discharge to 20% again, the generator comes on again.

7.9.6 Generator Operating Hours and Operating Cost

The more the generator runs, the more it will cost to operate. Generator operating cost consists of fuel cost, oil changes, tune-ups, and rebuilding. To determine the operating cost, then, it is necessary to determine the annual operating hours of the generator.

Table 7.21 lists the monthly kWh requirements of the batteries from the generator. The generator kWh output (kWh_{Gen}) needed to supply the battery kWh (kWh_{Batt}) needs is found from

$$kWh_{Gen} = \frac{kWh_{Batt} \times 1.2}{\eta}, \tag{7.8}$$

where 1.2 is the ratio of charging voltage to nominal battery voltage and η is the efficiency of conversion of ac to dc in the battery charger.

The monthly hours of generator operation can then be determined from the monthly kWh_{Gen} by the output power at which the generator is set to operate. Because most generators operate at maximum efficiency at approximately 90% of their rated output power, it will be assumed that the generator will be set at the winter $C/10$ rate of 4531 W. It is also reasonable to assume a battery charger conversion efficiency of 90%, as was done during the process of calculating the needed generator power.

The monthly generator fuel consumption can be determined from knowledge of either the hourly fuel consumption or the fuel consumed per kWh of generation. A utility-grade generator will generate approximately 12 kWh per gallon of fuel [13], but a small generator will generate considerably less, because the combustion efficiency of a small gasoline, propane, or diesel engine is much less than the efficiency of a large utility steam turbine. Table 7.22 estimates fuel consumption on the basis of a generator output of 5 kWh/gal of fuel. With a generator output of 4531 W, it will take 1.10 h to generate 5 kWh_{Gen}, resulting in a fuel consumption rate of 0.906 gal/h. Assuming an ac/dc conversion efficiency for the battery charger of 0.9, use of Equation (7.8) results in 5 kWh_{Gen} providing 3.75 kWh_{Batt}. This means that in 1 h, the generator will produce 4.53 kWh_{Gen}, which results in $3.75 \times 4.53 \div 5 = 3.40$ KWh_{Batt}, which means that it will take 0.294 h to deliver 1 kWh to the batteries. The generator running time is thus equal to 0.294 times the kWh needed for the batteries from the generator. The generator fuel use is thus 0.906 times the running time, and the generator fuel cost follows by multiplying by the price/gal of fuel.

Generator annual operating costs can now be estimated from the maintenance information of Table 3.4, Chapter 3, along with known fuel cost. Table 7.23 summarizes the required generator maintenance for the six PV arrays previously considered.

Table 7.22 Monthly Generator Operating Hours and Annual Fuel Consumption for Six PV Array Sizes

Month	Battery kWh for Loads	Array					
		18 Mod	15 Mod	12 Mod	9 Mod	6 Mod	3 Mod
Jan	347	0.0	0.0	20.3	40.9	61.2	81.4
Feb	313	0.0	0.0	9.7	30.3	50.9	71.4
Mar	303	0.0	0.0	0.0	14.7	39.4	64.4
Apr	293	0.0	0.0	2.4	23.2	44.4	65.3
May	303	0.0	0.0	9.4	29.1	49.1	69.1
Jun	232	0.0	0.0	0.0	12.9	31.2	49.7
Jul	239	0.0	0.0	0.0	11.2	30.9	50.6
Aug	239	0.0	0.0	0.0	7.6	28.5	49.4
Sep	293	0.0	0.0	5.6	25.6	45.9	66.2
Oct	303	0.0	0.0	10.6	30.3	50.0	69.4
Nov	293	0.0	10.6	25.6	40.9	55.9	71.1
Dec	347	0.0	15.6	32.9	50.3	67.3	84.7
Ann gen op h/yr		0.0	26	116	317	554	793
Ann gen fuel in gal		0	24	105	287	502	718
Ann gen fuel cost[a]		0	$53	$237	$646	$1130	$1616

[a] @ $2.30/gal.

The table entries are determined by the quotient of the annual operating hours and the maintenance intervals of 25 h per oil change, 300 h per tune-up, and 3000 h per rebuild. Once the cost of each of these items is known, the annual maintenance cost can be determined. As will be seen in Chapter 8, annual maintenance costs can be an important component of the system life-cycle cost, depending on the frequency of maintenance operations.

7.9.7 Charge Controller and Inverter Selection

The control for a hybrid system is somewhat more complicated than the control for a conventional PV system. It must control battery charge and discharge by both the PV array and the generator. It must provide a starting signal/voltage for the generator when the batteries have discharged to a preset level and must shut down the

Table 7.23 Annual Generator Maintenance Frequency for Six PV Arrays

Item	18 Mod	15 Mod	12 Mod	9 Mod	6 Mod	3 Mod
Op h/yr	0	26	116	317	554	793
Oil changes/yr	0	1	4	13	22	32
Tune-ups/yr	0	0	0	1	2	2
Yr/rebuild	>20	>20	>20	9.5	5.4	3.8

generator when the batteries reach a preset level of charge. Proper setting of these levels is important. Too low a setting on discharge may render the batteries unable to provide starting current for the generator. Too high a setting may result in the generator's unnecessarily replacing energy that might be available from the PV array, with the PV array then being shut off with energy to spare. Fortunately, all of these control functions can be accomplished with an MPPT charge controller and an off-grid version of the inverter used previously in a grid-connected battery-backup configuration. The MPPT charge controller provides efficient battery charging and maximizes PV array output, while the inverter provides the interface between generator and batteries by using either the AC IN port or the GEN (AC2) IN port (if available) as the generator connection port, with all the programmable charging features of the grid-connected version except for the "sell" option and its corresponding tight controls on voltage and frequency at the AC IN port. The inverter is capable of monitoring battery voltage and providing a generator start command when battery voltage drops too low and a generator stop command when the battery voltage is adequate.

Obviously, in addition to the battery-charging function, regardless of the array size, the inverter must supply all the loads of the house, which have been previously tabulated at 5110 W in Table 7.17. So, an inverter rated at a minimum of 5110 W is needed. Fortunately, an inverter that supplies 6000 W @ 120/240 V is now available that has a separate generator input and operates on a 48 V battery system. Depending on the array size, either one or two charge controllers will be needed, if it is desired to have the inverter and charge controllers communicate with each other. Perhaps by the time this text reaches publication, a larger charge controller will be available that will handle an array power of 3600 W. The controller presently available works best with array power ratings below 3200 W.

So, if it is decided to use the 3000 W array rather than the 3600 W array, the cost of one charge controller as well as the cost of 3 modules will be avoided. The trade-off, however, is either using a kWh or two less energy during November and December, or including a generator in the system. For smaller arrays, the generator will be needed in order to prevent a significant reduction in daily kWh use.

Although the peak efficiency of the inverter is very close to 95% for $1000 < P_{out} < 2000$ W, the efficiency drops off at low power levels and at high power levels as shown in Figure 7.12. When $P_{out} < 300$ W, the efficiency drops below 90%.

Because the average power consumption of the loads is 414 W or less, depending on the month of the year, this means there may be times when the inverter is operating at efficiencies less than 90%. This means higher percentage losses in the inverter when the load on the inverter is small, which means the overall system load, including inverter losses, will be somewhat higher than calculated. For example, if the inverter delivers 100 W at 70% efficiency, this means the inverter input power must be 143 W, rather than the 105 W that would result if the inverter efficiency were 95% as assumed in the connected load calculation. Over a 10 h period, this additional 38 W loss amounts to 0.38 kWh, which must be added to the daily load on the batteries. With a daily kWh consumption of approximately 10 kWh, this amounts to an additional 3.8% load on the system. The bottom line is that, depending on the number of modules chosen for the system, there will be less excess kWh/mo for

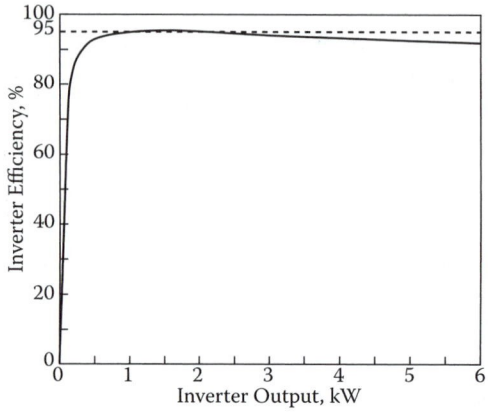

Figure 7.12 Inverter efficiency versus output power.

months where the PV delivers excess kWh, and the generator will run slightly longer during the months when the PV does not provide excess kWh. To provide this additional 0.38 kWh/day, the generator will need to run an additional 6.7 min per day. Fortunately, the inverter incorporates an adjustable search mode control, so that the inverter will "sleep" if the connected loads are below the sleep threshold.

Because the inverter runs on a real-time clock, it is possible to program the hours when generator operation will be permitted. It is also possible to program battery-charging current so that the generator will run at the design output of 4531 W.

7.9.8 Wire, Circuit Breaker, and Disconnect Selection

All wiring on the load side of the inverter can be done in a manner consistent with conventional 120/240 V residential wiring. The *NEC* requires a distribution panel no smaller than 100 A to be installed, and the panel will need to have sufficient circuit breaker pole spaces to accommodate all the loads plus a few spare spaces for possible future loads.

Table 7.24 shows the wire sizes required for array to source circuit combiner, combiner to charge controller, charge controller to inverter, battery to inverter, generator to inverter, and inverter to distribution panel, assuming distances of 40, 6, 4, 10, 30, and 10 ft, respectively. As usual, the source circuit to combiner ampacity and the combiner to charge controller ampacity must be at least 156% of the rated module short-circuit current. The current used for voltage drop calculations is I_{mp}, because this is the expected operating current of a source circuit under full sun. One might argue that the array will never be exposed to full sun, because, in winter, the altitude of the sun is low enough that the air mass increase causes a decrease in intensity, and in spring, summer, and fall, the sun is never perpendicular to the array.

Table 7.24 Summary of Wiring and Fusing for the Hybrid Residence Based on Use of 3600 W PV Array

Wire Location	VD A[a]	NEC A[b]	1-Way Length (ft)	System Volts[a] (V) A[b]	Wire Max (Ω/kft)	NEC Wire (A)	Wire size	C.B. Size (A)
Source circuits to combiner	7.61	12.8	40	78.9	2.59	33.7	#10	15
Combiner to controller[c]	22.8	38.4	6	78.9	5.76	38.4	#8	40
Controller to inverter[c]	32.3	50.3	6	48	2.48	50.3	#8	60
Batteries to inverter	130	163	10	48	0.369	163	#4/0	250
Generator to inverter	20.8	23.9	30	240	3.85	23.9	#10	25
Inverter to panel	37.5	46.9	10	120	3.20	46.9	#6	60

[a] Used for voltage drop calculations.
[b] Used for ampacity calculations.
[c] Based on use of two charge controllers, so half of array to each controller.

Thus, using I_{mp} gives a conservative estimate for the voltage drop calculation. The voltage used for voltage drop calculations is V_{mp} of a source circuit.

The minimum source circuit wire ampacity is determined by

$$I(30\text{C}) = \frac{1.56 I_{SC}}{D_T D_C}, \tag{7.9}$$

where I(30C) is the ampacity at a temperature of 30°C of the wire to be used, that is, the ampacity as tabulated in Table 4.2 in Chapter 4; I_{SC} is the module I_{SC}, D_T is the NEC temperature derating factor, and D_C is the NEC conduit fill derating factor. Using a maximum ambient temperature of 55°C on the roof, along with a conduit fill of 12 conductors, results, for wire with 90°C insulation (THWN-2), in values of D_T = 0.76 and D_C = 0.5. The necessary ampacity of the source circuit wiring in conduit must therefore be at least 33.7 A. Because the 30°C ampacity of #10 THWN-2 is 40 A, this is the minimum allowable wire size. Thus, even though the voltage drop calculation would allow #12 to be used, the ampacity calculation requires the use of #10 for the individual source circuits.

The charge controller output currents are determined by the portion of the rated array power that is connected to each controller. It can be assumed that 97% of this power will appear at the charge controller output at a battery-charging voltage of approximately 54 V.

The battery-to-inverter current is the rated inverter output power divided by the system dc voltage at its lowest expected value (44 V), which is normally considered to be the lowest rated inverter input voltage, and then divided by the inverter efficiency.

Alternatively, it is acceptable to use the maximum input current as listed on manufacturer's specification sheet when this number is available.

It might seem that the current from inverter to distribution panel should be determined from the rated inverter output power, divided by 240 V. This would be the case if the inverter output were balanced, but the inverter allows a maximum imbalance of 75% between Line 1 and Line 2 currents, with a maximum line-to-neutral output current of 37.5 A. All ac wiring must be sized to carry 125% of these rated currents, except for the wiring from generator to inverter/charger, where the wire must be sized to carry 115% of the rated generator output current, per *NEC*. The rated generator output current will be $5000 \div 240 = 20.8$ A, and 115% of 20.8 is 23.95 A. Thus, wiring from the generator to inverter will need to be #10, protected by a 25 A circuit breaker, unless the generator manufacturer will allow a 30 A circuit breaker to be used.

In a few previous examples, the wire size was determined on the basis of voltage drop for longer runs of wire. Table 7.24 shows that in this system, all distances are sufficiently short that wire sizes are determined on the basis of ampacity as opposed to voltage drop.

7.9.9 BOS Component Selection

The BOS components will include a battery storage container, an array mount, surge protection, and provisions for proper grounding of the system. Of course, the wiring from the distribution panel is also a part of the balance of the system, but its cost will not change as the mix of PV versus propane generation is varied. The cost of the array mount and the cost of wire from array to combiner box are the only items in the BOS that will vary with the number of modules.

7.9.10 Total System Design

Figure 7.13 shows the schematic diagram of the hybrid dwelling 3600 W PV system. It is assumed that the batteries will be placed in a reasonably well-insulated location so they will remain reasonably warm in the winter when they are needed the most. The array is located as close as practical to the batteries, but free of any objects that may shade the array. Although calculations show that the generator will not be needed for this system, it is included to show how it will be connected into the system. For the smaller-sized systems, all major components remain the same with the exception of fewer modules and only one charge controller for 3000 W and less. The 6 kW inverter is still needed for supplying adequate power to the loads, even though the array size may drop to 600 W.

The other components that will be included in the system will be an automatic generator start module and a system controller. The generator start module is controlled by the inverter, but it also monitors generator output to verify that the generator is operating properly. The inverter is programmable via the system controller so that generator start and run parameters can be varied as needed. This is a case where reading the manual is essential to proper operation of the generator.

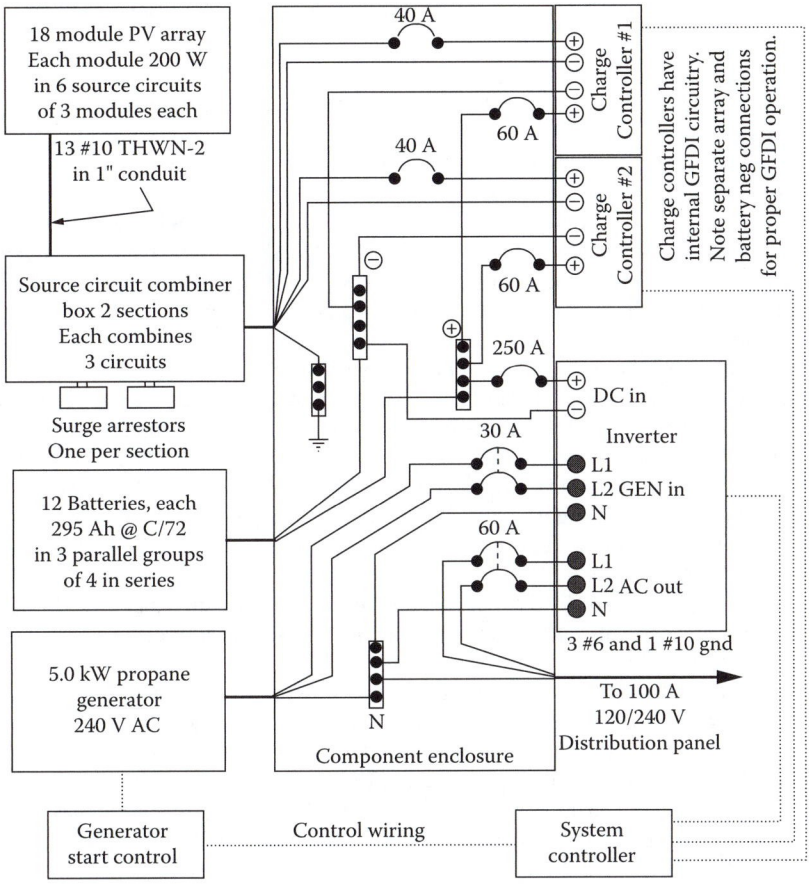

Figure 7.13 Hybrid residence electrical system.

7.10 SUMMARY OF DESIGN PROCEDURES

At this point, a relatively wide range of stand-alone PV systems have been designed in this chapter. An effort has been made to present alternate, commonsense design procedures to illustrate that no single methodology is necessarily the best one for any given problem. However, there are many commonalities among the designs, including the procedure for determining loads, the procedure for sizing batteries, and the procedure for sizing arrays. These procedures are summarized in the following listings. The designer should remember, however, that these procedures are intended to be used as guidelines. Specific design requirements may result in the need to modify these procedures.

DETERMINATION OF AVERAGE DAILY PV SYSTEM LOAD

1. Identify all loads to be connected to the PV system.
2. For each load, determine its voltage, current, power, and daily operating hours. For some loads, the operation may vary on a daily, monthly, or seasonal basis. If so, this must be accounted for in calculating daily averages.
3. Separate ac loads from dc loads.
4. Determine average daily Ah for each load from current and operating hours data. If operating hours differ from day to day during the week, the daily average over the week should be calculated. If average daily operating hours vary from month to month, then the load calculation may need to be determined for each month.
5. Add up the Ah for the dc loads, being sure all are at the same voltage.
6. If some dc loads are at a different voltage, which will require a dc-to-dc converter, then the converter input Ah for these loads needs to account for the conversion efficiency of the converter.
7. For ac loads, the dc input current to the inverter must be determined, and the dc Ah is then determined from the dc input current. The dc input current is determined by equating the ac load power to the dc input power and then dividing by the efficiency of the inverter.
8. Add the Ah for the dc loads to the Ah for the ac loads, then divide by the wire efficiency factor and the battery efficiency factor to obtain the corrected average daily Ah for the total load.
9. The total ac load power will determine the required size of the inverter. Individual load powers will be needed to determine wire sizing to the loads. Total load current will be compared with total array current when sizing wire from battery to controller. The power lost in the wiring is determined by the product of the current and the voltage drop. Depending on the nature of the load, a drop in voltage may result in either a drop or a rise in current. Normally the current will drop slightly, but, in some motors with constant loads, the current will rise. If the load is an MPPT, the current will also rise if the voltage drops. However, for a voltage drop of only 2%, it is reasonable to assume the load current will remain essentially constant. With this assumption, the power loss can be approximated by the product of load current at nominal load voltage and the voltage drop. Hence, a 2% voltage drop implies a 2% power loss, or, in terms of Ah, a 2% Ah loss. This loss must be added to the PV array power output requirements. Thus, a 2% voltage drop implies a wiring efficiency of 98%, and the load Ah per day must be increased by dividing by 0.98 to account for the wiring loss.

BATTERY SELECTION PROCEDURE

1. Determine the number of days storage required, depending on whether the load will be noncritical or critical. The designer may use discretion for special cases such as weekend occupancy or seasonal variations. It is possible that different seasons may have different numbers of storage days.
2. Determine the amount of storage required in Ah. This is the product of the corrected Ah per day and the number of days of storage required. This amount may vary with season. If so, list all values.

3. Determine the allowable level of discharge. Divide the required Ah by the level of allowed discharge, expressed as a fraction. For example, using 80% of total charge requires dividing by 0.8. This result is the total corrected Ah required for storage.
4. Check to see whether an additional correction for discharge rate will be needed. If so, apply this correction to the result obtained in (3).
5. Check to see whether a temperature correction factor is required. If so, apply this to the result of (3) or (4).
6. Check to see whether the rate of charge exceeds the rate specified by the battery manufacturer. If so, multiply the charging current by the rated number of hours for charging. If this number is larger than the result of Equation (7.2), use this as the required battery capacity.
7. Divide the final corrected battery capacity by the capacity of the chosen battery. The result may be rounded up or down, depending on the judgment of the system designer. Often the result can be rounded down, provided that the system receives at least some diffuse sunlight over prolonged overcast periods.
8. If more than four batteries are required in parallel, it is generally better to consider higher-capacity batteries to reduce the number of parallel batteries to provide for better balance of battery currents.

ARRAY SIZING AND TILT PROCEDURE

1. Determine the design current for each month of the year by dividing the corrected Ah load for the system each month by the monthly average peak sun hours at each array tilt angle.
2. Determine the worst-case (highest monthly) design current for each tilt angle.
3. For a fixed mount, select the tilt angle that results in the lowest worst-case design current.
4. If tracking mounts are considered, then determine the design current for one- and two-axis trackers. Note that the one-axis tracker tracks from east to west and needs to have its design currents checked for the three angles of fixed trackers.
5. Determine the derated array current by dividing the design current by the module derating factor.
6. Select a module that meets the illumination and temperature requirements of the system as well as having a rated output current and voltage at maximum power consistent with system needs.
7. Determine the number of modules in parallel by dividing the derated array current by the rated module current. Round up or down as deemed appropriate.
8. Determine the number of modules in series by dividing the nominal system voltage by the lowest anticipated module voltage of a module supplying power to the system. In this case, it is almost always necessary to round up.
9. The total number of modules is the product of the number in parallel and the number in series.

Problems

7.1 For the fan example, indicate applications for which it would be desirable to have an oversized PV module and indicate when it would be satisfactory to have a smaller module that would produce a significantly lower fan speed at lower light levels. Can you envision an application for which use of an MPPT/LCB would be advantageous?

7.2 How much power could be recovered at low light levels by using an MPPT on the fan of the first example in the text, assuming the MPPT to be 95% efficient? Estimate the additional maximum power output that would be required of a PV array that would produce the same low-light-level power as the system with the MPPT. If the additional PV cost is $4/W, how much could you spend on the MPPT to produce the same effect?

7.3 In a water-pumping system, it is desired to pump 2500 gal/day from a well that is 150 ft deep. The pump will have a wire-to-water efficiency of 30% and will be installed in a battery-backup system where the pump is set to run 12 h/day. The system will also have an MPPT charge controller. If the PV array receives 6 psh/day, determine

a. The necessary pump horsepower.
b. The size of the battery bank if one day of storage is desired.
c. The size of the PV array needed to power the pump if the efficiency of the charge controller is 96%.

7.4 For the pumping example in Section 7.3.3, show that the values obtained for n and for V_2 are correct under the assumptions made in the example.

7.5 Show that the module temperature of Section 7.3.3 can be expected to reach 69.5°C.

7.6 By plotting the composite *I-V* characteristic for five of module A in series (see Figure 7.4), and the approximate *I-V* characteristic of the pump as obtained from the data in Table 7.1, on the same set of coordinates, determine the overall operating point of the system when the module temperature is 50°C at full sun.

7.7 Search the Web for a suitable set of batteries that will provide 275 Ah @ 24 V when the discharge rate is $C/10$. Sketch the wiring diagram.

7.8 Find a suitable module and charge controller to charge the batteries used in the parking lot lighting example in Section 7.4.

7.9 Using the module and charge controller of Problem 7.8, determine the wire size between array and charge controller if the distance between the two is 25 ft and the voltage drop is to be limited to 1%.

7.10 Recalculate the battery size and array size for the parking lot fixture example if an inverter is used to power the lamp and the inverter has an efficiency of 94%. Check the Web to see if a suitable inverter is available, and list its specifications.

7.11 Determine the lamp wattage required to obtain an illumination level of 50 f-c over a 100 ft^2 area if a fixture is used with a CU of 0.75 and 80% of the available light reaches the work surface, the rest being absorbed by walls and other items in the space. Assume a luminous efficacy of 70 lumens/watt.

7.12 For a soil with a resistivity of 12,000 Ω-cm, and a cathode that requires 200 mA of current for protection, configure an anode system that will allow the use of a 12 V system. Then select a battery to provide 5 days of storage and one or more modules to supply the system energy needs if the minimum irradiation is 3 h/day.

7.13 For a soil with resistivity of 20,000 Ω-cm, and a cathode that requires 500 mA of current for protection, configure an anode system that will allow the use of a 12 V system. Then select a battery to provide 5 days of storage and one or more modules to supply the system energy needs for minimum peak sun of 2.5 h/day.

7.14 To test the soil resistivity after a large steel tank has been buried, a 6-ft-long, 4-in.-diameter anode is buried about 40 ft from the tank. When a 12 V battery is connected between the anode and the tank, a current of 0.5 A flows. What is the resistivity of the soil?

7.15 The highway information sign is to be designed for use in Miami, Florida, using the same PV modules. The sign still requires 5 days of autonomy under worst-case sun conditions, so the battery requirements will need to be recalculated. Then tabulate the average daily power available to the sign for each of the 12 months of the year.

7.16 Redo the highway information sign example for use in Phoenix, Arizona, again calculating the battery requirements and then showing the monthly average daily power availability.

7.17 For the highway sign of Section 7.6, determine the Ah rating of batteries that will provide 5 days of backup power for the month with the highest average daily sun hours.

7.18 How many days of storage will the batteries of the highway sign example of Section 7.6 provide at summer average daily power levels?

7.19 Offer an explanation for why a collector tilt of latitude + 15° gives better summer performance in Angola. Check the peak sun hours compilations in Appendix A for other locations to see whether this phenomenon is characteristic of any other locations. Comment on the meaning of "summer" and "winter" in the tables in Appendix A.

7.20 Assume 10 days of autonomy are desired for a battery system, but the battery size chosen only allows for 9.2 days of storage, with a maximum depth of discharge of 80%. Then assume 12 consecutive days occur during which peak sun averages only 10% of the predicted worst-case average. What will be the state of charge of the battery system at the end of the 12th day?

7.21 Show that the wire-to-water efficiency of the well pump used for the cabin is approximately 37%.

7.22 Determine the additional water that can be pumped with excess summer electricity produced by the cabin array.

7.23 Show why the hybrid residence will take 1.7 winter days to use up 5 h of charging at 4531 W, as claimed in the chapter text.

7.24 Show that the average power consumption in the hybrid residence is 414 W or less.

7.25
a. At what temperature will lead-acid batteries hold 80% of their rated capacity?
b. At what temperature will lead-acid batteries hold 90% of their rated capacity?
c. What percentage of their rated capacity will lead-acid batteries hold if their temperature is 0°C?

Design Projects

7.26 Write a program or develop a spreadsheet that will size a pump and wire to the pump for a PV-powered pumping system if the pumping height and daily volume are known along with the distance from the array to the pump.

7.27 Design your own little off-grid hideaway. Specify your own loads, occupancy, peak sun, and storage requirements and determine the number of batteries and the number of modules you would need to implement the system. Then specify BOS components.

7.28 Design your own slightly larger, off-grid hideaway for a location at a latitude higher than 50°, where winter peak sun hours are significantly less than summer peak sun hours. Use winter loads such that your system will end up as a hybrid system.

7.29 Complete the design of the hybrid residence electrical system of Section 7.8 using an array of 15 modules. Be sure to show the changes in wiring that will be made in the source circuit combiner box and in the component enclosure.

REFERENCES

1. *1997–98 Design Guide and Catalog*, Alternative Energy Engineering, Redway, CA, 1997.
2. *Lighting Handbook: Reference and Application,* 8th ed., Illuminating Engineering Society of North America, New York, 1993.
3. *IES Lighting Handbook*, Illuminating Engineering Society of North America, New York, 1984.
4. *9200 Lamp Catalog*, 22nd ed., GE Lighting, General Electric Co., 1995.
5. *Lamp Specification and Application Guide*, Philips Lighting Co., Somerset, NJ, 1999.
6. www.cetsolar.com for information on LED arrays.
7. U.S. Naval Observatory, Sun or Moon Rise/Set Table for One Year, http://aa.usno.navy.mil/data/docs/RS_OneYear.php.
8. *Stand-Alone Photovoltaic Systems: A Handbook of Recommended Design Practices*, Sandia National Laboratories, Albuquerque, NM, 1995.
9. U.S. Environmental Protection Agency and U.S. Department of Energy, Energy Star Products, Refrigerators and Freezers http://www.energystar.gov/index.cfm?c=refrig.pr_refrigerators.
10. ES-A Series photovoltaic panels, Evergreen Solar Inc., Marlboro, MA 01752, www.evergreensolar.com.
11. Solar Water Pumps, submersible and surface pumps, www.sun-pumps.com.
12. National Renewable Energy Laboratories, *The PV Watts Solar Calculator*, www.PVWatt.com.
13. Danley, D. R., Orion Energy Corporation, Ijamsville, MD, personal communication regarding fuel efficiency of fossil fueled generators, August 1999.

SUGGESTED READING

NFPA 70 National Electrical Code, 2008 Edition, National Fire Protection Association, Quincy, MA, 2008.

www.batteries4everything.com for battery information.

www.unirac.com for information on array mounts.

www.power-fab.com for additional information on array mounts.

www.zomeworks.com for information on tracking array mounts.

CHAPTER 8

Economic Considerations

8.1 INTRODUCTION

The costs of a PV system include acquisition costs, operating costs, maintenance costs, and replacement costs. At the end of the life of a system, it may have a salvage value or it may have a decommissioning cost. This chapter introduces the method of life-cycle costing, which accounts for all costs associated with a system over its lifetime, taking into account the time value of money. Life-cycle costing is used in the design of the PV system that will cost the least amount over its lifetime. Life-cycle costing, in general, constitutes a sensible means of evaluating any purchase options when a decision has been made to purchase. Sometimes a potential PV system owner wants to know how long it will take for the system to pay for itself, and sometimes a potential PV system owner simply wants to know what can be expected from the system, independent of cost. This chapter explores some of the economic considerations that enter the decision process when someone expresses an interest in owning a PV system.

If it is necessary to borrow money to purchase an item, the cost of the loan may also need to be incorporated into the total cost of a system. If incentive programs are available in any of a number of forms, they may significantly affect the life-cycle cost (LCC), or payback of a system. Often, the only question in the mind of the potential PV system owner is how long it will take to pay for itself. So some attention will also be paid to the payback question as well as various forms of incentive programs currently in place in certain areas.

Another economic option is third-party ownership, in which a for-profit company offers to install and own a PV system on a property and then sell electricity from the system to the property owner at a preferred rate. As the cost of PV systems continues to fall and as tax incentives continue to be adopted, this option is becoming more attractive, because it involves minimal risk for the property or system owner, as long as the sun rises and as long as the equipment functions properly with high reliability.

This chapter also introduces the concept of externalities. Externalities are costs that are not normally directly associated with an item. For example, it is generally agreed that acid rain can be caused by sulfur emissions from smokestacks. It is also generally agreed that acid rain can cause damage to buildings and lakes. Yet, the cost of these damages is generally not paid for directly by the entity that generates the emissions. Externalities will be considered in more detail in Chapter 9.

8.2 LIFE-CYCLE COSTING

8.2.1 The Time Value of Money

The life-cycle cost of an item consists of the total cost of owning and operating an item over its lifetime. Some costs involved in the owning and operating of an item are incurred at the time of acquisition, and other costs are incurred at later times. In order to compare two similar items, which may have different costs at different times, it is convenient to refer all costs to the time of acquisition. For example, one refrigerator may be initially less expensive than another, but it may require more electrical energy and more repairs over its lifetime. The additional costs of electrical energy and repairs may more than offset the lower acquisition cost.

Two phenomena affect the value of money over time. The inflation rate, i, is a measure of the decline in value of money. For example, if the inflation rate is 3% per year, then an item will cost 3% more next year. Because it takes more money to purchase the same thing, the value of the unit of currency, in effect, is decreased. Note that the inflation rate for any item need not necessarily follow the general inflation rate. Recently, health care costs have exceeded the general inflation rate in the United States, while the cost of most electronic goods has fallen far below the general inflation rate.

The discount rate, d, relates to the amount of interest that can be earned on principal that is saved. If money is invested in an account that has a positive interest rate, the principal will increase from year to year. The real challenge, then, in investing money, is to invest at a discount rate that is greater than the inflation rate.

As an example, assume an initial amount of money is invested at a rate of $100d\%$ per year, where d is the percentage rate expressed as a fraction. After n years, the value of the investment will be

$$N(n) = N_o(1+d)^n. \tag{8.1}$$

However, in terms of the purchasing power of this investment, $N(n)$ dollars may not purchase the same amount that this amount of money would have purchased at the time the investment was made. In order to account for inflation (or deflation), note that if the cost of an item at the time the investment was made is C_o, then the cost of the item after n years if the inflation rate is $100i\%$ per year, will be

$$C(n) = C_o(1+i)^n. \tag{8.2}$$

One might argue that if the cost of an item increases at a rate that exceeds the rate at which the value of saved money increases, that the item should be purchased right away. Similarly, if the cost of the item increases more slowly, or, perhaps, actually decreases over time, then one should wait before making the purchase, because the cost will be less at a later time. The disadvantage of this purchasing algorithm, of course, is that the item to be purchased will not be available for use until it is purchased. Hence, the new computer that becomes less and less expensive while the invested money continues to increase in value is not available for computing when it should be purchased. In other words, economics may not be the only consideration in making a purchase. Sometimes people buy things simply because they want them.

It is important to remember that choosing values for d and i is tantamount to predicting the future, because d and i fluctuate over time. Depending on the saving mechanism, the rate of return may be fixed or may be variable. Inflation is, at best, unpredictable. The worldwide recession of 2009 also surprised many observers and supports the observation that the future will be different from the past. The important point is to be able to distinguish between predictable events, such as death, taxes, sunrise, and loan repayment amounts, and unpredictable events, such as the value of the stock market, the inflation rate, and available interest rates on savings plans.

Figure 8.1 shows how the consumer price index, as a measure of inflation, the Dow Jones Industrial Average, as one possible measure of d, and the government prime lending rate, as a measure of the minimum borrowing interest rate, have varied over the period between 1980 and 2008. It is interesting to note how the prime lending rate is adjusted with the intent of either controlling inflation by discouraging borrowing or stimulating the economy by encouraging borrowing. The high prime lending rate in the early 1980s reflects the attempt to control high inflation in the late 1970s, resulting from significant increases in energy prices during this period. The low rates in 2002 and 2008 reflect the attempt to stimulate the economy and bring about a recovery of the stock market.

8.2.2 Present Worth Factors and Present Worth

If $C_o = N_o$, the ratio of $C(n)$ to $N(n)$ becomes a dimensionless quantity, Pr, which represents the present worth factor of an item that will be purchased n years later, and is given by

$$Pr = \left(\frac{1+i}{1+d}\right)^n. \tag{8.3}$$

The present worth of an item is defined as the amount of money that would need to be invested at the present time with a return of $100d\%$ in order to be able to purchase the item at a future time, assuming an inflation rate of $100i\%$. Hence, for the item to be purchased n years later, the present worth (PW) is given by

$$PW = (Pr)C_o. \tag{8.4}$$

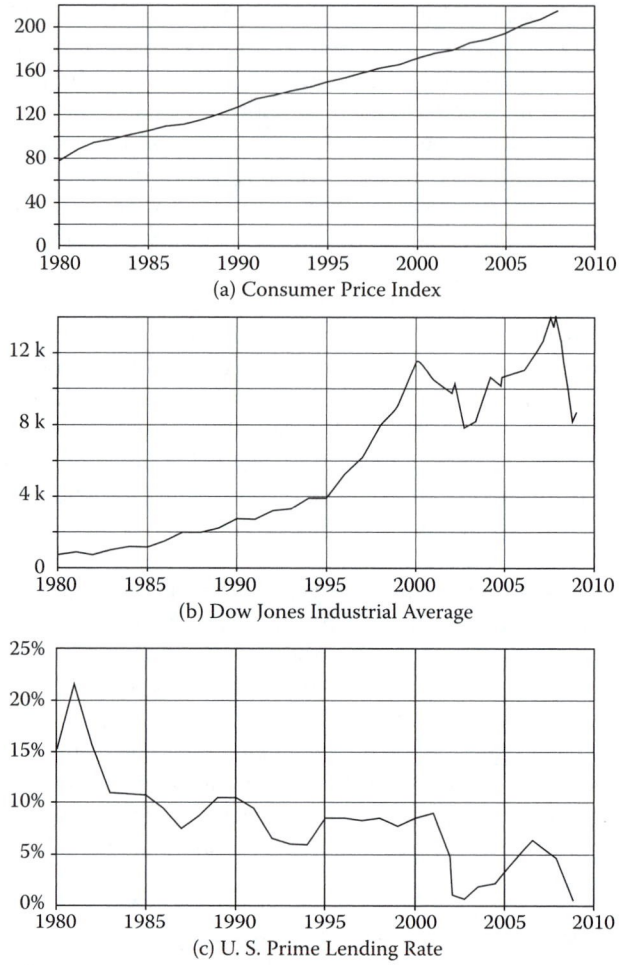

Figure 8.1 Comparison of consumer price index, Dow Jones Industrial Average, and prime lending rate, 1980–2008. (Based on data from U.S. Department of Labor, Bureau of Labor Statistics, Consumer Price Index, Washington, D.C. ftp://ftp.bls.gov/pub/special.requests/cpi/cpiai.txt; Dow Jones Indexes, http://www.djaverages.com; Historical Changes of the Target Federal Funds and Discount Rates http://newyorkfed.org/markets/statistics/dlyrates/fedrate.html)

Sometimes it is necessary to determine the present worth of a recurring expense, such as fuel cost. Because recurring expenses can be broken down into a series of individual expenses at later times, it is possible to determine the present worth of a recurring expense by simply summing up the present worth of each of the series. For example, suppose a commodity such as diesel fuel is to be used over the lifetime of a diesel generator. It is desired to determine how much money must be invested at present at an annual interest rate of $100d\%$, under conditions of $100i\%$ annual inflation in order to purchase fuel for n years. If the first year's supply of fuel is purchased at the time the system is put into operation, and each successive year's fuel supply is purchased at the beginning of the year, the present worth of the fuel acquisitions will be

$$\mathrm{PW} = C_o + C_o\left(\frac{1+i}{1+d}\right) + C_o\left(\frac{1+i}{1+d}\right)^2 + C_o\left(\frac{1+i}{1+d}\right)^3 + \cdots + C_o\left(\frac{1+i}{1+d}\right)^{n-1}. \tag{8.5}$$

Letting $x = (\frac{1+i}{1+d})$, Equation (8.5) becomes

$$\mathrm{PW} = C_o(1 + x + x^2 + \cdots + x^{n-1}). \tag{8.5a}$$

This expression can be simplified by observing that

$$\frac{1}{1-x} = 1 + x + x^2 + x^3 + \cdots = \sum_{i=0}^{\infty} x^i. \tag{8.6}$$

Now, the cumulative present worth factor can be defined as

$$Pa = PW/C_o = \frac{1}{1-x} - \sum_{i=n}^{\infty} x^i = \frac{1}{1-x} - x^n \sum_{i=0}^{\infty} x^i,$$

or, finally,

$$Pa = \frac{1-x^n}{1-x}. \tag{8.7}$$

It is important to recognize that Equation (8.7) is based on the assumption that the first year's supply is purchased at the beginning of the year at a time when the fuel is at its present value. The fuel is then purchased annually, with the last purchase occurring 1 year before the system lifetime has expired. In other words, there are n purchases of fuel, each at the beginning of the nth year.

If the recurring purchase does not begin until the end of the first year, and if the last purchase occurs at the end of the useful life of the system, there will still be n purchases, but, using x again, the cumulative present worth factor becomes

$$\begin{aligned} Pa_1 &= x + x^2 + x^3 + \cdots + x^n \\ &= x(1 + x + x^2 + \cdots + x^{n-1}) = xPa. \end{aligned} \tag{8.8}$$

Because x will typically be in the range $0.95 < x < 1.05$, and because determination of i and d are, at best, good guesses, and because it would be unusual to purchase an entire year's supply of many things all at once at the beginning of the year or at the end of the year, either Equations (8.7) or (8.8) will provide a good estimate of the present worth of a cumulative expenditure. Often, the values for *Pr* and for *Pa* are tabulated. As most engineers have programmable calculators and computers, there is little point in repeating tabulated values for these expressions. Once the values for the variables have been decided upon, the present worth of quantities can be calculated. For that matter, if an engineer wanted to assume different values for *d* and *i* for different years, the same methodology could be used to determine the present worth of a quantity.

8.2.3 Life-Cycle Cost

Once the PW is known for all cost categories relating to the purchase, maintenance, and operation of an item, the life-cycle cost (LCC) is defined as the sum of the PWs of all the components. The LCC may contain elements pertaining to original purchase price, replacement prices of components, maintenance costs, fuel and/or operation costs, and salvage costs or salvage revenues. Calculating the LCC of an item provides important information for use in the process of deciding which choice is the most economical. The following example demonstrates the use of LCC.

Example 8.1

Refrigerator A costs \$600 and uses 150 kWh of electricity per month. It is designed to last 10 years with no repairs. Refrigerator B costs \$800 and uses 100 kWh of electricity per month. It is also designed to last 10 years with no repairs. Assuming all the other features of the two refrigerators are the same, which is the better buy if the cost of electricity is \$0.07/kWh? What if the cost of electricity is \$0.15/kWh? Assume a discount rate of 3%, and assume an inflation rate of 5% for the electrical costs. Also, note that this example is being written out in early 2009, a time where estimating either *i* or *d* is highly speculative.

SOLUTION

The solution, of course, is to perform life-cycle cost analyses on each refrigerator. First, note that $x = 1.05 \div 1.03 = 1.019$. For a 10-year period, because electricity purchase begins at the time of purchase of the refrigerator, using Equation (8.7) gives $Pa = 11.16$.

Then, note that for refrigerator A, the electrical cost for the first year will be (12 mo) × (150 kWh/mo) × (\$0.07/kWh) = \$126, and for refrigerator B, the electrical cost for the first year will be \$84. Multiplying the first year cost by *Pa*, yields the PW of the electrical cost. A simple table may be constructed to compare the two refrigerators. Note that separate columns are used for the PW associated with electricity at \$0.07/kWh and electricity at \$0.15/kWh.

From Table 8.1 it is evident that the \$800 refrigerator has a lower LCC than the \$600 refrigerator. It is also evident that as the price of electricity is increased, the LCC of the more expensive first cost refrigerator becomes more and more attractive. Federal law requires that

Table 8.1 Life-Cycle Cost Analysis for Two Refrigerators at $0.07/kWh and $0.15/kWh

	Refrigerator A			Refrigerator B		
	First Year	PW	PW	First Year	PW	PW
Purchase price ($)	600	600	600	800	800	800
Electrical cost ($) @ $.07/kWh	126	1406		84	937	
Electrical cost ($) @ $.15/kWh	270		3013	180		2008
LCC ($)		2006	3613		1737	2808

certain appliances, including refrigerators, have labels that disclose the energy consumption. It is not necessarily the case that the more expensive units use less electricity. One should check the labels carefully when contemplating a purchase.

Example 8.2

Compare the life-cycle cost of a highway construction warning sign that is PV powered versus using a gasoline generator to power the same sign. The system is to be capable of 24-h-per-day operation with minimal downtime. Assume the load to be 2 kWh per day with a 20-year lifetime.

To power this load with a PV system, it will take a 500 W array of PV modules at a cost of $4 per watt; $900 worth of storage batteries, which need to be replaced every 5 years; and a $300 charge controller. Assume a system maintenance cost of $100 per year.

Although the average power requirement is only 83 W, it is unlikely that an 83 W generator will be used. For the purposes of this example, assume that a 500 W gasoline generator can be purchased for $250. Because it is running well under rated load, a generous efficiency estimate is 2 kWh per gallon and will thus use about 365 gallons of gasoline per year at an initial cost of $2/gal, and will require frequent maintenance with an annual cost of about $1500 for oil changes, tune-ups, and engine rebuilds. Because of the heavy use, after 5 years the generator must be replaced. Assume an inflation rate of 3% and a discount rate of 5%.

SOLUTION

For the PV system, Pr is needed for 5 years, 10 years, and 15 years, using Equation (8.3). For the generator, Pr is also needed for the generator replacement after 5 years, 10 years, and 15 years. For the PV system, Pa_1, using Equation (8.8) is needed for maintenance costs, and Pa, using Equation (8.7), is needed for generator fuel and maintenance costs. For the given inflation and discount figures, $x = 0.981$, $Pa = 16.81$, and $Pa_1 = 16.49$. So Table 8.2 can now be completed.

Hence, even though the initial cost of the PV system is significantly higher, its LCC is significantly lower. Could this possibly explain the rapid deployment of these signs?

Table 8.2 Comparison of LCCs for PV System and Generator System for Highway Sign

PV System			Generator System			
Component	**Initial Cost ($)**	**PW ($)**	**Component**	**Initial Cost ($)**	**Annual Cost ($)**	**PW ($)**
Array	2,000	2,000	Generator	250		250
Controller	300	300				
Batteries	900	900	Fuel		730	12,271
Batteries 5 yr	900	818	Generator 5 yr	250		227
Batteries 10 yr	900	743	Generator 10 yr	250		206
Batteries 15 yr	900	675	Generator 15 yr	250		188
Annual maintenance	100	1,649	Annual maintenance		1,500	25,215
LCC		7,085	LCC			38,357

Example 8.3

Compare the life-cycle costs of the hybrid residence electrical system of Chapter 7, Section 7.8, for the six different PV array options. Assume $x = 0.981$. This comparison has been done in the previous two editions of this text. In the first edition, the more PV that was used, the higher was the LCC. By the time the second edition was written, prices of PV had decreased and fuel costs had increased such that the LCC was optimized when PV supplied approximately 95% of the annual energy mix. So, naturally, the tradition must be carried on using 2009 pricing. It should also be noted that the overall system size has increased for this edition, making it even more interesting to compare the cost with the smaller system. The second edition maximum PV system consisted of 24 modules, each rated at 110 W, for a maximum PV array rated power of 2640 W, compared with the maximum array size this time of 3600 W, using 200 W modules.

SOLUTION

Much of the information needed for the solution of this example is available in Chapter 7, Tables 7.17–7.23. In the ideal case, a price should be affixed to the relative environmental costs of each type of generation as well as to the hardware associated with each type. While this will not be done for this example, environmental costs will be discussed in Chapter 9. In this chapter, it will simply be assumed that the less the generator runs, the less noise and air pollution will be created.

Table 8.3 presents the elements that contribute to the LCC of the various systems. Table entries are only shown for systems of 18 modules and 9 modules. It should also be noted that the component prices shown in this table are somewhere between wholesale and retail cost. The reader is encouraged to obtain the best price quotes for the individual items as a check on the prices in the table.

Reasonable assumptions are made for fuel cost and maintenance cost for the generator. It is assumed that propane costs $2.25 per gallon, an oil change will cost $6, a tune-up will cost $50, and a rebuild will cost $400. Array mount costs are based on the cost of typical

Table 8.3 Comparison of LCC for 18-Module Hybrid System and 9-Module Hybrid System

	18-Module System			9-Module System		
Item	**Cost ($)**	**Present Worth ($)**	**% Total LCC ($)**	**Cost ($)**	**Present Worth ($)**	**% Total LCC ($)**
		Capital Costs				
Array	15,120	15,120	29.3	7,560	7,560	14.4
Batteries	6,000	6,000	11.6	6,000	6,000	11.4
Array mount	1,512	1,512	2.9	756	756	1.4
Charge controller	1,000	1,000	1.9	500	500	1.0
Inverter/charger	3,000	3,000	5.8	3,000	3,000	5.7
Source circuit combiner	220	220	0.4	175	175	0.3
Installation	4,536	4,536	8.8	2,268	2,268	4.3
Generator	2,500	2,500	4.8	2,500	2,500	4.8
BOS	3,878	3,878	7.5	2,624	2,624	5.0
		Recurring Costs				
Annual inspection	75	1,258	2.4	75	1,258	2.4
Generator fuel	0	0	0	646	10,835	20.7
Generator maintenance	0	0	0	128	2,147	4.1
		Replacement				
Batteries 8 yr	6,000	5,146	10.0	6,000	5,146	9.8
Batteries 16 yr	6,000	4,414	8.6	6,000	4,414	8.4
Charge cont 15 yr	1,000	750	1.5	500	375	0.7
Inverter 15 yr	3,000	2250	4.4	3,000	2,250	4.3
Generator rebuild 9.5 yr	400	0		400	333	0.6
Generator rebuild 19 yr	400	0		400	278	0.5
TOTAL		51,584	100		52,418	100

commercial array mounts. It is assumed that the inverter for the hybrid system will be the same as the inverter for the nonhybrid system to allow for addition of a generator at a later date if desired. The installation cost is assumed to be $1/W for the PV system plus 20% of the cost of the generator.

Figure 8.2a is a plot of LCC versus the number of modules in the system, and Figure 8.2b is a plot of the LCC versus the percentage of annual kWh provided by the PV array. It is now up to the system owner to decide whether to spend the additional $4102 for the convenience of the 100% PV system over the 89% PV system, or, perhaps to spend an additional $417 to achieve 97.5% instead of 89%. In any case, it makes no economic sense to install a system with fewer than 12 modules. For the LCC of the generator-only system, it must be remembered that the system also includes an inverter and batteries, so the generator need not run continuously and can run at its most efficient output power level. If the batteries and inverter are eliminated, the generator must run continuously and will run well below its maximum efficiency most of the time, thus significantly increasing annual fuel and maintenance costs and requiring additional cost analysis.

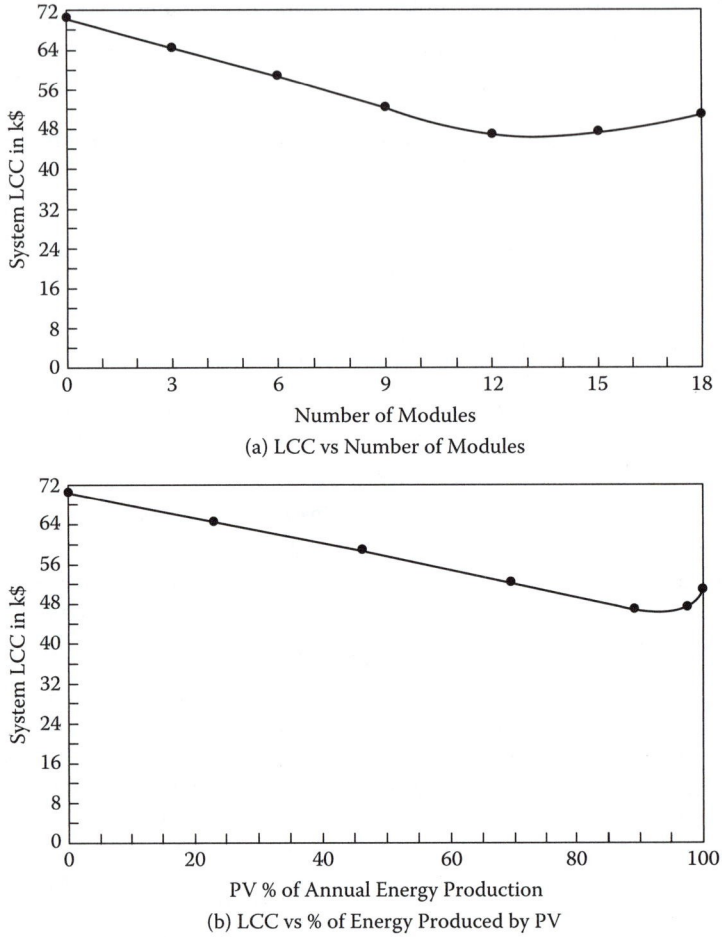

Figure 8.2 Optimizing LCC by varying mix of PV and propane generator.

8.2.4 Annualized Life-Cycle Cost

It is sometimes useful to compare the LCC of a system on an annualized basis. Dividing the system LCC by the expected lifetime of the system may appear to be the way to arrive at an annual cost. This, of course, would assume the cost per year to be the same for every year of operation of the system, which is assumed not to be the case in the original set of assumptions. Hence, to find the annualized LCC (ALCC) in present-day dollars, it is necessary to divide the LCC by the value of Pa or Pa_1 used in the PW analyses for the system components.

For the refrigerators in Example 8.1, this means dividing the LCC by 11.16 for each LCC evaluated. For the PV system of Example 8.2, the ALCC = \$7085 ÷ 16.81 = \$421, and for the gasoline generator system of Example 8.2, the ALCC = \$38,357 ÷ 16.81 = \$2,282. The ALCC for the hybrid residence system depends on which mix of PV versus generator is chosen.

8.2.5 Unit Electrical Cost

One especially valuable use of the ALCC is to determine the unit cost of electricity produced by an electrical generating system. Obviously, if the electricity is to be sold, it is necessary to know the price for which it should be sold to either earn a profit or to at least know how much will be lost in the process. Once the ALCC is known, the unit electrical cost is simply the ALCC divided by the annual electrical production. If the annual electrical production is measured in kWh, then the unit electrical cost will be measured in \$/kWh.

For the PV system of Example 8.2, the unit electrical cost is ALCC/kWh = \$421 ÷ 730 = \$0.577/kWh. For the gasoline generator system, the unit electrical cost is \$2282 ÷ 730 = \$3.13/kWh. Clearly, both of these costs far exceed the cost of utility-generated electricity, but because this is a portable application, where the sign is moved around from day to day, the cost of hooking up and disconnecting the system from utility power is impractical at best.

8.3 BORROWING MONEY

8.3.1 Introduction

Sometimes the desire to own something causes the potential owner to realize that money does not grow on trees. While some money comes from paychecks, usually, larger sums of money come from borrowing from banks or other lending institutions. The question at this point for the engineer-turned-economist is whether borrowed money is any different from paycheck money or money from a savings account, a mattress, or other form of liquid asset. For example, if the money to purchase one of the refrigerators of Example 8.1 had to be borrowed rather than taken from a wallet, would that affect the LCC of the refrigerator?

Once again, tables are readily available for looking up the annual payments on a loan of C_o dollars taken out at $100i\%$ annual interest over a period of n years. For that matter, it is possible to purchase a calculator that will automatically yield the answer when it has been given the terms of the loan. An engineer, of course, will want to know how the numbers are obtained.

8.3.2 Determination of Annual Payments on Borrowed Money

To satisfy this curiosity, consider Table 8.4, representing the principal, interest, total payments, and principal balance at the end of the kth year for repayment of

Table 8.4 Breakdown of Portions of Loan Payment Allocated to Principal and Interest

Yr	Payment on Principal	Interest Payment	Total Payment	Balance of Principal
1	A_1	iC_o	$A_1 + iC_o$	$C_o - A_1$
2	A_2	$i(C_o - A_1)$	$A_2 + i(C_o - A_1)$	$C_o - A_1 - A_2$
3	A_3	$i(C_o - A_1 - A_2)$	$A_3 + i(C_o - A_1 - A_2)$	$C_o - A_1 - A_2 - A_3$
n	A_n	$i(C_o - \ldots - A_{n-1})$	$A_n + i(C_o - \ldots - A_{n-1})$	0

an n-year loan at $100i\%$ annual interest, where C_o represents the amount borrowed. Note that during any of the years, it is not yet known how much will be paid on the principal in order to repay all of the principal in n years. The challenge is to arrive at an appropriate formula for equal total annual payments over the period of the loan.

In Table 8.4, A_k represents the amount paid on the principal after the kth year. To pay the principal fully in n years requires that the sum of the annual payments on principal add up to the loan amount, C_o.

Setting the total payments of each year to be equal yields a solution for A_1. For example,

$$A_1 + iC_o = A_2 + iC_o - iA_1,$$

which yields

$$A_2 = A_1(1+i).$$

In general,

$$A_n = A_{n-1}(1+i) = A_1(1+i)^{n-1}.$$

Next, letting $x = (1 + i)$ and summing all the payments toward principal yields the amount borrowed.

$$A_1 + A_1x + A_1x^2 + \cdots + A_1x^{n-1} = A_1(1 + x + \cdots + x^{n-1}) = C_o.$$

But Equations (8.5a) and (8.7) have shown that

$$1 + x + x^2 + \cdots + x^{n-1} = \frac{1-x^n}{1-x},$$

which yields

$$A_1 = \frac{C_o(1-x)}{(1-x^n)}. \tag{8.9}$$

Now all that remains is to add the interest payment for the first year to the principal payment, given by Equation (8.9), to get the total payment for the first year, which will be equal to the total payment for each succeeding year. Proceeding in this way yields

$$\text{ANN PMT} = \frac{C_o(1-x)}{(1-x^n)} + C_o i = C_o\left(\frac{i}{(1+i)^n - 1} + i\right).$$

Simplifying this result yields, finally,

$$\text{ANN PMT} = C_o i\left(\frac{(1+i)^n}{(1+i)^n - 1}\right). \tag{8.10}$$

Usually payments are made monthly rather than annually. In this case, one need only note that rather than n payments, there will be $12n$ payments, and the monthly interest rate will simply be the annual rate divided by 12. Doing so converts Equation (8.10) into an equation for monthly payments. Obviously, if one wants to split hairs even more finely, Equation (8.10) could be modified for weekly, daily, or hourly payments. Equation (8.10) is also an equation that might be conveniently tabulated for various values of i and n, but again, it is left to the reader to use either a computer or a programmable calculator to generate the numbers that apply to the problem at hand.

8.3.3 The Effect of Borrowing on Life-Cycle Cost

Depending on the nature of a purchase, it is interesting to compare whether the cost of borrowing money will render the purchase undesirable. For example, if money is borrowed to purchase something that will provide a return on the investment, it may make economic sense to borrow the money. This is the standard criterion for commercial loans. The better the return, the better the reason to borrow the money. But what if it is necessary to borrow the money for the initial cost of a system that does not have an obvious return on investment? Does it make sense to borrow for something having a greater first cost if it results in higher loan repayment costs? Although there are several ways to evaluate the worthiness of a purchase, once again, people sometimes borrow money to purchase things that do not have a measurable monetary return on investment, such as automobiles, simply because they want them.

As a simple example, suppose it is possible to spend \$2000 on a maintenance-free system that will last for 25 years and will reduce an electric bill by a certain amount per year. The exact system might be insulation, good windows, a solar water heater, or any number of energy efficiency measures. Suppose also that money is borrowed to purchase the system and the period of the loan is 25 years. This would be the case if the system is purchased for a dwelling at the time of construction and the item is included in the 25-year mortgage. If the mortgage rate is 9%, then Equation (8.10) shows the additional annual mortgage payments would be \$203.61.

If the annual savings on the electric bill exceed $203.61, then it is worth borrowing the money.

One should note, however, that the additional $203.61 mortgage payment will be constant over the life of the loan. In an inflationary environment, this means that even though the electric bill savings may not be $203.61 the first year, after a certain number of years, the annual savings may well exceed $203.61, making the investment worth considering.

What if the money is not borrowed? What if the $2000 is at hand and available for investing? What would be the return on investment if it is spent on one of the items of the previous example? The simple answer would say the return is equal to the resulting savings less the resulting operating or maintenance costs. This works fine for the first year, but for successive years to yield a more precise estimate, the time value of money must be taken into account. Assuming the value of the quantity saved increases at the inflation rate, i, and the discount rate is d, the present worth of the savings accrued over n years will be given by Equation (8.5). As a result, the PW of the annual savings over n years is given by Equation (8.7). Hence, Equation (8.7) can be applicable in either an expenditure mode or in a savings mode. By converting costs and savings to LCC, it is easy to compare savings with costs to determine whether to make the purchase.

Assuming that a system must be acquired to do something, and assuming that the system with the least LCC has been identified, it is simply a matter of deciding whether to borrow money or to use money at hand, if, indeed, the money is at hand. If the money is not at hand, then the only option is to borrow. If the money is at hand, then the criterion is whether the lending rate is less than or more than the discount rate. If money can be borrowed at a rate less than what it can earn, then it makes sense to borrow for the acquisition and to invest the money that might have been used for the purchase. This is a choice often made by the buyer of a new automobile. Should the savings account be used or should the money be borrowed? The answer depends on the relative interest on savings versus the interest on the loan.

8.4 PAYBACK ANALYSIS

As mentioned in the introduction to the chapter, some people are not interested in a renewable energy system unless they know how long it will take the system to pay for itself. Often, these same people are owners of expensive automobiles who did not even think about how long it would take the expensive automobile to pay for itself. In other cases, the potential owners are businesspeople who are interested in "greening" their business, but only if the "greening" has a reasonable payback period.

Once again, determining a payback period depends on predicting the future, because the idea is to install a system that will generate enough electricity in the first few years to offset all of the installation costs that remain after all incentive payments have been accounted for. The question is, what will be the value of the electricity? The part of the future that is relatively predictable is the annual psh and the system performance. The hard part is to figure out what will happen with electric rates over the life of the system. So the process involves some guessing.

One way to take some of the uncertainty out of the guessing is to establish lower and upper limits of assumed utility cost escalation rates and then analyze the resulting payback periods. It is also possible to attempt to take into account the time value of money into the analysis, but this adds even more uncertainty to the analysis, so this analysis will be based on as many certainties as possible.

Example 8.4

Assume that the installed cost of a 50 kW PV system is $7/W. Also assume that this system will qualify for a 30% investment tax credit and that the cost of the system can be depreciated at the rate of 20%/yr for 5 years. For system performance, assume 2000 annual psh on the array and an 80% dc:ac conversion ratio for the system. Assume the money is borrowed at 5% over 15 years to pay for the system and that the overall utility cost/kWh is $0.12. A more sophisticated analysis would separate out the energy and the demand portions of the electric bill, but this analysis simply divides the total bill by the total kWh. Perform a payback analysis on the system.

SOLUTION

First, determine the annual loan repayment from Equation (8.10). Also, note that the interest on the loan will be tax deductible for a business, but this will need to be handed over to the accountant at the end of the year. It will not be considered in this analysis. Using Equation (8.10), the annual loan repayment amount is found to be $33,720.

Next, the annual system kWh production is determined to be 80,000, which has a first-year value of $9600. At this rate, it might appear that the system will never pay for itself, but it is still useful to construct a table that looks at income versus expenses on an annual basis over the lifetime of the system, which will be assumed to be 30 years. In the analysis, utility escalation rates of 4%/yr and 10%/yr will be used. Note that this exercise is perfectly suited to a spreadsheet analysis, where as many variables can be included as suits the fancy of the engineer and/or the accountant, including such items as renewable energy credits (see Section 8.5.1), tax refunds on interest paid, etc. Table 8.5 shows the results. In the table, note that the annual depreciation is considered an operating expense, so the assumption is that at the end of the year after all expenses are included, the owner will still have tax liability.

An examination of Table 8.5 shows that over the first 5 years, the combination of 30% investment tax credit plus rapid depreciation more than offset the loan repayment value, so the fact that the value of the system electricity is less than the loan repayment is offset by the tax credit and rapid depreciation. By the end of year 5, after the system has been fully depreciated, the cumulative value of savings is about $340,000, depending on the utility escalation rate. Then, beginning in year 6, with no further depreciation or tax credits, the system begins to lose money each year for a while. So, the cumulative value of savings begins to fall, but never goes below zero (meaning that in the table, the cumulative value of cost always is negative). After year 15, there are no further loan payments, so the value of the electricity is now subtracted from owner expenses for the rest of the life of the system. Also, note that in year 15, the value of the electricity in the 10% utility escalation scenario exceeds the value of the loan payment.

Table 8.5 Payback Analysis for Example 8.4

Year	1	2	3	4	5	6
			Costs			
Loan payment	33,720	33,720	33,720	33,720	33,720	33,720
			Credits			
Util @ 4%	9,600	9,984	10,383	10,799	11,231	11,680
Util @ 10%	9,600	10,560	11,616	12,778	14,055	15,461
Inv tax cr	105,000	0	0	0	0	0
Ann depr	70,000	70,000	70,000	70,000	70,000	0
Net @ 4%	−150,880	−46,264	−46,663	−47,079	−47,511	22,040
Cum @ 4%	−150,880	−197,144	−243,807	−290,886	−338,397	−316,357
Net @ 10%	−150,880	−46,840	−47,896	−49,058	−50,335	18,259
Cum @ 10%	−150,880	−197,720	−245,616	−294,674	−345,009	−326,750

Year	7	8	9	10	11	12
			Costs			
Loan payment	33,720	33,720	33,720	33,720	33,720	33,720
			Credits			
Util @ 4%	12,147	12,633	13,138	13,664	14,210	14,779
Util @ 10%	17,007	18,708	20,578	22,636	24,900	27,390
Inv tax cr	0	0	0	0	0	0
Ann depr	0	0	0	0	0	0
Net @ 4%	21,573	21,087	20,582	20,056	19,510	18,941
Cum @ 4%	−294,784	−273,697	−253,115	−233,059	−213,549	−194,608
Net @ 10%	16,713	15,012	13,142	11,084	8,820	6,330
Cum @ 10%	−310,037	−295,025	−281,883	−270,799	−261,979	−255,649

Year	13	14	15	16	17	18
			Costs			
Loan payment	33,720	33,720	33,720	0	0	0
			Credits			
Util @ 4%	15,370	15,985	16,624	17,289	17,981	18,700
Util @ 10%	30,129	33,142	36,456	40,102	44,112	48,523
Inv tax cr	0	0	0	0	0	0
Ann depr	0	0	0	0	0	0
Net @ 4%	18,350	17,735	17,096	−17,289	−17,981	−18,700
Cum @ 4%	−176,258	−158,522	−141,426	−158,715	−176,696	−195,396
Net @ 10%	3,591	578	−2,736	−40,102	−44,112	−48,523
Cum @ 10%	−252,058	−251,480	−254,216	−294,317	−338,429	−386,952

8.5 EXTERNALITIES

8.5.1 Introduction

What happens if something owned and operated by one party causes damage to something owned by another party? A quick response would be that the entity causing the damage would be liable, and that the cost of repair of the damaged property should be the responsibility of the party doing the damage. The problem is complicated in many instances, however, when more than one party is responsible for creating the cause of the damage, and when more than one party suffers damage as a result of the cause. And things get even more complicated when debate ensues over whether the alleged cause is really the cause. Classic examples of such situations are the link between smoking and cancer and the link between burning of fossil fuels, acid rain, and global warming.

Much debate has taken place regarding the liability of tobacco companies for alleged first-hand, smoke-induced cancer and even regarding the alleged incidence of second-hand, smoke-induced cancer. Until recent large judgments against the tobacco industry, the cost of producing cigarettes did not include a component to underwrite the cost of paying the judgments. Yet, it has been argued that tobacco is the cause of billions of dollars in medical bills, none of which has been paid out of tobacco revenues. Thus, in the past, the medical bills incurred by those exposed to tobacco smoke have been treated as externalities by the tobacco industry, while presently, these bills have become direct cost components of doing business.

There is now reasonable agreement that one cause of acid rain is the burning of fuel that contains sulfur, such as petroleum or coal. As a result, many power plants are now equipped with elaborate scrubbers that remove sulfur and other pollutants from smokestack emissions.

The U.S. Environmental Protection Agency (EPA) has established limits on the amounts of various pollutants that may be contained in smokestack emissions. Limits also exist for regions, so that if a region has met its limit, then no further burning may take place unless one of the burners can be made cleaner. In some cases, when a company has not reached its emission limit, the company will sell or trade the remaining allowed emissions with another burner. As a result, monetary value is evolving for certain emissions, albeit in a bit of a roundabout way. The monetary value, however, does not relate directly to the damage done in either the form of acid rain damage or the cost of respiratory diseases. The cost of scrubbers, however, appears as a system cost during the LCC process.

Recently, the concept of Renewable Energy Credit (REC) has been introduced to the marketplace. The REC, once called "green tag," is equivalent to 1000 kWh of electricity produced by renewable sources, and RECs are now traded among energy producers. Once a renewable energy source has sold its RECs, no further claims can be made for the elimination of greenhouse gases, as selling the RECs allows another party to produce the greenhouse gases instead. Sometimes it is less expensive for a generation facility to purchase RECs from a renewable generator than to clean up the CO_2 emissions of its own generation facility.

Another factor that needs consideration in establishing cost is the effect of various forms of subsidies. Some subsidies may enter the picture as direct costs, while others may appear as externalities.

8.5.2 Subsidies

When performing an LCC, sometimes the cost of a component, a fuel, or the operation of a system may be affected by a subsidy or subsidies. For example, the cost of military presence in a region to ensure the steady flow of a fuel from the region is never included in the selling price of the product. Mineral depletion allowances, however, can be factored into the selling price of a product. In other cases, governments have been known to offer price supports in order to ensure competitive prices in a world market. Tariffs, in effect, are a form of subsidy, because they ensure that domestic production will be sold at a profit, thus not competing directly with less costly products from outside a country.

Green pricing is a form of subsidy for the acquisition of clean energy sources. An example is when the customers of a utility express a willingness to pay extra every month to ensure that a part of their energy mix comes from renewable sources, such as photovoltaics.

Subsidies in the form of tax breaks tend to come and go on a year-by-year basis. For a few years during the 1970s, homeowners could deduct a fraction of the cost of a domestic solar hot water system from their federal income tax. Some states initiated grant programs to encourage homeowners to install solar systems and some utilities established rebate programs for part of the cost of installing various energy conservation measures, such as more efficient air conditioning or attic insulation. Recently, several states have created buy-down programs in which rebates are offered toward the installation of PV and other renewable energy systems.

The argument is often set forth that all competing interests must compete on a level playing field. The meaning is simply that with so many subsidies, some of which are obvious and others of which are hidden, it is difficult for two competing interests to engage in fair competition. This is true in many industries and particularly in the energy industry.

The reader is thus reminded that no economic analysis or comparison is complete until all forms of subsidy have been considered.

8.5.3 Externalities and Photovoltaics

The cost of an electrical generation source often excludes externalities. Subsequently, if a source is cleaner from an environmental viewpoint but has a higher LCC based on parameters considered, that source may not be chosen. A proper treatment of externalities includes not only the operating externalities, but the externalities associated with the construction and salvage or decommissioning of the facility. In both categories, photovoltaics show significant advantages over nonrenewable sources, as will be shown in Chapter 9.

Problems

8.1 Obtain the data for consumer price index, Dow Jones Industrial Average, and prime lending rate from References [1–3]. Plot the data, and attempt to fit curves to the data to show trends. For example, use Excel graphs with trend lines, equations, and R^2 values added to show the "goodness of fit." Compare the R^2 values for linear, exponential, and polynomial fits to your curves.

8.2 Determine the present electricity cost for which the \$600 refrigerator of Example 8.1 will have the same LCC as the \$800 unit, assuming all other parameters to be the same.

8.3 The next time you are in an appliance store, record the first cost and annual operating costs of several refrigerators that are comparable. Then, making reasonable assumptions about discount rates, inflation rates, and appliance lifetimes, compare the LCCs of the units. You might want to ask a salesperson for information on expected lifetime and repair costs of the units, including annual maintenance contracts.

8.4 Use Equation (8.10) to compare 12 equal monthly payments to 1 annual payment by modifying the equation to account for monthly interest rate versus annual rate and monthly payments versus annual payments. How does the sum of 12 monthly payments compare with a single annual payment?

8.5 Rework Example 8.2 for a system that uses 4 kWh per day. This will require double the PV array and double the batteries, as well as double the fuel. However, the cost of the generator will remain the same, and the cost of the maintenance will remain the same. Assume the cost of the charge controller will increase by 50%.

8.6 Calculate a set of reasonable conditions on interest rate, term of loan, and cost per installed kW for a PV system that generates power for an average of 5 peak sun hours per day so the annual loan repayment can be recovered if the value of the electricity generated is \$0.20/kWh.

8.7 The value of electricity during utility peaking hours is \$0.25/kWh, and the money for a utility-interactive system with a 30-year expected lifetime can be borrowed at an interest rate of 7%. Calculate the installed cost per kW for the system that will result in annual loan payments equal to the value of electricity produced by the system if the electricity is produced during utility peak hours.

8.8 The Feed-in Tariff (FIT) is a means of providing an incentive for consumers to install PV systems that has been widely accepted, particularly in Germany and Spain. The idea is to pay the PV electric producer a rate per kWh produced that is equal to (or, sometimes, greater than) the loan repayment on the system. At the time of this writing, the City of Los Angeles, California, is seriously adopting this system. If the installed cost of a 5 kW grid-connected system is \$7/W, if the federal investment tax credit of 30% of the system cost is applied to the system cost, if the system is installed south-facing with a 25° array tilt, the dc-to-ac conversion efficiency of the system is 80%, and if the balance of the system cost after the investment tax credit can be borrowed for the system at 4%/yr for 20 years,

a. Use PVWatts to determine the expected annual kWh generation.
b. Determine the payment in \$/kWh needed to enable the system to earn enough annually to equal the value of the loan repayment.
c. Look up the present Los Angeles electric rate (or assume it to be \$0.15/kWh). Then assume the rate will increase at 5%/yr. Determine the year when the electric rate in \$/kWh will equal the rate of payment to the PV system owner.

8.9 Net metering is when a utility pays the same rate for electricity sold back to the utility as is charged to the consumer for purchased electricity. Assume a straight grid-connected 5 kW PV system can be installed for \$4/W after all incentive payments are accounted for. Also assume the system is installed in an area that receives 1900 psh/yr of sunlight at the orientation of the array. Then assume that the present utility electric rate is \$0.14/kWh and that the rate appears to be increasing at 6%/yr. If the dc:ac conversion efficiency of the system is 80%, and money can be borrowed at 5%/yr:

a. Calculate the annual loan repayment rate if the loan is for 20 years.
b. Calculate the annual kWh production of the system and the value of this electricity at the present utility rate.
c. Construct a bar graph that shows the difference between the annual value of the electricity generated and the annual loan repayment rate for a 30-year period. Keep in mind that the loan payments cease after 20 years.
d. Construct the equivalent of the integral of the bar graph of part c that shows the cumulative return on investment for the system.

8.10 Assume that the system of Problem 8.9 is installed using a single-axis tracking array to increase the annual psh on the array to 2400. The additional cost of the tracking array mount increases the cost after incentives to $5/W.

a. If all the other conditions of Problem 8.9 are the same for this case, evaluate parts a–d for the tracking system.
b. Which system appears to be the best choice?

8.11 A battery storage system is to be designed to provide a storage capacity of somewhere between 440 and 555 Ah @ 48 V at a *C*/20 discharge rate. Four battery types are under consideration:

Battery	Volts	Capacity	Type	Lifetime	Weight	Cost Each
A	6	220 Ah	Flooded	5 yr	56 lb	$ 66
B	12	255 Ah	agm	8 yr	168 lb	$ 356
C	12	180 Ah	agm	8 yr	135 lb	$ 301
D	12	555 Ah	gel	12 yr	564 lb	$ 1941

Assume $i = 2\%$, $d = 5\%$, and a 24-year system lifetime. Perform an LCC for the four battery types and discuss other considerations that may influence the choice of batteries. If the site were a home, what would you recommend? Why? If the site were a remote communication system, what would you recommend? Why?

REFERENCES

1. U.S. Department of Labor, Bureau of Labor Statistics, Consumer Price Index, Washington, D.C. ftp://ftp.bls.gov/pub/special.requests/cpi/cpiai.txt.
2. Dow Jones Indexes, http://www.djaverages.com.
3. Historical Changes of the Target Federal Funds and Discount Rates http://newyorkfed.org/markets/statistics/dlyrates/fedrate.html.

SUGGESTED READING

42 U.S.C. § 7401 et seq. (Clean Air Act).

Markvart, T., Ed., *Solar Electricity*, John Wiley & Sons, Chichester, U.K., 1994.

"National Appliance Energy Conservation Act of 1987" (PL 100–12, March 17, 1987, 101 § 103).

Stand-Alone Photovoltaic Systems: A Handbook of Recommended Design Practices, Sandia National Laboratories, Albuquerque, NM, 1995.

CHAPTER 9

Externalities and Photovoltaics

9.1 INTRODUCTION

Pollution is an insidious by-product of civilization that threatens the quality and integrity of our air, our soils, our water, and our infrastructure. The planet has a remarkable ability to cleanse itself, with significant feedback loops in place for restoration of equilibrium. One example is photosynthesis, in which the oxygen–carbon dioxide cycle is sustained by regeneration of oxygen from the carbon dioxide of respiration and oxidation. Another is digestion, in which bacteria process organic matter into a form that enables the matter to enrich soils or to nourish living bodies.

In electrical engineering terms, the feedback processes can be characterized in control terms by observing the locations of the poles of the processes. If the poles are in the left-half plane, then the process is stable. If the poles are in the right-half plane, then the process is unstable. Stable implies that the process will tend to correct itself. *Unstable* implies unbounded increase.

While many processes appear to be stable, examples of unstable processes have also been noted. In particular, increased CO_2 from nonpolar regions causes warming in the polar regions, resulting in thawing of permafrost regions, in which large quantities of CH_4, another greenhouse gas, are trapped. Liberation of the CH_4 causes additional warming, resulting in additional liberation of CH_4, and so forth. The question is whether another natural process exists to counteract this cycle. The process itself is unstable.

Human intervention is at a point where serious questions are now being asked with regard to the effects of human-generated pollutants in terms of the stability of the solutions of the process equations. One might argue that as long as the poles are kept in the left-half plane, the pollution causing the appearance of the poles is at acceptable levels for the planet to sustain life. The challenge becomes one of defining acceptable limits on pollution and the resulting changes in natural cycles caused by human intervention.

9.2 EXTERNALITIES

Externalities were mentioned briefly in Chapter 8, Section 8.5, as factors resulting from energy or other manufacturing that are not incorporated directly into the selling price of a product. That is, once the smoke is out of the smokestack, it is no longer the fiscal responsibility of the smoke producer. In this chapter, externalities will be explored in greater detail. In particular, environmental effects, health and safety issues, and subsidies will be discussed in the context of comparing energy produced by photovoltaic means with energy produced by other means. In general, the externalities discussed are not associated with costs that appear in a life-cycle cost analysis of an energy source. The reader can expect that, in the future, greater attention will be paid to those issues currently considered as externalities, as means for attaching monetary value to these issues are developed and adopted. Some readers may find the research necessary to affix monetary value to externalities of sufficient interest and challenge that they may choose to pursue this interesting and important area further.

Perhaps one of the more challenging considerations relating to externalities is the debate as to whether they really merit consideration. Acid rain and global warming, for example, are two phenomena that have been the subjects of a great deal of discussion over the past few decades.

Acid rain is claimed to result from the emissions of sulfur and nitrogen oxides by fossil-fuel burners. As these oxides are dissolved in raindrops, they turn into weak acids, which have been blamed for the changes in pH in lakes and soils and for the etching away of building structures. The scientific community has reached consensus on specific greenhouse gases that cause global warming. While it is generally accepted among the scientists that these effects really do exist, there remain skeptics who either do not believe in the phenomena or otherwise claim that insufficient information is available to confirm the cause-effect relationships with 100% certainty.

For example, Moore [1], an economist, argues that policies to reduce the emission of greenhouse gases "may be unnecessary, would be inordinately expensive and would lead to worldwide recession, rising unemployment, civil disturbances, and increased tension between nations ..." He notes that "even if significant warming were to occur, public policymakers could, at the time it became evident, launch programs to adapt to the change, such as building dikes, increasing air conditioning, and aiding farmers and ecosystems to adjust to the new weather." He goes on to observe that during colder periods, more people die from exposure to cold than the number who die from exposure to intense heat during warmer periods, suggesting that the potential consequences may not be as dire as predicted by those concerned about global warming.

The history of climate concern, with an extensive bibliography, is nicely documented in Paterson [2]. The greenhouse properties of the atmosphere were noted as early as 1827 by Fourier, who is perhaps better known to the electrical engineer for the Fourier series. During the 19th century, discussions continued about the causes of climate change, and in 1872, the International Meteorological Organization was formed. In 1896, Arrhenius published an article in which he calculated that the temperature of the earth would rise about 5°C if atmospheric carbon dioxide were to double. In 1908, he suggested that industrial carbon dioxide emissions might cause a noticeable change in atmospheric carbon dioxide levels within the next few centuries. He continued to observe that perhaps this would be good, because it would warm up some of the cold regions of the planet and make them more useful for agricultural and other purposes. An alternative viewpoint was that there was no need for concern over carbon dioxide emissions, because the oceans would remove excess carbon dioxide and maintain the delicate balance between atmospheric oxygen and carbon dioxide.

The first formal significant consensus on global warming and its relation to the production of specific greenhouse gases probably came out of the Conference on the Physical Basis of Climate and Climate Modeling, held in 1974. The Intergovernmental Panel on Climate Change (IPCC), in its 1990 report, identified carbon dioxide, methane, chlorofluorocarbons (CFCs), and nitrous oxide as gases that will enhance the greenhouse effect and result, on the average, in an additional warming of the surface of the earth [2].

In this chapter, a general overview of relative environmental effects of energy sources is presented, followed by observations of specific areas of concern for a variety of PV sources and suggestions on how best to minimize these potential adverse effects. The reader is encouraged to acknowledge that externalities must be viewed from economic as well as from political perspectives and that scientific, economic, and political considerations often cause the decision-making process to merge into a complicated maze. One need only observe the controversy over what to do with nuclear waste to appreciate the interrelationships of these three areas.

9.3 ENVIRONMENTAL EFFECTS OF ENERGY SOURCES

9.3.1 Introduction

Although the methodology of assigning precise dollar values to carbon dioxide emission, particulate emission, nitrogen oxide emission, sulfur emission, and, in general, the adverse environmental effects of the construction, operation, and decommissioning of various energy sources is complex, it is generally possible to assign relative comparative values to most of these areas. Table 9.1 shows a matrix compiled by Baumann and Hill [3] that compares 10 negative effects for 12 different energy sources, with relative significances indicated. The observations should come as no particular surprise to anyone familiar with the sources listed.

9.3.2 Air Pollution

9.3.2.1 The Clean Air Act and the U.S. Environmental Protection Agency

The U.S. Environmental Protection Agency (EPA) is required by the 1990 amendment of the Clean Air Act [4] to regulate emissions of toxic air pollutants from a published list of source categories. The EPA is charged with setting standards and

Table 9.1 Relative Environmental Effects of a Variety of Renewable and Nonrenewable Energy Sources

Negligible/Significant = 1 Significant = 2 Significant/Large = 3 Large = 4	SO_x and NO_x	CO_2	CH_4	Health	Particulates	Heavy Metals	Catastrophes	Waste Disposal	Visual Intrusion	Noise	Land Requirements
Passive solar energy									1		
Photovoltaics					1	1		1	1		1
Wind									3	1	1
Biomass	1		3	1	1	1		1	1	1	3
Geothermal	1	1	1	1		1		2	1	1	
Hydro							2		3		3
Tidal							1		3		1
Water waves							1		1		
Coal	4	4	2	1	2	2	1	2	2	1	3
Oil	3	4	1	1	2	1	2	1	1		1
Natural gas	1	4	3	1			2		1		1
Nuclear	1	1		1			2	3	2		1

Source: Adapted from Baumann and Hill, *Proc. 10th EC Photovoltaics Solar Energy Conference*, Kluwer, Dordrecht, 1991, 834–837.

with establishing and promulgating rules for reducing toxic emissions to levels that meet the standards for nitrogen oxides, ozone, sulfur oxides, lead, carbon monoxide, and particulates.

Over the years, the EPA has presented data on these six pollutants in several formats, such as nonattainment maps. Currently, the EPA publishes a wealth of data, including national and regional summaries of historic and daily levels of these substances as compared to the standards [5]. The data are encouraging, because they show that for all but ground-level ozone, the atmospheric concentrations in the United States have been falling steadily since 1990. For example, NO_x concentrations have fallen 35%, CO concentrations have fallen 67%, and Pb has fallen 80%. Ozone has fallen 9% but is still above the national air quality standard level. Note, however, that greenhouse gases are not considered to be pollutants by these standards. And, of course, the identification of six pollutants does not imply that these six are the only ones that merit concern.

Table 9.2 shows the top U.S. polluters in the six categories of EPA pollutants as of 2002. Note that ground-level ozone is not included in the table, but, rather, the two components that, when exposed to sunlight, combine to produce ozone are shown.

Previous EPA listings have shown the top 25 sources of U.S. air pollution in each of the six categories. A steel producer was at the top of the list of 25 emitters of NO_2 on the March 1999 list. The remaining 24 entries were various electrical utilities, located predominantly along the Ohio River, where coal-fired generation is a major source of electricity.

Table 9.2 Top Polluters in the Six EPA Categories in 2002

Category	Source	% of Total
Nitrogen oxides[a]	On-road vehicles	38.3
	Electricity generation	22.0
Sulfur oxides	Electricity generation	71.4
	Fossil fuel combustion	13.9
Carbon monoxide	On-road vehicles	57.6
	Nonroad equipment	20.5
Volatile organic compounds[a]	On-road vehicles	24
	Solvent use	18
Lead	Industrial processes	42.8
	Nonroad equipment	27.9
Small particulates (PM 2.5)	Fires	26.8
	Road dust	18.6

Source: U.S. Environmental Protection Agency data on current and historic levels of pollutants regulated by the Clean Air Act www.epa.gov/airtrends.

[a] Ground-level ozone is not produced directly. It is produced by a reaction between VOCs and nitrogen oxides.

Carbon monoxide was once attributed mostly to steel companies, where carbon is used as a reducing agent to remove oxygen from the oxides of iron. Carbon monoxide is now predominantly produced by internal combustion engines in automobiles, trucks, and nonroad equipment.

In March 1999, the top 25 producers of SO_2 in the United States were all electric utilities, with a total estimated production of 4.672 million tons per year. Every state in the union except Utah had SO_2 emitters.

January 2003 ozone nonattainment areas covered a significant fraction of California and a strip along the eastern seaboard from Washington, D.C., to southern Maine. Ozone is primarily a by-product of internal combustion engines. Improving these engines or replacing much of the fleet, such as commuting vehicles with electric vehicles that run on PV-generated electricity from PV arrays in parking lots, are two suggestions for reducing ozone emissions. Fuel cells as energy sources for vehicles may be another promising solution. The hydrogen for the fuel cells can be produced with PV-generated electricity.

Particulates with diameters less than 10 μm, designated as PM-10 particles, tend to remain airborne for relatively long periods of time, causing respiratory problems in anyone with sensitivities to certain particulates. Again, these contaminants were at high levels in southern California, but also, surprisingly, in the Grand Canyon area, in southern Oregon, and in northern Idaho and western Montana. The 25 major fixed sources of PM-10 particles in the United States include steel mills and electric utility generators.

Although CO_2 is recognized as a greenhouse gas, its emission is not officially monitored or regulated at this time, although a move is under way to reduce CO_2 emissions in the United States to 1990 levels as a result of the 1998 Kyoto Conference [6].

Readers are encouraged to visit the EPA Web site for details of current EPA programs. The World Resources Institute (WRI) maintains another Web site that provides timely information on pollutants, including valuation methodologies and economic cost discussions [7]. According to WRI, the Intergovernmental Panel on Climate Change (IPCC) has calculated that to stabilize CO_2 levels at 1997 levels would require a 60% cut in emissions, maintained at that level for the next century. Readers may also wish to review their solutions to Problem 1.9 in Chapter 1, which dealt with CO_2 emissions from burning coal and petroleum to put the CO_2 from these sources into perspective.

Air quality is not a problem limited to the United States. As economies in developing countries continue to expand, pollution controls often take a back seat to development and the desires of the populations to have more energy at their disposal. Many of the world's larger cities have severe air pollution problems.

The World Bank has tabulated pollution indices for large cities around the world for the most recent years, available as of 2009 [8]. In 1995, the highest levels of total suspended particulates were found in Delhi, India (415 $\mu g/m^3$), with Beijing, China, and Bombay, India, following in close second and third place at 377 and 375 $\mu g/m^3$, respectively. In contrast, Stockholm, Sweden, had a level of 9 $\mu g/m^3$. By 2004, the leader was Cairo, Egypt, with 169 $\mu g/m^3$ followed by Delhi and Kolkata, India, with 150 and 128 $\mu g/m^3$, respectively, and then Tianjin and Chongqing, China, with 125 and 123 $\mu g/m^3$, respectively. In comparison, Los Angeles, California, had 34 $\mu g/m^3$, and Manchester, England, had 14 $\mu g/m^3$.

The most recent data on SO_2 show Guiyang and Chongqing, China, with 424 and 340 $\mu g/m^3$, respectively; Tehran, Iran, with 209 $\mu g/m^3$; Rio de Janeiro, Brazil, with 129 $\mu g/m^3$; Los Angeles, California, with 9 $\mu g/m^3$; and Havana, Cuba, with 1 $\mu g/m^3$.

The winner of the NO_2 competition was Milan, Italy, with a concentration of 248 $\mu g/m^3$. Mexico City, Mexico, came in at 130 $\mu g/m^3$, while Beijing, China, and Sofia, Bulgaria, both reported 122 $\mu g/m^3$. New York City won the honors for the United States with 79 $\mu g/m^3$. Havana, Cuba, was reported at 5 $\mu g/m^3$.

For the reader who wishes to spend a year or two in the further collection of information on air pollution, on May 1, 2009, a Web search for air pollution data netted 15,500,000 hits.

9.3.2.2 Greenhouse Gases

At this point in history, most people have heard of greenhouse gases and most people have heard somewhere that CO_2 is the major culprit. The common characteristics of greenhouse gases are that they are, for the most part, nontoxic, and they produce the greenhouse effect by being transparent to visible radiation, but reflective to infrared radiation. This means, simply, that as sunlight passes through the atmosphere and is absorbed by something on the surface of the earth, the object

Table 9.3 Summary of Potent Greenhouse Gases Showing Potency Compared to CO_2

Gas	$\times CO_2$	Gas	$\times CO_2$	Gas	$\times CO_2$
CO_2	1.0	CFC-113	4500	CH_4	21
CFC-13	13,000	CFC-114	7000	CCl_4	1300
N_2O	310	CFC-115	7000	CCl_2H_2	15
Halon-1211	4900	Halon-1301	4900	HFC-125	3400

Source: Based on data from Battisti and Corrado, *Energy* 30 (2005), p. 959.

that absorbed the relatively short-wavelength sunlight will then emit relatively longer wavelength energy in the infrared region of the spectrum. As the infrared radiation encounters a greenhouse gas molecule, the radiation will be absorbed or reflected back to the surface of the earth. Ideally, to maintain temperature equilibrium, the earth will emit the same amount of energy that it absorbs. Not doing so causes heat to accumulate or dissipate such that the temperature of the earth rises or falls enough to be able to emit the stabilizing amount of energy according to the blackbody radiation formula that was previously discussed in Chapter 2.

Less commonly known, however, is that there are greenhouse gases that exhibit a much higher ability to hold heat than CO_2. The only reason CO_2 is better known is because there is more of it, and the amount increases every year. But some of the other culprits are also increasing, so it should be of interest to know what they are. Table 9.3 summarizes the more potent greenhouse gases, indicating their relative ability to retain heat compared to that of CO_2 [9].

9.3.3 Water and Soil Pollution

Water pollution has perhaps received the most attention in the United States in terms of the lowering of pH of lakes in the Finger Lakes region of New York, presumably caused by acid rain from SO_2 and NO_2 emissions in the Ohio River region. The lowering of the pH as a result of the inability of the lakes to buffer the effect of the acid rain has been blamed for the loss of significant fish populations. This loss has, in turn, resulted in a loss of revenue to other sectors of society such as the sport fishing and tourist industries.

Other water pollution with perhaps more insidious consequences is the pollution of aquifers, lakes, rivers, and streams with solid and liquid pollutants. While aquifer pollution has not been blamed on SO_2 or NO_2 or other atmospheric gases, it has been blamed on industrial solvents and other chemicals, many of which are used to support an economy with a large energy appetite. Perhaps one of the best-known incidents in this category is the Love Canal, with many other similar sites as identified for cleanup by the EPA. At this point, the cleanup costs of these toxic waste dumps is being quantified, so these costs can be related back as externalities resulting from past waste-dumping practices.

It has been reported that soil acidification leads to permanently reduced productivity resulting from slower decomposition of humus and diminishing humus quality.

Other important effects of soil acidification include leaching of alkali and alkaline metals and release of heavy metals into soils. With changing quality of the soil surrounding the roots of trees, it is believed that the trees may become more sensitive to gaseous air pollutants [10].

9.3.4 Infrastructure Degradation

A number of studies have shown that pollution causes premature degradation of infrastructure [11]. Stonework, carbon steel, nickel and nickel-plated steel, zinc, and galvanized steel are all greatly affected by sulfur dioxide in rain. It has been estimated that in unpolluted environments, galvanized coatings last up to three times longer, and in polluted areas, galvanized transmission towers require repainting nearly twice as often. Hence, a cost figure can be attached to the restoration of structural finishes. However, assigning responsibility for the degradation is somewhat more problematic in that it is difficult to determine the mix of pollution sources that causes a specific amount of degradation to any particular site, because the origin of any pollutant at any particular time will depend on the wind direction.

9.3.5 Quantifying the Cost of Externalities

9.3.5.1 The Cost of CO_2

The cost of CO_2 can be considered in two ways: the cost of controlling CO_2 and the benefit of controlling CO_2. This method is applicable to other pollutants as well, but only CO_2 will be discussed in this section.

According to the World Resources Institute [7], holding CO_2 levels at 1990 levels will cost approximately 1% to 2% of the gross domestic product for developed countries over the long term, with reductions below 1990 levels cost increasing to 3%. An alternate analysis, however, suggests that by reducing CO_2 levels significantly by incorporation of energy conservation measures, the cost to reduce the CO_2 levels will be less than the savings in energy costs.

If action is not taken to stop the increase of CO_2 atmospheric concentration, it has been estimated that doubling of atmospheric CO_2 will lead to an average warming of 2.5°C [7]. According to current projections, this doubling will occur during the next century if no curtailment measures are taken. In this case, it is estimated that the damage due to the warming will reach approximately 1% to 1.5% of GDP per year in developed countries, with substantially more in island nations. If CO_2 levels continue to rise beyond the double point, the damage may come closer to 6% of GDP.

Hence, it appears certain that damage will occur if nothing is done, while it is debatable whether there will be cost or cost savings if a program to reduce CO_2 levels is pursued. Considering that CO_2 production is also generally accompanied by the production of NO_2, SO_2, and particulates, if reduction of CO_2 is also accompanied by reduction of other pollutants, then the benefits become subject to a multiplier effect. What is also being supported here is that it may be a better strategy to focus on reducing the production of CO_2 rather than sequestering the CO_2 already produced. Perhaps a carefully planned combination of these two strategies may produce the most cost-effective results.

9.3.5.2 Sequestering CO_2 with Trees

In order to quantify the cost of externalities, it is necessary to arrive at a methodology for determining the value of a ton of SO_2 or NO_2 or CO_2 or other substance. Because trees, other green plants, and oceans have maintained the balance between O_2 and CO_2 over the millennia, it would seem that a tree might be valued in terms of the number of tons of CO_2 it will remove from the atmosphere in its lifetime. Because SO_2 and NO_2 generally are removed from the atmosphere by rain, which dissolves them as weak acids, presumably a price per ton can be assigned in terms of the monetary value of any damage that may result to structures or the environment from acid rain. For example, if a building requires repair more often as a result of acid rain, this is a measurable cost, provided that the increased repair costs can be documented. If the fish die in a lake as a result of acid rain and, as a result, cause a decrease in tourism to the area, this also may constitute a documented cost if the source or sources can be identified.

The cost of CO_2 and other pollutants can be measured in several ways, including the cost of repairing the damage, the cost of controlling the damage by reducing emission of pollutants, and the cost of mitigating the damage, such as by absorbing the additional CO_2 by trees. Hodas (1990) [12] reported costs of $240 per ton of carbon for removal of CO_2 from exhaust gases, with significant technical difficulty. The scrubbed CO_2 would then be liquefied and pumped to a depth of about 500 ft in the ocean, where it would then dissolve and become available for plankton and ocean vegetation. The cost of such an operation would likely be close to $1000/kW for scrubbing and disposal. There would also be a 25% energy penalty, because the energy for the scrubbing and disposal would not be available for other end users, as well as a 22% capacity penalty and additional annual maintenance costs.

Supply- and demand-side efficiency improvements are estimated to cost in the range of $20 per ton of carbon avoided. Automobile efficiency increases also save 22 lb of CO_2 per gallon not burned, with an estimated cost of $0.53 per gallon for the effort required to increase automobile efficiency to 44 miles per gallon, assuming continued use of gasoline as a fuel [12].

It is estimated that the United States would require 1.5 billion hectares of forest to absorb its annual CO_2 emissions [12]. The problem is that the total land area of the United States is only 913 million hectares, and forest in Death Valley, California, and quite a few other areas appear unlikely prospects.

The idea of planting trees to mitigate the effects of CO_2 generation was first attempted in a project proposed by Applied Energy Service. This project involved planting approximately 50 million trees in Costa Rica to mitigate the CO_2 generated by a fossil fuel electric generator in Connecticut. The project also involved a fire protection program to save 2400 ha of forest from fires, with an overall estimated sequestering of 387,000 t of carbon per year [12].

When trees are used for CO_2 mitigation, it is necessary to consider what will be done with the trees after they mature. If they are used for firewood, then the sequestered carbon is returned to its CO_2 form. If they are used for construction, then the carbon remains sequestered. However, even if they are burned, on the assumption that something will be burned, the burning of these trees displaces burning of other trees. If the trees spared from fire are genetically diverse, then an additional value can be assigned to preservation of diversity in the biosphere.

Perhaps an even more valuable location for trees is the urban environment, particularly in tropic or near-tropic regions. Hodas [12] observes that by planting trees near buildings, the microclimate of the buildings is cooled by several degrees, thus reducing air conditioning needs and the corresponding energy required for cooling. If the energy for cooling comes from fossil-fueled generation, then the trees reduce atmospheric CO_2 directly as a result of their photosynthesis as well as indirectly by reducing the output demands and corresponding fuel requirements of the fossil-fueled generator.

9.3.5.3 The Clean Power Estimator

The PVWatts Web site also makes available the Clean Power Estimator, which has been licensed by the California Energy Commission from the Clean Power Institute for use in California [13]. Fortunately, Kyocera has made the Clean Power Estimator available throughout the United States via the Kyocera Web site [14]. The Clean Power Estimator estimates the annual amount of CO_2 that can be avoided as a result of using PV-generated electricity in place of electricity from fossil-fuel sources. The estimation procedure follows EPA guidelines. The program also produces an estimated net annual cost of a system on the basis of information keyed in by the person running the program. Parameters used in the calculation include local electrical rates, income taxes, incentive programs, weather data, financing details, and utility load profiles.

The net annual return on investment is calculated by taking the difference between loan payments and tax and utility bill savings. While the program does not take into account an economic value for the CO_2 avoided, if an economic cost for CO_2 were to be established, the program could readily include such cost toward the annual net cost of the PV system.

At the time of this writing, NO_2 and SO_2 credits can now be purchased in the United States and carbon credits can also be purchased or sold in the form of renewable energy credits, as discussed in Chapter 8. If these costs, along with particulates and other air and water pollutants, could be included in the Clean Power Estimator, PV installations would more than likely become more economically attractive than most other energy sources. In fact, as the cost of conventional sources continues to increase and the cost of PV continues to decrease, it is highly likely that before this book is out of print the cost of PV, taking all externalities into consideration, will fall below the cost of present conventional energy sources.

The Clean Power Estimator has existed since 1994, and, because it is very labor intensive to keep it up to date, its survival as an important Web tool depends on licensing fees. If the reader happens to go to the reference at the end of this chapter and finds that it no longer brings up the tool, the reader is encouraged to continue searching for it by keying in *Clean Power Estimator.*

9.3.5.4 Attainment Levels as Commodities

Another interesting method of assigning value to pollutants comes as a consequence of the Clean Air Act [4] and regulated attainment levels. Companies are granted a certain allowable number of pollution units for their facilities. For example, an electrical generating utility may be allowed a certain number of tons of SO_2 emissions per year. If the utility does not generate as many tons as allowed, it may transfer the remainder of its allowed emissions to another entity. In a free enterprise society, this is not normally done as a simple, friendly gesture. Rather, air quality points are traded on the commodity exchange.

It is now possible for an individual, an organization, or a company to purchase allowances for atmospheric SO_2 and NO_2 [15]. At the end of March each year, the U.S. Environmental Protection Agency holds an auction for SO_2 allowances. Anyone can bid on the allowances, and the highest bidder gets the allowance. It is also possible to purchase allowances through a commodity broker or various environmental groups that purchase the allowances and then do not use them. Not using the allowances results in cleaner air. It is also possible to purchase NO_2 allowances from brokers or from environmental groups, but they are not currently auctioned off by the EPA.

While this practice does not affix a price to emissions based on specific environmental costs, it at least shows that emissions do have some economic cost, and the reduction of emissions can result in economic benefit. What is interesting is that the law of supply and demand dictates the price of air quality points. If strict laws are enacted to reduce pollutants, companies end up investing in pollution control technologies, such as electrostatic precipitators and scrubbers. If these investments result in sufficient reduction in emission of pollutants, then a smaller number of air quality points will be needed and a larger number will be available. This means the price will go down.

Other related actions have also been offered by energy producers in trade for permission to operate, such as improving navigation channels to reduce the likelihood of accidental fuel spills. These added costs represent the attachment of a price to certain externalities, and, as a result, end up incorporating the costs of the externalities into the cost of the product.

9.3.5.5 Subsidies

Subsidies were included as components of direct costs of doing business in Chapter 8, because they are usually included in the reduction of the cost of a product, such as energy, to the consumer. However, unless a level playing field exists, the costs of two energy sources cannot be properly compared. A level playing field exists when all energy sources are subsidized equally on a kWh-to-kWh basis. The difficulty is the identification of all subsidies, because some are direct and others are indirect.

A direct subsidy, such as a depletion allowance, can be quantified in terms of dollars per kWh. An indirect subsidy, such as military presence, can also sometimes be quantified, but will normally not appear in the financial records of a company. The effects of favorable tax considerations and tariffs can also be quantified in either a direct or an indirect manner.

Government-sponsored research and government purchase of systems or rebates or tax credits on systems are also forms of subsidy. Generally, government involvement is justified on the basis that once the subsidized industry is able to compete without subsidy, it will generate sufficient revenues to result in sufficient taxes to reimburse the government for its initial investment. In any case, hardly an energy source exists that does not benefit to some extent from subsidies. To date, however, no one has requested a depletion allowance on sunlight, although solar access rights have been challenged in the case where one party shades another party's solar system.

9.3.6 Health and Safety as Externalities

Recent litigation against tobacco companies, resulting in the award of billions of dollars in damages to states to help reimburse the costs of tobacco-related illnesses, has shown that the cost of public health is becoming a recognized externality, with a significant price tag. Other lawsuits also have been successfully prosecuted against a variety of forms of pollution resulting from careless or indiscriminate disposal of many forms of toxic wastes. These cases confirm the linkage between public health and its economic cost.

Public safety is also at issue when energy sources are considered. The Chernobyl and Three Mile Island nuclear accidents both brought to the forefront public concern over nuclear safety. The location of large fuel tanks in highly populated areas also is of concern to public advocacy groups. Transport of various fuels is also problematic, especially when large tanker trucks are involved in accidents that result in spilling of their contents and subsequent major efforts to clean up the damage before aquifers are affected. Even in the case of renewable energy sources, concern has been expressed about the possibility of construction accidents, because installation of renewable sources tends to be more labor intensive per installed kW than conventional sources. Manufacture of the materials used in conventional generation facilities carries with it both energy costs and materials costs, along with certain levels of exposure to hazardous or toxic materials. Similar costs are associated with the manufacture of photovoltaic cells and system components. Although these costs tend to be incorporated into the production cost of the materials, when toxic waste by-products result, they are often treated as externalities.

The rest of this chapter explores the externalities associated with the production, deployment, operation, and decommissioning of PV power systems. Particular attention is paid to the environmental concerns associated with each of the phases, recognizing that in order to produce PV systems, initially other energy sources must be exploited.

9.4 EXTERNALITIES ASSOCIATED WITH PV SYSTEMS

9.4.1 Environmental Effects of PV System Production

The production phase of each PV technology can be described in terms of a production cycle, which includes potential pollutants or hazardous waste associated with each production step. Table 9.4 summarizes areas of environmental or health concern for current PV technologies that have promising possibilities for large-scale production. The associated CO_2 for each technology represents the CO_2 resulting

Table 9.4 Environmental Concerns Associated with PV System Production

Technology	Concern	Relative Significance[a]	Relative Control Cost[b]	Control Strategy
All	Mining	Low	Lower	MSA[c]
All	Cleaning solvents	High	Higher	Recycle
All	Steel	Medium	Lower	OSHA, EPA
All	Aluminum	Medium	Lower	OSHA, EPA
All	Concrete	Medium	Lower	OSHA, EPA
All	Glass/breakage	Low	Lower	n/a
All	Encapsulants	Low	Lower	n/a
Crystalline Si	CO_2	0.02[d]	Lower	EPA
Multicrystalline Si	CO_2	0.01	Lower	EPA
Thin-film Si	CO_2	0.005	Lower	EPA
Other thin films	CO_2	0.004	Lower	EPA
Future cells	CO_2	0.002	Lower	EPA
All Si cells	Silica dust	Low	Lower	OSHA, EPA
All Si cells	CH_4	Low	Lower	Confine and recycle (C and R)
All Si cells	B_2H_6	Low	Lower	C and R
All Si cells	PH_3	Low	Lower	C and R
All Si cells	AsH_3	Low	Lower	C and R
CIS	H_2Se	Low	Lower	C and R
CIS	Cd[e]	Low	Lower	C and R
CIS	Fire (Cd)	Low	Lower	None
CdTe	Cd and CdO	Low	Lower	C and R
CdTe	Te	Low	Lower	C and R
CdTe	Fire (Cd, Te)	Low	Lower	None

[a] Compared to amount produced from equivalent amount of energy from coal-fired generation. Some items are not present in coal-fired, but are compared to other by-products of coal-fired generation such as SO_2, NO_2 and particulates.
[b] Generally lower control costs because generally less to control.
[c] Mine Safety Administration.
[d] These numbers represent ratios of CO_2 to produce a kWh from PV over the system lifetime to CO_2 per kWh from burning coal.
[e] It is estimated that coal-fired stack emissions produce as much Cd per kWh as is contained in CIS per lifetime kWh. CIS Cd is recycled at the end of CIS life cycle.

from the production of the primary energy used to produce the PV systems, so, in effect, it represents an energy cost of production. In the future, it is conceivable that energy for PV production will come from PV sources, thus further reducing the CO_2 and other pollutants associated with the production of PV cells. This concept is feasible because current PV systems, including cells, module materials, and balance of system, will produce, over their useful lifetime, between 8 and 10 times the energy used in their fabrication [16]. It is expected that new technologies and processing techniques will lead to energy returns on investment ranging between 14 and 27.

Aside from the energy cost, each technology has its own specific toxic waste areas of concern. Some concerns are common to all technologies, such as production of the materials for support structures and encapsulants for the modules. Perhaps the most important common denominator associated with all technologies is the material used for cleaning the cells. In all cases, high levels of cleanliness are required to maximize performance, and in some cases the solvents used are highly toxic. Fortunately, these solvents have been in common use in the semiconductor industry and careful means of control are well understood within the industry. The common means of dealing with these solvents is to confine them and then recycle them.

In the case of Si PV cells, the most significant concerns are the proper handling of the dopants for n-type and p-type. Because the semiconductor industry has been handling these components effectively, the processes are well defined. Once the impurities are in the Si, the amounts are so insignificant that the doped Si is considered to be benign.

For the CIS cells and the CdTe cells, the Cd content is the primary item of concern, but because the films in these devices are generally less than 2 μm thick, the total amount of Cd and Te in these cells is very small. Another concern for CIS is the use of H_2Se for the deposit of Se into the cell, but again, the use of this highly toxic substance can be kept well under control with minimal risk of having it escape into the environment.

9.4.2 Environmental Effects of PV System Deployment and Operation

The deployment of PV systems has associated environmental and health costs similar to the deployment of other energy technologies [17]. Steel, aluminum, and concrete can be expected to be part of the structures on which the PV arrays and associated BOS components will be mounted. Hence, the production costs and associated environmental costs, such as CO production in the reduction of iron oxides, become associated with PV deployment.

Just as construction accidents occur in any construction project, it is anticipated that PV project construction will also result in construction accidents. For the most part, materials used in construction of PV facilities are nontoxic. It is thus the responsibility of the installation contractor to ensure that the workplace and work practices follow Occupational Safety and Health Administration (OSHA) rules.

Once the PV system is installed, it quietly generates pollution-free electricity any time the sun is shining. The main environmental concern associated with the operation of PV systems is if a system containing Cd or Se or Te should be exposed to fire. Analysis has shown, however, that because these materials appear in such minute quantities in thin-film PVs, anyone approaching close enough to encounter significant exposure to any of the toxins would face significantly more danger from the fire itself. For example, only 400 g of Cd are present in a 1 MW CIS PV system, whereas 5 g/m^2 of Cd are present in CdTe systems, or about 25 times the amount in CIS systems.

Other dangers associated with the installed system include, of course, the possibility of electrical shock or fire. Systems installed in accordance with the *National Electrical Code* [18] will reduce any shock or fire hazard to a minimum, both for the general public as well as maintenance personnel. The fact that PV systems are inherently current limiting tends to reduce their potential to produce high-current arcing if they are inadvertently shorted. Incorporation of ground fault protection in systems adds further protection.

A final safety consideration for installed PV systems relates to their ability to withstand high winds. Inadequately mounted PV modules in high winds may tear loose from their mounting and become projectiles. Installations compliant with local building code wind-loading requirements present minimal risk.

9.4.3 Environmental Effects of PV System Decommissioning

Certain regulatory requirements apply to the decommissioning of PV systems. Assuming that the system will be disassembled at the end of its useful lifetime, it is then necessary to consider the destination of the disassembled components. Some of the components will require disposal according to toxic waste regulations unless they can be recycled, and other components will not be subject to quite as stringent requirements.

The Resource and Conservation Recovery Act constitutes the primary set of rules governing wastes containing Cd, Se, Pb, Cu, or Ag, provided that these wastes are considered to be discarded material and are not included in any specific exclusions [19]. The EPA defines the Toxicity Characteristic Leaching Procedure (TCLP) for 39 materials to determine whether waste products containing them may be classified as hazardous. Any waste product containing any of the 39 materials that yields a soluble concentration in excess of its TCLP limits is considered hazardous.

The California Hazardous Waste Control Law introduces an additional test for toxicity characterization, with two additional indicators. The Waste Extraction Test (WET) is used to identify non-RCRA toxic wastes. The WET threshold limits are defined as the Soluble Threshold Limit Concentration (STLC) and the Total Threshold Limit Concentration (TTLC), which is the total concentration of listed materials, in any form.

In terms of STLC and TTLC, it appears that CdTe modules may be characterized as hazardous in terms of TTLC and STLC for Cd. CIS modules may be subject to TTLC for Se and STLC for Se and Cd. Polycrystalline Si modules may have a problem with TTLC and STLC for Ag or Pb. In some cases, Pb and Cu may be a problem, and a-Si:H modules do not seem to have any toxicity problems [19,20]. If CdTe and CIS modules are recycled, they are exempt from the disposal regulations.

Studies have projected recycling costs of approximately \$0.01 to \$0.04 per watt for both materials, using relatively standard separation techniques.

Problems

9.1 Look up the U.S. nonattainment regions on the U.S. Environmental Protection Agency Web site.

9.2 If recycling of CdTe PV modules will cost approximately \$0.04/W and the lifetime of a CdTe module is 20 years, what is the present value of the recycling cost for a 10 kW array?

9.3 Determine the prices of a ton of SO_2 and a ton of NO_2 on the commodity exchange. The EPA Web site lists Web sites and phone numbers of commodity brokers and environmental organizations that purchase and sell them.

9.4 If burning a gallon of gasoline produces 22 lb of CO_2, estimate the CO_2 that your vehicle emits in a year. Compare this with the amount that would have been generated had you been driving an electric vehicle charged from a PV array or a hybrid auto or a fuel-cell-powered auto. Take into account the origin of the H_2 used in the fuel cell.

9.5 In the CO_2 scrubbing and liquefaction scheme reported by Hodas, estimate the increase in electrical cost that would be necessary to offset the cost of the process.

9.6 Use the Clean Power Estimator [14] to determine for your residential zip code area:

a. A combination of system cost per PV watt (at STC), interest rate, and loan duration that will result in a positive net annual cash flow for a PV system owner of your income level and marital status.

b. Assume the cost of a PV system to be \$3/W and the annual interest rate on a 25-year loan for the PV system to be 6%. Determine the necessary value of a ton of avoided CO_2 to bring the annual system net cost to zero for your personal income and tax situation (or your hypothetical income and tax situation).

REFERENCES

1. Moore, T. G., *Climate of Fear: Why We Shouldn't Worry about Global Warming*, the Cato Institute, Washington, D.C., 1998.
2. Paterson, M., *Global Warming and Global Politics*, Routledge, London, 1996.
3. Baumann, A. and Hill, R., *Proc. 10th EC Photovoltaics Solar Energy Conference*, Kluwer, Dordrecht, 1991, 834–837.
4. 42 U.S.C. § 7401 et seq. (Clean Air Act).
5. U.S. Environmental Protection Agency data on current and historic levels of pollutants regulated by the Clean Air Act www.epa.gov/airtrends.
6. 3rd Conference of the Parties to the U.N. Framework Convention on Climate Change, Kyoto, Japan, December 1997.
7. World Resources Institute data on economic cost of pollutants http://www.wri.org/.
8. http://siteresources.worldbank.org/DATASTATISTICS/Resources/table3_13.pdf (World Bank pollution data).
9. Battisti, R. and Corrado, A., Evaluation of technical improvements of photovoltaic systems through life cycle assessment methodology, *Energy* 30 (2005), p. 959.

10. Guderian, R., *External Environmental Costs of Electric Power*, Springer-Verlag, Berlin, 1990, 9–24.
11. Weltschev, M., *External Environmental Costs of Electric Power*, Springer-Verlag, Berlin, 1990, 25–35.
12. Hodas, D. R., *External Environmental Costs of Electric Power*, Springer-Verlag, Berlin, 1990, 59–78.
13. The PVWatts Solar Calculator, National Renewable Energy Laboratory, Golden, CO, 2008, http://www.pvwatt.com/.
14. Access to Clean Power Estimator for locations outside California http://kyocerasolar.cleanpowerestimator.com/default.aspx.
15. Information on SO_2 and NO_2 commodity trading. http://www.epa.gov/airmarkt/trading/buying.html.
16. Information on energy return on investment for current and new PV technologies http://www.nrel.gov/docs/fy04osti/35489.pdf.
17. Markvart, T., Ed., *Solar Electricity*, John Wiley & Sons, Chichester, U.K., 1994.
18. *NFPA 70 National Electrical Code, 2008 ed.*, National Fire Protection Association, Quincy, MA, 2008.
19. Fthenakis, V. M. and Moskowitz, P. D., Emerging photovoltaic technologies: environmental and health issues update, *NREL/SNL Photovoltaics Program Review*, AIP Press, New York, 1997.
20. Fthenakis, V.M., Kim, H.C., and Alsema, E., Emissions from photovoltaic life cycles, *Environ. Sci. Technol.* 42(6), pp. 2168–2174, March 2008.

CHAPTER 10

The Physics of Photovoltaic Cells

10.1 INTRODUCTION

By this time, it is possible that the reader, now highly skilled at photovoltaic (PV) system design, would be interested in what goes on inside the PV cell during the process of converting light energy into electrical energy. This chapter is presented with the purpose of enabling the reader to become familiar with the challenges facing those engineers and physicists who spend their lives working on processes and materials aimed at reducing the cost and increasing the efficiency of PV cells. Initially, the basics of PV energy conversion are presented, followed by discussion of present limitations of cell production and some of the ideas that have emerged toward overcoming these limitations. Most readers will have already had at least an exposure to the theory of operation of the pn junction and the semiconductor diode in an electronics course, so it will be assumed that the reader will at least be familiar with the basic diode equation.

10.2 OPTICAL ABSORPTION

10.2.1 Introduction

When light shines on a material, it is either reflected, transmitted, or absorbed. Absorption of light is simply the conversion of the energy contained in the incident photon to some other form of energy, typically heat. Some materials, however, happen to have just the right properties needed to convert the energy in the incident photons to electrical energy.

When a photon is absorbed, it interacts with an atom in the absorbing material by giving off its energy to an electron in the material. This energy transfer is governed by the rules of conservation of momentum and conservation of energy. Because the zero-mass photon has very small momentum compared to the electrons and holes, the transfer of energy from a photon to a material occurs with inconsequential momentum transfer. Depending on the energy of the photon, an electron may be raised to a higher energy state within an atom or it may be liberated from the atom.

Liberated electrons are then capable of moving through the crystal in accordance with whatever phenomena may be present that could cause the electron to move, such as temperature, diffusion, or an electric field.

10.2.2 Semiconductor Materials

Semiconductor materials are characterized as being perfect insulators at absolute zero temperature, with charge carriers being made available for conduction as the temperature of the material is increased. This phenomenon can be explained on the basis of quantum theory, by noting that semiconductor materials have an energy band gap between the valence band and the conduction band. The valence band represents the allowable energies of valence electrons that are bound to host atoms. The conduction band represents the allowable energies of electrons that have received energy from some mechanism and are now no longer bound to specific host atoms.

At $T = 0$ K, all allowable energy states in the valence band of a semiconductor are occupied by electrons, and no allowable energy states in the conduction band are occupied. Because the conduction process requires that charge carriers move from one state to another within an energy band, no conduction can take place when all states are occupied or when all states are empty. This is illustrated in Figure 10.1a.

As the temperature of a semiconductor sample is increased, sufficient energy is imparted to a small fraction of the electrons in the valence band for them to move to the conduction band as indicated in Figure 10.1b. In effect, these electrons are leaving covalent bonds in the semiconductor host material. When an electron leaves the valence band, an opening is left that may now be occupied by another electron, provided the other electron moves to the opening. If this happens, of course, the electron that moves in the valence band to the opening, leaves behind an opening in the location from which it moved. If one engages in an elegant quantum mechanical explanation of this phenomenon, it must be concluded that the electron moving in the

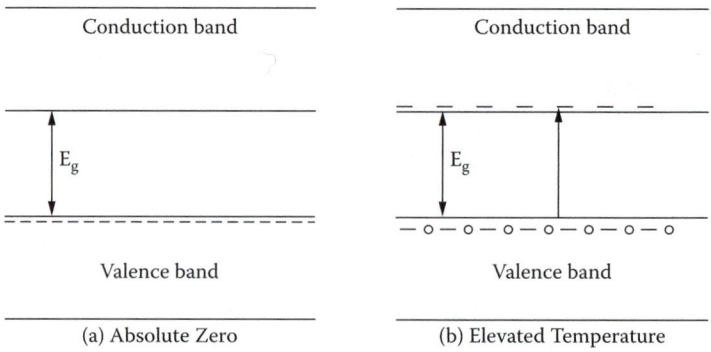

Figure 10.1 Illustration of availability of states in valence band and conduction band for semiconductor material.

valence band must have either a negative effective mass and a negative charge, or, alternatively, a positive effective mass and a positive charge. The latter has been the popular description, and, as a result, the electron motion in the valence band is called hole motion, where *holes* is the name chosen for the positive charges, because they relate to the moving holes that the electrons have left in the valence band.

What is important to note about these conduction electrons and valence holes is that they have occurred in pairs. Hence, when an electron is moved from the valence band to the conduction band in a semiconductor by whatever means, it constitutes the creation of an electron–hole pair (EHP). Both charge carriers are then free to become a part of the conduction process in the material.

10.2.3 Generation of EHP by Photon Absorption

The energy in a photon is given by the familiar equation,

$$E = h\upsilon = \frac{hc}{\lambda} \text{ (joules)}, \tag{10.1}$$

where h is Planck's constant ($h = 6.63 \times 10^{-34}$ J-s), c is the speed of light ($c = 2.998 \times 10^8$ m/s), ν is the frequency of the photon in Hz, and λ is the wavelength of the photon in meters. Because energies at the atomic level are typically expressed in electron volts (1 eV = 1.6×10^{-19} J) and wavelengths are typically expressed in micrometers, it is possible to express hc in appropriate units so that if λ is expressed in μm, then E will be expressed in eV. The conversion yields

$$E = \frac{1.24}{\lambda} \text{ (eV)}. \tag{10.1a}$$

The energy in a photon must exceed the semiconductor bandgap energy, E_g, to be absorbed. Photons with energies at and just above E_g are most readily absorbed because they most closely match bandgap energy and momentum considerations. If a photon has energy greater than the bandgap, it still can produce only a single EHP. The remainder of the photon energy is lost to the cell as heat. It is thus desirable that the semiconductor used for photoabsorption have a bandgap energy such that a maximum percentage of the solar spectrum will be efficiently absorbed.

Now, referring back to the Planck formula for blackbody radiation in Chapter 2, Equation (2.1), note that the solar spectrum peaks at $\lambda \approx 0.5$ μm. Equation (10.1a) shows that a bandgap energy of approximately 2.5 eV corresponds to the peak in the solar spectrum. In fact, because the peak of the solar spectrum is relatively broad, bandgap energies down to 1.0 eV can still be relatively efficient absorbers, and in certain special cell configurations to be discussed later, even smaller bandgap materials are appropriate.

The nature of the bandgap also affects the efficiency of absorption in a material. A more complete representation of semiconductor bandgaps must show the relationship between bandgap energy and bandgap momentum. As electrons make

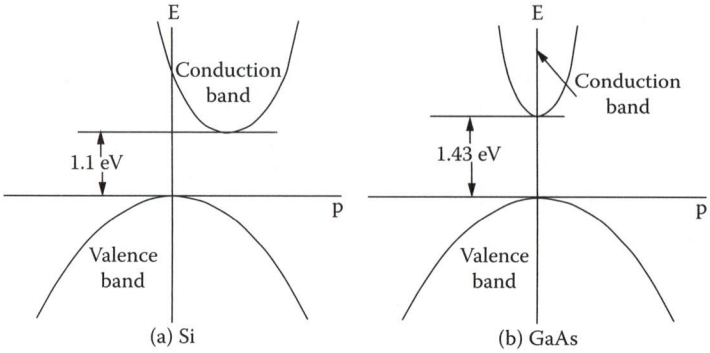

Figure 10.2 The energy and momentum diagram for valence and conduction bands in Si and GaAs.

transitions between conduction band and valence band, both energy and momentum transfer normally take place, and both must be properly balanced in accordance with conservation of energy and conservation of momentum laws.

Some semiconducting materials are classified as direct bandgap materials, while others are classified as indirect bandgap materials. Figure 10.2 shows the bandgap diagrams for two materials, considering momentum as well as energy. Note that for silicon, the bottom of the conduction band is displaced in the momentum direction from the peak of the valence band. This is an indirect bandgap, while the GaAs diagram shows a direct bandgap, where the bottom of the conduction band is aligned with the top of the valence band.

What these diagrams show is that the allowed energies of a particle in the valence band or the conduction band depend on the particle momentum in these bands. An electron transition from a point in the valence band to a point in the conduction band must involve conservation of momentum as well as energy. For example, in Si, even though the separation of the bottom of the conduction band and the top of the valence band is 1.1 eV, it is difficult for a 1.1 eV photon to excite a valence electron to the conduction band because the transition needs to be accompanied by sufficient momentum to cause displacement along the momentum axis, and photons carry little momentum. The valence electron must thus simultaneously gain momentum from another source as it absorbs energy from the incident photon. Because such simultaneous events are unlikely, absorption of photons at the Si bandgap energy is several orders of magnitude less likely than absorption of higher-energy photons.

Because photons have so little momentum, it turns out that the direct bandgap materials, such as gallium arsenide (GaAs), cadmium telluride (CdTe), copper indium diselenide (CIS), and amorphous silicon (aSi:H) absorb photons with energy near the material bandgap energy much more readily than do the indirect materials, such as crystalline silicon. As a result, the direct bandgap absorbing material can be several orders of magnitude thinner than indirect bandgap materials and still absorb a significant part of the incident radiation.

The absorption process is similar to many other physical processes, in that the change in intensity with position is proportional to the initial intensity. As an equation, this becomes

$$\frac{dI}{dx} = -\alpha I, \tag{10.2}$$

with the solution

$$I = I_o e^{-\alpha x}, \tag{10.3}$$

where I is the intensity of the light at a depth x in the material, I_o is the intensity at the surface, and α is the absorption constant. The absorption constant depends on the material and on the wavelength. Equation (10.3) shows that the thickness of material needed for significant absorption needs to be several times the reciprocal of the absorption constant. This is important information for the designer of a PV cell, because the cell must be sufficiently thick to absorb the incident light. In some cases, the path length is increased by causing the incident light to reflect from the front and back surfaces while inside the material until it ultimately generates an EHP. Figure 10.3 shows the dependence of absorption constant on wavelength for several materials. Observe that at energies below the bandgap energy, no absorption takes place. The material is transparent to these low-energy photons. At energies above the bandgap, the absorption constant increases relatively slowly for indirect bandgap semiconductors and increases relatively quickly for direct bandgap materials.

In any case, when the photon is absorbed, it generates an EHP. The question, then, is what happens to the EHP?

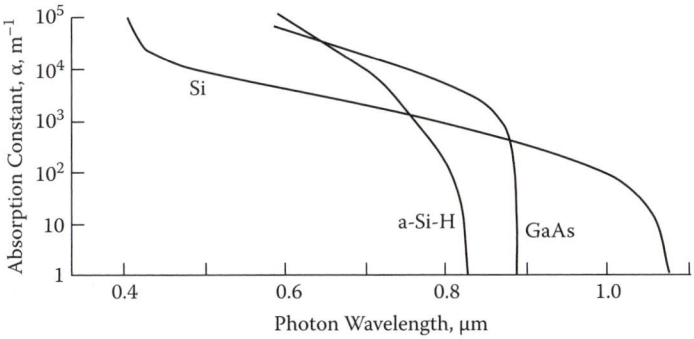

Figure 10.3 Dependence of absorption constant on wavelength for several semiconductors. (Adapted from Yang, E. S., *Microelectronic Devices*, 1988, McGraw-Hill.)

10.2.4 Photoconductors

Once an EHP is generated, it becomes a question of how long the EHP lasts before the conduction electron returns to the valence band. Remember that the creation of an EHP does not imply that the electron will remain in the conduction band and the hole will remain in the valence band. Thermal equilibrium in a semiconductor comprises a constant generation and recombination of EHPs, so that, on the average, the population of electrons and the population of holes remain constant. In fact, the product of the concentration of holes and the concentration of electrons in thermal equilibrium is a constant, which depends on temperature and the bandgap energy of the semiconductor, along with a few other parameters unique to each semiconductor. If n_o represents the thermal equilibrium concentration of electrons per cm^3 and if p_o represents the thermal equilibrium concentration of holes per cm^3, then, in thermal equilibrium, with n_o and p_o at any single location in the material,

$$n_o p_o = n_i^2(T) = KT^3 e^{-\frac{E_g}{kT}}, \tag{10.4}$$

where n_i represents the intrinsic carrier concentration; K is semiconductor-material dependent, as will be discussed later; E_g is the bandgap energy; k is Boltzmann's constant; and T is the temperature in K. Because of the temperature dependence of n_i, it is customary to plot n_i versus $1000/T$ as in Figure 10.4, which shows how n_i varies with temperature for several semiconductor materials. Note that materials with smaller bandgap energies have higher intrinsic carrier concentrations. Table 10.1 shows bandgap energies and several other properties of some common semiconductors used in PV cells.

Electrons and holes generated as a result of optical absorption bring the material into a state of nonthermal equilibrium. The photon-generated excess electrons and

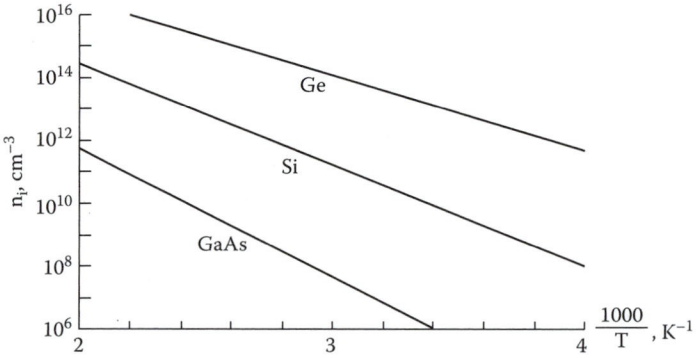

Figure 10.4 Temperature dependence of intrinsic carrier concentration for several common semiconductors. (Adapted from Streetman, B., *Solid State Electronic Devices*, 4th ed., 1995, Prentice-Hall.)

Table 10.1 Properties of Several Common Semiconductors at Room Temperature

Material	E_g (eV)	n_i (cm^{-3})	μ_n (cm^2/Vs)	μ_p (cm^2/Vs)	ε_r	Melting Point (°C)
Si	1.11	1.5×10^{10}	1350	480	11.8	1415
Ge	0.67	2.5×10^{13}	3900	1900	16	936
GaAs	1.43	2×10^6	8500	400	13.2	1238
CdS	2.42		340	50	8.9	1475
CdTe	1.48		1050	100	10.2	1098

Sources: Based on data from Yang, *Microelectronic Devices*, McGraw-Hill, New York, 1988; Streetman, *Solid State Electronic Devices*, 4th ed., Prentice-Hall, Englewood Cliffs, NJ, 1995.

holes remain, on the average, for a time, τ, which is defined as the excess carrier lifetime. The recombination of EHPs is a statistical process in which some EHPs recombine in times shorter than the carrier lifetime, and some take longer to recombine.

EHPs, then, are subject to a generation rate, measured in number per cm^3 per second, which is proportional to the incident photon flux, and are subject to a recombination rate that is proportional to the departure of np from its equilibrium value. Under steady-state conditions, these two rates are equal.

The conductivity of a material is proportional to the density of free charge carriers, and is given by

$$\sigma = q(\mu_n n + \mu_p p), \tag{10.5}$$

where μ_n and μ_p represent, respectively, the mobility of electrons and the mobility of holes in the material. Mobility is simply a measure of how easily the particles can move around in the material when an electric field is present. Mobilities are included in the material characteristics presented in Table 10.1.

Equation (10.5) clearly indicates that, if shining a light on a piece of material can create excess charge carriers, then the conductivity of the material will increase, causing a corresponding decrease in the resistance of the material. The material becomes a light-sensitive resistor. A common material used for such devices is cadmium sulfide (CdS), which is used in many of the light sensors that turn lights on after dark. The most sensitive photoconductors are materials that have long lifetimes of excess EHPs. However, if the problem is to detect short bursts of light occurring at a high repetition rate, it is necessary to have any excess population of electrons and holes quickly die out as soon as the light source is removed. The trade-off between speed and sensitivity is similar to the trade-off between bandwidth and gain for an amplifier as described by the familiar gain-bandwidth product.

Note that even though EHPs are generated in the host material, the material remains passive because the generated EHPs have random thermal velocities. This means that no net current flow results from their creation and, because no separation of charges occurs, no voltage is produced. With no resulting voltage or current, the only effect of the creation of the additional charge carriers is the reduction in resistance of the host material. The next step, then, is to figure out a way to get work out of these photon-generated charges.

10.3 EXTRINSIC SEMICONDUCTORS AND THE PN JUNCTION

10.3.1 Extrinsic Semiconductors

Up to this point, the semiconductors discussed have been intrinsic semiconductors, meaning that the populations of holes and electrons have been equal. A somewhat more formal definition of an intrinsic semiconductor takes into account differences in electron and hole mobilities and defines intrinsic semiconductors as materials for which the Fermi level energy is at the center of the bandgap. Because Fermi levels have not been discussed, and because electron and hole mobilities are generally close enough to keep the Fermi level quite close to the center of the bandgap, the equal carrier definition will suffice for the following discussion. The reader is referred to a text on semiconductor device physics for more details on Fermi levels [1,2].

At $T = 0$ K, intrinsic semiconductors have all covalent bonds completed with no leftover electrons or holes. If certain impurities are introduced into intrinsic semiconductors, there can be leftover electrons or holes at $T = 0$ K. For example, consider silicon, which is a group IV element, which covalently bonds with four nearest neighbor atoms to complete the outer electron shells of all the atoms. At $T = 0$ K, all the covalently bonded electrons are in place, whereas at room temperature, about 1 in 10^{12} of these covalent bonds will break, forming an EHP, resulting in minimal charge carriers for current flow.

If, however, phosphorus, a group V element, is introduced into the silicon in small quantities, such as 1 part in 10^6, four of the valence electrons of the phosphorus atoms will covalently bond to the neighboring silicon atoms, while the fifth valence electron will have no electrons with which to covalently bond. This fifth electron remains weakly coupled to the phosphorus atom, readily dislodged by temperature, because it requires only 0.04 eV to excite the electron from the atom to the conduction band. At room temperature, sufficient thermal energy is available to dislodge essentially all of these extra electrons from the phosphorus impurities. These electrons thus enter the conduction band under thermal equilibrium conditions, and the concentration of electrons in the conduction band becomes nearly equal to the concentration of phosphorus atoms, because the impurity concentration is normally on the order of 10^8 times larger than the intrinsic carrier concentration.

Because the phosphorus atoms donate electrons to the material, they are called donor atoms and are represented by the concentration, N_D. Note that the phosphorus, or other group V impurities, do not add holes to the material. They only add electrons. They are thus designated as n-type impurities.

On the other hand, if group III atoms such as boron are added to the intrinsic silicon, they have only three valence electrons to covalently bond with the nearest silicon neighbors. The missing covalent bond appears the same as a hole, which can be released to the material with a small amount of thermal energy. Again, at room temperature, nearly all of the available holes from the group III impurity are donated to the conduction process in the host material. Because the concentration of impurities will normally be much larger than the intrinsic carrier concentration, the concentration of holes in the material will be approximately equal to the concentration of impurities.

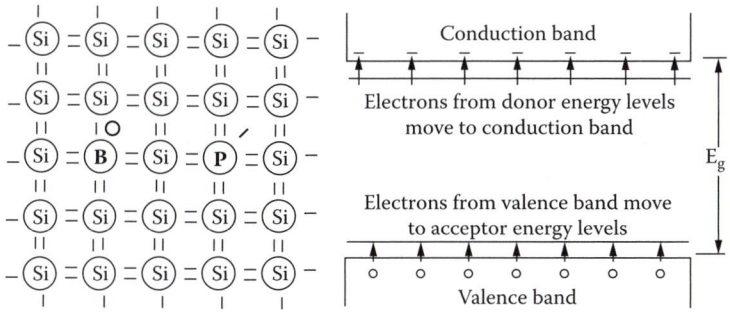

Figure 10.5 Acceptor and donor impurities in Si.

Historically, group III impurities in silicon have been viewed as electron acceptors, which, in effect, donate holes to the material. Rather than being termed hole donors, however, they have been called acceptors. Thus, acceptor impurities donate holes, but no electrons, to the material, and the resulting hole density is approximately equal to the density of acceptors, which is represented as N_A. Figure 10.5 shows the effects of donor and acceptor impurities on the intrinsic material along with the positions of the energy levels of the impurities in the bandgap of the material.

Equation (10.5) shows that adding a mere one part in a million of a donor or acceptor impurity can increase the conductivity of the material by a factor of 10^8 for silicon. Equation (10.4) shows that in thermal equilibrium, if either an n-type or a p-type impurity is added to the host material, that the concentration of the other charge carrier will decrease dramatically, because it is still necessary to satisfy Equation (10.4). In extrinsic semiconductors, the charge carrier with the highest concentration is called the majority carrier, and the charge carrier with the lowest concentration is called the minority carrier. Hence, electrons are majority carriers in n-type material, and holes are minority carriers in n-type material. The opposite is true for p-type material.

If both n-type and p-type impurities are added to a material, then whichever impurity has the higher concentration will become the dominant impurity. However, it is then necessary to acknowledge a net impurity concentration that is given by the difference between the donor and acceptor concentrations. If, for example, $N_D > N_A$, then the net impurity concentration is defined as $N_d = N_D - N_A$. Similarly, if $N_A > N_D$, then $N_a = N_A - N_D$.

10.3.2 The pn Junction

10.3.2.1 Drift and Diffusion

When charged particles are placed in an electric field, they are exposed to an electrostatic force. This force accelerates the particles until they undergo a collision with another component of the material that slows them down. They then accelerate once again and collide again. The process continues, with the net result that the charge carrier achieves an average velocity, either in the direction of the electric field for positive charges, or opposite the electric field for negative charges. It should be noted that this average, or *drift* velocity, is superimposed on the thermal velocity of the charge carrier. Normally the thermal velocity is much larger than the drift velocity, but the thermal velocity is completely random, so the net displacement of the charge carriers is zero. Another way to consider the thermal velocity is that at any instant, it is equally likely that a particle will be moving in any direction.

Drift current, then, is simply the component of current flow due to the presence of an electric field, and is described by the familiar equation,

$$\vec{J} = \sigma \vec{E}. \tag{10.6}$$

This is simply the vector form of Ohm's law, where J is the current density in A/cm^2, when σ is expressed in Ω^{-1}cm^{-1} and E is measured in V/cm.

Diffusion is that familiar process by which random thermal motion of particles causes them to ultimately distribute themselves uniformly within a space. Whenever particles are in thermal motion, they will tend to move from areas of greater concentration to areas of lesser concentration, simply because at any point, the probability of motion in all directions is equal. Suppose, for example, that regions A and B are adjoining, as shown in Figure 10.6. Suppose also that all particles in both regions are experiencing random thermal motion and that the concentration of certain particles, z, in region A is greater than the concentration of z-particles in region B. At any instant, half the z-particles in region A will be moving toward B, and half the z-particles in B will be moving toward A. Because there are more z-particles in A, the net motion of z-particles from A to B will continue until the concentrations in A and B are equal. If the particles are holes, this net movement from regions of greater

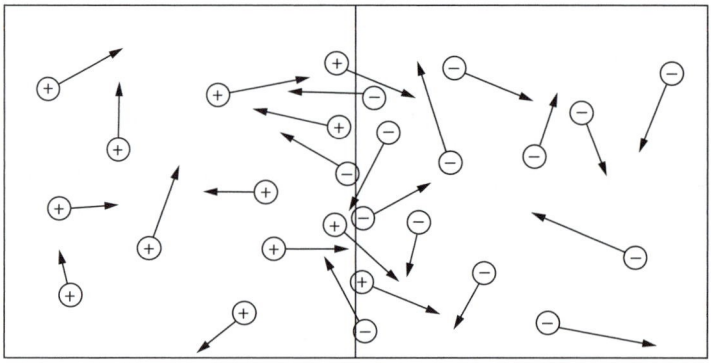

⊕= Hole ⊖= Electron

(a) More Holes Than Electrons (b) More Electrons Than Holes

Figure 10.6 Random thermal motion and diffusion for electrons and holes.

concentration to regions of lesser concentration constitutes a flow of current that can be described in one dimension by the equation

$$J_p = -qD_p \frac{dp}{dx}, \tag{10.7a}$$

and if the particles are electrons,

$$J_n = qD_n \frac{dn}{dx}. \tag{10.7b}$$

In Equations (10.7a) and (10.7b), the change in concentration with position is known as the concentration gradient, and D_p and D_n are the hole diffusion constant and the electron diffusion constant, respectively. The minus sign in Equation (10.7a) accounts for the fact that if the gradient is negative, the holes flow in the positive x-direction. The lack of the minus sign in Equation (10.7b) accounts for the fact that if electrons flow in the positive x-direction, the associated current is in the negative x-direction. In each case, $q = 1.6 \times 10^{-19}$ C.

At this point, all that is necessary is to make two additional observations. The first observation is that donor and acceptor atoms become donor and acceptor ions when they give up their electron or hole to the host material. These ions are fixed in position in the host by covalent bonds. The second observation is that in either n-type material or in p-type material, any point in the material will have charge neutrality. That is, the net charge present at any point is zero due to positive charges being neutralized by negative charges. However, when n-type and p-type materials are joined to form a pn junction, something special happens at the boundary.

10.3.2.2 Junction Formation and Built-In Potential

Although n-type and p-type materials are interesting and useful, the real fun starts when a junction is formed between n-type and p-type materials. The pn junction is treated in gory detail in most semiconductor device textbooks. Here, the need is to establish the foundation for the establishment of an electric field across a pn junction and to note the effect of this electric field on photo-generated EHPs.

Figure 10.7 shows a pn junction formed by placing p-type impurities on one side and n-type impurities on the other side. There are many ways to achieve this structure. The most common is the diffused junction.

To form a diffused pn junction, the host material is grown with impurities, so it will be either n-type or p-type. The material is either grown or sliced into an appropriate thickness. Then the material is heated in the presence of the opposite impurity, which is usually in the form of a gas. This impurity will diffuse into the host material at a level that exceeds the host impurity level, but will only penetrate a small distance into the host material, depending on how long the host material is left at the elevated temperature. The result is a layer of material of one dominant impurity on top of the remainder of the material, which is doped with the other dominant impurity.

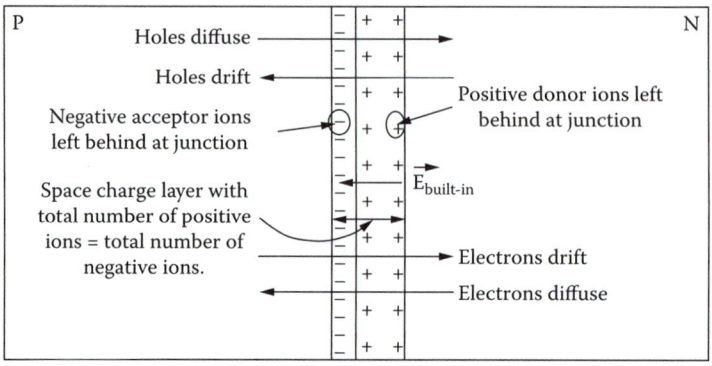

Figure 10.7 The pn junction showing electron and hole drift and diffusion.

When the two materials are brought together, the first thing to happen is that the conduction electrons on the *n*-side of the junction "notice" the scarcity of conduction electrons on the *p*-side, and the valence holes on the *p*-side "notice" the scarcity of valence holes on the *n*-side. Because both types of charge carrier are undergoing random thermal motion, they begin to diffuse to the opposite side of the junction in search of the wide open spaces. The result is diffusion of electrons and holes across the junction, as indicated in Figure 10.7.

When an electron leaves the *n*-side for the *p*-side, however, it leaves behind a positive donor ion on the *n*-side, right at the junction. Similarly, when a hole leaves the *p*-side for the *n*-side, it leaves a negative acceptor ion on the *p*-side. If large numbers of holes and electrons travel across the junction, large numbers of fixed positive and negative ions are left at the junction boundaries. These fixed ions, as a result of Gauss' law, create an electric field that originates on the positive ions and terminates on the negative ions. Hence, the number of positive ions on the *n*-side of the junction must be equal to the number of negative ions on the *p*-side of the junction.

The electric field across the junction, of course, gives rise to a drift current in the direction of the electric field. This means that holes will travel in the direction of the electric field, and electrons will travel opposite the direction of the field, as shown in Figure 10.7. Note that for both the electrons and for the holes, the drift current component is opposite the diffusion current component. At this point, one can invoke Kirchhoff's current law to establish that the drift and diffusion components for each charge carrier must be equal and opposite, because there is no net current flow through the junction region. This phenomenon is known as the law of detailed balance.

By setting the sum of the electron diffusion current and the electron drift current equal to zero and recalling from electromagnetic field theory that

$$E = -\frac{dV}{dx}, \tag{10.8}$$

it is possible to solve for the potential difference across the junction in terms of the impurity concentrations on either side of the junction. Proceeding with this operation yields

$$-q\mu_n n\frac{dV}{dx}+qD_n\frac{dn}{dx}=0, \tag{10.9a}$$

that can be rewritten as

$$dV=\frac{D_n}{\mu_n}\frac{dn}{n}. \tag{10.9b}$$

Finally, recognizing the Einstein relationship, $D_n/\mu_n = kT/q$, which is discussed in solid-state physics textbooks, and integrating both sides from the n-side of the junction to the p-side of the junction, yields the magnitude of the built-in voltage across the junction to be

$$V_j=\frac{kT}{q}\ln\frac{n_{no}}{n_{po}}. \tag{10.10a}$$

It is now possible to express the built-in potential in terms of the impurity concentrations on either side of the junction by recognizing that $n_{no} \cong N_D$ and $n_{po} \cong (n_i)^2/N_A$. Substituting these values into Equation (10.10a) yields, finally,

$$V_j=\frac{kT}{q}\ln\frac{N_A N_D}{n_i^2}. \tag{10.10b}$$

At this point, a word about the region containing the donor ions and accepter ions is in order. Note first that outside this region, electron and hole concentrations remain at their thermal equilibrium values. Within the region, however, the concentration of electrons must change from the high value on the n-side to the low value on the p-side. Similarly, the hole concentration must change from the high value on the p-side to the low value on the n-side. Considering that the high values are really high, that is, on the order of $10^{18}/\text{cm}^3$, while the low values are really low, that is, on the order of $10^2/\text{cm}^3$, this means that within a short distance of the beginning of the ionized region, the concentration must drop to significantly below the equilibrium value. Because the concentrations of charge carriers in the ionized region are so low, this region is often termed the depletion region, in recognition of the depletion of mobile charge carriers in the region. Furthermore, on account of the charge due to the ions in this region, the depletion region is also often referred to as the space charge layer. For the remainder of this text, this region will simply be referred to as the junction.

The next step in the development of the behavior of the pn junction in the presence of sunlight is to let the sun shine in and see what happens.

10.3.2.3 The Illuminated pn Junction

Equation (10.3) governs the absorption of photons at or near a pn junction. Noting that an absorbed photon releases an EHP, it is now possible to explore what happens after the generation of the EHP. Those EHPs generated within the pn junction will be considered first, followed by the EHPs generated outside, but near, the junction.

If an EHP is generated within the junction, as shown in Figure 10.8 (points B and C), both charge carriers will be acted upon by the built-in electric field. Because the field is directed from the *n*-side of the junction to the *p*-side of the junction, the field will cause the electrons to be swept quickly toward the *n*-side and the holes to be swept quickly toward the *p*-side. Once out of the junction region, the optically generated carriers become a part of the majority carriers of the respective regions, with the result that excess concentrations of majority carriers appear at the edges of the junction. These excess majority carriers then diffuse away from the junction toward the external contacts, because the concentration of majority carriers has been enhanced only near the junction.

The addition of excess majority charge carriers to each side of the junction results in either a voltage between the external terminals of the material, or a flow of current in the external circuit, or both. If an external wire is connected between the *n*-side of the material and the *p*-side of the material, a current, I_l, will flow in the wire from the *p*-side to the *n*-side. This current will be proportional to the number of EHPs generated in the junction region.

If an EHP is generated outside the junction region, but close to the junction (with "close" yet to be defined, but shown as point A in Figure 10.8), it is possible that, due to random thermal motion, either the electron, the hole, or both will end up moving into the junction region. Suppose, for example, that an EHP is generated in the *n*-region close to the junction. Then suppose the hole, which is the minority carrier in the *n*-region, manages to reach the junction before it recombines. If it can do this, it will be swept across the junction to the *p*-side, and the net effect will be the same as if the

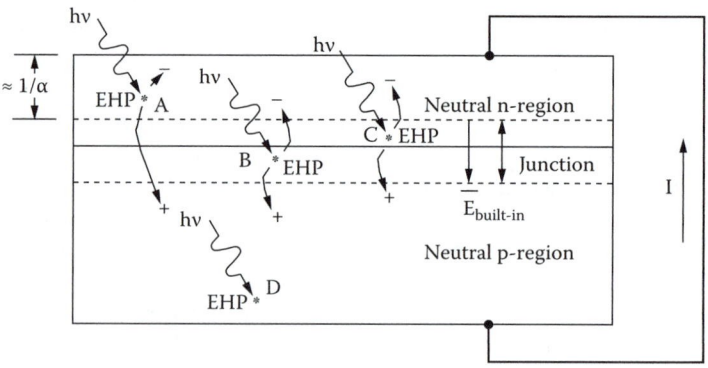

Figure 10.8 The illuminated pn junction showing desirable geometry and the creation of electron–hole pairs.

EHP had been generated within the junction, because the electron is already on the n-side as a majority carrier. Similarly, if an EHP is generated within the p-region, but close to the junction, and if the minority carrier electron reaches the junction before recombining, it will be swept across to the n-side, where it is a majority carrier. So what is meant by "close"?

Clearly, the minority carriers of the optically generated EHPs outside the junction region must not recombine before they reach the junction. If they do, then, effectively, both carriers are lost from the conduction process, as in point D in Figure 10.8. Because the majority carrier is already on the correct side of the junction, the minority carrier must therefore reach the junction in less than a minority carrier lifetime, τ_n or τ_p.

To convert these times into distances, it is necessary to note that the carriers travel by diffusion once they are created. Because only the thermal velocity has been associated with diffusion, but because the thermal velocity is random in direction, it is necessary to introduce the concept of minority carrier diffusion length, which represents the distance, on the average, that a minority carrier will travel before it recombines. The diffusion length can be shown to be related to the minority carrier lifetime and diffusion constant by the formula

$$L_m = \sqrt{D_m \tau_m}, \tag{10.11}$$

where m has been introduced to represent n for electrons or p for holes. It can also be shown that on average, if an EHP is generated within a minority carrier diffusion length of the junction, the associated minority carrier will reach the junction. In reality, some minority carriers generated closer than a diffusion length will recombine before reaching the junction, while some minority carriers generated farther than a diffusion length from the junction will reach the junction before recombining.

Hence, to maximize photocurrent, it is desirable to maximize the number of photons that will be absorbed either in the junction or within a minority carrier diffusion length of the junction. The minority carriers of the EHPs generated outside this region have a higher probability of recombining before they have a chance to diffuse to the junction. If a minority carrier from an optically generated EHP recombines before it crosses the junction and becomes a majority carrier, it, along with the opposite carrier with which it recombines, is no longer available for conduction. Furthermore, the combined width of the junction and the two diffusion lengths should be several multiples of the reciprocal of the absorption constant, α, and the junction should be relatively close to a diffusion length from the surface of the material upon which the photon impinges, to maximize collection of photons. Figure 10.8 shows this desirable geometry. The engineering design challenge then lies in maximizing α, as well as maximizing the junction width and minority carrier diffusion lengths.

10.3.2.4 The Externally Biased pn Junction

In order to complete the analysis of the theoretical performance of the pn junction operating as a photovoltaic cell, it is useful to look at the junction with external bias. Figure 10.9 shows a pn junction connected to an external battery with the

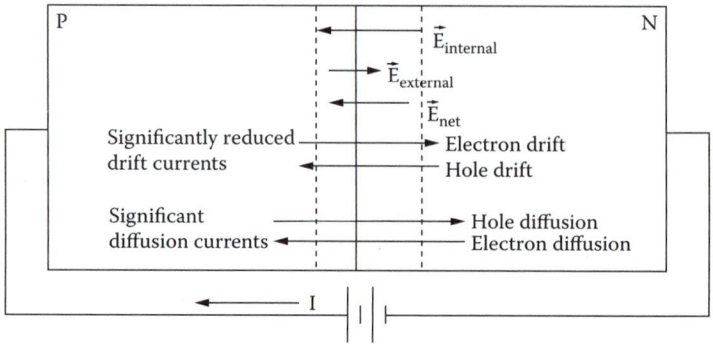

Figure 10.9 The pn junction with external bias.

internally generated electric field direction included. If Equation (10.10a) is recalled, taking into account that the externally applied voltage, with the exception of any voltage drop in the neutral regions of the material, will appear as opposing the junction voltage, the equation becomes

$$V_j - V = \frac{kT}{q} \ln \frac{n_n}{n_p}. \tag{10.12}$$

Note that the only difference between Equations (10.12) and (10.10a) is that the electron concentrations on the n-side and on the p-side of the junction are no longer expressed as the thermal equilibrium values. This will be the case only when the externally applied voltage is zero. However, under conditions known as low injection levels, it will still be the case that the concentration of electrons on the n-side will remain close to the thermal equilibrium concentration. For this condition, Equation (10.12) becomes

$$V_j - V = \frac{kT}{q} \ln \frac{N_d}{n_p}. \tag{10.13}$$

Because V_j can be calculated from Equation (10.10b), the quantity of interest in Equation (10.13) is n_p, the concentration of minority carriers at the edge of the junction on the p-side. Equation (10.13) can thus be solved for n_p with the result

$$n_p = N_D e^{-\frac{q(V_j - V)}{kT}} = N_D e^{\frac{-qV_j}{kT}} e^{\frac{qV}{kT}}. \tag{10.14}$$

Next, note that the thermal equilibrium value of the minority carrier concentration occurs when $V = 0$. If the excess minority carrier concentration is now defined as $n'_p = n_p - n_o$, and if n_o is subtracted from Equation (10.14), the result is

$$n'_p(0) = n_p(0) - n_o(0) = N_D e^{\frac{-qV_j}{kT}} \left(e^{\frac{qV}{kT}} - 1 \right) = n_{po} \left(e^{\frac{qV}{kT}} - 1 \right). \tag{10.15}$$

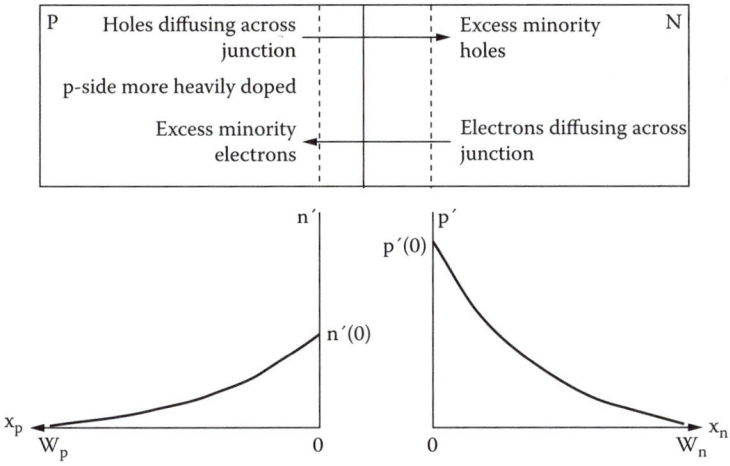

Figure 10.10 Excess minority carrier concentrations in neutral regions of pn junction device.

What happens to these excess minority carriers? They diffuse toward a region of lesser concentration, which happens to be away from the junction toward the contact, as shown in Figure 10.10.

If the contact is a few diffusion lengths away from the junction, it can be shown, by solving the well-known continuity equation, that the distribution of excess minority electrons between the junction and the contact will be

$$n'(x_p) = n'_p(0)\cosh\frac{x_p}{L_n} - n'_p(0)\operatorname{ctnh}\frac{w_p}{L_n}\sinh\frac{x_p}{L_n}. \tag{10.16}$$

A similar expression can be obtained for the concentration of excess minority holes in the neutral region of the n-side of the device. To obtain an expression for the total device current, Equation (10.7) is used for each side of the device, noting that the gradient in the total carrier concentration is given by the gradient of the excess carrier concentration, because the spatial variation in the excess concentration will far exceed any spatial variation in the equilibrium concentration. Combining the results for electron and hole currents yields the total current in the device as it depends on the externally applied voltage along with the indicated device parameters. The result is, if A represents the cross-sectional area of the pn junction and adjoining regions,

$$I = I_n + I_p = qA\left(\frac{D_n n_{po}}{L_n}\operatorname{ctnh}\frac{w_p}{L_n} + \frac{D_p p_{no}}{L_p}\operatorname{ctnh}\frac{w_n}{L_p}\right)\left(e^{\frac{qV}{kT}} - 1\right). \tag{10.17}$$

Equation (10.17) is, of course, the familiar diode equation that relates diode current to diode voltage. Note that the current indicated in Equation (10.17) flows in the direction opposite to the optically generated current described earlier. Letting qA(nasty expression) $= I_o$ and incorporating the photocurrent component into Equation (10.17) finally yields the complete equation for the current in the PV cell to be

$$I = I_\ell - I_o\left(e^{\frac{qV}{kT}} - 1\right). \tag{10.18}$$

Those readers who remember everything they read will recognize Equation (10.18) to be the same as Equation (3.1) in Chapter 3. Note that the current of Equation (10.18) is directed out of the positive terminal of the device, so that when the current and voltage are both positive, the device is delivering power to the external circuit.

10.4 MAXIMIZING PV CELL PERFORMANCE

10.4.1 Introduction

Equation (10.18) indicates, albeit in a somewhat subtle manner, that to maximize the power output of a PV cell, it is desirable to maximize the open-circuit voltage, short-circuit current, and fill factor of a cell. Recalling the plot of Equation (10.18) and Figure 3.3, it should be evident that maximizing the open-circuit voltage and the short-circuit current will maximize the power output for an ideal cell characterized by Equation (10.18). Real cells, of course, have some series resistance, so there will be power dissipated by this resistance, similar to the power loss in a conventional battery due to its internal resistance. In any case, recalling that the open-circuit voltage increases as the ratio of photo current to reverse saturation current increases, a desirable design criterion is to maximize this ratio, provided that it does not proportionally reduce the short-circuit current of the device.

Fortunately, this is not the case, because maximizing the short-circuit current requires maximizing the photocurrent. It is thus instructive to look closely at the parameters that determine both the reverse saturation current and the photocurrent. Techniques for lowering series resistance will then be discussed.

10.4.2 Minimizing the Reverse Saturation Current

Beginning with the reverse saturation current as expressed in Equation (10.17), the first observation is that the equilibrium minority carrier concentrations at the edges of the pn junction are related to the intrinsic carrier concentration through Equation (10.4). Hence,

$$p_{no} = \frac{n_i^2}{N_D} \quad \text{and} \quad n_{po} = \frac{n_i^2}{N_A}. \tag{10.19}$$

So far, no analytic expression for the intrinsic carrier concentration has been developed. Such an expression can be obtained by considering Fermi levels, densities of states, and other quantities that are discussed in solid-state devices textbooks. Because the goal here is to determine how to minimize the reverse saturation current, and not to go into the detail of quantum mechanical proofs, the result is noted here with the recommendation that the interested reader consult a good solid-state devices text for the development of the result. The result is

$$n_i^2 = 4\left(\frac{2\pi kT}{h^2}\right)^3 (m_n^* m_p^*)^{\frac{3}{2}} e^{-\frac{E_g}{kT}}, \tag{10.20}$$

where m_n^* and m_p^* are the electron effective mass and hole effective mass in the host material and E_g is the bandgap energy of the host material. These effective masses can be greater than or less than the rest mass of the electron, depending on the degree of curvature of the valence and conduction bands when plotted as energy versus momentum as in Figure 10.2. In fact, the effective mass can also depend on the band in which the carrier resides in a material. For more information on effective mass, the reader is encouraged to consult the references listed at the end of the chapter.

Now, using Equation (10.11) with Equations (10.19) and (10.20) in Equation (10.17), the following final result for the reverse saturation current is obtained:

$$I_o = \left(4qA\left(\frac{2\pi kT}{h^2}\right)(m_n^* m_p^*)^{\frac{3}{2}} e^{-\frac{E_g}{kT}}\right)$$
$$\times\left(\frac{1}{N_A}\sqrt{\frac{D_n}{\tau_n}}\operatorname{ctnh}\frac{\ell_p}{\sqrt{D_n\tau_n}} + \frac{1}{N_D}\sqrt{\frac{D_p}{\tau_p}}\operatorname{ctnh}\frac{\ell_n}{\sqrt{D_p\tau_p}}\right). \tag{10.21}$$

Because the design goal is to minimize I_o while still maximizing the ratio I_l:I_o, the next step is to express the photocurrent in some detail so that the values of appropriate parameters can be considered in the design choices.

10.4.3 Optimizing Photocurrent

In Section 10.3.2, the photocurrent optimization process was discussed qualitatively. In this section, the specific parameters that govern the absorption of light and the lifetime of the absorbed charge carriers will be discussed, and a formula for the photocurrent will be presented for comparison with the formula for the reverse saturation current. In particular, minimizing reflection of the incident photons, maximizing the minority carrier diffusion lengths, maximizing the junction width, and minimizing surface recombination velocity will be discussed. The PV cell designer will then know exactly what to do to make the perfect cell.

10.4.3.1 Minimizing Reflection of Incident Photons

The interface between air and the semiconductor surface constitutes an impedance mismatch, because the electrical conductivities and the dielectric constants of air and a PV cell are different. As a result, part of the incident wave must be reflected

Table 10.2 Some Antireflective Coatings Useful in PV Cell Production

Material	Index of Refraction
Al_2O_3	1.77
Glasses	1.5–1.7
MgO	1.74
SiO	1.5–1.6
SiO_2	1.46
Ta_2O_5	2.2
TiO_2	2.5–2.6

Source: Adapted from Hu, C. et al., *Solar Cells: From Basic to Advanced Systems*, McGraw-Hill, 1983.

in order to meet the boundary conditions imposed by the solution of the wave equation on the electric field, E, and the electric displacement, D.

Readers who know electromagnetic field theory well will recognize that this problem is readily solved by the use of a quarter-wave matching coating on the PV cell. If the coating on the cell has a dielectric constant equal to the geometric mean of the dielectric constants of the cell and of air and if the coating is one-quarter-wavelength thick, it will act as an impedance-matching transformer and minimize reflections. Of course, the coating must be transparent to the incident light. This means that it needs to be an insulator with a bandgap that exceeds the energy of the shortest-wavelength light to be absorbed by the PV cell. Alternatively, it needs some other property that minimizes the value of the absorption coefficient for the material, such as an indirect bandgap. Several of these coatings are listed in Table 10.2.

It is also important to realize that a quarter wavelength is on the order of 0.1 μm. This is extremely thin, and may pose a problem for spreading a uniform coating of this thickness. And, of course, because it is desirable to absorb a range of wavelengths, the antireflective coating will be optimized at only a single wavelength. Despite these problems, coatings have been developed that meet the requirements quite well.

An alternative to antireflective coatings now commonly in use with Si PV cells is to manufacture the cells with a textured front surface, as shown in Figure 10.14. Textured front and back surfaces and their contribution to the capture of photons will be discussed later in this section in conjunction with Figure 10.14. The bottom line is that a textured surface acts to enhance the capture of photons and also acts to prevent the escape of captured photons before they can produce EHPs. Furthermore, the textured surface is not wavelength dependent as is the antireflective coating.

10.4.3.2 Maximizing Minority Carrier Diffusion Lengths

Because the diffusion lengths are given by Equation (10.11), it is necessary to explore the factors that determine the diffusion constants and minority carrier lifetimes

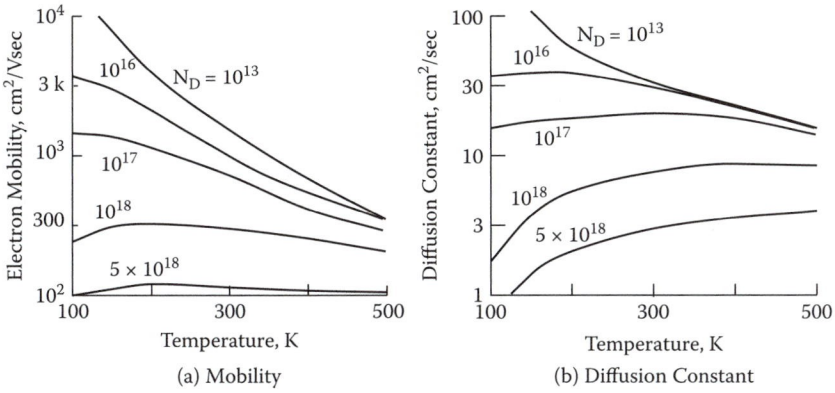

Figure 10.11 Temperature and impurity dependence of electron mobility and diffusion constant in silicon.

in different materials. It needs to be recognized that changing a diffusion constant may affect the minority carrier lifetime, so the product needs to be maximized.

Diffusion constants depend on scattering of carriers by host atoms as well as by impurity atoms. The scattering process is both material dependent and temperature dependent. In a material at a low temperature with a well-defined crystal structure, scattering of charge carriers is relatively minimal, so they tend to have high mobilities. Figure 10.11a illustrates the experimentally determined dependence of the electron mobility on temperature and on impurity concentration in Si. The Einstein relationship shows that the diffusion constant is proportional to the product of mobility and temperature. This relationship is shown in Figure 10.11b. So, once again, there is a trade-off. While increasing impurity concentrations in the host material *increases* the built-in junction potential, increasing impurity concentrations *decreases* the carrier diffusion constants.

The material of Figure 10.11 is single-crystal material. In polycrystalline or amorphous material, the lack of crystal lattice symmetry has a significant effect on the mobility and diffusion constant, causing significant reduction in these quantities. However, if the absorption constant can be made large enough for these materials, the corresponding decrease in diffusion length may be compensated for by the increased absorption rate.

When an electron and a hole recombine, certain energy and momentum balances must be achieved. Locations in the host material that provide for optimal recombination conditions are known as recombination centers. Hence, the minority carrier lifetime is determined by the density of recombination centers in the host material.

One type of recombination center is a crystal defect, so that as the number of crystal defects increases, the number of recombination centers increases. This means that crystal defects reduce the diffusion constant as well as the minority carrier lifetime in a material.

Impurities also generally make good recombination centers, especially those impurities with energies near the center of the bandgap. These impurities are thus different from the donor and acceptor impurities that are purposefully used in the host material, because donor impurities have energies relatively close to the conduction band, and acceptor impurities have energies relatively close to the valence band.

Minority carrier lifetimes also depend on the concentration of charge carriers in the material. An approximation of the dependence of electron minority carrier lifetime on carrier concentration and location of the trapping energy within the energy gap is given by

$$\tau_n = \frac{n'\left[n + p + 2n_i \cosh \frac{E_t - E_i}{kT}\right]}{CN_t\left(np - n_i^2\right)}, \tag{10.22a}$$

where C is the capture cross section of the impurity in cm^3/s, N_t is the density of trapping centers, and E_t and E_i are the energies of the trapping center and the intrinsic Fermi level. In most materials, the intrinsic Fermi level is very close to the center of the bandgap. Under most illumination conditions, the hyperbolic term will be negligible compared to the majority carrier concentration, and the excess electron concentration as minority carriers in p-type material will be much larger than the electron thermal equilibrium concentration. Under these conditions, for minority electrons in p-type material, Equation (10.22a) reduces to

$$\tau_n \cong \frac{1}{CN_t}. \tag{10.22b}$$

Hence, to maximize the minority carrier lifetime, it is necessary to minimize the concentration of trapping centers and to be sure that any existing trapping centers have minimal capture cross sections.

10.4.3.3 Maximizing Junction Width

Because it has been determined that it is desirable to absorb photons within the confines of the pn junction, it is desirable to maximize the width of the junction. It is therefore necessary to explore the parameters that govern the junction width. Perhaps the reader recalls similar discussions in a previous electronics class.

An expression for the width of a pn junction can be obtained by solving Gauss' law at the junction, because the junction is a region that contains electric charge. Solution of Gauss' law, of course, is dependent on the ability to express the spatial distribution of the space charge in mathematical, or, at least, in graphical form. Depending on the process used to form the junction, the impurity profile across the junction can be approximated by different expressions. Junctions formed by epitaxial growth or by ion implantation can be controlled to have impurity profiles to meet the discretion of the operator. Junctions grown by diffusion can be reasonably approximated by a linearly graded model. The interested reader is encouraged to consult a reference on semiconductor devices for detailed information on the production of various junction impurity profiles.

The junction with uniform concentrations of impurities is convenient to use to obtain a feeling for how to maximize the width of a junction. Solution of Gauss' law for a junction with uniform concentration of donors on one side and a uniform concentration of acceptors on the other side yields solutions for the width of the space charge layer on each side of the junction. The total junction width is then simply the sum of the widths of the two sides of the space charge layer. The results for each side are

$$W_n = \left[\frac{2\varepsilon N_A}{qN_D(N_A + N_D)} \right]^{\frac{1}{2}} (V_j - V)^{\frac{1}{2}} \tag{10.23a}$$

and

$$W_p = \left[\frac{2\varepsilon N_D}{qN_A(N_D + N_A)} \right]^{\frac{1}{2}} (V_j - V)^{\frac{1}{2}}. \tag{10.23b}$$

Before combining these two results, it is interesting to note that the width of the junction on either the p-side or on the n-side depends on the ratio of the impurity concentrations on each side. Again, because Gauss' law requires equal numbers of charges on each side of the junction, the side with the smaller impurity concentration will need to have a wider space charge layer to produce enough impurity ions to balance out the impurity ions on the other side. The overall width of the junction can now be determined by summing Equations (10.23a) and (10.23b) to get

$$W = \left[\frac{2\varepsilon(N_D + N_A)}{qN_A N_D} \right]^{\frac{1}{2}} (V_j - V)^{\frac{1}{2}}. \tag{10.24}$$

At this point, it should be recognized that the voltage across the junction due to the external voltage across the cell, V, will never exceed the built-in voltage, V_j. The reason is that, as the externally applied voltage becomes more positive, the cell current increases exponentially and causes voltage drops in the neutral regions of the cell, so only a fraction of the externally applied voltage actually appears across the junction. Hence, there is no need to worry about the junction width becoming zero or imaginary. In the case of PV operation, the external cell voltage will hopefully be at the maximum power point, which is generally between 0.5 and 0.6 V for silicon.

Next, observe that, as the external cell voltage increases, the width of the junction decreases. As a result, the absorption of photons decreases. This suggests that it would be desirable to design the cell to have the largest possible built-in potential to minimize the effect of increasing the externally applied voltage. This involves an interesting trade-off, because the built-in junction voltage is logarithmically dependent on the product of the donor and acceptor concentrations [see Equation (10.10b)], and the junction width is inversely proportional to the square root of the product of the two quantities. Combining Equations (10.10b) and (10.24) results in

$$W = \left[\frac{2\varepsilon(N_D + N_A)}{qN_D N_A}\right]^{\frac{1}{2}} \left[\left(\frac{kT}{q}\ln\frac{N_A N_D}{n_i^2}\right) - V\right]^{\frac{1}{2}}. \tag{10.25}$$

Note now that maximizing W is achieved by making either $N_A >> N_D$ or by making $N_D >> N_A$. For example, if $N_A >> N_D$, then Equation (10.25) simplifies to

$$W = \left[\frac{2\varepsilon}{qN_D}\right]^{\frac{1}{2}} \left[\left(\frac{kT}{q}\ln\frac{N_A N_D}{n_i^2}\right) - V\right]^{\frac{1}{2}}, \tag{10.26a}$$

or, if $N_D >> N_A$, then

$$W = \left[\frac{2\varepsilon}{qN_A}\right]^{\frac{1}{2}} \left[\left(\frac{kT}{q}\ln\frac{N_A N_D}{n_i^2}\right) - V\right]^{\frac{1}{2}}. \tag{10.26b}$$

Another way to increase the width of the junction is to include a layer of intrinsic material between the p-side and the n-side as shown in Figure 10.12. In this pin junction, there are no impurities to ionize in the intrinsic material, but the ionization still takes place at the edges of the n-type and the p-type material. As a result, there is still a strong electric field across the junction, and there is still a built-in potential across the junction. Because the intrinsic region could conceivably be of any width, it is necessary to determine the limits on the width of the intrinsic region.

The only feature of the intrinsic region that degrades performance is the fact that it has a width. If it has a width, then it takes time for a charge carrier to traverse this width. If it takes time, then there is a chance that the carrier will recombine. Thus, the width of the intrinsic layer simply needs to be kept short enough to minimize

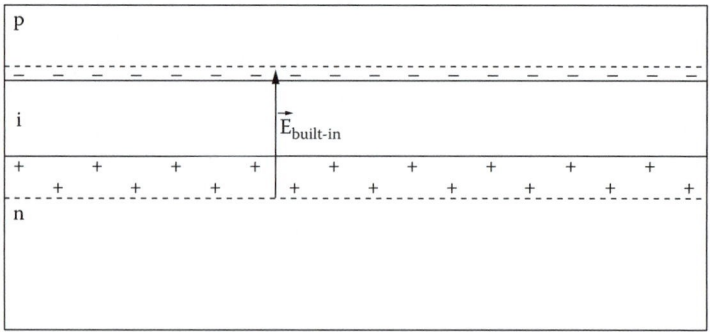

Figure 10.12 The pin junction.

recombination. The particles travel through the intrinsic region with a relatively high drift velocity due to the built-in electric field at the junction. Because the thermal velocities of the carriers still exceed the drift velocities by several orders of magnitude, the width of the intrinsic layer needs to be kept on the order of about one diffusion length.

10.4.3.4 Minimizing Surface Recombination Velocity

If an EHP is generated near a surface, it becomes more probable that the minority carrier will diffuse to the surface. Because photocurrent depends on minority carriers' diffusing to the junction and ultimately across the junction, surface recombination of minority carriers before they can travel to the junction reduces the available photocurrent. When the surface is within a minority carrier diffusion length of the junction, which is often desirable to ensure that generation of EHPs is maximized near the junction, minority carrier surface recombination can significantly reduce the efficiency of the cell.

Surface recombination depends on the density of excess minority carriers, in this case, as generated by photon absorption, and on the average recombination center density per unit area, N_{sr}, on the surface. The density of recombination centers is very high at contacts and is also high at surfaces in general, because the crystal structure is interrupted at the surface. Imperfections at the surface, whether due to impurities or to crystal defects, all act as recombination centers.

The recombination rate, U, is expressed as number/cm^2/s and is given by

$$U = cN_{sr}m', \tag{10.27}$$

where m' is used to represent the excess minority carrier concentration, whether electrons or holes, and c is a constant that incorporates the lifetime of a minority carrier at a recombination center. Analysis of the dimensions of the parameters in Equation (10.27) shows that the units of cN_{sr} are cm/s. This product is called the surface recombination velocity, S. The total number of excess minority carriers recombining per unit time, and subsequent loss of potential photocurrent, is thus dependent on the density of recombination centers at the surface and on the area of the surface. Minimizing surface recombination thus may involve reducing the density of recombination centers or reducing the density of minority carriers at the surface.

If the surface is completely covered by a contact, then little can be done to reduce surface recombination if minority carriers reach the surface, because recombination rates at contacts are very high. However, if the surface is not completely covered by a contact, such as at the front surface, then a number of techniques have been discovered that will result in passivation of the surface. Silicon oxide and silicon nitrogen passivation are two methods that are used to passivate silicon surfaces.

Another method of reducing surface recombination is to passivate the surface and then only allow the back contact to contact the cell over a fraction of the total cell area. While this tends to increase series resistance to the contact, if the cell material near the contact is doped more heavily, the ohmic resistance of the material is decreased and the benefit of reduced surface recombination offsets the cost of somewhat higher series resistance. Furthermore, an E-field is created that attracts majority carriers to the contact and repels minority carriers. This concept will be explored in more detail in Section 10.5.2.

10.4.3.5 A Final Expression for the Photocurrent

An interesting exercise is to calculate the maximum obtainable efficiency of a given PV cell. Equation (10.3) indicates the general expression for photon absorption. Because the absorption coefficient is wavelength dependent, the general formula for overall absorption must take this dependence into account. The challenge in design of the PV cell and selection of appropriate host material is not only to avoid absorption before the photon is close enough to the junction, but also to ensure absorption when the photon is within a minority carrier diffusion length of either side of the junction. Figure 10.13 shows a typical photon absorption profile for the n-region and the p-region of a cell with the p-region at the surface.

Because direct bandgap materials tend to have larger absorption constants than indirect materials such as Si, it is relatively straightforward to capture photons in these materials. In Si, however, several clever design practices are used to increase absorption. Because some photons will travel completely across the cell to the back of the cell without being absorbed, if the back of the cell is a good reflector, the photons will be reflected back toward the junction. This is easy to do, because the back contact of the Si cell covers the entire back of the cell. However, rather than having a smooth back surface that will reflect photons perpendicular to the surface, the back surface is textured so the incident photons will be scattered at angles, thus increasing the path length.

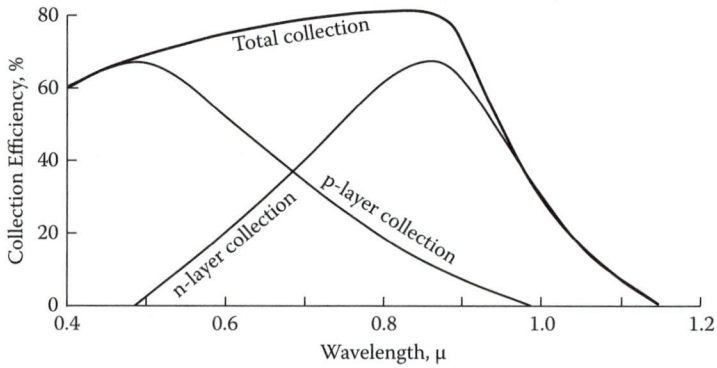

Figure 10.13 Photon collection efficiency versus wavelength and location in cell for a typical PV cell with p-layer at the surface. (Adapted from Yang, E. S., *Microelectronic Devices*, 1988, McGraw-Hill.)

The front of the cell, however, can be covered with an antireflective coating to maximize transmission of photons into the material. Hence, a similar scheme is necessary to capture the photons in the material if, after bouncing off the back surface, they are still not absorbed. Once again, a textured surface will enhance the probability that a photon will undergo internal reflection, because the dielectric coefficient, and, hence, the index of refraction of the host material, is greater than that of the antireflective coating. This is analogous to when ripples on water prevent a person below the surface from seeing anything above the surface. When the surface is smooth, it is possible to see objects above the water, provided that the angle of view is sufficiently close to the perpendicular. Figure 10.14 shows a cell with textured front and back surfaces and the effect on the travel of photons that enter the host material.

The foregoing discussion can be quantified in terms of cell parameters in the development of an expression for the photocurrent. Considering a monochromatic photon flux incident on the p-side of a p^+n junction (the + indicates strongly doped), the following expression for the hole component of the photocurrent can be obtained. The expression is obtained from the solution of the diffusion equation in the neutral region on the n-side of the junction for the diffusion of the photon-created minority holes to the back contact of the cell.

$$\Delta I_{\ell p} = \frac{qAF_{ph}\alpha L_p}{\alpha^2 L_p^2 - 1}\left[\frac{S\cosh\frac{w_n}{L_p} + \frac{D_p}{L_p}\sinh\frac{w_n}{L_p} + (\alpha D_p - S)e^{-\alpha w_n}}{S\sinh\frac{w_n}{L_p} + \frac{D_p}{L_p}\cosh\frac{w_n}{L_p}} - \alpha L_p\right]. \tag{10.28}$$

It is assumed that the cell has a relatively thin p-side and that the n-side has a width, w_n. In Equation (10.28), F_{ph} represents the number of photons per cm^2 per second per unit wavelength incident on the cell. The effect of the surface recombination velocity on the reduction of photocurrent is more or less clearly demonstrated by Equation (10.28). The mathematical whiz will immediately be able to determine that small values of S maximize the photocurrent, and large values reduce the photocurrent, while the person with average math skills may need to plug in some numbers.

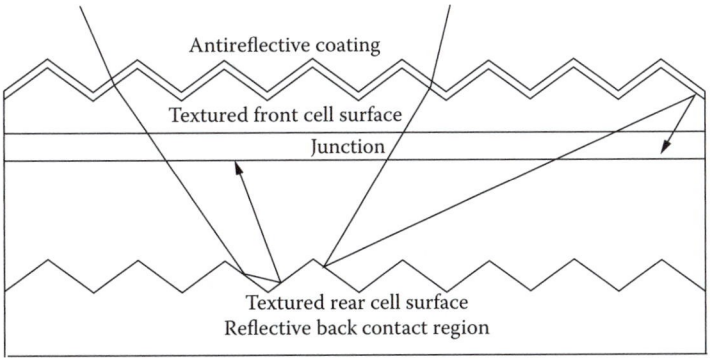

Figure 10.14 Maximizing photon capture with textured surfaces.

Table 10.3 Theoretical Conversion Efficiency Limits for Several PV Materials at 25°C

Material	E_g	η_{max} (%)
Ge	0.6	13
CIS	1.0	24
Si	1.1	27
InP	1.2	24.5
GaAs	1.4	26.5
CdTe	1.48	27.5
AlSb	1.55	28
a-Si:H	1.65	27
CdS	2.42	18

Sources: Based on data from Rappaport, *RCA Review*, Vol. 20, September 1959, 373–379; Zweibel, *Harnessing Solar Power*, Plenum Press, New York, 1990.

Equation (10.28) is thus maximized when α and L_p are maximized and S is minimized. The upper limit of the expression then becomes

$$\Delta I_{\ell p} = -qAF_{ph}, \tag{10.29}$$

indicating that all photons have been absorbed and all have contributed to the photocurrent of the cell.

Because sunlight is not monochromatic, Equation (10.28) must be integrated over the incident photon spectrum, noting all wavelength-dependent quantities, to obtain the total hole current. An expression must then be developed for the electron component of the current and integrated over the spectrum to yield the total photocurrent as the sum of the hole and electron currents. This mathematical challenge is clearly a member of the nontrivial set of math exercises and is not included as a homework problem. Yet, some have persisted at a solution to the problem and have determined the maximum efficiencies that can be expected for cells of various materials. Table 10.3 shows the theoretical optimum efficiencies for several different PV materials.

10.4.4 Minimizing Cell Resistance Losses

Any voltage drop in the regions between the junction and the contacts of a PV cell will result in ohmic power losses. In addition, surface effects at the cell edges may result in shunt resistance between the contacts. It is thus desirable to keep any such losses to a minimum by keeping the series resistance of the cell at a minimum and the shunt resistance at a maximum. With the exception of the cell front contacts, the procedure is relatively straightforward.

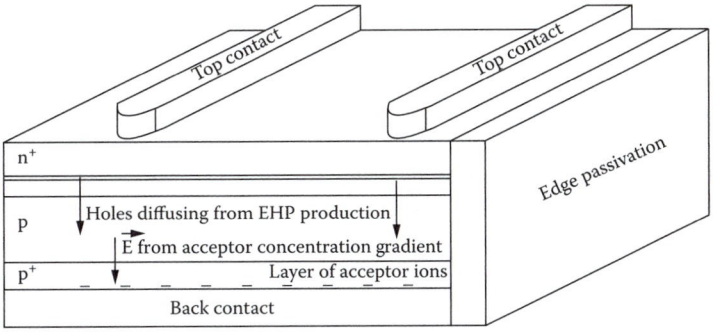

Figure 10.15 PV cell geometry for minimizing losses from cell series and shunt resistance.

Most cells are designed with the front layer relatively thin and highly doped, so the conductivity of the layer is relatively high. The back layer, however, is generally more lightly doped in order to increase the junction width and to allow for longer minority carrier diffusion length to increase photon absorption. There must therefore be careful consideration of the thickness of this region in order to maximize the performance of these competing processes.

If the back contact material is allowed to diffuse into the cell, the impurity concentration can be increased at the back side of the cell, as illustrated in Figure 10.15. This is important for relatively thick cells, commonly fabricated by slicing single crystals into wafers. The contact material must produce either *n*-type or *p*-type material if it diffuses into the material, depending on whether the back of the cell is *n*-type or *p*-type.

In addition to reducing the ohmic resistance by increasing the impurity concentration, the region near the contact with increased impurity concentration produces an additional electric field that increases the carrier velocity, thus producing a further equivalent reduction in resistance. The electric field is produced in a manner similar to the electric field that is produced at the junction.

For example, if the back material is *p*-type, holes from the more heavily doped region near the contact diffuse toward the junction, leaving behind negative acceptor ions. Although there is no source of positive ions in the *p*-region, the holes that diffuse away from the contact create an accumulated positive charge that is distributed through the more weakly doped region. The electric field, of course, causes a hole drift current, which, in thermal equilibrium, balances the hole diffusion current. When the excess holes generated by the photoabsorption process reach the region of the electric field near the contact, however, they are swept more quickly toward the contact. This effect can be viewed as the equivalent of moving the contact closer to the junction, which, in turn, has the ultimate effect of increasing the gradient of excess carriers at the edge of the junction. This increase in gradient increases the diffusion current of holes away from the junction. Because this diffusion current strongly dominates the total current, the total current across the junction is thus increased by the heavily doped layer near the back contact.

At the front contact, another balancing act is needed. Ideally, the front contact should cover the entire front surface. The problem with this, however, is that if the front contact is not transparent to the incident photons, it will reflect them away. In most cases, the front contact is reflecting. Because the front/top layer of the cell is generally very thin, even though it may be heavily doped, the resistance in the transverse direction will be relatively high because of the thin layer. This means that if the contact is placed at the edge of the cell to enable maximum photon absorption, the resistance along the surface to the contact will be relatively large.

The compromise, then, is to create a contact that covers the front surface with many tiny fingers, as shown in Figure 10.15. This network of tiny fingers, which, in turn, are connected to larger and larger fingers, is similar to the configuration of the capillaries that feed veins in a circulatory system. The idea is to maintain more or less constant current density in the contact fingers, so that as more current is collected, the cross-sectional area of the contact must be increased. This subject is covered in more detail in Chapter 11.

Finally, shunt resistance is maximized by ensuring that no leakage occurs at the perimeter of the cell. This can be done by nitrogen passivation or simply by coating the edge of the cell with insulating material to prevent contaminants from providing a current path across the junction at the edges.

10.5 EXOTIC JUNCTIONS

10.5.1 Introduction

Thus far, only relatively simple junctions have been considered. PV cell performance can be enhanced significantly by incorporating a variety of more sophisticated junctions, including, but not necessarily limited to, graded junctions, heterojunctions, Schottky junctions, multijunctions, and tunnel junctions. This section provides an introduction to these junctions so that the reader will better understand the various junction types in the specific devices to be presented in the next chapter.

10.5.2 Graded Junctions

All the formulations up to this point have related to the abrupt junction, in which the impurity concentration is constant to the junction and then abruptly changes to the opposite impurity concentration. While such junctions do exist, graded junctions are more common in materials for which the junction is fabricated by diffusion of impurities from the surface. The fabrication process for obtaining a linearly graded impurity profile as shown in Figure 10.16 will be discussed in some detail in Chapter 11.

The significance of the graded junction is that majority carrier transport beyond the junction is improved by an additional electric field component resulting from the decreasing impurity concentration from surface to junction, as shown in Figure 10.16. The origin of this electric field component is the same as the origin of the additional

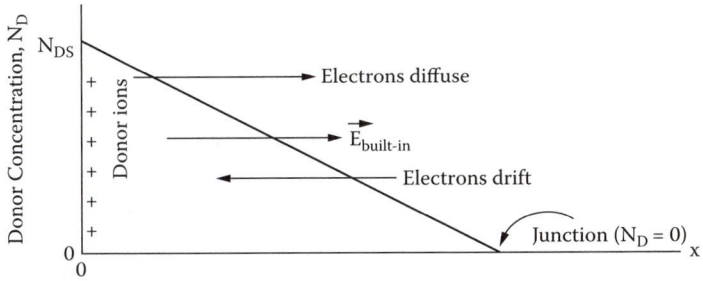

Figure 10.16 Diffusion and drift directions and electric field resulting from linearly decreasing impurity profile.

field near the back contact when the back metallization is allowed to diffuse into the back of the cell. Each impurity atom donates either an electron or a hole and becomes ionized. If there is a gradient in impurity concentration, then the mobile carriers will diffuse in the direction of the gradient, leaving behind fixed impurity ions that serve either as the origin or termination of electric field lines.

An expression for the electric field in terms of the impurity concentration can be obtained by equating the diffusion current to the resulting drift current needed to balance the net current to zero when no external circuit is available for current flow. Assuming the majority carriers to be electrons in a heavily doped surface region, setting the sum of electron drift and diffusion current to zero results in

$$q\mu_n nE + qD_n \frac{dn}{dx} = 0, \tag{10.30}$$

which can be solved for E with the result

$$E = -\frac{D_n}{\mu_n}\frac{1}{n}\frac{dn}{dx} = -\frac{kT}{q}\frac{1}{N_D}\frac{dN_D}{dx}. \tag{10.31}$$

This additional E-field provides a drift component to the electron flow away from the junction toward the contact that effectively increases the velocity of the electrons toward the contact, resulting in lower resistive losses in the electron transport process toward the contact. Problem 10.12 provides an opportunity for calculation of the E-field present if the impurity concentration decreases linearly from surface to junction.

10.5.3 Heterojunctions

One of the problems identified with the optical absorption process is the absorption of higher-energy photons close to the surface of the cell. Most of the EHPs generated by these higher-energy photons are lost to recombination when they are created more than a diffusion length from the junction.

This phenomenon can be mitigated to some extent by the use of a heterojunction. A heterojunction is simply a composite junction comprising two materials with closely matched crystal lattices, so the bandgap near the surface of the material is greater than the bandgap near the junction. The higher-bandgap region will appear transparent to photons with lower energies, so these photons can penetrate to the junction region where the bandgap is less than the incident photon energy. In the region of the junction, they can generate EHPs that will be collected before they recombine.

Heterojunctions are sometimes made between two *n*-regions or two *p*-regions as well as between *n*-region and *p*-region. The behavior of the heterojunction is dependent on the crystal lattices, work functions, impurity doping profiles, and energy band properties of each semiconductor material, to the extent that discussion of any particular junction would probably not be applicable to a different junction. Readers interested in specific junctions, such as Ge:GaAs or AlGaAs:GaAs, will be able to obtain more information from journal and conference publications.

In some cases, materials cannot be made either *n*-type or *p*-type. Use of a heterojunction with a material that can be made to complete a pn junction is another important use of heterojunctions. For example, *p*-type CdS is yet to be produced. But a thin *n*-type CdS layer on top of a CIS structure will produce a pn junction effect, so CdS is often used as a part of a thin-film structure to produce a thin, heavily doped *n*-region near the front surface of the cell.

10.5.4 Schottky Junctions

Sometimes, when a metal is contacted with semiconductor material, an ohmic contact is formed and sometimes a rectifying contact is formed. It all depends on the relative positions of the work functions of the two materials. Figure 10.17 shows situations where the work function of the metal, $q\phi_m$, is greater than that of the *n*-type semiconductor, $q\phi_s$, and where the work function of the metal is less than that of the *n*-type semiconductor material. The work function is simply the energy difference between the Fermi level and the vacuum level, and the Fermi level is that energy where the probability of occupancy by a mobile carrier is 0.5. For the semiconductor, the energy difference between the bottom of the conduction band and the vacuum level is designated as $q\psi_s$.

When the metal and semiconductor are joined, electrons flow either from metal to semiconductor or from semiconductor to metal, depending on which has the higher Fermi level. In the case where $q\phi_s < q\phi_m$, as indicated in Figure 10.17a, electrons will flow from semiconductor to metal in a manner similar to diffusion across a pn junction from *n*-type to *p*-type. As the electrons leave, they leave behind positive donor ions, just as in the pn junction. Because the materials are in contact, at the point of contact the probability of electron occupancy must be the same for each material, which requires the Fermi levels of each material to align. Because the semiconductor surface in contact with the metal is depleted of electrons, the Fermi level must move farther away from the conduction band, because, as the Fermi level moves closer to the conduction band, more electrons will appear in the conduction band and vice versa.

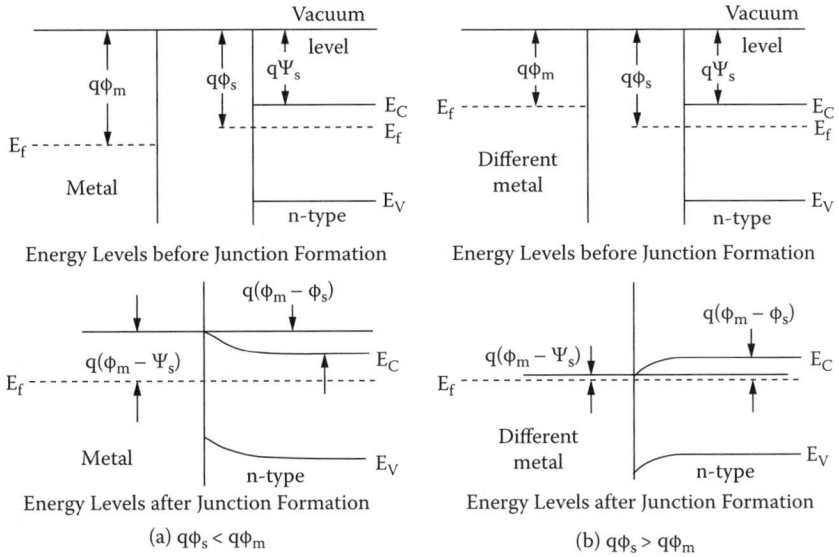

Figure 10.17 Energy levels in metal semiconductor junctions.

The donor ions on the *n*-side of the contact thus become the origin of an electric field directed from the semiconductor to the metal. This field causes a drift of electrons from metal to semiconductor to balance the diffusion from semiconductor to metal. Note that holes cannot diffuse from metal to semiconductor, because the metal supports only electrons. Thus, the built-in potential is only about half what it might be if holes could also diffuse across the junction.

In Figure 10.17b, where $q\phi_s > q\phi_m$, the electrons diffuse from the metal to the semiconductor, thus increasing the concentration of electrons on the semiconductor side of the junction. This diffusion, in effect, causes the equivalent of heavier doping of the *n*-type semiconductor and enables current to flow easily in either direction across the junction. This is therefore an ohmic junction, whereas the junction of Figure 10.17a is a rectifying junction. External contacts to a PV cell need to be ohmic to prevent unnecessary voltage drop at the contact, whereas the rectifying contacts are useful for other purposes.

For the rectifying junction, a positive external voltage from metal to semiconductor reduces the built-in field, enabling the electrons that have diffused to the metal to continue flowing in the external circuit, as shown in Figure 10.18a. Note that as the Fermi level on the semiconductor side is drawn closer to the conduction band, more electrons can diffuse to the metal, resulting in significant current flow from metal to semiconductor. The reduced built-in field reduces drift components in the opposite direction in a manner similar to the forward bias condition of a conventional pn junction.

When the external voltage is negative, the built-in field is as shown in Figure 10.18b. Although this might be expected to result in significant electron flow

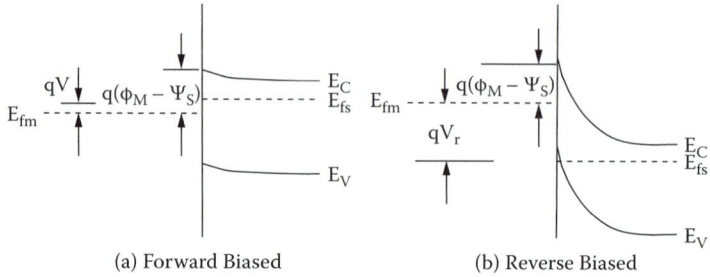

Figure 10.18 The rectifying metal-*n*-type semiconductor junction under forward-biased and reverse-biased conditions.

from metal to semiconductor, this does not occur, because the metal electrons must overcome the barrier between the metal Fermi level and the semiconductor conduction band. Only a few of the metal electrons are sufficiently energetic to do so, so the result is a relatively small reverse saturation current, and thus a rectifying contact. The contact is rectifying even though the relatively strong E-field in the semiconductor region would be capable of enhancing drift current if replacement mobile charges were available.

For PV action, photons need to generate EHPs on the semiconductor side of the junction within a minority carrier diffusion length of the junction, so that the minority carrier can be swept out of the region to the other side of the junction by the built-in field.

It has been shown that, if the semiconductor material is *n*-type and $q\phi_m < q\phi_s$, the junction becomes ohmic, with bidirectional current conduction. When the semiconductor is *p*-type, the opposite is true. That is, for $q\phi_m > q\phi_s$, the contact is ohmic, and for $q\phi_m < q\phi_s$, the contact is rectifying.

These conditions must thus be met when ohmic contacts are to be made to semiconductor materials. This is why certain metals such as Mo, Al, and Au are acceptable for ohmic contacts on some materials but not on others.

The Schottky barrier junction is relatively straightforward to fabricate, but is not very efficient as a PV cell because it has a relatively smaller open-circuit voltage than a conventional pn junction due to diffusion currents flowing in only one direction across the junction. On the other hand, the lower voltage drop across the junction when it is conducting results in lower power dissipation, making the Schottky diode convenient for use as a bypass diode.

10.5.5 Multijunctions

Because photon energy is most efficiently absorbed when it is near the bandgap, a clever way to absorb more photons is to stack junctions of different bandgaps. By starting near the front surface with a relatively larger bandgap material, the higher-energy photons can be absorbed relatively efficiently at this junction. Then a smaller bandgap junction, perhaps followed by an even smaller bandgap junction, will enable

Figure 10.19 Three junctions in tandem (series), showing opposing pn junctions.

lower-energy photons to be absorbed more efficiently. Because the junctions are in series, they must produce equal currents.

Figure 10.19 shows three junctions in series, each of somewhat different material in order to achieve three different bandgaps. The illustration may seem very logical, but there is a bit of a problem that is encountered when more than one junction is connected in series. That is simply the fact that, although the p-to-n direction of the first junction may be forward-biased, this makes the n-to-p junction between the first and second pn junctions become reverse-biased. The reverse bias across these junctions causes unnecessary voltage drop equivalent to the drop across blocking diodes that are sometimes inserted in strings of PV modules. Hence, a means must be devised to eliminate the effect of these reverse-biased junctions. Fortunately, the tunnel junction can eliminate this problem.

10.5.6 Tunnel Junctions

Tunnel junctions take advantage of the Heisenberg uncertainty principle, that is,

$$\Delta p \Delta x \geq \hbar, \tag{10.32}$$

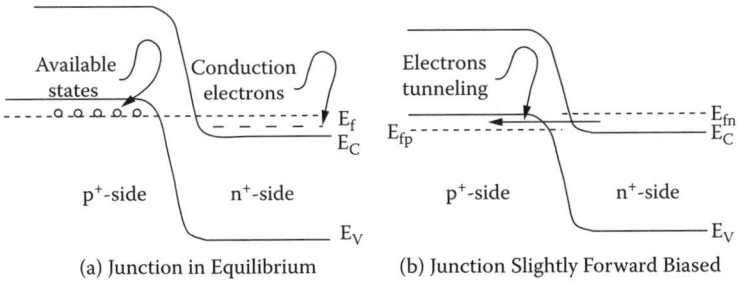

Figure 10.20 Tunnel junction showing conditions necessary for tunneling.

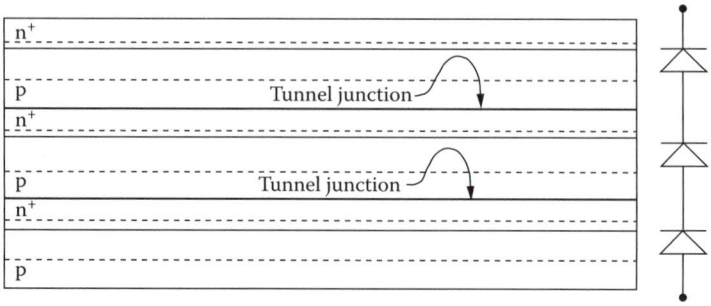

Figure 10.21 Separation of multiple pn junctions with tunnel junctions.

where Δp and Δx are, respectively, the uncertainty in particle position and the uncertainty in particle momentum, and $\hbar$ is Planck's constant divided by 2π. In a tunnel junction, the junction width is so extremely narrow, that with relatively small uncertainty in particle momentum, it cannot be specified with certainty that the particle is on one side or the other side of the junction. This phenomenon is known as quantum mechanical tunneling. The process takes place without any change in energy, because the particle essentially tunnels through the potential barrier produced by the tunnel junction. Noting Equation (10.24), it is evident that if impurity concentrations are very large on each side of the junction, the junction width will be quite small and the built-in potential will be relatively large, according to Equation (10.10b). As the junction is forward-biased, the junction width becomes even smaller, per Equation (10.24). Tunneling occurs when the electrons in the conduction band on the *n*-side rise to energy levels adjacent to empty states on the *p*-side, provided that the junction is sufficiently narrow, as illustrated in the energy band diagram of Figure 10.20.

Thus, the multiple pn junctions of the multijunction configuration can be accomplished by incorporating p^+n^+ tunnel junctions between each junction to eliminate the reverse-biased pn junctions. This is shown in Figure 10.21.

Problems

10.1 What can be said about the bandgap of glass? Why is silicon opaque? For what range of wavelengths will silicon appear transparent?

10.2 Given the arguments about direct versus indirect bandgap materials and the ability of these materials to absorb photons, offer an argument about the optimal bandgap structure of a material that will *emit* photons efficiently. Note that photon emission involves a transition of an electron from the conduction band to the valence band.

10.3 What would the bandgap need to be for an infrared LED? What about a red LED? What about a green LED? What about a blue LED? Would you expect the materials with these bandgaps to be direct bandgap materials or indirect bandgap materials?

10.4 Compare the average thermal velocity of an electron, v_{th}, at room temperature, if $\frac{1}{2}mv_{th}^2 = \frac{1}{2}kT$, where m is the electron mass, k is Boltzmann's constant, and

T is the absolute temperature, with the drift velocity of an electron having a mobility of 10^3 cm^2/V-s in an electric field of 1000 V/cm. The drift velocity equals the product of the mobility and the electric field. At what electric field strength will the drift velocity equal the thermal velocity?

10.5 Use Equation (10.9b) to solve for the concentration of electrons as a function of position between the neutral n-side edge of the junction and the edge of the neutral p-side of the junction. Plot this concentration function on a logarithmic scale along with the concentration of ionized impurities in the space charge layer.

10.6 Solve for the concentration of holes in the space charge layer. Show that in the space charge layer in thermal equilibrium $np = n_i^2$ at every position, x.

10.7 Show that Equation (10.10b) can also be obtained by setting the net hole current across the junction equal to zero.

10.8 If a material has $\mu_n = 8600$ cm^2/V-s, $\mu_p = 400$ cm^2/V-s, and $\tau_n = \tau_p = 10$ ns, would you design a PV cell with the light incident on the n-side or the p-side of the material? Approximately what should be the distance between the edge of the space charge layer and the back contact? Explain why.

10.9 Show that Equation (10.24) results from adding Equation (10.23a) to Equation (10.23b).

10.10 Show that combining Equations (10.10b) and (10.24) results in Equation (10.25).

10.11 Assume $N_A = 10^{20}$ cm^{-3} and $N_D = 10^{17}$ cm^{-3}. Assume room temperature.

a. Calculate the built-in junction potential for Si and GaAs.

b. Calculate the junction width for Si and for GaAs under short-circuit conditions.

10.12 Use Equation (10.30) to calculate the electric field in the n-region adjacent to the junction if the donor concentration is given by $N_D(x) = N_S(1 - bx)$, where x is distance measured from the surface of the cell and b is a constant.

10.13 Calculate the widths of the following abrupt junctions in Si at room temperature with no external bias applied.

a. $N_A = 10^{16}$ and $N_D = 10^{20}$.

b. $N_A = 10^{21}$ and $N_D = 10^{21}$.

Next, calculate the widths of each junction if a forward bias of 0.3 V is applied, with all the external voltage appearing across the junction. Sketch the energy band picture for part b.

REFERENCES

1. Yang, E. S., *Microelectronic Devices*, McGraw-Hill, New York, 1988.
2. Streetman, B. G., *Solid State Electronic Devices*, 4th ed., Prentice-Hall, Englewood Cliffs, NJ, 1995.
3. Hu, C. and White, R. M., *Solar Cells: From Basic to Advanced Systems*, McGraw-Hill, New York, 1983.
4. Rappaport, P., *RCA Review*, Vol. 20, September 1959, 373–379.
5. Zweibel, K., *Harnessing Solar Power*, Plenum Press, New York, 1990.

CHAPTER 11

Present and Proposed PV Cells and Systems

11.1 INTRODUCTION

In Chapter 10, the basic theory of PV cells was presented without regard to any specific cell technology. This chapter will cover some of the fabrication processes associated with current cells along with discussions of the operation of a variety of cells, some of which are commonly in use and others that are still in the experimental phases. But the world of PV is experiencing more than changes in cell technology. It is also in the midst of a systems paradigm shift as large-scale system deployment appears to be well under way.

Progress in PV research and development is moving so rapidly that by the time this chapter is in print, much of it will likely be outdated. As an example of the extent of PV research currently under way, the 2009 IEEE PV Specialists Conference had a total of more than 600 oral and poster presentations. Hence, in addition to introducing the reader to technologies of 2009, it is also the goal of this chapter to provide the reader with the intellectual tools needed to read and understand current and future literature. Many of the sources of current information on PV technology are listed in the reference section of the chapter.

In general, cell fabrication begins with the refining and purification of the cell base material. After extremely pure material is available, then the pn junction or its equivalent must be formed. In some multiple-layer cells, more than one junction is formed along with various fascinating isolation steps. The single-crystal silicon cell has made its mark on history. Whether it will continue to make its mark on the future will depend on whether the amount of energy consumed in the production of the cell can be reduced. Its fabrication and characteristics will be discussed first. During the discussion of the single-crystal silicon cell, important processes such as crystal growth and diffusion will be discussed. These basic processes in many cases are applicable to the fabrication of other types of PV cells.

Perhaps the most promising near-term PV cells will consist of thin films. Crystalline cells are generally considered to be the first generation of PV cells, and thin films are considered to be the second generation. The third generation, yet to be commercialized, will have production costs in the range of \$0.20–\$0.50/W and will have efficiencies between 31% and 74%, according to Martin Green [1]. Certain materials have direct bandgaps with energies near the peak of the solar spectrum, along with relatively high absorption constants and the capability of being fabricated with pn junctions. These films are not single-crystal devices, so there are limitations to carrier mobilities and subsequent device performance. However, in spite of the non-single-crystal structures, laboratory conversion efficiencies close to 20% have been achieved for laboratory cells [2].

Thin films are advantageous because only a minimal amount of material is required to deposit a film with a thickness of 1 or 2 μm. Figure 11.1 shows the photon current versus optical path length for common thin-film materials. Note that within 1 to 2 μm, all the materials approach photon current saturation, whereas crystalline Si requires significantly greater thickness for full photon absorption. Examples of thin-film PV cells that are currently in commercial production are amorphous silicon (a-Si), copper indium (gallium) diselenide (CIGS), and cadmium telluride (CdTe). Gallium arsenide has been used for high-performance cells and has also been the subject of thin-film experimentation. And, with the dawn of the age of nanotechnology, PV applications have been given more than a cursory bit of attention.

Availability of cell component materials is also of interest if many gigawatts of PV power are to be obtained from any technology. Discussion of each technology includes commentary on the availability and refining of the cell component materials. Of all the thin-film materials, availability of indium may become a limiting factor in the production of copper indium diselenide (CIS) PV cells, but this limit will not be approached until 200 GW of cells have been produced [4]. Annual production of In would limit CIS cell production to about 4 GW per year. For comparison purposes, 2010 worldwide electrical generation capacity is projected to be approximately 4593 GW [5]. Other thin-film materials, although scarce, are sufficiently abundant

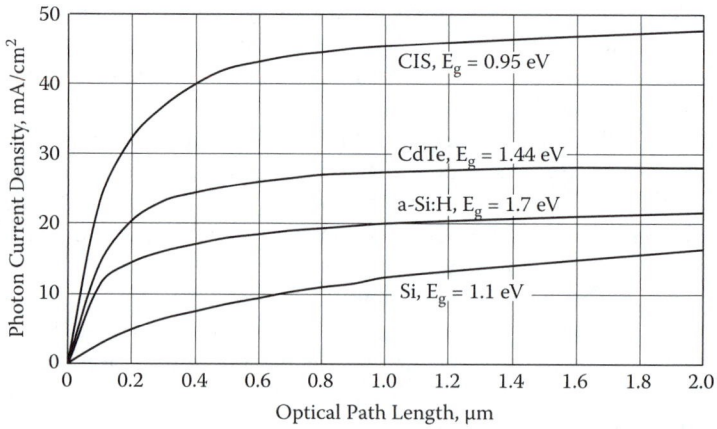

Figure 11.1 Photon current versus optical path length for thin-film materials, compared with crystalline silicon, under standard test conditions. (Adapted from Tarrant, and Gay, *Research on High Efficiency, Large Area CuInSe2-Based Thin Film Modules, Final Technical Report*, NREL, Golden, CO, April 1995.)

for the production of many gigawatts of PV cells at reasonable cost, provided that certain production shortcomings that limit cell performance can be overcome.

11.2 SILICON PV CELLS

The first silicon PV cells were of the single-crystal variety. Single-crystal cells are the most efficient and most robust of the silicon PV cell family, but are also the most energy intensive in their production. For this reason, other varieties of silicon cells have been developed. Polycrystalline cells are somewhat less efficient, but are less energy intensive. Amorphous silicon cells are even less energy intensive, but have lower conversion efficiencies. However, because amorphous silicon is a direct bandgap material with an energy gap larger than pure silicon, it has a much larger absorption constant than crystalline silicon, with peak absorption at a wavelength closer to the peak of the solar spectrum, and is thus a suitable material for thin-film cells. The instability mechanism present in early versions of amorphous silicon is reasonably well understood and means for overcoming the instability are now in common use. Additional efficiency increases have been achieved for amorphous silicon by the use of multijunction cell structures.

Thin Si cells, such as crystalline silicon on glass (CSG), represent a compromise between crystalline cells and amorphous cells. These relatively recent introductions to the Si cell family use novel techniques for photon capture and minimization of top surface area blocked by contacts. Good efficiencies have been achieved in laboratory-scale devices.

11.2.1 Production of Pure Silicon [6]

The first thing that must be done in the production of a silicon cell is to find some silicon. Although nature has not seen fit to leave behind large quantities of relatively pure silicon in a form similar to the large deposits of carbon that have been mined for many years and burned in furnaces and worn on fingers, silicon has been provided in abundant quantities mixed with oxygen. Indeed, more than half the crust of the earth is composed of this silica compound, so all that is needed is to mine it and remove the oxygen and any other impurities.

Of course, this is not done with zero energy. Separation of oxygen from silicon is quite energy intensive, requiring the reduction of SiO_2 with carbon in a carbon arc furnace. Regrettably, the CO_2 by-product of this process is a greenhouse gas, so the production of pure Si for PV cells begins with the production of a greenhouse gas along with the silicon. Fortunately, over the lifetime of the cell, the amount of CO_2 from the reduction process is significantly less than the amount of CO_2 that would be produced if fossil fuels were used instead of PV cells to generate the same amount of electrical energy. Furthermore, as carbon capture technology advances, it is possible that the CO_2 from the production of Si will ultimately be sequestered and kept out of the atmosphere. The energy cost of this step, once at approximately 50 kWh/kg of metallurgical-grade (99.0% pure) silicon, continues to decrease as more efficient processing steps are developed.

A purity of 99% means impurity levels of 1 part in 100. This means that additional refining needs to take place, because the purity of PV-grade silicon needs to be in the neighborhood of 1 part in 10^8 or so, depending on how efficient the cell is to be. Unwanted impurities lead to crystal defects and trapping levels that affect the cell output. Hence, the next step in the refining process is to react the metallurgical-grade silicon with either hydrogen or chlorine to form either SiH_4 (silane) or $SiHCl_3$ (trichlorosilane).

The more common reaction is to combine the Si with HCl in a fluidized bed reaction. This reaction, however, produces a combination of Si-H-Cl compounds, such as SiH_2Cl_2, $SiHCl_3$, and $SiCl_4$, along with chlorides of impurities. Each of these liquids has its own boiling point, so it is relatively straightforward to use fractional distillation to separate the $SiHCl_3$ from the other components. Although it looks simple on paper, the energy required in this step is relatively high. The next step is to remove the silicon from the trichlorosilane. This is done by reacting the trichlorosilane with hydrogen at a high temperature, producing polycrystalline silicon and HCl. Once again, improvements in process efficiency have led to reductions in the energy cost, which in 1999 was close to 200 kWh/kg of silicon. At this point, the silicon achieves the electronic-grade level of purity (99.999999%).

11.2.2 Single-Crystal Silicon Cells

11.2.2.1 Fabrication of the Wafer

To produce single-crystal silicon from the polycrystalline material, the silicon must be melted and recrystallized. This is done by dipping a silicon seed crystal into the melt and slowly withdrawing it from the melt with a slight twisting motion. As the silicon is pulled from the melt, it reaches a somewhat higher purity level because the remaining impurities tend to remain behind in the melt. Under properly controlled conditions for solidification and replenishment of the melt, crystals as large as 8 in. in diameter and 6 ft long can be readily grown. To produce *n*-type or *p*-type crystals, boron or arsenic can be introduced to the melt in whatever quantities are needed to produce the desired doping levels. Other type III or type V materials may also be introduced, but generally the diffusion constants or activation energies of these materials are less desirable from a PV performance perspective. Indium, for example, is a group III impurity, but at room temperature, because of the higher activation energy of indium, a significant percentage of the indium atoms does not ionize and donate holes to the host crystal. Phosphorus is a group V impurity that has satisfactory ionization properties, but also has a higher diffusion constant in silicon, so that if the host material is heated, phosphorus atoms will migrate in the host material at a faster rate than arsenic atoms.

Special saws are then used to cut the ingots into wafers. Because circular wafers mounted in a module leave a large amount of empty space between the wafers, often the edges of the wafers are trimmed to make the wafers closer to square. The wafers are approximately 0.01 in. thick and are quite brittle. The sawing causes significant surface damage, so the wafers must next be chemically etched to restore the surface.

In order to achieve the textured finish as described in Chapter 10, it is possible to use a preferential etching process, so that after etching, the wafer surface is textured and relatively defect free. Again, the wafering process is quite energy intensive.

11.2.2.2 Fabrication of the Junction [7]

The next step in the production of the single-crystal silicon cell is to create the pn junction. In electronic semiconductors, pn junctions can be created by diffusion, epitaxial growth, and ion implantation. These methods work fine for junctions having areas in the 10^{-12} m^2 range, but for PV cells with junction areas in the 10^{-2} m^2 range, the time it takes to produce a given area of junction becomes very important. Hence, due to cost constraints, pn junctions in crystalline silicon PV cells are made primarily by diffusion.

Impurity atoms diffuse into silicon in a manner similar to the diffusion of holes and electrons across regions of nonuniform impurity density. The concentration as a function of distance from the surface, x, and time is determined by the solution of the familiar diffusion equation

$$\frac{\partial N}{\partial t} = D\frac{\partial^2 N}{\partial x^2}, \tag{11.1}$$

where N represents the density of the impurity and D is the diffusion constant associated with the specific impurity. It should be noted that D is highly temperature dependent, especially close to the melting temperature of the host material. At room temperature, diffusion is negligible, but in the neighborhood of 1000°C, diffusion becomes appreciable.

The solution to Equation (11.1) depends on the boundary conditions imposed on the impurity. Most diffusion processes begin by holding the surface concentration of impurities constant for a fixed time. Then the impurity source is removed so that the total number of impurities now remains constant, equal to the number of impurity atoms that diffused into the host during the constant surface concentration step of the process. The first step, known as the predeposition, leads to the well-known complementary error function solution, that is,

$$N(x,t) = N_o erfc\frac{x}{2\sqrt{Dt}}, \tag{11.2}$$

where N_o is the concentration of impurities at the surface of the wafer. Suppose the substrate is uniformly doped p-type material and the impurities being diffused are n-type. The net impurity concentration at any depth, x, then, is the difference between the n-type and the p-type concentrations or

$$N(x,t) = N_o erfc\frac{x}{2\sqrt{Dt}} - N_A. \tag{11.3}$$

When the net impurity concentration is positive, the material is n-type, and when the net impurity concentration is negative, the material is p-type. The junction is located where the net impurity concentration is zero, which represents the transition between n-type and p-type. The depth of the junction can thus be determined by setting $N(x,t) = 0$ and solving for x, yielding the result

$$x = 2\sqrt{Dt}\,erfc^{-1}\frac{N_A}{N_o}. \tag{11.4}$$

This will not normally be the final junction depth, because after the predeposition step, the impurity concentration at the wafer surface is close to the solid solubility limit for the host material, which shortens the minority carrier diffusion length in this region. This preliminary junction depth estimate is useful, however, because it leads to a determination of the amount of time needed in the predeposition step to produce a junction at a depth less than the final desired junction depth, which will ultimately be 5 to 10 times the initial junction depth.

To reduce the surface concentration, a **drive-in** diffusion is carried out next. The drive-in diffusion is simply the continued application of heat to the material after the supply of the surface impurity is removed. The result is that the impurity continues to diffuse from the region of greater concentration to the region of lesser concentration. The impurity concentration is given, to a good approximation, by the solution of the diffusion equation under conditions of constant impurities per unit area. The assumption is that all of the impurities diffused into the material during the predeposition step remain at the surface of the material. The number of impurities per unit area can be determined by integrating Equation (11.3) from the surface into the material. While the actual distance of travel of the impurities is less than 1 μm, the integration is performed from 0 to infinity to yield

$$Q(t) = \int_0^\infty N(x,t)dx = 2N_o\sqrt{\frac{Dt}{\pi}}. \tag{11.5}$$

After the drive-in step, the net impurity concentration is represented fairly accurately by the Gaussian distribution function,

$$N(x,t) = \frac{Q}{\sqrt{\pi Dt}}e^{-\frac{x^2}{4Dt}} - N_A. \tag{11.6}$$

This expression can now be solved for the junction depth by again setting $N(x,t) = 0$. The solution is left as an exercise for the reader (Problem 11.2). It is thus a straightforward matter to control the diffusion of impurities into the wafer to produce a junction at any desired depth from the surface. Because it is desirable to have the junction within a minority carrier diffusion length of the surface, the depth can be set to achieve this goal. Note that Equations (11.3) through (11.6) also enable the cell designer to control the impurity concentration profile on the n-side of the junction. Figure 11.2 shows the impurity concentrations after the two diffusion steps have been completed.

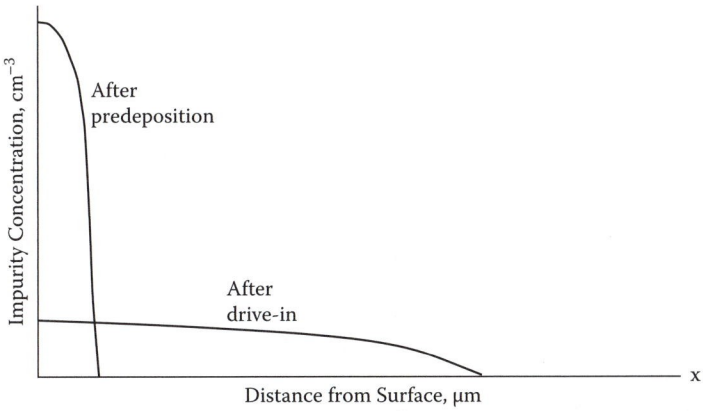

Figure 11.2 Concentrations of impurities after two diffusion steps.

11.2.2.3 Contacts

After creating the pn junction, the next step is to affix contacts to the cell. If the intent is to fabricate contacts that will last as long as the rest of the cell, then alligator clips are somewhat inadequate. It is necessary to affix contacts that are low-resistance ohmic contacts that will remain in good contact with the cell through extreme temperature cycling. The contact material must lend itself to bonding of connecting wires/ribbons so the cell can be integrated into a module.

For many cells, the back contact covers the entire cell and is commonly made of evaporated aluminum that is annealed by heating the material after evaporation. The annealing process causes the aluminum to diffuse slightly into the silicon, creating a strong bond that will not break under thermal cycling. Because aluminum is a group III element, it also acts as a *p*-type donor and produces a heavily doped p-region adjoining the contact. This heavily doped p-region creates an impurity gradient that produces a resulting electric field that accelerates holes toward the back contact.

The front contact is considerably more challenging for several reasons. First of all, if the front contact is opaque to light, then it will block photons from entering the cell and being absorbed. This means the area of the front contact must be minimized. On the other hand, if the area is too small, then the resistance of the contact increases. Because the junction is only a small distance below the front surface, if the contact material is annealed, it is possible that it will diffuse into the silicon and short out the junction. For that matter, even if the contact does not diffuse across the junction initially, this can occur over time under operation at higher temperatures and shorten the lifetime of the cell. Hence, the design constraints on the front contact are that it must be small, ohmic, and present low series resistance while not threatening the junction.

The series resistance at the front contact of a cell depends on the distance the charge carrier must travel through the host material to the contact in addition to the

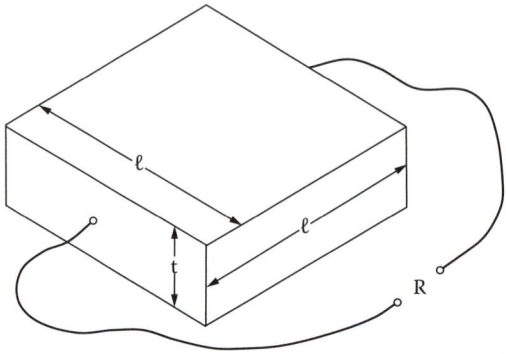

Figure 11.3 Determination of resistance per square.

resistance of the contact itself. The conductivity of a material is given by

$$\sigma = q(\mu_n n + \mu_p p). \tag{11.7}$$

It is convenient to consider the resistance between two points on an ohms-per-square basis. Figure 11.3 shows a piece of material with dimensions $\ell \times \ell \times t$ (i.e these are the dimensions of the object in Fig. 11.3.) The resistance between the two ends of the material is given by

$$R = \frac{1}{\sigma}\frac{\ell}{\ell\, t} = \frac{1}{q(\mu_n n + \mu_p p)t} \text{ ohms/square.} \tag{11.8}$$

For a typical thin n-type silicon region with $n \approx 10^{20} >> p$, $\mu_n \approx 100$ and a junction depth of approximately 1 μm (= t), the resistance per square becomes $R = 6.25\ \Omega$.

Now suppose two contacts are each 2 cm long and spaced 0.5 cm apart, as shown in Figure 11.4. The problem is to approximate the resistance between bulk and contact. Because the worst-case resistance is from the midpoint between the two

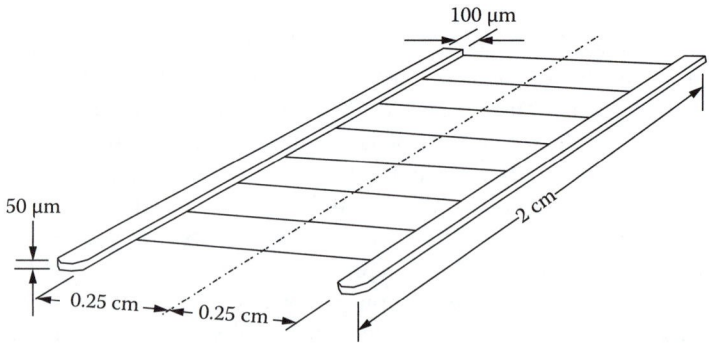

Figure 11.4 Determination of resistance between bulk and contacts.

contacts, the longest distance to a contact is 0.25 cm. Over the 2 cm distance, a total of eight squares, each 0.25 cm on a side, can be inserted between the midpoint and one of the contacts, and another eight can be inserted between the midpoint and the other contact. Hence, there are the equivalent of 16 squares in parallel, each of which has a resistance of 6.25 Ω. The resulting resistance of the 16 parallel squares is thus $6.25 \div 16 = 0.39\ \Omega$.

Suppose further that the cell is approximately 10 cm in diameter and generates a total of 2.5 A under standard illumination. Assuming uniform generation of current, this means that each square cm generates approximately 0.032 A. Because the area between the two contacts is 1 cm^2, approximately 32 mA is generated in this region that needs to be carried by the contacts after it reaches the contacts. In general, a contact will be collecting current from both sides, so that each cm of contact length will collect approximately 16 additional mA of current, if the spacing is maintained at 0.5 cm.

An estimate of the power loss to the series cell resistance can be obtained from the worst-case resistance and the total current flowing through this resistance, with the result that $P = I^2R = 0.4$ mW. The power generated in this region, assuming a maximum power voltage of approximately 0.55 V, is $P = IV = 0.032 \times 0.55 = 16.5$ mW. Hence, approximately 2.4% of the power generated is lost to the ohmic resistance of the cell between the generation point and the contact in this worst-case example. Because most of the charge carriers have a smaller distance of travel to the contact, the resistance experienced will be less, and the overall power loss will be less than 2% between bulk and contact.

To determine power losses in the contacts, it is necessary to determine the resistance of the contact. Aluminum has a bulk resistivity, ρ, of 2.7×10^{-6} Ω-cm. Assuming a contact with typical thickness of 50 μm and width of 100 μm, the resistance per centimeter length is $R = \rho/A = 0.054$ Ω/cm. Hence, the contact with 2 cm length will have a resistance of 0.108 Ω.

Because the current in the contact increases linearly from zero at the beginning to 32 mA at the end, the average current in the contact is 16 mA. However, to more accurately express the ohmic losses in the contact, the losses should be integrated over the length of the contact. Choosing any point, x, along the contact, the current at point x will be $I(x) = 0.016x$, and the differential resistance of this portion of the contact will be 0.054 dx ohm, because the resistance of the contact is 0.054 Ω/cm of length. The result of this calculation is

$$P = \int_0^2 d(I^2R) = \int_0^2 (.016x)^2 .054\, dx = 3.7 \times 10^{-5}\ \text{W}. \tag{11.9}$$

This is only about 10% of the power loss over the cell surface. Hence, most of the series resistance is due to the surface resistivity of the bulk cell material. However, if the contact is only 5 μm thick, the resistance increases by a factor of 10, and the power loss is comparable to the bulk losses. These calculations illustrate a means that may be used to determine optimal spacing between top contacts as well as the dimensions of the top contacts to keep power losses to a minimum while still enabling maximum photon absorption.

As the current in the contacts continues to increase, the cross-sectional area of the contact must increase to enable the contact to carry the total current within acceptable voltage drop limitations. This is analogous to the circulatory system of plants, in which leaves contain capillaries that extend over the leaf to provide transport for nutrients.

The front contacts can be fabricated by several means, including evaporation in a vacuum chamber or silk screening with a paste. Fine lines may be defined with photoresist as is done in the production of small geometry electronic semiconductors.

If aluminum is the contact material, in order to make an ohmic contact to the n-type material, it is necessary to anneal the contact for approximately 5 min at a temperature of approximately 450°C. To prevent transport of the aluminum through to the junction, it is advisable to use a small percentage of silicon in the aluminum or to first evaporate a very thin ($\approx 0.01\ \mu m$) layer of titanium or chromium to act as a barrier to the aluminum diffusing into the junction.

11.2.2.4 Antireflective Coating

After affixing the contacts to the material, an antireflective coating may be applied to the cell, normally by evaporation, because the coating must be so thin. A quarter-wavelength coating has a thickness of approximately 0.15 μm. These coatings are commonly used for photographic lenses to increase their speed by reducing reflection and simultaneously increasing transmission. Typical coatings are listed in Table 10.2 in Chapter 10. As long as the photon wavelength is relatively close to the quarter-wavelength constraint, transmission in excess of 90% can be achieved for the cell surface.

Because antireflective coatings optimize transmission at only one wavelength, textured cell surfaces are becoming more common for enhancing light trapping over the full spectrum.

11.2.2.5 Modules

The cells must then be mounted in modules, with interconnecting wires or metal foil "ribbons." The interconnects are generally either ultrasonically bonded or soldered to the cell contacts. The cells are then mounted to the module base and encapsulated with a glass or composite. The encapsulant must be chosen for long life in the presence of ultraviolet radiation as well as possible degradation from other environmental factors. Depending on the specific module environment, such as blowing sand, salt spray, acid rain, or other not-so-friendly environmental component, the encapsulant must be chosen to minimize scratching, discoloration, cracking, or any other damage that might be anticipated. Earlier encapsulants, such as ethylene vinyl acetate (EVA), tended to discolor as a result of exposure to high levels of ultraviolet radiation and higher temperatures. Other encapsulants had a tendency to delaminate under thermal stresses. Modern materials now seem to have overcome these problems [8].

After the cells are encapsulated in modules, they are ready to produce electricity very reliably for a long time, normally in excess of 20 years.

11.2.3 A High-Efficiency Si Cell with All Contacts on the Back [9]

Figure 11.5 shows the cross section of the SunPower A-300 monocrystalline Si cell. At the time of this writing, cell efficiencies of 22% have been achieved, and 20.3% efficiency has been achieved for modules [1]. Basically, the cell consists of a substrate of *n*-type Si that has alternating positive and negative contacts to alternating n^+- and *p*-type contact areas that are diffused into the substrate. As EHPs are generated in the substrate, with a few being generated in the pn junction regions, the electrons are attracted to the *n*-type contact and the holes are attracted to the *p*-type contact as a result of the E-fields across the pn junction and between the n^+ contact and the substrate.

Note that the region between the n^+ contact and the weaker *n*-type substrate essentially constitutes a graded junction, where electrons flow from the more heavily doped contact region to the less heavily doped substrate. As a result, net positive charges are left behind in the n^+ region such that an electric field is created that is directed toward the substrate from the contact region. This field enhances the drift of electrons from the photon-generated EHPs from the substrate to the contact region.

The p-regions create pn junctions with the substrate with resulting E-fields directed from the substrate toward the p-region, as is the case for conventional pn junctions. The substrate doping is weaker than the p-region doping, so that the junction region is predominantly in the substrate. This field enhances the drift of minority holes from photon-generated EHPs in the substrate, across the junction into the p-regions. The cell surface is textured to optimize photon capture. The lack of front contacts on the cell allows a greater level of photon penetration into the substrate layer, because the surface is not reflecting.

In 2005, the surface polarization effect was announced for these cells. When modules that used these cells were installed in the field with negative grounds, it was noticed that the module performance degraded over time much faster than anyone

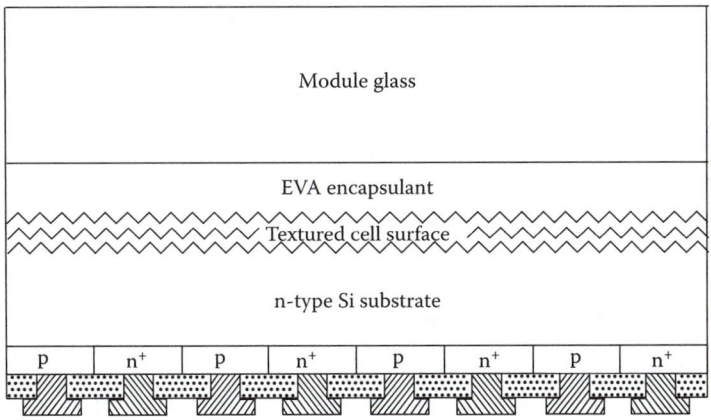

Figure 11.5 SunPower all back contact A-300 cell showing encapsulant and module glass.

would expect for a crystalline Si technology. Further analysis revealed that, even though the glass and EVA encapsulant are essentially insulators, it was still possible for minute currents to flow *from* the more positive EVA *to* the grounded module frames via the EVA and glass pathway, equivalent to charging a capacitor. Current flowing *from* the EVA means electrons flowing *to* the EVA from the module frame.

If the EVA layer is considered to be analogous to the gate region of a MOSFET, then the accumulation of negative charges in the EVA layer is equivalent to a negative charge on the gate of the MOSFET. This negative charge, in turn, attracts positive charge from the substrate, namely, minority carrier holes, to the cell surface rather than to the *p*-type contact. These holes are then lost to recombination with substrate electrons, which is the equivalent of an EHP being generated more than a minority carrier diffusion length from the pn junction.

Fortunately, this phenomenon is easily avoided, and reversible, if the module is operated with a positive ground rather than a negative ground. This results in positive charge buildup in the EVA layer rather than negative charge buildup. The resulting E-field from EVA to substrate repels holes rather than attracting them, so the hole collection efficiency is actually enhanced somewhat by operating the module with a positive ground. So, the system designer simply needs to be sure to use system power-conditioning components that can operate with a positive grounded array.

11.2.4 Multicrystalline Silicon Cells

At this point, it should be clear that the production of single-crystal Si cells is highly energy intensive. The large amount of energy used in the wafering process includes the single-crystal Si lost when the crystals are sawed into wafers. Further loss occurs if the round wafers are trimmed to approximate a square to fill a greater percentage of a module with cells. The question arises whether a means can be conceived for growing single-crystal wafers or perhaps a compromise can be achieved by manufacturing wafers that approximate single-crystal material. To date, at least three methods have been explored and developed for the production of multicrystalline PV cells: crucible growth, the EFG process, and string ribbon technology.

One compromise involves pouring molten Si into a crucible and controlling the cooling rate. The result is not single-crystalline material, because no seed crystal is used, but the multicrystalline Si obtained by this process has a square cross section and is sufficiently close to the single-crystal ideal that small cell efficiencies in the range of 20% have been obtained and module efficiencies of 15.5% have been achieved [1]. It is still necessary to saw the ingots into wafers, but the wafers are square, so no additional sawing is needed, as was the case with the round Si ingots. This process increases the production rate per kg of material by reducing the kerf loss.

Another method of producing multicrystalline Si is the edge-defined film-fed growth (EFG) method [10]. The EFG process involves pulling an octagon tube, 6 m long, with a wall thickness of 330 μm, directly from the Si melt. The octagon is then cut by laser along the octagonal edges into individual cells. Module efficiencies of 13.2% have been achieved.

The third method of producing multicrystalline Si cells involves pulling a ribbon of Si, or dendritic web, from the melt. The difficulty in this process is controlling the width of the ribbon. To do so, high-temperature string materials are used to define the edges of the ribbon. The string materials are pulled through a crucible of molten Si in an Ar atmosphere after the attachment of a seed crystal to define the crystal structure of the ribbon. The nonconducting string material has a coefficient of thermal expansion close to that of Si, so during the cooling process, the string material will not affect the Si crystallization process. A laboratory cell efficiency of 16.2% has been achieved, and an efficiency of 15.4% has been achieved for an 80 cm^2 cell [11].

Essentially the same processes are used with the multicrystalline wafers that are used with the single-crystal wafers. Multicrystalline modules are characterized by the cells completely filling the modules, with the cells having a sort of speckled appearance resulting from the departure from single-crystal structure. Because the multicrystalline cells still maintain the basic crystalline properties, the 1.1 eV indirect bandgap results in the need for thicker cells with surface texturing to provide for maximum photon capture, as in the case of single-crystal cells.

Because multicrystalline cells are currently in the production phase, it is reasonable to project efficiency increases and cost per watt reductions as research and development progresses.

11.2.5 Other Thin Silicon Cells

As this text goes to press in 2009, significant advances have been made in the area of thin Si cells. The same technologies that were reported as having promise at the IEEE 29th Photovoltaic Specialists Conference in 2002 were also included on the 2009 IEEE 34th Photovoltaic Specialists Conference list of promising technologies. Thin crystalline cell technology has been pursued with the hope that cell efficiency can be improved through the use of crystalline Si and that cell manufacturing cost can be lowered by using less material in the construction of the cell. A number of processes, including thin Si on ceramics [12], thin-film crystalline silicon on glass (CSG) [13], and epitaxial growth of Si on existing crystalline Si with subsequent removal of the epitaxially grown cell from the existing Si substrate (the PSI Process) [14] are described in the Conference Proceedings.

The CSG process has produced modules of areas in the range of 480–900 cm^2 since 1998. Efficiencies rose from 2% in 1999 to 8% in 2002. The structure consists of textured glass, antireflective coating, n^+pp^+ structure, resin insulator, and metallization that dips into the structure from the back surface to form back and front cell contacts. The front contact does not require a transparent conducting oxide (TCO), so losses from the TCO are eliminated. TCOs will be discussed in more detail later in this chapter.

The modules consist of cells that are monolithically connected, and due to isolated metal interconnects that produce a fault-tolerant structure, performance of the module is not degraded as the size is scaled up. The production cost of this module in May of 2002 was $1.95/W.

In 2009, other emerging thin Si technologies included microcrystalline and nanocrystalline materials.

11.2.6 Amorphous Silicon Cells

Amorphous silicon has no predictable crystal structure. Its atoms are located at more or less random angles and distances from each other. As a result, many of the potential covalent bonds in the silicon are not completed. These incomplete bonds cause a large number of equivalent impurity states in the bandgap, and the noncrystalline nature of the material results in very low values for electron and hole mobilities. The impurity states would typically result in trapping of mobile carriers, so the combination of impurity states and diminished transport properties at first rendered amorphous silicon as a rather poor semiconductor material.

But persistent solid-state physicists confirmed that the incomplete bonds could be passivated with hydrogen, thus significantly reducing the number of impurity states in the bandgap, and that with n-type and p-type materials, a pn junction could be formed. Furthermore, by incorporating an intrinsic layer between p^+ and n-type material, a reasonable EHP generation region could be created with reasonable transport properties resulting from minimized impurity scattering in the passivated intrinsic material. Another positive feature of the a-Si:H system is that it has a direct bandgap close to 1.75 eV, resulting in a high absorption coefficient and qualifying a-Si:H as a good potential candidate for a thin-film photovoltaic material.

11.2.6.1 Fabrication

Figure 11.6 shows the structure of a basic a-Si:H cell. Fabrication of the cell begins with the deposition of a transparent conducting oxide layer on a glass substrate. The TCO, typically n^+ SnO, constitutes the front contact of the cell. Next, a very thin layer of p^+ a-Si:H is deposited, usually by plasma decomposition of SiH_4. The degenerate n-type TCO and the degenerate p-layer form a tunnel heterojunction. After the p^+ layer, a slightly n-type intrinsic layer is deposited, followed by a stronger n-layer and finally a back contact, usually of Al, is deposited.

Initial operation of the basic cell with a relatively wide intrinsic layer was found to result in cell degradation when the cell was operated under sunlight conditions. The degradation was accounted for by the Staebler–Wronski effect, which explains

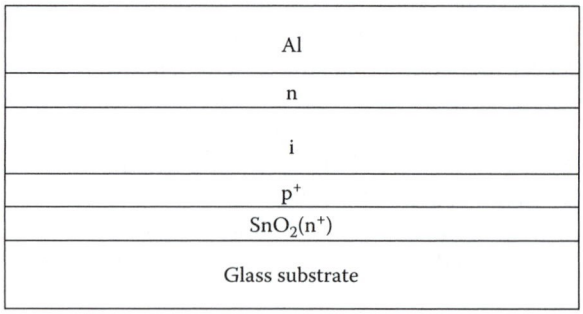

Figure 11.6 Basic a-Si:H cell structure.

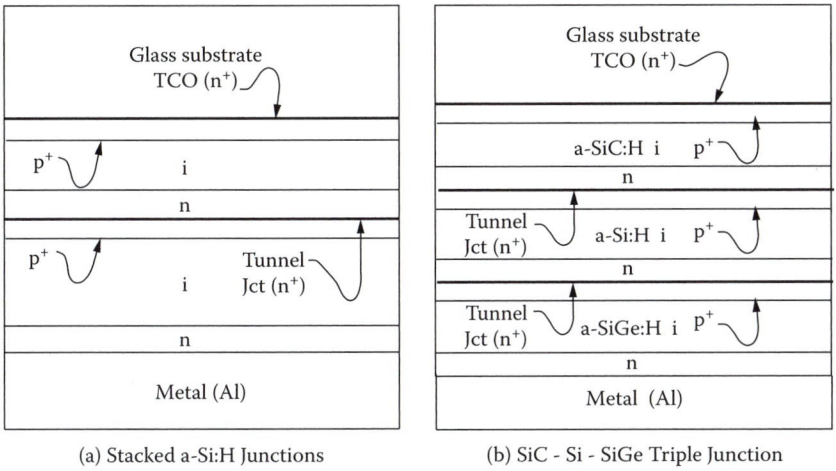

(a) Stacked a-Si:H Junctions (b) SiC - Si - SiGe Triple Junction

Figure 11.7 Stacked a-Si:H with Si only and with SiGe and SiC layers.

the degradation in terms of increased density of scattering and trapping states in the intrinsic layer in proportion to photon exposure.

To mitigate the effects of the wider i-layer, cells were designed with narrower i-layers, but stacked, as shown in Figure 11.7a. These stacked cells require inclusion of tunnel junctions to prevent the blocking action of the pn junctions of adjacent cells, as discussed in Section 10.5.6. These cells have shown improved long-term performance.

The concept of stacked cells has been extended one step further by recognizing that C and Ge also bond reasonably well with a-Si to produce either a-SiC:H or a-SiGe:H. The C alloy has a higher bandgap, thus enhancing absorption in the blue range, while the Ge alloy has a smaller bandgap, thus enhancing absorption in the infrared range. The net result is an overall performance improvement over the a-Si:H cell. The structure is shown in Figure 11.7b, noting again the need for tunnel junctions between cell layers.

Another interesting variation in cell structure is shown in Figure 11.8, where a stainless steel substrate has been used to produce a flexible cell. Rolls of a-Si cells for use in roofing and other building integrated applications are currently on the market [15]. Figure 11.8 also shows a-Si on a very lightweight polymer substrate, about 2 mil thick, that has been proposed for extraterrestrial applications where minimal weight is important and where stresses on the structure are minimal [16]. Both types of cell have been reported to have efficiencies in the 10% range.

The cell is produced on stainless steel by a proprietary roll-to-roll process. Note that the back surface incorporates an Al/ZnO textured film between the n-layer and the stainless steel to enhance photon trapping. The bandgap of an a-SiGe:H cell is dependent on the fraction of Ge in the mix. In this structure, the top a-Si:H intrinsic layer has a bandgap of approximately 1.8 eV. The middle a-SiGe:H cell has a Ge fraction of approximately 15% in the intrinsic layer, resulting in a bandgap of approximately 1.6 eV, and the bottom a-SiGe:H cell has an intrinsic layer Ge fraction of approximately 45%, resulting in a bandgap for this layer of approximately 1.4 eV [17].

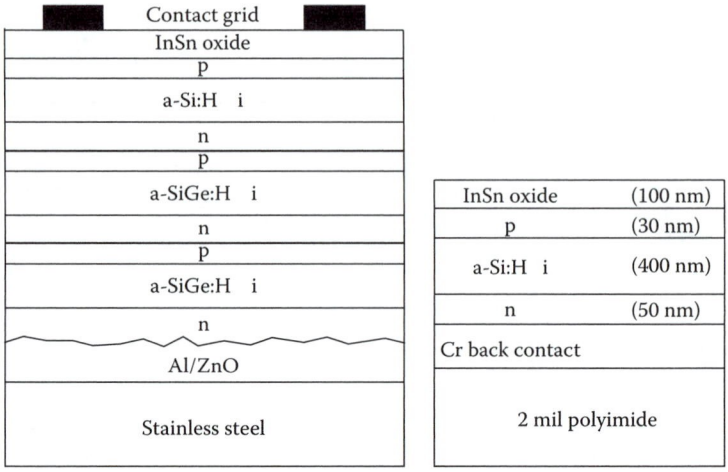

Figure 11.8 a-Si:H cell on stainless steel and polymer substrates. (Adapted from Guha et al. and Huang et al., 699, *Proc. 26th IEEE PV Spec. Conf.*, 1997, ©1997 IEEE.)

The challenge in creating the PV-on-polymer structure is keeping the polymer from deforming during heating portions of the processing. To do so, the process involves using a silicone gel between the bottom of the polymer and a heat sinking material, so that the polyimide can be protected from heat stresses.

11.2.6.2 Cell Performance

Table 10.3 in Chapter 10 shows a theoretical maximum efficiency for a-Si of 27%, whereas in the late 1990s, large-scale efficiencies for a-Si devices were reported in the 10% range, with lab cell stable efficiencies for triple junction devices of approximately 14% [18]. As of 2009, these records remain in place [1]. Considering that initial efficiencies of only a few percent were achieved, these efficiencies represent significant progress.

With the worldwide efforts in place to continually improve cell processing, along with the incorporation of novel and practical applications that can justify the use of cells with efficiencies in the 8% to 10% range, continued progress can be expected with amorphous and thin-film silicon cells.

When cells are stacked, they must be designed so that absorbing layers produce equal photocurrent, because the layers are approximately ideal current sources connected in series. The increased efficiency of stacked (series) cells is thus due to more complete photon capture, with EHPs being effectively separated at the junctions, resulting in higher overall cell open-circuit voltages resulting from the combined series junctions.

The efficiency achieved by increasing the number of series junctions is also affected by the air mass to which the cell is exposed, because certain wavelengths are subjected to greater atmospheric absorption as the air mass changes. Thus, a terrestrial cell may have high efficiency at noon, but earlier or later, when the sun travels through more atmosphere, thus increasing the air mass, the performance of one or more sections of a stacked cell may be degraded, thus reducing overall cell current. Optimal performance of stacked cells is thus achieved when they are designed for extraterrestrial use.

Overall cell efficiency is also limited by the wavelength range over which the ARC will effectively minimize reflections, because the ARC optimal thickness is 1/4 wavelength. This limits the number of effective series junctions.

11.3 GALLIUM ARSENIDE CELLS

11.3.1 Introduction

The 1.43 eV direct bandgap, along with a relatively high absorption constant, makes GaAs an attractive PV material. Historically high production costs, however, have limited the use of GaAs PV cells to extraterrestrial and other special-purpose uses, such as in concentrating collectors. Recent advances in concentrating technology, however, enable the use of significantly less active material in a module, such that cost-effective, terrestrial devices may soon be commercially available.

Production of pure gallium arsenide requires first the production of pure gallium and pure arsenic. The two materials are then combined to form GaAs. Most modern GaAs cells consist of thin films of GaAs grown on substrates such as Ge by an assortment of film growth processes.

11.3.2 Production of Pure Cell Components

11.3.2.1 Gallium

Gallium [19] was predicted by Mendeleev and first discovered by spectroscopic analysis by Lecoq de Boisbaudran in 1875. De Boisbaudran then separated out Ga from its hydroxide by electrolysis in the same year. Ga is a metal that liquefies just above room temperature, but has a very high boiling point. It is found as a trace element in diaspore, sphalerite, germanite, bauxite, and coal, all of which contain Zn, Ge, or Al, in addition to the Ga. In fact, coal flue dusts can contain up to 1.5% Ga, although most contain less than 0.1%. The relative abundance of Ga is comparable to the abundance of Pb and As, but the percentage composition of Ga in any naturally occurring mineral rarely exceeds 1%. The most important source of Ga is bauxite, even though this ore only contains 0.003 to 0.01% Ga.

Gallium can be extracted by many different methods, depending on the host material. For example, in bauxite, the weight ratio of Al to Ga is approximately 8000:1. In the Bayer process for Al extraction, bauxite is first mixed with a NaOH solution. The solution is autoclaved, diluted, and decanted, at which point a red mud is removed from the material. Decomposition follows, which results in removal of hydrated alumina along with extracts having Al:Ga ratios of approximately 200:1. This Al:Ga mixture is then subjected to evaporation and is again mixed with NaOH and fed back into the mixing, autoclave, dilution, decantation, and decomposition steps again.

When Al:Ga ratios of about 200:1 are reached in the sodium aluminate, the sodium aluminate solution can be subjected to processes such as the Pechiney process. In the Pechiney process, which was invented by Beja and came into use after World War II, the sodium aluminate solution is reacted with CO_2 in several stages. In the first stage, much of the soda and alumina are precipitated. This leaves a liquor enriched in Ga, from which Ga-rich alumina is obtained. This alumina is again dissolved in NaOH, and fractional carbonation is repeated. The final product, after enough fractional carbonation steps, is a mixed oxide with adequate Ga content to enable electrolytic deposition of Ga from the NaOH solution.

Semiconductor-grade Ga, with a purity of 99.999+%, is obtained by a variety of physical and chemical processes. These methods include chemical treatments with acids or gases at high temperatures, physicochemical methods, such as filtration of fused metal, heating in vacuum, dissolving again, and subjecting to further electrolysis or crystallization as monocrystals. It is also possible to react Ga with Cl, fractionally distill the solution until the desired purity is reached, then recover the Ga and reconvert it to metal.

11.3.2.2 Arsenic

Arsenic [20] has been known since 1250 A.D. In 1649, Schroeder published two methods of preparing the element. Arsenic is found in many forms in nature, including sulfides, arsenides, sulfarsenides, oxides, and arsenates. The most common source of As is FeSAs. When FeSAs is heated, the As sublimes, leaving FeS behind. Arsenic oxidizes rapidly if heated and, along with its compounds, is very poisonous. Arsenic is typically marketed in its arsenic trioxide form, which can be obtained at various purity levels by resublimation. This form is most commonly used in the manufacture of insecticides.

Semiconductor-grade As can be obtained by reducing a chemically purified compound with a highly pure solid or gas. One such highly pure form is arsine (AsH_3), which is also highly poisonous and requires very special precautions. If elemental As is desired, the AsH_3 can be decomposed by heating, resulting in highly pure elemental As. A large percentage of use of As in semiconductors involves the decomposition of AsH_3 at high temperatures.

11.3.2.3 Germanium

The existence of Ge [19] was predicted by Mendeleev in 1870, and it was isolated by Winkler in 1886. Only a small number of minerals contain Ge in appreciable quantities, including (AgGe)S, (AgSnGe)S, (CuZnAsGe)S, and (CuFeAsGe)S. Most Ge, however, is recovered from Pb-Zn-Cu ores in Africa.

The recovery process involves heating the ore under reducing conditions. This vaporizes Zn and Ge so they can be oxidized and collected. The fumes are then leached with H_2SO_4, and the pH is then gradually increased. As the pH increases to 3, Ge is 90% precipitated, whereas Zn begins to precipitate at a pH of 4. If Zn and Ge are both precipitated at a pH of 5, a 50:1 Zn:Ge ratio solution will produce a Zn-Ge precipitate containing close to 10% Ge. If MgO is used as the base, then a Mg-Ge precipitate containing about 10% Ge is obtained.

The next step is to react the precipitate with strong HCl, which causes $GeCl_4$ to form. The $GeCl_4$ can then be fractionally distilled and reacted with H_2O to form GeO_2. The GeO_2 can then be reduced with H_2 to obtain pure Ge. The resulting Ge can then be zone-refined to obtain semiconductor-grade Ge.

11.3.3 Fabrication of the Gallium Arsenide Cell

Crystalline gallium arsenide is somewhat more difficult to form than silicon, because gallium and arsenic react exothermally when combined. The most common means of growing GaAs crystals is the liquid-encapsulated Czochralski (LEC) method. In this method, the GaAs crystal is pulled from the melt. The melted GaAs must be confined by a layer of liquid boric oxide. The biggest challenge is to create the GaAs melt in the first place. Several methods have been developed, such as first melting the Ga, then adding the boric oxide, and finally injecting the As through a quartz tube [21].

Most modern GaAs cells, however, are prepared by growth of a GaAs film on a suitable substrate. Figure 11.9 shows one basic GaAs cell structure. The cell begins with the growth of an n-type GaAs layer on a substrate, typically Ge. Then a p-GaAs layer is grown to form the junction and collection region. The top layer of p-type GaAlAs has a bandgap of approximately 1.8 eV. This structure reduces minority carrier surface recombination and transmits photons below the 1.8 eV level to the junction for more efficient absorption.

A number of other GaAs structures have been reported, including cells of other III-V compounds. Figure 11.10 and Table 11.1 show a cascaded AlInP/GaInP/GaAs structure grown by molecular beam epitaxy (MBE) [22] and an InP cell fabricated with the organometallic vapor phase epitaxy process (OMVPE) [23].

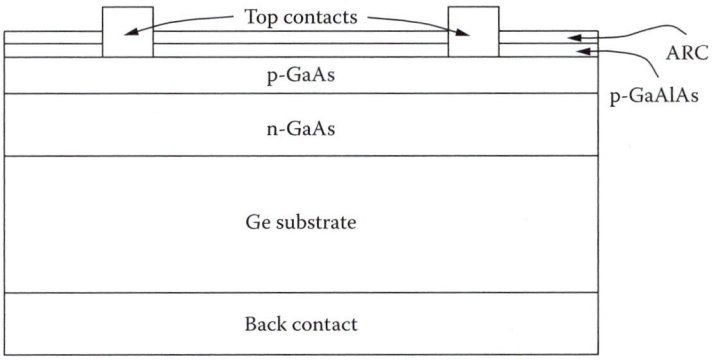

Figure 11.9 Structure of a basic GaAs cell with GaAlAs window and Ge substrate.

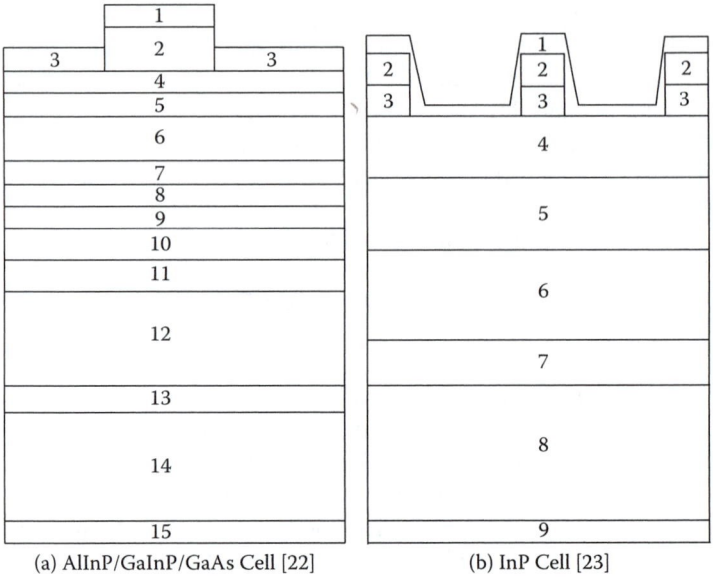

(a) AlInP/GaInP/GaAs Cell [22] (b) InP Cell [23]

Figure 11.10 Cascaded AlInP/GaInP/GaAs cell and InP cell. Properties of the regions of the cells are summarized in Table 11.1. (Adapted from Lammasniemi, et al., 823, and Hoffman, et al., 815, *Proc 26th IEEE PV Spec. Conf.*, 1997. ©1997 IEEE.)

Table 11.1 Summary of Composition of Regions of Cells of Figure 11.10

	AlInP/GaInP/GaAs Cell			InP Cell		
Region	**Material**	**Thickness**	**Doping**	**Material**	**Thickness**	**Doping**
1	Au/Ni/Ge	Contact		MgF_2/ZnS	110–55 nm	ARC
2	SiO_2/SiN_x	ARC		Au-Ge	2–3 μm	Front grid
3	GaAs	600 nm	n = 8E18	InGaAs	0.1–0.5 μm	
4	AlInP	25 nm	n = 2E18	InP	50–100 nm	p = 1E18
5	GaInP	75 nm	n = 1–4E18	InP	100–200 nm	p = 1E17
6	GaInP	400 nm	p = 5–500E16	InP	1.5–4.0 μm	n = 1E17
7	AlInP	25 nm	p = 5E18	InP	250–500 nm	n = 1E18
8	GaAs	10 nm[a]	p = 1E20	InP	400 μm	n > 1E18
9	GaAs	10 nm[a]	n = 8E18	Au-Ge	Contact	
10	GaInP	50 nm	n = 2E18			
11	GaAs	100 nm	n = 1E18			
12	GaAs	3500 nm	p = 1E17			
13	GaInP	100 nm	p = 1E19			
14	GaAs	Substrate				
15	Au/Pt/Ti	Contact				

[a] Tunnel junction.

The epitaxial growth process involves passing appropriate gases containing the desired cell constituents over the surface of the heated substrate. As the gases contact the substrate, the H or CH_3 attached to the In or P or Ga are liberated and the In, P, or Ga attaches to the substrate. Hence, to grow a layer of p-type InP, a combination of trimethyl indium, phosphine (for P), and diethyl zinc (for acceptor impurities) are mixed in the desired proportions and passed over the heated substrate for a predetermined time until the desired layer thickness is obtained. The process is repeated with different mixes of gases to form the other layers at the desired thickness.

GaAs cells remain expensive to fabricate and are thus used primarily for extraterrestrial applications and in concentrating systems.

11.3.4 Cell Performance

Cells fabricated with III-V elements are generally of extraterrestrial quality. In other words, they are expensive, high-performance units. Efficiencies in excess of 20% are common, and efficiencies of cells fabricated on more expensive GaAs substrates have exceeded 34% [24].

An important feature of extraterrestrial-quality cells is the need for them to be radiation resistant. Cells are generally tested for their degradation resulting from exposure to healthy doses of 1 MeV or higher-energy protons and electrons. Degradation is generally less than 20% for high exposure rates.

Extraterrestrial cells are sometimes exposed to temperature extremes, so the cells are also cycled between −170°C and +96°C for as many as 1600 cycles. The cells also need to pass a bending test, a contact integrity test, a humidity test, and a high-temperature vacuum test, in which the cells are tested at a temperature above 140°C in vacuum for 168 h [25].

Fill factors in excess of 80% have been achieved for GaAs cells. Single-cell open-circuit voltages are generally between 0.8 and 0.9 V.

The design of stacked cells depends on the air mass under which the device is intended to operate. In order to ensure equal photon-generated current in each absorption layer, layer thickness needs to be adjusted for the air mass under which operation is anticipated, because air masses do not attenuate the entire spectrum proportionately, as shown in Figure 2.2 in Chapter 2.

In particular, if a cell absorbs efficiently at $\lambda = 0.9$ μm, significantly more photons are available at this wavelength at AM0 than at AM1. Thus, for operation at AM1, the absorber width would need to be increased to generate a photocurrent comparable to that which a narrower layer would generate at AM0. This consideration is particularly important when optimizing cell performance at AM0, but can also be relevant for cells designed for use in regions where exposures to AM 1.5 or AM 2.0 may occur for long periods of cell operation.

The bottom line is that III-V cells are excellent extraterrestrial performers, and it appears that continued research will improve performance, reduce mass, and reduce cost of the cells. Whether III-V cells will be able to compete with other technologies for terrestrial applications will depend on the degree of future improvements in all technologies. This mystery will likely unfold before the eyes of any reader born after the fabrication of the first commercial PV module. The first commercial terrestrial module will very likely be a concentrating module, because concentrations up to 2000 suns can increase cell efficiency beyond 30% and such concentrations also minimize the use of expensive cell material and use less expensive concentrating and cooling materials instead.

11.4 COPPER INDIUM (GALLIUM) DISELENIDE CELLS

11.4.1 Introduction

The first CIS PV cell was reported in 1974 by a group at Bell Laboratories [26]. Copper indium diselenide was chosen as a potential PV material because of its attractive direct bandgap (1.0 eV), its very high optical absorption coefficient, and its potentially inexpensive preparation. Furthermore, the cell components are available in adequate quantities and the manufacturing, deployment, and decommissioning of the technology fall within acceptable environmental constraints.

While current technology involves the use of Cd as a cell component, the total quantity is relatively small, and efforts are under way to identify alternate, less toxic materials to replace the Cd. The challenge with CIS, as with other thin-film technologies, is to prepare a cell with appropriate electric field for collection of photon-generated, electron–hole pairs. Then ohmic contacts need to be affixed, with the front contact being highly conducting but transparent to incident photons. The final device structure must be stable and must be encapsulated to ensure long module lifetime.

Unlike Si, which has been studied intensely for decades and is well understood by the scientific community, the fundamentals of CIS are less well understood. For example, energy bands in Si have been studied extensively, and detailed explanations are available in nearly any solid-state devices text. CIS, on the other hand, has only recently received significant attention. As basic research on the properties of CIS leads to increased understanding of the fundamental properties of the material, chances for significant device improvement will be enhanced.

After more than 10 years of field testing and continued work on product development, commercial CIS modules are now available. One creative development has been the encapsulation of the CIS cells in cylindrical geometry, in sizes similar to standard 4 ft fluorescent lamps [27]. To add to the utility of these novel modules, the module frames are designed to impose minimal wind load so they can be installed without penetration on flat roof surfaces. Furthermore, on reflective roof surfaces, the cylindrical cells can also collect sunlight reflected off the roof surface.

11.4.2 Production of Pure Cell Components

11.4.2.1 Copper

Of all the elements used in the production of semiconductors, except for silicon, copper [28] is probably the most abundant. It is known from prehistoric times. It occasionally occurs in pure form in nature, but also occurs in many minerals, such as cuprite, malachite, azurite, chalcopyrite, and bornite. The sulfides, oxides, and carbonates are the most important sources of copper. The refining process involves smelting, leaching, and electrolysis. Because copper is used in so many applications, and is found in so many locations in so many different compounds, many different methods of copper refining have been developed. The reader is referred to Reference [28] for examples.

Copper can be refined to semiconductor grade by electrolysis of a solution of copper sulfate and sulfuric acid. The anode is made of approximately 99% pure Cu, and the solution is maintained so that almost everything that plates out on the cathode is Cu. The chemical reactions at anode and cathode are

$$\text{anode} \qquad Cu \rightarrow Cu^{2+} + 2e^{-}, \tag{11.10a}$$

$$\text{cathode} \qquad Cu^{2+} + 2e^{-} \rightarrow Cu. \tag{11.10b}$$

Although it appears that this process is exclusively moving Cu from anode to cathode, it is also moving elements such as oxygen, sulfur, arsenic, antimony, bismuth, lead, nickel, selenium, tellurium, gold, and silver from anode into solution. Oxygen is generally the most common impurity in the anode Cu, in the form of Cu_2O. As the Cu_2O moves into the solution, it reacts with the sulfuric acid to form $CuSO_4$ and H_2O along with Cu, which precipitates into the slime along with most of the other elements from the anode that do not dissolve. This loses half the Cu to the slime, but the slime then can be reprocessed.

Only the silver will dissolve, but addition of HCl or NaCl to the solution will precipitate the silver. Hence, only the $CuSO_4$ is left in solution for migration to the cathode, with the net result that the Cu plates out onto the cathode.

The slime thus contains compounds such as Cu_2S, Cu_2Se, Ag_2S, and Ag_2Se, all of which may be of interest in further refining operations for recovery of any of the other elements. In the meantime, the Cu on the cathode will be very pure if the solution is properly controlled.

11.4.2.2 Indium

Indium [29] is most commonly found with zinc, but is also found with iron, lead, and copper ores. Little use was made of In until it was found to be useful in certain semiconductors. Most commercial In is now obtained from the flue dusts and residues from smelting lead and zinc. As late as 1924, less than an ounce of pure In existed, even though it is rather widely scattered throughout the world, albeit in very small concentrations in host minerals.

Several means of extracting In from its natural states are available. In the presence of lead and zinc, the material can be melted and treated with chlorine gas or other chlorine source. This removes the Zn and In as chlorides, provided that the temperature is low enough to prevent evaporation of the In. The chloride slag is then leached with dilute sulfuric acid, which causes the In to precipitate along with Zn dust. The next step is to melt the In–Zn mixture and remove the Zn with Cl.

Indium is refined to semiconductor grade by further physical or chemical separation techniques, such as zone refining and fractional distillation of liquid In compounds.

11.4.2.3 Selenium

Selenium [30] is a group VI element, and, as a result, is very similar to sulfur in many of its chemical properties, which accounts for its common occurrence with CuS. In its natural form, it combines naturally with 16 other elements and is a major component of 39 mineral species and a minor component of another 37 mineral species. Because Se is always combined with other elements, such as S, there are no identified "reserves" of Se. Certain native plant species preferentially absorb Se, and are sometimes indicators of the presence of the element. Elemental Se is relatively nontoxic, but many Se compounds are very poisonous. Selenium is recovered primarily from the anode muds from the refining of copper. It is also recovered from flue dusts from processing copper sulfide ores.

A number of methods of recovery of Se are available, depending on the starting material and the desired end products. One method is to first eliminate the Cu in Cu ores through either aeration with H_2SO_4 or by first oxidizing the mud and then reacting with H_2SO_4 in a leaching process. The H_2SO_4 causes the Cu to precipitate out of solution, leaving the Se behind in the mud for smelting with soda and silica. The initial product of this smelting process contains only about 1% Se, but most of the Fe, As, Sb, and Pb are eliminated.

The molten charge is then oxidized with air. This volatilizes the Se, which is then caught in a scrubber and combined again with soda to produce a Se-rich soda slag. The slag is then subjected to a precious metal recovery process, leaving a slag that is leached with water and filtered, yielding sodium selenite and sodium tellurite alkaline liquor.

When the pH of the liquor is reduced to 6.2 by adding acid, the Te precipitates as tellurous acid, leaving the selenium behind in the sludge. Additional acid, followed by sulfur dioxide, precipitates Se as an amorphous sludge. This Se is washed, dried, and reduced to a powder after boiling with steam. At this point, the Se is relatively pure, but still contains some Te. The chemical reactions take place with Se and Te both included in the first three reactions, prior to precipitation of the Te in the fourth reaction. After separation of the Se, the Te can be prepared in a similar fashion. The process will yield both Se and Te, both of which will require further refining for purities in excess of 99.99%.

$$Se + O_2 \rightarrow SeO_2 \tag{11.11a}$$

$$SeO_2 + H_2O \rightarrow H_2SeO_3 \tag{11.11b}$$

$$SeO_2 + Na_2CO_3 \rightarrow Na_2SeO_3 + CO_2 \tag{11.11c}$$

$$Na_2TeO_3 + H_2SO_4 \rightarrow Na_2SO_4 + H_2TeO_3\downarrow \tag{11.11d}$$

$$H_2SeO_3 + 2SO_2 + H_2O \rightarrow 2\ H_2SO_4 + Se\downarrow \tag{11.11e}$$

$$H_2TeO_3 + 2SO_2 + H_2O \rightarrow 2\ H_2SO_4 + Te\downarrow \tag{11.11f}$$

One method of achieving high-purity Se is to dissolve the Se from the previous process in hot sodium sulfite, wherein Te and many other impurities do not dissolve. The solution is then filtered and acidified with sulfuric acid to again liberate the Se. This distillation process is repeated several times until the Se is ultimately highly purified, at which point it is formed into shot for transport and storage. The two final reactions are

$$Se + Na_2SO_3 \leftrightarrow Na_2SeSO_3, \tag{11.12a}$$

$$Na_2SeSO_3 + H_2SO_4 \rightarrow Na_2SO_4 + Se + SO_2 + H_2O. \tag{11.12b}$$

The purity of the higher-purity final Se products is generally verified spectrographically, with commercial grade having purity of better than 99.0%, high-purity grade having purity in excess of 99.99%, and ultra-high-purity grade having a purity as high as 99.9999%.

11.4.2.4 Cadmium

Cadmium [31] was discovered in 1817 [19] by Friedrich Stromeyer in zinc carbonate and by K. S. L. Hermann in a specimen of zinc oxide. The common denominator is that cadmium is commonly found embedded in zinc compounds, although cadmium sulfide is found in Scotland, Bohemia, and Pennsylvania.

Because zinc is routinely refined, the leftover cadmium from the process is generally recovered, mixed with carbon, and redistilled to yield an enriched dust. This process is repeated several times. Then, hydrochloric acid is mixed with the cadmium dust, and zinc is added, resulting in the precipitation of the cadmium. After several repetitions of this process, electrolysis is finally used to deposit the final material. Cadmium is a group II metallic element, and along with solutions of its compounds, is highly toxic.

11.4.2.5 Sulfur

Sulfur [32] is abundant in the crust of the earth, both in elemental and combined forms. Elemental sulfur tends to occur in layers above salt domes. Abundant supplies of elemental sulfur have been discovered in many locations around the world. It has been used for thousands of years in many applications.

Elemental sulfur is generally recovered by pumping superheated water down a well to melt the sulfur, after which the molten sulfur is extracted from the well. The elemental sulfur is then generally reacted with a suitable element to enable the sulfur to be carried in a gaseous state for vapor deposition onto a substrate.

11.4.2.6 Molybdenum

Molybdenum [33], a group VI metal, is an important metal for making ohmic contacts, particularly as the back contact in a CIS cell. The metal was first prepared by P. J. Hjelm in 1782. It has high strength, high corrosion resistance, and is often used in alloys, particularly in the production of stainless steel. It retains its strength at high temperatures better than most other metals.

Most Mo is mined in the United States, Canada, and Chile. Molybdenum occurs principally in MoS_2, but the concentration of MoS_2 in ores is rather small. About a ton of ore must be mined, crushed, and milled to recover about 4 lb of Mo. After the ore is crushed, the MoS_2 is recovered in relatively high concentration by floatation.

Roasting then drives off the sulfur and oxidizes the Mo to MoO_3. Reduction with H_2 at high temperature yields Mo in powder form with purity in the 99% range. The powder can be reacted with halogens to form compounds such as MoF_6 and $MoCl_5$, which are suitable for vapor-phase deposition of Mo.

11.4.3 Fabrication of the CIS Cell

While it is possible to produce both n-type and p-type CIS, homojunctions in the material are neither stable nor efficient. A good junction can be made, however, by creating a heterojunction with n-type CdS and p-type CIS.

The ideal structure uses near-intrinsic material near the junction to create the widest possible depletion region for collection of generated EHPs. The carrier diffusion length can be as much as 2 μm, which is comparable with the overall film thickness. Figure 11.11 shows a basic ZnO/CdS/CIGS/Mo cell structure, which was in popular use in 2004. Again, CIGS technology is advancing rapidly as a result of the Thin-Film Photovoltaics Partnership Program, so by the time this paragraph is read, the structure of Figure 11.11 may be suitable only for history books and general discussion of the challenges encountered in thin-film cell development.

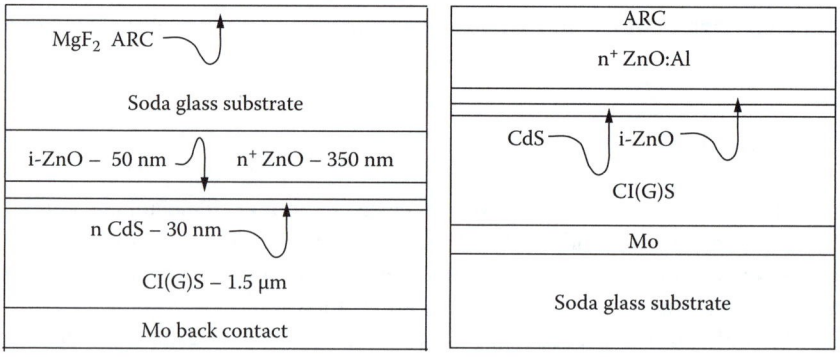

Figure 11.11 Typical CIGS thin-film structures. (Adapted from Ullal et al., *Proc 26th IEEE Spec. Conf.*, 1997, 301. ©1997 IEEE.)

Nearly a dozen processes have been used to achieve the basic cell structure of Figure 11.11. The processes include rf sputtering, reactive sputtering, chemical vapor deposition, vacuum evaporation, spray deposition, and electrodeposition. Sometimes these processes are implemented sequentially, and sometimes they are implemented concurrently.

In the physical vapor deposition (PVD) process, which was used to achieve a record laboratory cell efficiency, the constituent elements are deposited under a relatively high vacuum of 10^{-6} torr. In the PVD process, the four elements can be simultaneously evaporated, they can be sequentially evaporated followed by exposure to Se, or they can be sequentially evaporated in the presence of Se. The soda lime glass substrate is maintained between 300°C and 600°C during the evaporation process.

The front ohmic contact is straightforward, and ZnO generally works well. The trick is to achieve sufficiently high conductivity without absorbing any of the incident photons. Often, the ZnO is applied in two layers. The layer on the glass is fairly strongly n-type, with a very thin intrinsic layer in contact with the CdS. The more heavily doped layer has high conductivity, in the neighborhood of 4 Ω/square, and the narrow intrinsic layer acts as a passivation layer between the thin CdS layer and the TCO, but is narrow enough to allow efficient transport of electrons to the TCO.

11.4.4 Cell Performance

When new-technology PV cells are developed, normally small area cells are fabricated first to determine whether it appears practical to extend the technology to larger area cells and modules. At the time of this writing, the highest performance achieved for a CIGS laboratory cell was 19.4%. Scaling up to the production of modules with areas up to 3459 cm^2 resulted in decrease in efficiency to 13.5% [1]. Clearly, the challenge in module development is to overcome the factors that result in degradation of cell performance, and it is equally clear that these challenges are being undertaken ambitiously. Doing so requires an understanding of the factors that cause the degradation. Some of these considerations include general device design, contact grid design, antireflection coatings, and the sheet resistance of window layers.

An example of the trade-offs involved in scaling up a technology is the ZnO transparent contact. For a laboratory-scale device with an area of 1 cm^2 or so, the ZnO layer can be relatively thin, with a sheet resistivity of 15 Ω/square, because the relatively small current from the cell will experience minimal voltage drop through this contact. The thin window absorbs a minimal amount of incident radiation so the CIS absorber layer can achieve maximum conversion efficiency. However, as the cell is made larger, the sheet resistance of the TCO must be reduced to prevent voltage drop at the contact and corresponding degradation of fill factor and cell efficiency. The price paid for lower sheet resistance is a greater amount of absorption of incident photons by the transparent contact.

Series connection of cells can also cause cell performance degradation. Figure 11.12 shows how individual cells have been replaced by minimodules that are

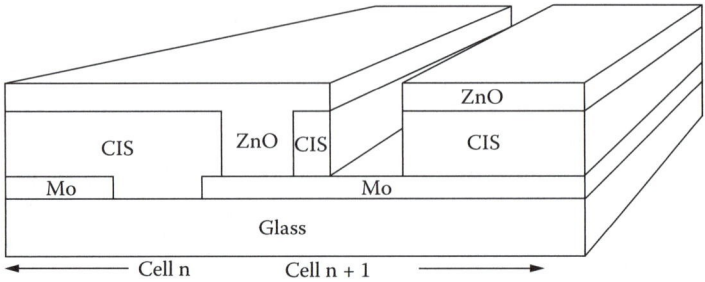

Figure 11.12 Siemens monolithic method of series cell connection. (Adapted from Tarrant, D. and Gay, *Thin-Film Photovoltaic Partnership Program—CIS-Based Thin Film PV Technology. Phase 2 Technical Report, October 1996–October 1997*, NREL report NREL/SR-520-24751, Golden, CO, May 1998.)

monolithically connected. This monolithic connection process eliminates the need for separate fabrication processes for interconnecting cells.

After deposition of the Mo, it is scribed to separate adjacent cells, creating cells with a length of about 4 ft and a width of a fraction of an inch. Next, after deposition of the CIS cell components, the CIS is scribed. Finally, after deposition of the TCO, the TCO and CIS are scribed. While this process may appear to be straightforward, it can also be appreciated that the first cut needs to remove all of the Mo, but none of the glass. The second and third cuts must leave the Mo intact. Considering thickness of a few micrometers or less for these layers, the process falls somewhat short of being classified as trivial.

By incorporating Ga into the CIS mixture, the bandgap of the material can be increased beyond 1.1 eV. This movement of the bandgap energy closer to the peak of the solar spectrum increases conversion efficiency in this wavelength region and leaves lower-energy photons for capture by the free carrier absorption process in the transparent conducting oxide (TCO) layer while the higher-energy photons are converted to EHPs. The result is increasing the cell open-circuit voltage from approximately 0.4 V to as high as 0.68 V, with fill factors approaching 80% for laboratory cells. Experiments with adding sulfur to the selenium have also resulted in cell performance improvements.

11.4.4.1 Transient Effects

Unlike the a-Si cells, the CIS cells have shown good long-term outdoor stability, with interesting recovery behavior for cells that have been light starved. Earlier modules (1990) had aperture efficiencies in the neighborhood of 7%; a 2002 module achieved an efficiency of 7.4% on a metal foil substrate [37], and a conventional CIGS module attained an efficiency of 13.6% [38].

Exposure to more intense light has actually caused efficiencies to increase. Exposure to elevated temperatures has resulted in loss of efficiency, but light soaking restored the modules to original efficiency levels.

11.5 CADMIUM TELLURIDE CELLS

11.5.1 Introduction

In theory, CdTe cells have a maximum efficiency limit close to 25%. The material has a favorable direct bandgap and a large absorption constant, allowing for cells with thickness of a few micrometers. By 2001, efficiencies approaching 17% were being achieved for laboratory cells, and module efficiencies had reached 11% for the best large-area (8390 cm^2) module [37]. Efforts were then focused on scaling up the fabrication process to mass-produce the modules, with the result that a production cost of less than $1.00/W was achieved in 2008. This cost included an escrow account to be used for recycling the materials at the end of module life [39].

Although tellurium is not as abundant as other cell components, cadmium and tellurium are both available in sufficient quantities for the production of many gigawatts of array. The amount of cadmium poses a possible fire hazard and some concern at the time of decommissioning of the modules, as indicated in Chapter 9. Analysis of the concerns, however, has shown that the Cd of the cells would be recycled at decommissioning time and that the danger of burns from any fire far exceeds the danger of contact with any Cd released from heating of the modules. Means for recycling CdTe modules exist that result in recovery of glass, $CdCO_3$, electrolytically refined Te, and clean EVA at a cost of less than $0.04 per watt [40].

Purification of all of the components of the CdTe cell has been discussed so far except Te. After a brief summary of the extraction and purification of Te, cell fabrication and cell performance will be discussed.

11.5.2 Production of Pure Tellurium [32]

Tellurium is a group VI metallic element, which was discovered by Muller in Transylvanian gold ore in 1782 and first extracted and identified by M. H. Klaproth in 1798. It is found as tellurides of copper, lead, silver, gold, iron, and bismuth and is widely distributed over the surface of the earth, although its percentage in the earth's crust is very small. The primary sources of tellurium for production are leftovers from copper and lead refining, where tellurium and selenium both appear in very small quantities.

When copper is produced by electrolysis, the tellurium is precipitated along with other impurities in the copper to become a sludge at the anode. Several alternative chemical means are then used to separate the tellurium from the other impurities in the sludge. The first step is to produce tellurium dioxide, which precipitates out of solution while the other impurities remain dissolved. Alternatively, other impurities may first be precipitated, leaving behind higher concentrations of tellurium in the form of tellurous acid. If the oxide is produced, then the oxide is reduced to form elemental tellurium.

Elemental tellurium remains contaminated with iron, copper, tin, silver, lead, antimony, and bismuth and can be further purified by low-pressure distillation, where the heavier metals remain in the residue. Selenium, however, is volatile and remains a contaminant in the distilled tellurium. Further purification can be achieved by dissolving the tellurium in strong nitric acid. Diluting and boiling hydrolyzes the tellurium to a precipitate form. The precipitate is separated, washed, dissolved in hydrochloric acid, and reduced with sulfur dioxide.

This relatively pure tellurium can be brought to the ultra-high-purity state by zone refining in an inert gas atmosphere, and single crystals can be grown by either the Czochralski or Bridgman methods.

Tellurium is classified as "probably" toxic, and reasonable care is recommended in its handling.

11.5.3 Production of the CdTe Cell

Figure 11.13 shows a typical CdTe cell structure. The cell begins with a glass superstrate with a transparent, conducting oxide layer about 1 μm thick, a thin CdS buffer layer about 0.1 μm thick, a CdTe layer a few μm thick, and a rear contact of Au, Cu/Au, Ni, Ni/Al, ZnTe:Cu, or (Cu, HgTe).

The TCO layer has been fabricated with SnO, InSnO, and Cd_2SnO_4. The Cd_2SnO_4 has been shown to exhibit better conductivity and better transparency [41] than the other TCOs and may end up as the preferred TCO. The Cd_2SnO_4 layer is deposited by combining CdO and SnO_2 in a 2:1 proportion in a target material, which is then deposited by means of radio frequency magnetron sputtering onto the glass superstrate.

The thin *n*-type CdS layer has been deposited by the metal organic chemical vapor deposition (MOCVD) process as well as by other thin-film deposition techniques. The layer needs to be annealed prior to deposition of the CdTe in order to reduce CdS surface roughness, thus reducing defects on the CdS/CdTe boundary. This is normally accomplished in air at approximately 400°C for about 20 min.

An interesting challenge existed with regard to the CdS layer, because the layer is so thin. During deposition of the CdTe or subsequent heat treatment of the cell, intermixing of the CdS and CdTe can occur at the boundary. This can result in junctions between the CdTe and the TCO layer, which causes significant reductions in

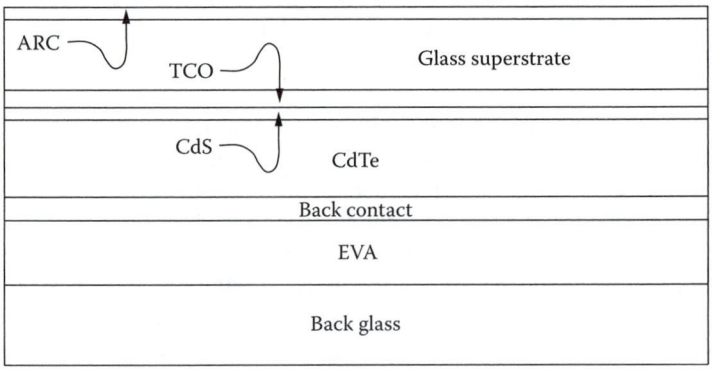

Figure 11.13 Basic structure of a CdTe PV cell. (Adapted from Ullal et al., *Proc. 26th IEEE PV Spec. Conf.*, 1997, 301, ©1997 IEEE.)

cell open-circuit voltage. Several methods of minimizing this mixing were proposed [42], and this problem has now been overcome.

Numerous methods have been used to deposit the CdTe layer, including atmospheric pressure chemical vapor deposition (APCVD), atomic layer epitaxy (ALE), close-spaced sublimation (CSS), electrodeposition (ED), laser ablation, physical vapor deposition (PVD), screen printing (SP), spray, sputtering, and MOCVD [24]. The CdTe layer is subjected to a heat treatment in the presence of $CdCl_2$ for about 20 min at about 420°C. This treatment enhances grain growth in the CdTe layer to reduce grain boundary trapping effects on minority carriers. Non-heat-treated CdTe cells tend to have open-circuit voltages less than 0.5 V, while after heat treating, V_{OC} can exceed 0.8 V.

Another important step in optimizing cell performance is to ensure stability of the back contact. Before the back contact is applied, the CdTe is etched with nitric-phosphoric (NP), resulting in a layer of elemental Te at the back CdTe surface. The elemental Te produces a more stable contact between the *p*-type CdTe and the back metal [43].

Final processing of modules involves encapsulation of the back of the cell with a layer of EVA between the metallization and another layer of glass.

11.5.4 Cell Performance

No fewer than nine companies have shown an interest in commercial applications of CdTe. As of 2001, depending on the fabrication methodology, efficiencies of close to 17% had been achieved for small-area cells (≈1 cm^2), and 11% on a module with an area of 8390 cm^2 [24]. Furthermore, sufficient experience had been logged with large-scale production to identify areas in which improvement was needed [44]. These areas included improving the design, operation, and control of a CdTe reactor; increasing the understanding of the fundamental properties of CdTe films; maintaining uniformity of materials and device properties over large areas, including the interdiffusion of CdTe and CdS and the back contact; obtaining stability of the back contact; and addressing any environmental concerns over Cd. Because Te availability may limit cell production, increasing performance with thinner layers of CdTe is also desirable. Reducing the thickness of the CdTe layer to 0.5 μm will allow for four to five times the cell area, provided that efficiency can be maintained or increased. Just as in the case of other thin-film arrays, area-related costs limit the minimal cost per watt of CeTe arrays, so increase in efficiency remains important to minimize the overall cost per watt.

Fill factors for lab cells have been obtained in the 65% to 75% range [45], with the slope of the cell J-V curve at V_{OC} in the range of 5 Ω-cm^2. Short-circuit current densities upward from 25 mA/cm^2 were obtained by using a thin buffer layer of CdS along with a thin insulating TCO layer between the heavily doped TCO and the CdS layer [24].

Experimental CdTe arrays in sizes up to 25 kW were deployed in California, Ohio, Colorado, and Florida, and in Tunisia, for testing purposes [24]. Soon after it was shown that no degradation was observable after 2 years, production-scale manufacturing began. CdTe modules are now being manufactured and marketed at the rate of more than 1 GW annually for utility-scale projects and the magic $1.00/W production cost barrier has now been broken for these modules, with an announced 4th Quarter 2008 production cost of $0.93/W [39].

11.6 EMERGING TECHNOLOGIES

11.6.1 New Developments in Silicon Technology

While progress continues on conventional Si technology, new ideas are also being pursued for crystalline and amorphous Si cells. The goal of Si technology has been to maintain good transport properties, while improving photon absorption and reducing the material processing cost of the cells. When the thickness of Si is reduced, some sort of substrate material is required to maintain sufficient physical strength for the cell. Two substrates under investigation in 2002 were ceramic and graphite, both having led to cell efficiencies in the 10% range [46]. Diffusion of contaminants such as Mg, Mn, and Fe from ceramic substrates has been shown to limit performance of Si on ceramics, and methods of introducing barrier layers at the Si–ceramic interface have been shown to limit the introduction of contaminants into the Si [47]. It is likely that these and other versions of thin Si cells will continue to attract the attention of the PV community, including recent research on thin Si on glass.

Another interesting opportunity for cost reduction in Si cell production is to double up on processing steps. For example, a technique has been developed for simultaneously diffusing boron and phosphorus in a single step, along with growing a passivating oxide layer [46].

As an alternative to the pn junction approach to Si cells, MIS-IL (metal insulator semiconductor inversion layer) cells have been fabricated with 18.5% efficiency [48]. The cell structure is shown in Figure 11.14. The cell incorporates a point-contacted back electrode to minimize the rear surface recombination, along with Cs beneath

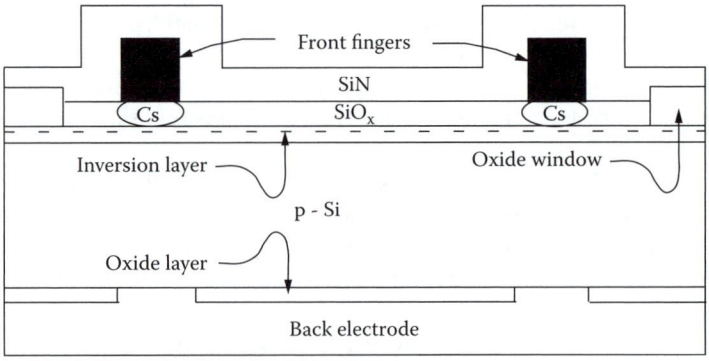

Figure 11.14 Structure of an MIS-inversion layer Si cell. (Adapted from Metz et al., *Proc. 26th IEEE PV Spec. Conf., 1997*, 31. ©1997 IEEE.)

the MIS front grid and oxide window passivation of the front surface to define the cell boundaries. Further improvement in cell performance can be obtained by texturing the cell surfaces.

The MIS-IL cell uses the top SiO_x layer as a tunnel junction. The presence of positive charges in the SiO_x layer creates the electric field from oxide to p-Si, thus creating the inversion layer at the top of the p-type material. This field then separates the EHPs just as the E-field at a pn junction separates the carriers.

Other groups have worked on cells with both contacts on the back, in order to eliminate the shadowing of the front surface by the front contacts [45]. The ACE Designs project, funded by the European Community, resulted in the development of three types of rear contact Si cells: metallization wraparound (MWA), metallization wrap-through (MWT), and emitter wrap-through (EWT). A laser-grooved, buried grid process (LGBG) applied to the MWA technology was estimated to have the best potential for lowest-cost rear-contacted cells [49].

New developments in surface texturing may also simplify the process and result in additional improvement in Si device performance. Discovery of new substrates and methods of growing good-quality Si on them is also an interesting possibility for performance improvement and cost reduction for Si cells.

Another idea that has been investigated is to combine crystalline and amorphous Si into a tandem cell arrangement to take advantage of the different bandgaps of the two materials in increasing absorption efficiency [50]. Work in Japan produced a 1 cm^2 tandem cell with an open-circuit voltage of 1.4 V, a fill factor of 71.9%, and an efficiency of 14.5% [51]. The cell is composed of a textured, transparent conducting oxide on a glass substrate, followed by an a-Si:H cell, an interlayer, a microcrystalline Si cell, and a back contact/reflector. The interlayer is incorporated to produce some reflection of the incident photons back into the a-Si to better match the current densities of the two cells. This Heterojunction with Intrinsic Thin layer (HIT) technology is now commercially available in two types of modules, one of which is a standard, single-sided module and the other of which is a bifacial module that absorbs photons from the front and the back sides. The cell efficiency approaches 19%, and the module efficiency is rated at 15.7% [52].

Work also continues on dendritic web Si. Standard dendritic web cell thicknesses were 100 μm in 2002. Recent reports describe 70 μm cell thickness, with cell efficiencies up to 14.1% [53].

Still another interesting process for producing crystalline, thin-film Si cells involves the epitaxial growth of very thin crystalline cells on existing crystalline cells [54]. The growth takes place at a temperature that does not melt the existing cell, and thus the epitaxially grown cell can be "peeled" off the existing cell and mounted on its own substrate, usually glass. By fabricating the new cell on a cell with a textured surface, the new cell also will have a textured surface. The overall thickness of the epitaxially grown cell is less than 20 μm, and the epitaxial growth process is convenient for adding *n*-type impurities and then switching to *p*-type impurities to produce the pn junction of the cell. The thin layers justify epitaxial growth as the mechanism for cell production.

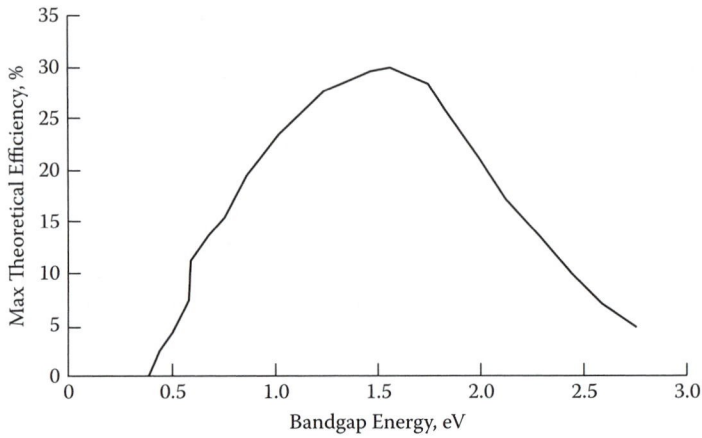

Figure 11.15 Theoretical maximum PV cell efficiency versus bandgap. (Adapted from Granata, Sites, and Tuttle, *Proc. 14th NREL PV Program Review*, AIP Conf. Proceedings 394, Lakewood, CO, 1996, 621–630.)

Because new ideas will continue to emerge as interest in Si PV technology continues to grow, the interested reader is encouraged to attend PV conferences and to read the conference publications to stay up to date in the field.

11.6.2 CIS-Family-Based Absorbers

Figure 11.15 shows the theoretical maximum efficiency of a solar absorber as a function of bandgap. Table 11.2 shows the bandgaps of a family of CIS-type

Table 11.2 Bandgaps of CIS-Related Materials

Material	Bandgap (eV)
$CuInSe_2$	1.05
$AgInSe_2$	1.24
$CuInS_2$	1.56
$CuGaSe_2$	1.67
$AgGaSe_2$	1.69
$AgInS_2$	1.95
$CuGaS_2$	2.33
$AgGaS_2$	2.56

Source: Based on data from Tarrant and Gay, *Research on High Efficiency, Large Area CuInSe2 Modules, Final Technical Report*, NREL, Golden, CO, April 1995.

materials. Much is yet to be learned about inhomogeneous absorbers and composite absorbers composed of combinations of these various materials. The possibility of multijunction devices is also being explored.

Efficiency of large-area devices is critically dependent on spatial uniformity of the absorber layers and electrodes as well as on the performance of cell interconnects. Hence, with any of the listed absorber possibilities, the key to performance will be in the engineering of reliable and reproducible module processing techniques.

Even an issue as mundane as the quality and cleanliness of the glass substrate can have a significant effect on module performance. Polishing the glass with CeO_2 has been shown to increase module performance [56]. Furthermore, it appears that if a certain amount of Na diffuses from the glass to the absorber, the absorber performance is increased [55].

Meanwhile, work is under way to reduce the material usage in the production of CIS modules in order to further reduce production costs. Examples of reduction of material use include halving the width of the Mo contact layer, reduction in the use of H_2S and H_2Se, and a reduction in ZnO, provided that a minimum thickness can be maintained [57].

11.6.3 Other III-V and II-VI Emerging Technologies

It appears that compound tandem cells will receive appreciable emphasis in the III-V family of cells over the next few years. For example, $Ga_{0.84}In_{0.16}As_{0.68}P_{0.32}$, lattice-matched to GaAs, has a bandgap of 1.55 eV and may prove to be an ideal material for use under AM0 conditions, because it also has good radiation resistance [58]. Cells have been fabricated with $Al_{0.51}In_{0.49}P$ and $Ga_{0.51}In_{0.49}P$ window layers, with the best 1 cm^2 cell having an efficiency of just over 16%, but having a fill factor of 85.4%. The best-performing window was the AlInP.

Mechanical stacking of materials having different lattice constants has also been proposed [59]. Performance modeling of a proposed 26-layer, 4-junction device based on GaInP, GaAs, GaInAsP, and GaInAs has shown the possibility of 35.3% conversion efficiency.

Cell efficiencies can be increased by concentrating sunlight on the cells. Although the homojunction cell efficiency limit under concentration is just under 40%, quantum well (QW) cells have been proposed to increase the concentrated efficiency beyond the 40% level [60]. In QW cells, intermediate energy levels are introduced between the host semiconductor's valence and conduction bands to permit absorption of lower-energy photons. These levels must be chosen carefully so that they will not act as recombination centers, however, or the gains of EHPs from lower-energy incident photons will be lost to the recombination processes. Laboratory cells have shown higher V_{OC} resulting from a decrease in dark current for these cells. Intense study is under way with the goal of understanding the electronic processes that take place within these cells [61].

11.6.4 Other Technologies

11.6.4.1 Thermophotovoltaic Cells

Up to this point, discussion has been limited to the conversion of visible and near-infrared spectrum to EHPs. The reason is simply that the solar spectrum peaks out in the visible range. However, heat sources and incandescent light sources produce radiation in the longer infrared regions, and in some instances, it is convenient to harness radiated heat from these processes by converting it to electricity. This means using semiconductors with smaller bandgaps, such as Ge. More exotic structures, such as InAsSbP, with a bandgap of 0.45–0.48 eV have also been fabricated. The InAsSbP can be fabricated as *p*-type and *n*-type. The pn junction is grown on a substrate of InAs [62].

11.6.4.2 Intermediate Band Solar Cells

In all cells described up to this point, absorption of a photon has resulted in the generation of a single EHP. If an intermediate band material is sandwiched between two ordinary semiconductors, it appears that it may be possible for the material to absorb two photons of relatively low energy to produce a single EHP at the combined energies of the two lower-energy photons. The first photon raises an electron from the valence band to the intermediate level, creating a hole in the valence band, and the second photon raises the electron from the intermediate level to the conduction band. The trick is to find an intermediate band material that will "hold" the electron until another photon of the appropriate energy impinges upon the material. Such a material should have half its states filled with electrons and half should be empty in order to optimally accommodate this electron transfer process. It appears that III-V compounds may be the best candidates for implementation of this technology. Theoretical maximum efficiency of such a cell is 63.2% [63].

11.6.4.3 Supertandem Cells

If a large number of cells are stacked with the largest bandgap on top and the bandgap of subsequent cells decreasing, the theoretical maximum efficiency is 86.8% [64]. A 1 cm^2 4-junction cell has been fabricated with an efficiency of 35.4%. The maximum theoretical efficiency of this cell is 41.6% [65]. Perhaps one day one of the readers of this paragraph (or one of their great-great grandchildren) will fabricate a cell with the maximum theoretical efficiency.

11.6.4.4 Hot Carrier Cells

The primary loss mechanism in PV cells is the energy lost in the form of heat when an electron is excited to a state above the bottom of the conduction band of a PV cell by a photon with energy greater than the bandgap. The electron will normally drop to the lowest-energy available state in the conduction band, with the energy lost in the process being converted to heat. Hence, if this loss mechanism can be overcome, the efficiency of a cell with a single junction should be capable of approaching that of a supertandem cell. One method of preventing the release of this heat energy by the electron is to heat the cell, so that the electron will remain at the higher-energy state. The process is called *thermoelectronics* and is currently being investigated [64].

11.6.4.5 Optical Up- and Down-Conversion

An alternative to varying the electrical bandgap of a material is to reshape the energies of the incident photon flux. Certain materials have been shown to be capable of absorbing two photons of two different energies and subsequently emitting a photon of the combined energy. Other materials have been shown to be capable of absorbing a single high-energy photon and emitting two lower-energy photons. These phenomena are similar to up-conversion and down-conversion in communications circuits at radio frequencies.

By the use of both types of materials, the spectrum incident on a PV cell can be effectively narrowed to a range that will result in more efficient absorption in the PV cell. An advantage of this process is that the optical up- and down-converters need not be a part of the PV cell. They simply need to be placed between the photon source and the PV cell. In tandem cells, the down-converter would be placed ahead of the top cell and the up-converter would be integrated into the cell structure just ahead of the bottom cell.

11.6.4.6 Organic PV Cells

Even more exotic than any of the previously mentioned cells is the organic cell. In the organic cell, electrons and holes are not immediately formed as the photon is absorbed. Instead, the incident photon creates an *exciton*, which is a bound EHP. In order to free the charges, the exciton binding energy must be overcome. This dissociation occurs at the interface between materials of high electron affinity and low ionization potential [65]. Photoluminescence is related to this process. Just to end this section with a little chemistry, the reader will certainly want to know that one material that is a candidate for organic PV happens to be poly{2,5-dimethoxy-1,4-phenylene-1,2-ethenylene-2-methoxy-5-(2-ethyl-hexyloxy)-1,4-phenylene-1,2-ethenylene}, which goes by the nickname M3EH-PPV. Whether M3EH-PPV will dominate the PV market one day remains to be seen. So far, efficiencies of this very challenging technology have been in the 1% range.

11.7 NEW DEVELOPMENTS IN SYSTEM DESIGN

11.7.1 Micro Grids

Currently, utility-interactive inverters must incorporate anti-islanding algorithms so they will shut down if the grid shuts down, regardless of whether other PV inverters are connected to the grid that might tend to fool the inverter into thinking they are the utility grid. But what if a situation might exist where it is important to maintain a PV island in the midst of a utility outage? One might imagine a remote town or village with a single feeder connection to the utility grid. If the grid goes down, should the entire town go down, or, alternatively, be transferred to individual or collective nonutility generation? What about setting up PV systems such that islanding is allowed for the systems in the town, but still disconnects the town or neighborhood from the utility feed until utility power is restored? The question, yet to be answered, includes determining the extent to which battery backup will be required, the means of connecting alternate, nonutility generation sources, whether inverters should act as current sources or, perhaps, voltage sources, and how to minimize any sort of system losses.

11.7.2 Smart Grids

The best sources of renewable energy tend to be in places that are not densely populated, such as sunny and dry regions in the southwestern United States to windy regions along the coasts and in the region between Texas and the Dakotas. As a result, the ability to move electricity generated in these regions to highly populated areas is lacking. In order to achieve goals of 20% or 25% of all electricity used coming from renewable sources, it will be necessary to develop a grid structure that will enable efficient transmission of power from regions where renewable energy is most efficiently generated to regions where it is needed for consumption. It has been reported that because of transmission congestion, consumers in the eastern United States pay $16.5 billion annually in higher electricity prices [66]. Furthermore, it is estimated that by updating our electricity infrastructure, Americans can save $80 billion in unnecessary power plant construction over the next 20 years [67].

Smart grids are a work in progress. Many of the characteristics that smart grids will need have been identified, such as updating regulations to discourage utilities from purchasing obsolete equipment, developing open standards, providing consumers with energy management tools and time-of-use pricing, and providing utilities with incentives for selling less electricity. Advanced load management to better match load profiles to electric generation capacity as well as using consumer generation and storage to supplement grid needs will also likely be components of the smart grid. The grid will likely be able to communicate with plug-in electric vehicles such that the vehicles will be able to deliver power to the grid when the grid needs power, while when the grid has excess power, it will be available for charging of electric vehicle batteries.

Communications will be the heart of the smart grid, as generation and consumption capacity and needs will be able to communicate, the result being more efficient management of the transfer of power from generation to load. It is likely that many of the lessons learned from the Internet and cell phone networks will be applicable to the smart grid communication network, with the difference that rather than multiplexing packets of information, large packets of power will be efficiently dispatched over extra-high-voltage transmission lines as well as over neighborhood distribution lines.

With advanced communications, the "smart garage" concept also becomes possible [68], in which plug-in hybrid cars can deliver energy to the grid during peak hours and then charge from the grid during off-peak hours.

Thus, as new ideas are brought to market in PV, wind, and other renewable sources, it is highly likely, if not absolutely certain, that communication with a smart grid will be an integral part of the design of the systems.

11.7.3 Inverter Performance Enhancement

As utility-scale PV systems grow larger and larger, interesting questions arise over these systems. For example, how will the grid react if all of a sudden, clouds cause the output of a 20 MW system to drop within seconds to 5 MW or less? Or, for that matter, what if the opposite should happen? Will it be necessary to incorporate some sort of storage at the array output so the inverter output will be more stable when cloud cover is anything but stable? Will the grid have sufficient damping capability to prevent rapid PV power changes from affecting grid stability? In effect, this amounts to applying either an impulse or a step function excitation to the grid and then observing the response. Will the grid be smart enough to deal with fluctuations in supply and load?

Then the question arises as to whether an inverter should always be required to operate at unity power factor. Generally, large-scale utility rotating generation must supply reactive power as well as real power. There is no question that inverter output power factor will undergo serious discussions among the ranks of utilities, inverter manufacturers, and government research laboratories. Fortunately, it should be a straightforward exercise for an inverter to be programmed to determine the optimal phase angle for current injection into the grid. This would include having the inverter output supply leading power factor current to cancel out lagging power factor loads when such a power factor adjustment would lower utility line losses. Not only is this an interesting electrical issue, but it will probably also lead to some interesting billing questions, such as whether the PV generator should be credited for any role it may play in reducing utility line losses.

On a smaller scale, present dc-coupled battery-backup utility-interactive systems encounter greater losses than straight grid-connected systems. Improving the overall electrical throughput efficiency of these systems will also present a challenge to the designers of inverters and charge controllers. The competition, of course, lies in ac-coupled systems that reduce losses to battery charging and discharging. The competition is not unlike the competition between crystalline silicon and thin-film technologies for the most cost-effective life-cycle energy production.

11.8 SUMMARY

Regardless of the technology or technologies that may result in low-cost, high-performance PV cells, it must be recognized that the life-cycle cost of a cell depends on the cell's having the longest possible, maintenance-free lifetime. Thus, along with the developments of new technologies for absorbers, development of reliable encapsulants and packaging for the modules will also merit continued research and development activity.

Every year engineers make improvements on products that have been in existence for many years. Automobiles, airplanes, electronic equipment, building materials, and many more common items see improvement every year. Even the yo-yo, a popular children's toy during the 1940s and 1950s, came back with better-performing models. Hence, it should come as no surprise to the engineer to see significant improvements and scientific breakthroughs in the PV industry well into the next millennium. The years ahead promise exciting times for the engineers and scientists working on the development of new PV cell and system technologies, provided that the massive planning and execution phases can be successfully undertaken.

Problems

11.1. Assume a PV module is to have dimensions of 37.5 × 63.2 in. Also assume that 8-in. round cells are available.
   a. Calculate the percentage of the module area that can be covered with circular cells.
   b. If the cells operate with 18% efficiency, and if there is no mismatch loss among cells, calculate the overall module efficiency if the round cells are used.
   c. Assume the cells are "squared up" by sawing off the edges to obtain cells that measure 6 in. between the straight sides. Calculate the percentage fill of these "squared up" cells, and repeat the module efficiency calculation.
   d. What is the percentage loss of material in the "squaring up" process?

11.2. Determine the expression for the depth of the junction after the drive-in diffusion step.

11.3. Sketch the impurity distribution profile between the junction and the back contact of a single-crystal silicon cell to show how the annealed aluminum of the back contact creates an accelerating E-field.

11.4. What volume and weight of tellurium are needed to produce a square meter of CdTe thin film with a thickness of 1.5 μm? Assume that the Cd and Te occupy equal volumes within the film.

11.5. What volume and weight of indium are required to fabricate a 1 MW PV CIS array if the CIS layer thickness is 1.5 μm, assuming that 18% of the layer volume is due to the In? Assume an array efficiency of 10% and standard test conditions.

11.6 The energy payback time (EPBT) for a PV technology is defined as the time needed for the completed cell to generate an amount of electrical energy equal to the energy required for producing the cell. Refer to current literature, and tabulate the EPBT for the most efficient crystalline Si cells, multicrystalline Si cells, thin Si (microcrystalline or nanocrystalline) cells, and amorphous Si cells.

11.7 Calculate or look up the EPBT for CIGS cells, CdTe cells, and GaAs or other III-V cell. If you do the calculation, list your assumptions.

REFERENCES

1. Green, M., Third generation photovoltaics: Assessment of progress over the last decade, *Proc. 34th IEEE PV Spec. Conf.*, 2009.
2. Green, M. et al., Solar cell efficiency tables (Version 33), *Progress in Photovoltaics Research and Applications*, Vol. 17, No. 1, John Wiley & Sons, Chichester, U.K., 2009, 85–93.

3. Tarrant, D. E. and Gay, R. R., *Research on High Efficiency, Large Area CuInSe2-Based Thin Film Modules, Final Technical Report*, NREL, Golden, CO, April 1995.
4. Fthenakis, V., Recycling workshop 2, *Proc. 34th IEEE PV Spec. Conf., 2009.*
5. U. S. Energy Information Administration, *Reference Case Projections for Electricity Capacity and Generation by Fuel Tables (2006–2030)*, www.eia.doe.gov/oiaf/ieo/ieoecg.html.
6. Markvart, T., Ed., *Solar Electricity*, John Wiley & Sons, Chichester, U.K., 1994.
7. Streetman, B. G., *Solid State Electronic Devices,* 4th ed., Prentice-Hall, Englewood Cliffs, NJ, 1995.
8. Bohland, J., Accelerated aging of PV encapsulants by high intensity UV exposure, *Proc. 1998 Photovoltaic Performance and Reliability Workshop*, Cocoa Beach, FL, November 3–5, 1998.
9. *Sunpower discovers the "Surface Polarization Effect" in high efficiency solar cells*, Sunpower Corp Internal Document, personal communication with SunPower representative, May 2009.
10. www.schottsolar.com (for information on EFG process).
11. Hanoka, J. I., *Proc. 29th IEEE PV Spec. Conf.*, 2002, 66–69.
12. DelleDonne, E. et al., *Proc. 29th IEEE PV Spec. Conf.*, 2002, 82–85.
13. Basore, P. A., *Proc. 29th IEEE PV Spec. Conf.*, 2002, 4952.
14. Brendel, R. et al., *Proc. 29th IEEE PV Spec. Conf.*, 2002, 86–89.
15. Guha, S., Yang, J. et al., *Proc. 26th IEEE PV Spec. Conf.*, 1997, 607–610.
16. Huang, J., Lee, Y. et al., *Proc. 26th IEEE PV Spec. Conf.*, 1997, 699–702.
17. Guha, S. and Yang, J., *Proc. 29th IEEE PV Spec. Conf.*, 2002, 1070–1075.
18. Yang, J., Banerjee, A. et al., *Proc. 26th IEEE PV Spec. Conf.*, 1997, 563–568.
19. Standen, A., Executive Ed., *Kirk-Othmer Encyclopedia of Chemical Technology*, 2nd ed., Vol. 10, John Wiley & Sons, New York, 1968.
20. Standen, A., Executive Ed., *Kirk-Othmer Encyclopedia of Chemical Technology*, 2nd ed., Vol. 2, John Wiley & Sons, New York, 1968.
21. Williams, R., *Modern GaAs Processing Methods*, Artech House, Norwood, MA, 1990.
22. Lammasniemi, J. et al., *Proc. 26th IEEE PV Spec. Conf.*, 1997, 823–826.
23. Hoffman, R. et al., *Proc. 26th IEEE PV Spec. Conf.*, 1997, 815–818.
24. Kazmerski, L. L., *Proc. 29th IEEE PV Spec. Conf.*, 2002, 21–27.
25. Brown, M. R. et al., *Proc. 26th IEEE PV Spec. Conf.*, 1997, 805–810.
26. Shay, J. L., Wagner, S., and Kasper, H. M., *Appl. Phys. Lett.*, 27, No. 2, 1975, 89.
27. Optimized Photovoltaic Systems for Commercial Rooftops http://www.solyndra.com/Products/Optimized-PV.
28. Standen, A., Executive Ed., *Kirk-Othmer Encyclopedia of Chemical Technology*, 2nd ed., Vol. 6, John Wiley & Sons, New York, 1968.
29. Standen, A., Executive Ed., *Kirk-Othmer Encyclopedia of Chemical Technology*, 2nd ed., Vol. 11, John Wiley & Sons, New York, 1968.
30. Standen, A., Executive Ed., *Kirk-Othmer Encyclopedia of Chemical Technology*, 2nd ed., Vol. 17, John Wiley & Sons, New York, 1968.
31. *The Encyclopedia Britannica*, 15th ed., Encyclopedia Britannica, Inc., Chicago, IL, 1997.
32. Standen, A., Executive Ed., *Kirk-Othmer Encyclopedia of Chemical Technology*, 2nd ed., Vol. 19, John Wiley & Sons, New York, 1968.
33. Standen, A., Executive Ed., *Kirk-Othmer Encyclopedia of Chemical Technology*, 2nd ed., Vol. 13, John Wiley & Sons, New York, 1968.
34. Ullal, H. S., Zweibel, K., and von Roedern, B., *Proc. 26th IEEE PV Spec. Conf.*, 1997, 301–305.

35. Carlson, D. E. et al., *Proc. 25th IEEE PV Spec. Conf*, 1996, 1023–1028.
36. Tarrant, D. E. and Gay, R. R., *Thin-Film Photovoltaic Partnership Program—CIS-Based Thin Film PV Technology. Phase 2 Technical Report, October 1996–October 1997*, NREL report NREL/SR-520-24751, Golden, CO, May 1998.
37. Kazmerski, L. L., *Proc 29th IEEE PV Spec. Conf.*, 2002, 21–27.
38. Konagai, M., *Proc. 29th IEEE PV Spec. Conf.*, 2002, 38–43.
39. Fthenakis session on recycling, First Solar person at *Proc. 34th IEEE PV Spec. Conf*, 2009, Fthenakis presentation.
40. Bohland, J. et al., *Proc. 26th IEEE PV Spec. Conf.*, 1997, 355–358.
41. Wu, X., Sheldon, P. et al, *Proc. 26th IEEE PV Spec. Conf.*, 1997, 347–350.
42. McCandless, B. E. and Birkmire, R. W., *Proc 26th IEEE PV Spec. Conf.*, 1997, 307–312.
43. Levi, D. H. et al., *Proc. 26th IEEE PV Spec. Conf.*, 1997, 351–354.
44. Birkmire, R. W., *Proc. 26th IEEE PV Spec. Conf.*, 1997, 295–300.
45. Zhou, C. Z. et al., *Proc. 26th IEEE PV Spec. Conf.*, 1997, 287–290.
46. Krygowski, T., Rohatgi, A., and Ruby, D., *Proc 26th IEEE PV Spec. Conf.*, 1997, 19–24.
47. Slaoui, A. et al., *Proc. 29th IEEE PV Spec. Conf.*, 2002, 90–93.
48. Metz, A. et al., *Proc. 26th IEEE PV Spec. Conf.*, 1997, 31–34.
49. Schönecker, A. et al., *Proc. 29th IEEE PV Spec. Conf.*, 2002, 106–109.
50. Kuznicki, Z. T., *Proc. 26th IEEE PV Spec. Conf.*, 1997, 291–294.
51. Yamamoto, K. et al., *Proc. 29th IEEE PV Spec. Conf.*, 2002, 1110–1113.
52. HIT Double™ Bifacial Photovoltaic Modules data sheet download from http://sanyo.com/solar/.
53. Meier, D. L. et al., *Proc. 29th IEEE PV Spec. Conf.*, 2002, 110–113.
54. Brendel, R. et al., *Proc. 29th IEEE PV Spec. Conf.*, 2002, 86–89.
55. Granata, J., Sites, J. R. and Tuttle, J. R., *Proc. 14th NREL PV Program Review*, AIP Conf. Proceedings 394, Lakewood, CO, 1996, 621–630.
56. Tarrant, D. E. and Gay, R. R., *Research on High Efficiency, Large Area CuInSe2 Modules, Final Technical Report*, NREL, Golden, CO, April 1995.
57. Weiting, R., *Proc. 29th IEEE PV Spec. Conf.*, 2002, 478–483.
58. Jaakkola, R. et al., *Proc. 26th IEEE PV Spec. Conf.*, 1997, 891–894.
59. Sharps, P. R. et al., *Proc. 26th IEEE PV Spec. Conf.*, 1997, 895–898.
60. Barnham, K. W. J. and Duggan, G., A new approach to high-efficiency multi-band-gap solar cells, *J. Appl. Phys.* 67, 1990, 3490–3493.
61. Corkish, R. and Honsberg, C., *Proc. 26th IEEE PV Spec. Conf.*, 1997, 923–926.
62. Khvostikov, V. P. et al., *Proc. 29th IEEE PV Spec. Conf.*, 2002, 943–946.
63. Luque, A. et al., *Proc. 29th IEEE PV Spec. Conf.*, 2002, 1190–1193.
64. Green, M. A., *Proc. 29th IEEE PV Spec. Conf.*, 2002, 1330–1334.
65. Zahler, J. M. et al., *Proc. 29th IEEE PV Spec. Conf.*, 2002, 1029–1032.
66. Hendricks, Bracken, Wired for Progress, *Solar Today*, Vol. 23, No. 4, May 2009, p. 26 (solartoday.org).
67. Berst, Jesse, For a Smart Grid, Look to Smart States, *Solar Today*, Vol. 23, No. 4, May 2009, p. 30 (solartoday.org).
68. Burns, Cameron M., The Smart Garage (V2G): Guiding the next big energy solution, *RMI Solutions J.*, Vol. 1, No. 1, July 2008, p. 22.

Appendix A

Average Daily Irradiation for Selected Cities

The U.S. data in the following tables have been compiled by the National Renewable Energy Laboratory. Data outside the United States have been compiled by Sandia National Laboratories. Average daily irradiation in kWh/m^2 is tabulated for each month of the year, including annual averages, for an assortment of orientations, including:

1. Fixed, south-facing arrays tilted at latitude – 15°, latitude, and latitude + 15°
2. Single-axis, east-west tracking mounts tilted at three angles
3. Horizontal and vertical mounts (U.S. data only)
4. Double-axis tracking mounts

The following cities are included, listed alphabetically by state, then by country. For additional listings, the reader is encouraged to consult *Stand-Alone Photovoltaic Systems: A Handbook of Recommended Design Practices* by Sandia National Laboratories, Albuquerque, New Mexico. Appendix B lists the NREL and several other Web sites that provide irradiation information.

1. Fairbanks, Alaska
2. Sacramento, California
3. Denver, Colorado
4. Miami, Florida
5. Atlanta, Georgia
6. Boston, Massachusetts
7. Albuquerque, New Mexico
8. Bismarck, North Dakota
9. Austin, Texas
10. Seattle, Washington
11. Luanda, Angola
12. Buenos Aires, Argentina
13. Melbourne, Australia
14. Shanghai, People's Republic of China
15. Paris-St. Maur, France
16. New Delhi, India
17. Tokyo, Japan
18. Nairobi, Kenya
19. Mexico D. F., Mexico
20. Stockholm, Sweden

Fairbanks, AK Average Daily Peak Sun Hours, kWh/m², Latitude: 64°49′ N, Longitude: 147°52′ W

	Array Tilted at Latitude – 15°		Array Tilted at Latitude		Array Tilted at Latitude + 15°				
Month	Fixed Array	1-Axis Tracker	Fixed Array	1-Axis Tracker	Fixed Array	1-Axis Tracker	Horizontal Array	Vertical South-Facing Array	2-Axis Tracking Array
Jan	0.7	0.7	0.7	0.8	0.8	0.8	0.1	0.8	0.8
Feb	2.2	2.6	2.4	2.7	2.5	2.8	0.8	2.5	2.8
Mar	4.5	5.7	4.7	5.9	4.7	5.8	2.3	4.5	5.8
Apr	5.6	7.7	5.6	7.7	5.3	7.4	4.0	4.9	7.7
May	5.7	8.2	5.3	8.0	4.6	7.6	5.1	4.1	8.4
Jun	5.7	8.3	5.2	8.0	4.5	7.6	5.6	3.9	8.7
Jul	5.4	7.6	4.9	7.4	4.3	6.9	5.1	3.7	7.9
Aug	4.5	6.0	4.2	5.8	3.8	5.5	3.7	3.3	6.0
Sep	3.4	4.4	3.4	4.4	3.2	4.2	2.3	3.0	4.4
Oct	1.9	2.2	2.0	2.3	2.0	2.3	1.0	2.0	2.3
Nov	1.0	1.1	1.1	1.2	1.1	1.2	0.3	1.1	1.2
Dec	0.2	0.2	0.3	0.3	0.3	0.3	0.0	0.3	0.3
Avg	3.4	4.6	3.3	4.5	3.1	4.4	2.5	2.8	4.7

Sacramento, CA Average Daily Peak Sun Hours, kWh/m², Latitude: 38°31′ N, Longitude: 121°30′ W

	Array Tilted at Latitude – 15°		Array Tilted at Latitude		Array Tilted at Latitude + 15°				
Month	Fixed Array	1-Axis Tracker	Fixed Array	1-Axis Tracker	Fixed Array	1-Axis Tracker	Horizontal Array	Vertical South-Facing Array	2-Axis Tracking Array
Jan	2.6	3.0	2.9	3.3	3.1	3.4	1.9	2.7	3.4
Feb	3.9	4.7	4.2	4.9	4.3	5.0	3.0	3.6	5.0
Mar	5.2	6.6	5.4	6.7	5.2	6.6	4.3	3.8	6.7
Apr	6.5	8.6	6.3	8.5	5.9	8.2	5.9	3.6	8.6
May	7.3	10.1	6.8	9.8	6.0	9.2	7.2	3.0	10.2
Jun	7.6	10.8	7.0	10.3	6.0	9.7	7.9	2.7	11.0
Jul	7.8	11.2	7.2	10.8	6.3	10.1	7.9	2.9	11.4
Aug	7.5	10.4	7.2	10.2	6.5	9.8	7.0	3.6	10.4
Sep	6.7	9.1	6.9	9.2	6.6	9.0	5.7	4.5	9.2
Oct	5.3	6.8	5.7	7.1	5.8	7.2	4.0	4.6	7.2
Nov	3.3	4.0	3.7	4.3	3.9	4.5	2.4	3.4	4.5
Dec	2.4	2.8	2.7	3.0	2.9	3.2	1.7	2.6	3.2
Avg	5.5	7.3	5.5	7.4	5.2	7.2	4.9	3.4	7.6

Denver, CO Average Daily Peak Sun Hours, kWh/m², Latitude: 39°45′ N, Longitude: 104°52′ W

	Array Tilted at Latitude – 15°		Array Tilted at Latitude		Array Tilted at Latitude + 15°				
Month	Fixed Array	1-Axis Tracker	Fixed Array	1-Axis Tracker	Fixed Array	1-Axis Tracker	Horizontal Array	Vertical South-Facing Array	2-Axis Tracking Array
Jan	3.8	4.8	4.4	5.2	4.8	5.5	2.4	4.5	5.6
Feb	4.6	5.9	5.1	6.2	5.3	6.4	3.3	4.6	6.4
Mar	5.4	7.0	5.6	7.2	5.6	7.1	4.4	4.3	7.2
Apr	6.1	8.1	6.0	8.0	5.6	7.7	5.6	3.6	8.1
May	6.2	8.4	5.9	8.1	5.2	7.7	6.2	2.8	8.5
Jun	6.6	9.1	6.1	8.8	5.2	8.2	6.9	2.6	9.4
Jul	6.6	9.1	6.1	8.7	5.3	8.2	6.7	2.7	9.2
Aug	6.3	8.6	6.1	8.4	5.5	8.0	6.0	3.2	8.6
Sep	5.9	7.9	6.0	7.9	5.8	7.8	5.0	4.0	8.0
Oct	5.1	6.7	5.6	7.1	5.7	7.1	3.8	4.6	7.1
Nov	4.0	5.0	4.6	5.5	4.8	5.7	2.6	4.4	5.7
Dec	3.5	4.4	4.2	4.9	4.5	5.2	2.1	4.3	5.3
Avg	5.4	7.1	5.5	7.2	5.3	7.1	4.6	3.8	7.4

Miami, FL Average Daily Peak Sun Hours, kWh/m², Latitude: 25°48′ N, Longitude: 80°16′ W

	Array Tilted at Latitude – 15°		Array Tilted at Latitude		Array Tilted at Latitude + 15°				
Month	Fixed Array	1-Axis Tracker	Fixed Array	1-Axis Tracker	Fixed Array	1-Axis Tracker	Horizontal Array	Vertical South-Facing Array	2-Axis Tracking Array
Jan	4.1	5.2	4.7	5.7	5.0	5.9	3.5	4.1	6.0
Feb	4.7	6.1	5.2	6.4	5.4	6.5	4.2	3.9	6.6
Mar	5.5	7.0	5.7	7.2	5.6	7.1	5.2	3.4	7.2
Apr	6.2	7.9	6.1	7.8	5.7	7.5	6.0	2.6	7.9
May	5.9	7.4	5.6	7.2	5.0	6.7	6.0	1.9	7.4
Jun	5.5	6.6	5.1	6.3	4.5	5.9	5.6	1.6	6.7
Jul	5.7	7.0	5.4	6.7	4.8	6.3	5.8	1.7	7.1
Aug	5.6	6.9	5.5	6.7	5.0	6.4	5.6	2.1	6.9
Sep	5.1	6.2	5.1	6.2	4.9	6.1	4.9	2.7	6.2
Oct	4.7	5.9	5.1	6.1	5.1	6.2	4.4	3.5	6.2
Nov	4.2	5.2	4.7	5.6	4.9	5.8	3.7	3.9	5.9
Dec	3.9	4.9	4.5	5.4	4.9	5.7	3.3	4.1	5.8
Avg	5.1	6.4	5.2	6.5	5.1	6.3	4.8	3.0	6.7

Atlanta, GA Average Daily Peak Sun Hours, kWh/m², Latitude: 33°39′ N, Longitude: 84°36′ W

	Array Tilted at Latitude – 15°		Array Tilted at Latitude		Array Tilted at Latitude + 15°				
Month	Fixed Array	1-Axis Tracker	Fixed Array	1-Axis Tracker	Fixed Array	1-Axis Tracker	Horizontal Array	Vertical South-Facing Array	2-Axis Tracking Array
Jan	3.4	4.2	3.8	4.5	4.1	4.7	2.6	3.5	4.8
Feb	4.2	5.3	4.6	5.5	4.7	5.6	3.4	3.7	5.7
Mar	5.1	6.5	5.3	6.6	5.1	6.5	4.5	3.5	6.6
Apr	6.0	7.7	5.8	7.6	5.4	7.3	5.7	3.0	7.7
May	6.2	7.9	5.8	7.7	5.2	7.2	6.2	2.4	8.0
Jun	6.3	8.0	5.8	7.6	5.1	7.1	6.4	2.2	8.1
Jul	6.1	7.6	5.7	7.3	5.0	6.8	6.2	2.2	7.7
Aug	5.9	7.4	5.7	7.2	5.2	6.9	5.7	2.7	7.4
Sep	5.3	6.6	5.4	6.7	5.1	6.5	4.8	3.2	6.7
Oct	4.9	6.2	5.2	6.4	5.3	6.5	4.1	4.0	6.5
Nov	3.8	4.7	4.2	5.0	4.5	5.2	2.9	3.8	5.3
Dec	3.2	3.9	3.7	4.3	3.9	4.5	2.4	3.5	4.5
Avg	5.0	6.3	5.1	6.4	4.9	6.2	4.6	3.1	6.6

Boston, MA Average Daily Peak Sun Hours, kWh/m², Latitude: 41°40′ N, Longitude: 71°10′ W

	Array Tilted at Latitude – 15°		Array Tilted at Latitude		Array Tilted at Latitude + 15°				
Month	Fixed Array	1-Axis Tracker	Fixed Array	1-Axis Tracker	Fixed Array	1-Axis Tracker	Horizontal Array	Vertical South-Facing Array	2-Axis Tracking Array
Jan	3.0	3.6	3.4	3.9	3.6	4.1	1.9	3.4	4.1
Feb	3.8	4.7	4.2	5.0	4.3	5.1	2.7	3.9	5.1
Mar	4.6	5.7	4.7	5.9	4.6	5.8	3.7	3.7	5.9
Apr	5.2	6.5	5.0	6.5	4.7	6.2	4.7	3.1	6.6
May	5.7	7.3	5.3	7.1	4.7	6.6	5.6	2.8	7.4
Jun	6.0	7.7	5.5	7.4	4.8	6.9	6.1	2.6	7.9
Jul	6.0	7.8	5.6	7.5	4.9	7.0	6.1	2.8	7.9
Aug	5.7	7.3	5.5	7.1	5.0	6.8	5.4	3.1	7.3
Sep	5.0	6.3	5.1	6.4	4.9	6.2	4.3	3.5	6.4
Oct	4.1	5.0	4.3	5.2	4.4	5.2	3.0	3.6	5.3
Nov	2.8	3.4	3.1	3.6	3.3	3.7	1.9	3.0	3.8
Dec	2.5	2.9	2.9	3.2	3.1	3.4	1.6	2.9	3.4
Avg	4.5	5.7	4.6	5.7	4.4	5.6	3.9	3.2	5.9

Albuquerque, NM Average Daily Peak Sun Hours, kWh/m², Latitude: 35°03′ N, Longitude: 106°37′ W

	Array Tilted at Latitude – 15°		Array Tilted at Latitude		Array Tilted at Latitude + 15°				
Month	Fixed Array	1-Axis Tracker	Fixed Array	1-Axis Tracker	Fixed Array	1-Axis Tracker	Horizontal Array	Vertical South-Facing Array	2-Axis Tracking Array
Jan	4.6	5.9	5.3	6.5	5.8	6.9	3.2	5.2	6.9
Feb	5.4	7.1	6.0	7.5	6.2	7.7	4.2	5.1	7.7
Mar	6.3	8.3	6.5	8.6	6.5	8.5	5.4	4.5	8.6
Apr	7.3	10.0	7.2	9.9	6.6	9.5	6.8	3.7	10.0
May	7.7	10.6	7.2	10.3	6.3	9.7	73.7	2.8	10.8
Jun	7.8	10.8	7.1	10.4	6.1	9.7	8.1	2.4	11.1
Jul	7.4	9.9	6.9	9.5	6.0	8.9	7.5	2.5	10.0
Aug	7.2	9.5	6.9	9.3	6.3	8.9	6.9	3.2	9.5
Sep	6.6	8.8	6.8	9.0	6.5	8.8	5.9	4.2	9.0
Oct	5.9	7.9	6.5	8.3	6.6	8.4	4.7	5.1	8.4
Nov	4.8	6.3	5.5	6.8	5.9	7.1	3.5	5.2	7.2
Dec	4.3	5.5	5.0	6.1	5.5	6.5	2.9	5.1	6.6
Avg	6.3	8.4	6.4	8.5	6.2	8.4	5.6	4.1	8.8

Bismarck, ND Average Daily Peak Sun Hours, kWh/m², Latitude: 46°46′ N, Longitude: 100°45′ W

	Array Tilted at Latitude – 15°		Array Tilted at Latitude		Array Tilted at Latitude + 15°				
Month	Fixed Array	1-Axis Tracker	Fixed Array	1-Axis Tracker	Fixed Array	1-Axis Tracker	Horizontal Array	Vertical South-Facing Array	2-Axis Tracking Array
Jan	3.1	3.7	3.5	4.0	3.7	4.2	1.7	3.7	4.2
Feb	4.0	4.9	4.4	5.2	4.5	5.3	2.6	4.2	5.3
Mar	5.0	6.3	5.2	6.4	5.1	6.4	3.8	4.3	6.5
Apr	5.6	7.3	5.5	7.3	5.1	7.0	4.9	3.7	7.4
May	6.1	8.3	5.7	8.0	5.1	7.6	6.0	3.3	8.4
Jun	6.5	8.9	5.9	8.6	5.1	8.0	6.6	3.1	9.2
Jul	6.8	9.5	6.3	9.2	5.5	8.6	6.8	3.4	9.7
Aug	6.4	8.7	6.1	8.5	5.5	8.1	5.8	3.7	8.7
Sep	5.3	7.0	5.4	7.0	5.1	6.8	4.2	3.9	7.0
Oct	4.2	5.3	4.5	5.5	4.5	5.5	2.8	3.9	5.5
Nov	2.9	3.4	3.2	3.7	3.4	3.9	1.7	3.2	3.9
Dec	2.6	3.0	3.0	3.3	3.2	3.5	1.4	3.1	3.6
Avg	4.9	6.4	4.9	6.4	4.7	6.3	4.0	3.6	6.6

Austin, Texas Average Daily Peak Sun Hours, kWh/m², Latitude: 30°18′ N, Longitude: 97°42′ W

	Array Tilted at Latitude – 15°		Array Tilted at Latitude		Array Tilted at Latitude + 15°				
Month	Fixed Array	1-Axis Tracker	Fixed Array	1-Axis Tracker	Fixed Array	1-Axis Tracker	Horizontal Array	Vertical South-Facing Array	2-Axis Tracking Array
Jan	3.7	4.6	4.2	5.0	4.4	5.2	3.0	3.8	5.2
Feb	4.4	5.6	4.8	5.9	5.0	6.0	3.8	3.8	6.0
Mar	5.2	6.6	5.4	6.7	5.3	6.6	4.7	3.4	6.7
Apr	5.6	7.1	5.5	7.0	5.1	6.7	5.4	2.7	7.1
May	5.8	7.3	5.5	7.1	4.9	6.7	5.9	2.1	7.4
Jun	6.4	8.3	5.9	8.0	5.1	7.4	6.6	1.9	8.5
Jul	6.7	8.7	6.2	8.4	5.4	7.8	6.8	2.0	8.9
Aug	6.5	8.5	6.3	8.3	5.7	7.9	6.3	2.6	8.5
Sep	5.7	7.3	5.8	7.3	5.5	7.2	5.2	3.3	7.4
Oct	5.0	6.5	5.4	6.8	5.5	6.8	4.4	4.0	6.9
Nov	4.1	5.1	4.6	5.5	4.8	5.7	3.3	4.0	5.7
Dec	3.5	4.4	4.0	4.8	4.3	5.0	2.8	3.8	5.1
Avg	5.2	6.7	5.3	6.7	5.1	6.6	4.9	3.1	7.0

Seattle, WA Average Daily Peak Sun Hours, kWh/m², Latitude: 47°27′ N, Longitude: 122°18′ W

	Array Tilted at Latitude – 15°		Array Tilted at Latitude		Array Tilted at Latitude + 15°				
Month	Fixed Array	1-Axis Tracker	Fixed Array	1-Axis Tracker	Fixed Array	1-Axis Tracker	Horizontal Array	Vertical South-Facing Array	2-Axis Tracking Array
Jan	1.5	1.6	1.6	1.8	1.7	1.8	1.0	1.5	1.8
Feb	2.3	2.7	2.5	2.8	2.5	2.8	1.7	2.2	2.9
Mar	3.5	4.2	3.6	4.3	3.5	4.2	2.8	2.8	4.3
Apr	4.6	5.6	4.4	5.5	4.1	5.3	4.1	3.0	5.6
May	5.4	6.9	5.1	6.7	4.5	6.3	5.3	3.0	7.0
Jun	5.7	7.3	5.2	7.0	4.5	6.6	5.8	2.8	7.5
Jul	6.1	8.2	5.7	7.9	4.9	7.4	6.1	3.1	8.3
Aug	5.6	7.3	5.4	7.2	4.9	6.8	5.2	3.4	7.4
Sep	4.7	5.8	4.7	5.9	4.5	5.7	3.8	3.4	5.9
Oct	3.0	3.6	3.2	3.7	3.2	3.7	2.2	2.7	3.7
Nov	1.7	1.9	1.8	2.0	1.8	2.0	1.2	1.7	2.0
Dec	1.3	1.4	1.4	1.5	1.4	1.5	0.8	1.3	1.5
Avg	3.8	4.7	3.7	4.7	3.5	4.5	3.3	2.6	4.9

Luanda, Angola, Average Daily Peak Sun Hours, kWh/m², Latitude: 8°49′ S, Longitude: 13°13′ W

	Array Tilted at Latitude – 15°		Array Tilted at Latitude		Array Tilted at Latitude + 15°		
Month	Fixed Array	1-Axis Tracker	Fixed Array	1-Axis Tracker	Fixed Array	1-Axis Tracker	2-Axis Tracking Array
Jan	5.92	7.62	5.56	7.20	4.94	6.28	7.67
Feb	6.07	7.83	5.87	7.66	5.40	6.96	7.84
Mar	5.43	7.02	5.49	7.19	5.30	6.89	7.20
Apr	4.89	6.19	5.19	6.68	5.27	6.76	6.79
May	4.60	5.61	5.11	6.34	5.42	6.70	6.73
Jun	4.18	5.01	4.75	5.80	5.14	6.27	6.34
Jul	3.36	4.17	3.71	4.78	3.93	5.11	5.14
Aug	3.70	4.75	3.95	5.21	4.04	5.36	5.37
Sep	4.57	5.96	4.68	6.21	4.60	6.08	6.23
Oct	5.06	6.66	4.97	6.60	4.66	6.11	6.69
Nov	5.60	7.27	5.31	6.93	4.77	6.11	7.30
Dec	6.16	7.87	5.72	7.36	5.02	6.33	7.95
Ann Avg	4.96	6.33	5.03	6.50	4.87	6.25	6.77

Buenos Aires, Argentina, Average Daily Peak Sun Hours, kWh/m², Latitude: 34°58′ S, Longitude: 58°48′ W

	Array Tilted at Latitude – 15°		Array Tilted at Latitude		Array Tilted at Latitude + 15°		
Month	Fixed Array	1-Axis Tracker	Fixed Array	1-Axis Tracker	Fixed Array	1-Axis Tracker	2-Axis Tracking Array
Jan	7.13	9.80	6.58	9.24	5.77	8.05	9.85
Feb	6.49	8.72	6.19	8.52	5.62	7.74	8.74
Mar	5.45	7.02	5.47	7.20	5.21	6.89	7.22
Apr	4.46	5.50	4.75	5.97	4.80	6.03	6.07
May	3.57	4.07	4.02	4.64	4.25	4.90	4.91
Jun	2.93	3.13	3.39	3.67	3.65	3.96	4.01
Jul	3.24	3.57	3.70	4.14	3.95	4.42	4.45
Aug	4.11	4.98	4.48	5.51	4.60	5.66	5.67
Sep	5.07	6.38	5.19	6.68	5.06	6.52	6.70
Oct	5.90	7.86	5.71	7.80	5.27	7.21	7.91
Nov	6.47	8.90	6.02	8.46	5.33	7.44	8.92
Dec	7.12	9.85	6.51	9.18	5.65	7.88	9.94
Ann Avg	5.16	6.65	5.17	6.75	4.93	6.39	7.03

Melbourne, Australia, Average Daily Peak Sun Hours, kWh/m², Latitude: 37°49′ S, Longitude: 144°58′ E

	Array Tilted at Latitude – 15°		Array Tilted at Latitude		Array Tilted at Latitude + 15°		
Month	Fixed Array	1-Axis Tracker	Fixed Array	1-Axis Tracker	Fixed Array	1-Axis Tracker	2-Axis Tracking Array
Jan	7.15	9.95	6.60	9.39	5.78	8.19	9.99
Feb	6.37	8.63	6.07	8.44	5.51	7.68	8.65
Mar	3.96	5.38	3.94	5.53	3.74	5.30	5.54
Apr	4.14	5.06	4.41	5.49	4.45	5.55	5.58
May	3.51	3.93	3.96	4.49	4.20	4.74	4.76
Jun	3.13	3.32	3.65	3.90	3.96	4.22	4.27
Jul	3.31	3.61	3.80	4.19	4.08	4.48	4.51
Aug	3.72	4.37	4.05	4.85	4.17	4.99	4.99
Sep	4.61	5.89	4.72	6.17	4.59	6.04	6.19
Oct	5.36	7.27	5.18	7.22	4.77	6.68	7.32
Nov	5.37	7.62	5.01	7.25	4.45	6.39	7.63
Dec	5.93	8.45	5.45	7.88	4.77	6.78	8.51
Ann Avg	4.71	6.21	4.74	6.23	4.54	5.92	6.50

Shanghai, China, Average Daily Peak Sun Hours, kWh/m², Latitude: 31°17′ N, Longitude: 121°28′ E

	Array Tilted at Latitude – 15°		Array Tilted at Latitude		Array Tilted at Latitude + 15°		
Month	Fixed Array	1-Axis Tracker	Fixed Array	1-Axis Tracker	Fixed Array	1-Axis Tracker	2-Axis Tracking Array
Jan	3.38	3.74	3.82	4.31	4.06	4.59	4.62
Feb	3.07	3.55	3.28	3.92	3.33	4.02	4.03
Mar	4.27	5.54	4.35	5.80	4.23	5.67	5.82
Apr	4.85	6.58	4.70	6.53	4.34	6.04	6.62
May	5.34	7.38	4.99	7.02	4.45	6.17	7.40
Jun	4.69	6.63	4.33	6.17	3.83	5.29	6.69
Jul	5.82	8.01	5.38	7.53	4.74	6.54	8.06
Aug	5.99	8.04	5.72	7.84	5.20	7.11	8.05
Sep	5.20	6.72	5.22	6.90	4.98	6.61	6.91
Oct	4.38	5.37	4.66	5.83	4.71	5.89	5.93
Nov	3.47	3.90	3.88	4.45	4.08	4.70	4.72
Dec	3.11	3.35	3.57	3.92	3.84	4.22	4.27
Ann Avg	4.46	5.73	4.49	5.85	4.32	5.57	6.09

Paris-St. Maur, France, Average Daily Peak Sun Hours, kWh/m², Latitude: 48°49′ N, Longitude: 2°30′ E

	Array Tilted at Latitude – 15°		Array Tilted at Latitude		Array Tilted at Latitude + 15°		
Month	Fixed Array	1-Axis Tracker	Fixed Array	1-Axis Tracker	Fixed Array	1-Axis Tracker	2-Axis Tracking Array
Jan	1.77	1.77	2.06	2.06	2.24	2.24	2.24
Feb	2.47	2.54	2.75	2.82	2.91	2.94	2.94
Mar	3.75	4.56	3.90	4.79	3.88	4.69	4.81
Apr	4.32	6.02	4.25	5.99	4.04	5.54	6.06
May	5.01	7.39	4.78	7.05	4.41	6.22	7.41
Jun	5.37	8.04	5.05	7.50	4.61	6.45	8.10
Jul	5.14	7.66	4.87	7.21	4.47	6.28	7.69
Aug	4.59	6.60	4.45	6.46	4.18	5.87	6.62
Sep	3.95	5.04	4.02	5.19	3.93	4.98	5.20
Oct	2.74	3.01	2.95	3.27	3.02	3.31	3.33
Nov	1.71	1.71	1.95	1.95	2.11	2.11	2.11
Dec	1.56	1.56	1.83	1.83	2.02	2.02	2.02
Ann Avg	3.53	4.66	3.57	4.68	3.49	4.39	4.88

New Delhi, India, Average Daily Peak Sun Hours, kWh/m², Latitude: 28°35′ N, Longitude: 77°12′ E

	Array Tilted at Latitude – 15°		Array Tilted at Latitude		Array Tilted at Latitude + 15°		
Month	Fixed Array	1-Axis Tracker	Fixed Array	1-Axis Tracker	Fixed Array	1-Axis Tracker	2-Axis Tracking Array
Jan	5.04	6.38	5.83	7.38	6.28	7.87	7.92
Feb	6.37	8.09	7.04	8.97	7.31	9.23	9.24
Mar	7.05	8.60	7.31	9.02	7.18	8.83	9.05
Apr	7.12	9.23	6.94	9.17	6.42	8.50	9.30
May	7.38	9.83	6.87	9.36	6.08	8.25	9.86
Jun	6.76	9.15	6.19	8.53	5.38	7.32	9.23
Jul	4.50	6.31	4.20	5.94	3.75	5.17	6.34
Aug	5.53	7.44	5.30	7.27	4.83	6.60	7.46
Sep	5.66	7.23	5.70	7.44	5.46	7.13	7.45
Oct	6.09	7.34	6.57	7.99	6.69	8.09	8.13
Nov	5.62	7.49	6.43	8.56	6.88	9.05	9.08
Dec	4.87	6.06	5.73	7.11	6.26	7.68	7.77
Ann Avg	6.00	7.76	6.18	8.06	6.04	7.81	8.40

Tokyo, Japan, Average Daily Peak Sun Hours, kWh/m², Latitude: 35°41′ N, Longitude: 140°14′ W

	Array Tilted at Latitude – 15°		Array Tilted at Latitude		Array Tilted at Latitude + 15°		
Month	Fixed Array	1-Axis Tracker	Fixed Array	1-Axis Tracker	Fixed Array	1-Axis Tracker	2-Axis Tracking Array
Jan	2.95	3.14	3.34	3.63	3.55	3.87	3.90
Feb	3.22	3.64	3.47	4.03	3.53	4.14	4.15
Mar	3.42	4.52	3.47	4.74	3.35	4.64	4.76
Apr	3.63	5.21	3.50	5.18	3.23	4.80	5.25
May	3.81	5.61	3.58	5.34	3.21	4.71	5.62
Jun	3.32	5.03	3.09	4.69	2.76	4.03	5.08
Jul	3.68	5.47	3.43	5.15	3.07	4.48	5.49
Aug	3.80	5.49	3.62	5.37	3.30	4.88	5.50
Sep	2.99	4.28	2.96	4.40	2.80	4.23	4.41
Oct	2.56	2.98	2.67	3.24	2.65	3.27	3.29
Nov	2.63	2.79	2.92	3.19	3.06	3.37	3.39
Dec	2.68	2.76	3.08	3.24	3.31	3.50	3.54
Ann Avg	3.22	4.24	3.26	4.35	3.15	4.16	4.53

Nairobi, Kenya, Average Daily Peak Sun Hours, kWh/m², Latitude: 1°18′ S, Longitude: 36°45′ E

	Array Tilted at Latitude – 15°		Array Tilted at Latitude		Array Tilted at Latitude + 15°		
Month	Fixed Array	1-Axis Tracker	Fixed Array	1-Axis Tracker	Fixed Array	1-Axis Tracker	2-Axis Tracking Array
Jan	6.93	8.57	6.46	8.08	5.67	7.02	8.62
Feb	7.14	8.95	6.89	8.73	6.29	7.92	8.96
Mar	6.41	8.17	6.49	8.35	6.26	7.98	8.37
Apr	5.32	6.78	5.65	7.29	5.75	7.36	7.40
May	4.40	5.51	4.86	6.21	5.13	6.55	6.57
Jun	4.13	5.09	4.66	5.88	5.02	6.34	6.41
Jul	3.46	4.37	3.81	4.98	4.02	5.32	5.35
Aug	4.02	5.19	4.30	5.68	4.42	5.83	5.84
Sep	5.26	6.80	5.42	7.08	5.33	6.91	7.09
Oct	5.80	7.44	5.69	7.37	5.32	6.81	7.48
Nov	5.93	7.49	5.60	7.12	5.01	6.26	7.52
Dec	6.52	8.06	6.03	7.52	5.24	6.44	8.15
Ann Avg	5.44	6.87	5.49	7.02	5.29	6.73	7.31

Mexico D. F., Mexico, Average Daily Peak Sun Hours, kWh/m², Latitude: 19°33′ N, Longitude: 99°18′ W

	Array Tilted at Latitude – 15°		Array Tilted at Latitude		Array Tilted at Latitude + 15°		
Month	Fixed Array	1-Axis Tracker	Fixed Array	1-Axis Tracker	Fixed Array	1-Axis Tracker	2-Axis Tracking Array
Jan	4.32	5.06	4.90	5.85	5.23	6.23	6.27
Feb	6.24	7.39	6.86	8.17	7.11	8.40	8.41
Mar	7.71	9.51	7.99	9.96	7.86	9.74	9.99
Apr	6.22	8.07	6.07	8.02	5.64	7.41	8.13
May	5.93	7.84	5.57	7.45	4.97	6.56	7.86
Jun	4.94	6.66	4.58	6.20	4.06	5.32	6.72
Jul	4.92	6.64	4.60	6.24	4.10	5.42	6.67
Aug	5.43	7.19	5.22	7.02	4.78	6.37	7.20
Sep	5.00	6.51	5.04	6.69	4.84	6.41	6.70
Oct	4.45	5.67	4.82	6.15	4.87	6.22	6.26
Nov	4.50	5.29	5.06	6.04	5.36	6.38	6.40
Dec	4.51	5.54	5.23	6.49	5.68	6.99	7.07
Ann Avg	5.36	6.78	5.50	7.04	5.38	6.79	7.31

Stockholm, Sweden, Average Daily Peak Sun Hours, kWh/m², Latitude: 59°21′ N, Longitude: 17°57′ E

	Array Tilted at Latitude – 15°		Array Tilted at Latitude		Array Tilted at Latitude + 15°		
Month	Fixed Array	1-Axis Tracker	Fixed Array	1-Axis Tracker	Fixed Array	1-Axis Tracker	2-Axis Tracking Array
Jan	1.43	1.43	1.67	1.67	1.81	1.81	1.81
Feb	2.46	2.47	2.76	2.76	2.91	2.91	2.91
Mar	3.85	4.63	4.02	4.85	3.99	4.74	4.86
Apr	4.12	5.82	4.05	5.77	3.82	5.34	5.86
May	5.17	8.16	4.91	7.76	4.52	6.83	8.18
Jun	5.45	8.94	5.12	8.33	4.67	7.14	9.03
Jul	5.27	8.51	4.98	8.00	4.56	6.95	8.56
Aug	4.57	6.79	4.42	6.62	4.13	6.00	6.80
Sep	3.46	4.42	3.52	4.53	3.42	4.34	4.54
Oct	2.09	2.20	2.25	2.38	2.30	2.41	2.43
Nov	1.09	1.09	1.25	1.25	1.34	1.34	1.34
Dec	1.05	1.05	1.24	1.24	1.35	1.35	1.35
Ann Avg	3.33	4.63	3.35	4.60	3.24	4.26	4.81

Appendix B

Design Review Checklist

The purpose of this textbook is to prepare the engineer for the design of photovoltaic systems. The purpose of this appendix is to provide a checklist against which the system designer can verify that sufficient information has been provided in the design to meet the needs of whoever is given the tasks of design review, installation, inspection, operation, and maintenance of the designed system. The following format, based on recommendations by the Florida Solar Energy Center, is used by VB Engineering, Inc., for preparation of permit application packages for design review and permitting. Course instructors may request electronic copies of sample permit packages from roger@vbengineering.com. Please include name of instructor, course number, name of institution, and class size with the request.

If design projects are assigned for a course that uses this book, it is recommended that the designs contain the information listed in this appendix.

RECOMMENDED INFORMATION TO SUBMIT FOR PV INSTALLATION PERMIT

1. Permit application forms for the code jurisdiction issuing the installation permit.
2. Notice of Commencement and Release of Lien (Florida requirement; check local requirements).
3. System description (about 1 letter-size page long), including
   a. Whether the system is utility interactive
   b. Whether the system has batteries
   c. Dc array rating
   d. Number of source circuits
   e. Number of modules per source circuit
   f. Number of inverters, inverter ratings, and inverter compliance with UL 1741, if applicable
   g. Whether fossil fuel generator is included in system
   h. Nominal battery storage in kWh
   i. System block diagram showing power flows
   j. Brief description of how system operates
   k. Statement of expected system performance, for example, average kWh/day from PVWatts
4. Site plan showing locations of all new equipment, existing electrical service equipment, and point of utility connection.
5. Electrical riser diagram showing how the new PV system connects into the existing service.
6. Standby load summary if battery-backup system.

7. Electrical schematic diagrams showing all wire sizes and types, junction boxes, conduit sizes, disconnects, overcurrent protection, and major equipment wiring connections.
8. Certification of component listings (UL, CE, etc.), including manufacturers' data sheets for all major equipment, such as modules, array mounts, charge controllers, inverters, and batteries.
9. Site information for wind loading, including of wind load in psf.
10. Structural drawings showing means of attachment of PV array to roof, if roof mounted and means of attachment of array to other structures if not roof mounted.
11. Electrical calculations showing ampacity, temperature ratings, conduit fill deratings, voltage drop calculations for all dc and ac conductors, ambient temperature effects on PV array output voltage. Sizes of disconnects and overcurrent protection should also be included, as well as input and output voltages and currents, and voltage and power specifications, for all major pieces of equipment.
12. Structural calculations to show necessary screw thread penetrations into trusses for roof-mounted arrays, along with calculated maximum pull on attachment points and safety factors included in calculations. If other than roof mount, show calculations to indicate that the structure can withstand all forces that are significant in the area, such as wind, snow, seismic, etc.
13. Code references for electrical and structural calculations (NEC, ASCE-7, etc.) and code references relating to the point of utility connection and instructions for labeling.

Index

"f" indicates material in figures. "p" indicates material in the problems. "t" indicates material in tables.

A

Absolute zero, 370, 370f, 376
Absorbed Glass Mat (AGM) batteries, 61, 223, 292, 350p
Absorption constant
  air mass and, 24
  amorphous vs. crystalline silicon, 409
  bandgap and, 373, 394
  of CdTe, 435
  diffusion length and, 383, 389
  of GaAs compounds, 423
  of next generation cells, 408
  pn junction width and, 383
  thickness of material and, 373
  wavelength and, 373, 373f
Absorption stage, 75, 75f, 233, 255
Acceptors
  conductivity and, 377
  at contacts, 397
  covalent bonding of, 379
  description of, 377
  energy of, 390
  for InP, 427
  pn junction and, 380, 380f
  vs. recombination centers, 390
  resistance and, 397f
  in silicon, 377f
  space charge layer and, 381, 391
  voltage and, 391
Accessibility, 178–179, 192
ac-coupled BBGC systems
  120 V with 240 V SGCS, 260–261
  120/240 V with 240 V SGCS, 258–260
  advantages/disadvantages of, 257
  demand for, 215
  losses in, 445
  sales by, 103, 261
  schematic diagram of, 216f
Accumulation, 9–10
ac/dc disconnects
  in 2.85 kW 1-phase system, 125, 130–131
  in 3.04 kW BBGC system, 229
  in 3.20 kW 1-phase system, 137–138
  in 120 V BBGC system, 229
  in 250 kW 3-phase system, 155–156
  lockable, 130
  main and, 138
  maintenance and, 155
  NEC on, 266
ACE Designs project, 439
Acid rain, 331–332, 347, 352, 357, 359
Adams, William, 14
Aesthetics of PV system, 115, 119, 192, 201–202
Africa
  germanium from, 424
  irradiation of, 33f–34f
  Sahara Desert, 25
AGM batteries, 61, 223, 292, 350p
AIC, 256, 266
Air
  in atmosphere; *See* Atmosphere
  above cells, 387–388, 422–423, 427
  pollution of; *See* Air pollution
  temperature of; *See* Air temperature
  velocity of; *See* Wind speed
Air conditioning, 238t, 348, 360
Air mass
  0 (AM 0), 24, 26, 427, 441
  1 (AM 1), 24, 24f, 26, 427
  1.5 (AM 1.5), 26, 50, 53, 282
  air temperature and, 25
  altitude and, 24
  cell efficiency and, 422–423, 427
  intensity of sunlight and, 24, 26
  PV standard test conditions, 26, 50
  at solar noon, 36
  solar spectrum and, 24f, 427
  zenith angle and, 24
Air pollution
  acid rain and, 331–332, 347
  Clean Air Act, 353–354, 361
  energy sources and, 354t
  EPA on, 347, 353–356
  greenhouse gases and, 356–357
  Olympic Games and, 2
  soil acidification and, 358
  top contributors, 355t
Air pressure
  on 3.20 kW 1-phase system, 208–209
  on 6 kW 1-phase system, 209–210
  air density and, 184
  altitude and, 184
  ASCE on, 207t
  Bernoulli's equation for, 184
  designing for, 183–191

dynamic vs. static, 184
free stream, 184
ground clearance and, 188
on Hughes modular array, 187–189, 188f, 189f
modeling, cylindrical, 184–185, 185f
roof slope and, 207t
roof zone and, 207t, 209–210
row configuration and, 188
at stagnation point, 184
streamlining and, 185–186, 186f
tilt and, 186–188, 186f
topography and, 190
uplift; *See* Uplifting force
velocity exposure coefficient, 190
wind direction and, 190
wind speed and, 184, 207t, 208
worst-case conditions for, 191
Air temperature
air mass and, 25
carbon dioxide and, 358
cell temperature and, 53
compression and, 172
conduit and, 106, 127–128, 128t, 137
copper wiring and, 106, 127, 128t
encapsulants and, 416
for NOCT, 53
stress and, 164, 170–172
tension and, 172
voltage and, 53
wiring and, 127, 128t
Air velocity, 274
AISC, 181
AISI, 181
Alaska
Fairbanks, 450t
petroleum production in, 13
time zones in, 28
Albany, microinverter system in, 140–142, 141t
Albedo radiation, 23
Albuquerque
average daily irradiation, 453t
cost-availability curve for, 71f
ALE, 437
Alkaline batteries, 55
Allowable stress, 169–170
Alloy steel, 175
Alternators, 59, 90
Aluminum
alloy series, 177, 177t
in batteries, 65
Bayer extraction process for, 423–424
cast, 176
coefficient of linear expansion for, 212p
conductivity of, 106
for contacts, 413, 415–416, 420
copper and, 176, 177t
corrosion resistance of, 176, 177, 179
cost of, 179
fabrication processes for, 176
fatigue failure in, 165, 165f
gallium and, 423–424
in galvanic series, 173t
magnesium and, 176, 177t
manganese and, 176, 177t
in manufacturing, 363t
modulus of elasticity for, 169t, 204
modulus of rigidity for, 169t
in mounts, 364
for ohmic contacts, 402
properties of, 176
resistivity of, 415
selective leaching of, 174
silicon and, 176, 177, 177t, 413
vs. steel, 169–170, 176, 211p
stress corrosion cracking of, 174
stress limits of, 164
tensile strength of, 170t, 176
terminating, 106
weight of, 106
for wiring, 104–106, 176
wrought, 176, 177t
yield strength of, 170t
zinc and, 173, 175, 176, 177t
Aluminum antimony, 396t
Aluminum Association, 181
Aluminum oxide, 106, 176, 388t
American Institute of Steel Construction (AISC), 181
American Iron and Steel Institute (AISI), 181
American National Standards Institute (ANSI), 105t
American Society of Civil Engineers (ASCE)
on mechanical forces, 116, 181
on seismic loads, 192
standard; *See* ASCE 7-05 standard
American Society of Mechanical Engineers (ASME), 181
American Society of Metals (ASM), 181
American Society of Testing and Materials (ASTM), 105t, 175, 181
Amorphous silicon (aSi)
absorption constant vs. photon wavelength for, 373f
bandgap energy of, 396t, 408f, 409, 420
bandgap of, 372, 409, 421
carbon and, 421
in cells, 408, 409, 420–423, 420f

crystalline silicon in tandem, 439
diffusion constant and, 389
efficiency of, 396t, 409, 421–423
germanium and, 421
mobility of EHPs and, 389
photon current density vs. optical path length in, 408f
solar spectrum and, 409, 421
stacked layers of, 421f, 422–423
stainless steel and, 421, 422f
structure of, 420
Ampere hours (Ah)
for 3.04 kW, 120 V BBGC system, 222–223
battery voltage and, 218, 326–327
to Btu conversion, 307
definition of, 58
for hybrid home, 316t
of lead-acid batteries, 60
for loads, 326
maximum charge rating and, 60
for mountain cabin, 303, 304t, 307t
of Ni-Cd batteries, 63
of Ni-Zn batteries, 64
to Watt hours conversion, 223, 294
Analemma, 28–29
Angle of attack, 186f
Angola, Luanda, 293–299, 296, 455t
Angular displacement, 30–31, 31f, 36–37, 37f, 39–40
ANSI, 105t
Antimony, 429, 430, 435
Antimony-lead batteries, 59, 60
Antireflective coating (ARC)
application of, 53, 416
diagram of, 395f
index of refraction of, 388t, 395
optimization of, 388, 416
purpose of, 395
reflection and, 52
vs. texture, 388
thickness of, 416, 423
wavelength range of, 423
Annualized life-cycle cost, 340–341
APCVD, 437
Applied Energy Service, 183
ARC; *See* Antireflective coating
Arc discharge, 97, 116
Arc fault circuit interrupter, 230
Argentina, Buenos Aires, 455t
Argon gas, 419
Arizona
Phoenix, 36
PM-10 particles in, 355
Array
battery backup for; *See* Battery-backup systems
cells in; *See* Cells, PV
controllers for; *See* Charge controllers
description of, 55
inverters for; *See* Inverters
modules; *See* Modules
mounts for; *See* Mounts
shade on; *See* Shade
spacing between; *See* Row spacing
straight; *See* Straight grid-connected systems
structural loading by, 182–183
utility and; *See* Utility interactive systems
wiring for; *See* Wiring
Arrhenius, S. A., 352–353
Arsenic, 410, 423, 424, 429, 430
Arsine, 363t
ASCE; *See* American Society of Civil Engineers
ASCE 7-05 standard
adherence to, 116
building codes and, 181
gust effect factors in, 189
load formulas in, 181, 182–183
on roof exposures, 205, 210, 210t
on roof height, 205, 210, 210t
on roof wind loads, 134, 189–191, 205–210
title of, 105t
on wind speed, 189, 207t
Ascension Technology, 140
Ash, 25
Asia; *See also specific countries in*
energy consumption in, 5f
irradiation of, 33f–34f
OECD, 5f
ASM, 181
ASME, 181
Aspect ratio, 197, 201
ASTM, 105t, 175, 181
ASTM A123 standard, 175
Asymmetrical interrupting capacity (AIC), 256, 266
Atlanta
average daily irradiation, 452t
highway signs for, 291–293
irradiance in, 292t
latitude/longitude of, 42
Sun's position over, 42–43
Atmosphere
density of, 183–184, 187
greenhouse gases in; *See* Greenhouse gases
ice fogs, 183
mass of; *See* Air mass
photosynthesis and, 25
properties of, 183

sunlight and, 23–25, 422–423
temperature of; *See* Air temperature
viscosity of, 183, 184
weight of, at earth surface, 17p
Atmospheric pressure chemical vapor deposition (APCVD), 437
Atomic layer epitaxy (ALE), 437
Attachment cap, 96
Austin, average daily irradiation, 454t
Australia
charge controller settings in, 234
GDP vs. kW per capita, 6f
irradiation of, 33f–34f
Melbourne, 456t
AutoCAD, 39
Automobiles
air pollution and, 355
altitude and, 93
aluminum alloys in, 177
batteries in, 59, 64, 90
carbon monoxide from, 355
efficiency of, 359
fasteners for, 180
health and safety concerns, 362
hydrogen for, 65
payback from, 343, 344
smart grids and, 444–445
starting in cold weather, 59
torsional shear stress in, 169
Autotransformers, 260–261, 261f
Availability, 69–72, 70f, 71f, 311; *See also* Reliability
Axial loading, 169
Azimuth angle
angular displacement and, 31
collector orientation and, 43
compass and, 29
declination and, 31
definition of, 29
diagram of, 30f
double-axis trackers and, 297
optimum, 195
row spacing and, 39–40, 144, 144f
shade and, 38, 38–40
solar altitude and, 31, 32f
at solar noon, 29
time of day and, 31, 32f
Azurite, 429

B

Backup systems
battery; *See* Battery-backup systems
generators; *See* Generators
UPS, 84, 289
utility as, 47
Bandgap
direct; *See* Direct bandgap
indirect, 372–373, 372f, 388, 394, 419
Bandgap energy
at absolute zero, 370f
for aluminum antimony, 396t
carrier concentration and, 374
for CdS, 375t, 396t
for CdTe, 375t, 396t
vs. cell efficiency, 440f
of CIS, 396t, 408f, 428, 440t
description of, 370–374
energy, momentum, and, 371–372
for GaAs compounds, 372, 372f, 375t, 396t
for germanium, 375t, 396t
heat and, 371, 442–443
impurities and, 390
for indium phosphide, 396t
for phosphorus, 376
reverse saturation current and, 387
for silicon, 372, 372f, 375t, 377f, 396t
temperature and, 370, 370f
trapping within, 390
Bardeen, John, 14
Barium, 62
Batteries
for 3.04 kW BBGC system, 221–223
for 7.20 kW BBGC system, 239
for 18 kW BBGC system, 251
for 120 V BBGC system, 221–223
for 120/240 V BBGC system, 239
AGM, 61, 223, 292, 350p
alkaline, 55
aluminum-based, 65
capacity of, 58
cell voltage, 72–73, 73f, 233
charge controllers for; *See* Charge controllers
charging of, 72–76, 74f, 75f, 232–233, 294, 295
connections to, 263–266
cost of, 218, 272
current
basics, 72–73, 73f
charge control and, 72–76, 75f
in lead-acid, 56–60, 56f, 264
MPPT and, 79
in Ni-Cd, 63
range of, 47
Cu-Zn, 64
discharge of, 72–76, 73f, 74f, 294–295
efficiency of, 223

equalization mode, 75–76, 233
function of, 47
gassing stage, 57
gel, 61, 223, 350p
generators and, 48f, 89–93, 234–235
housing for, 96, 256–257, 324
iron-based, 65
I-V curve and, 67f
lead-acid; *See* Lead-acid batteries
lithium, 55, 65
load on, 218
losses in, 251, 294
MPPT and, 79, 80f
Ni-Cd, 55, 62–64
NIMH, 55, 64
Ni-Zn, 64
parallel connection of, 218, 251, 263–266, 264f
performance of, 57–58
pn junction and, 383–384, 384f
rated capacity of, 294–295
replacement of, 266
resistance
  age of battery and, 266
  basics, 72–73, 73f
  in BBGC systems, 257
  charge/discharge rate and, 218
  controller and, 72–75
  corrosion and, 263
  in Ni-Cd, 63
  series-parallel connection options, 265f
  temperature and, 73
  from terminal lugs, 263
  voltage drop across, 61, 73
  from wiring, 263–264, 266
selection procedures for, 326–327
series connection of, 263–266, 264f
sodium-based, 65
space for, 218
system voltage regulation by, 47
temperature of
  backup system size and, 222
  capacity and, 218
  controller and, 75, 76
  derating factor, 294
  for hybrid home, 315t
  of lead-acid, 58–59, 59f, 61–63, 62t
  of lithium, 65
  for mountain cabin, 304, 304t
  of Ni-Cd, 63–64
  of Ni-Zn, 64
  resistance and, 73
  voltage and, 73
terminal voltage
  basics, 72–73, 73f
  bulk limit of, 233
  charge control and, 72–76, 75f
  for highway signs, 292–293
  hysteresis and, 74f, 76
  LBCI point, 234, 234t
  LBCO point, 234, 234t, 260
  of lithium, 65
  MPPT and, 80f
  of Ni-Cd, 63–64
  of Ni-Zn, 64
  nominal, 218
  refloat, 234
  resistance and, 73
  sell level, 233
  state of charge and, 74f
  temperature and, 73
  Thevenin circuits, 73f
Thevenin circuits for, 59, 72–73, 73f
transient suppression by, 47
units of measurement of, 58
venting of, 60, 76, 218
zinc-based, 64–65
Battery-backup grid-connected (BBGC) systems
3.04 kW, 220–235
7.20 kW, 235–243
18 kW, 244–257
100 kW, 244
120 V
  3.04 kW, 220–235
  with 240 V SGCS, 260–261
120/208 V 3-phase, 261–263
120/240 V
  7.2 kW, 235–243
  18 kW, 244–257
  with 240 V SGCS, 258–260
ac-coupled; *See* ac-coupled BBGC systems
array for, 219
batteries for, 217–218
charge/discharge rate of, 218
controller for, 219
conversion efficiency and, 117
dc-coupled; *See* dc-coupled BBGC systems
design of, 216–220
inverter bypass switches in, 98
inverters for, 89, 215, 217
losses in, 219, 445
payback from, 217
for standby loads, 89, 217–218
surge capacity of, 120
synchronization of, 89
Battery-backup systems (BBS)
for cathodic protection system, 289–290
controller for; *See* Charge controllers

for critical vs. noncritical systems, 70–71, 70f, 272
function of, 47, 69
generators and, 48f, 89, 89–93, 90, 234–235, 312
for highway signs, 292–293, 337, 338t
for hybrid home, 312, 314–315, 315t, 318–321, 339t
inverters and, 312
modules for, 52
for mountain cabin, 303–304, 304t
for outdoor lights, 280, 284
for refrigeration system, 294–296
schematic diagram of, 48f
utility interactive; *See* Battery-backup grid-connected systems
for water pumps, 274, 276
Bauxite, 423
Bayer extraction process, 423–424
BBGC; *See* Battery-backup grid-connected (BBGC) systems
Beam radiation; *See also* Direct radiation
collector orientation and, 35–37, 35f
definition of, 23
irradiance and, 26
measurements of, 32–35
seasons and, 36f
solar altitude and, 29
tracking collectors and, 36
zenith angle and, 24
Becquerel, Edmund, 14
Beijing
nitrogen oxide in, 356
Olympic Games in, 2
particulates in, 356
Beja, Maurice, 424
Bell Laboratories, 428
Bending, 164, 169, 427
Bernoulli's equation, 184
Biofuels, 4, 4f, 354t
BIPV, 105t, 115, 194, 195
Birds, 54, 179
Bismarck
average daily irradiation, 453t
hybrid home near, 311–324, 325f, 338–339
Bismuth, 429, 435
Black-and-white pyranometers, 32
Blackbody radiation, 22, 22f, 281–282, 371
Boca Raton
angular displacement in, 44p
clock vs. solar noon in, 29f
longitude of, 44p
Boeing, 188
Bohemia, 431
Boltzmann's constant, 22
Bombay, particulates in, 356
Boric oxide, 425
Bornite, 429
Boron, 376–377, 377f, 410, 438
Borrowing money; *See* Loans
Boston, average daily irradiation, 452t
Branch circuits
current in, 231f, 267p
definition of, 107
in microinverter system, 141, 142f
NEC on, 106, 302
neutral conductor in, 230–231, 267p
in PV systems, 107
standby power for, 231, 242
taps, 156
Brass, 174
Brattain, Walter, 14
Brazil
GDP vs. kW per capita, 6f
Rio de Janeiro, 356
British thermal unit (Btu), 2, 307
Bromine, 65
Bronze, 171–172, 171f, 173t
Btu, 2, 307
Btu standard, 13
Buckling, 170, 188, 203–205
Buenos Aires, average daily irradiation, 455t
Building codes, 96, 116, 179, 181–182
Building-integrated PV (BIPV), 105t, 115, 194, 195
Bulgaria, Sofia, 356
Bulk stage, 75, 75f
Buoys, 41
Burlington
7.20 kW BBGC system for, 235–243
120/240 V BBGC system for, 235–243
air temperature range in, 236
cost-availability curve for, 71f
Busbars
for 7.20 kW BBGC system, 237
for 21 kW 3-phase system, 147
for 120 V BBGC system, 227
for 120/240 V BBGC system, 237
for 250 kW 3-phase system, 155–156
circuit breakers and, 131, 148
cost-availability curve for, 130
current distribution on, 131
inverters and, 147, 254–255
labeling of, 131
NEC on, 130–131, 242
PUC circuit breaker in, 131, 131f
sizing of, 151
tap rules for, 156

Buy-down programs, 348
Bypass diodes, 52–53, 53f, 402

C

Cadmium, 363t, 364–365, 428, 431
Cadmium sulfide (CdS)
  bandgap energy of, 375t, 396t
  in cells
    CdTe, 436–437, 436f
    CIGS, 432f
    CIS, 400, 432–433
    for light sensors, 35, 375
  deposits of, 431
  efficiency of, 396t
  heterojunctions with, 400
  intrinsic carrier concentration in, 375t
  melting point of, 375t
  mobility of EHPs in, 375t
  relative dielectric constant of, 375t
Cadmium telluride (CdTe)
  bandgap energy of, 375t, 396t, 408f
  bandgap type, 372
  in cells, 408, 435, 436–438, 436f
  vs. CIS, 365
  efficiency of, 396t, 435
  environmental concerns, 363t, 364, 365, 435
  intrinsic carrier concentration in, 375t
  mobility of EHPs in, 375t
  photon current density vs. optical path length in, 408f
  relative dielectric constant of, 375t
Cairo, particulates in, 356
Calcium-lead batteries, 59, 60
California
  CdTe arrays in, 437
  CEC, 122, 150, 360
  fuel cell plants in, 66
  Hazardous Waste Control Law of, 365
  Los Angeles, 349p, 356
  MSRI, 1
  ozone and, 355
  PM-10 particles in, 355
  Sacramento, 42f, 450t
  Southern California Gas Company, 66
California Energy Commission (CEC), 122, 150, 360
Calorie, 2
Canada
  GDP vs. kW per capita, 6f
  molybdenum deposits in, 432
  Newfoundland, 28
  time zones in, 28
Cancer, 347
Capacitors, 55
Carbon dioxide
  air temperature and, 358
  Arrhenius on, 352–353
  carbon tetrachloride and, 357t
  CFCs and, 357t
  from coal, 17p, 354t
  cost of, 358–359
  dichlomethane and, 357t
  geothermal systems and, 354t
  halon and, 357t
  IPCC on, 353, 356
  in manufacturing, 363t
  methane and, 351, 357t
  natural gas and, 354t
  nitrogen oxide and, 357t
  nuclear energy and, 354t
  oceans and, 353
  oil and, 354t
  pentafluoroethane and, 357t
  permafrost and, 351
  from petroleum, 17p
  photosynthesis and, 25, 351, 359–360
  REC and, 347
  scrubbed and liquefied, 359
  sequestering, 359–360
  from silicon processing, 409
  sodium aluminate and, 424
  sunlight and, 23, 25
  U.S. emissions of, 356
Carbon monoxide, 354–355, 355t, 364
Carbon steel, 175, 358
Carbon tetrachloride, 357t
Cast iron, 169t, 170t, 173t
Cathodic protection system, 287–291
Caulking, 174
CdS; *See* Cadmium sulfide
CdTe; *See* Cadmium telluride
Cells, battery
  in AGM, 61
  alkaline, 55
  caps for, 60–61
  electrolyte specific gravity and, 61
  gel, 61, 223, 350p
  lead-acid; *See* Lead-acid batteries
  lithium, 55, 65
  Ni-Cd, 55, 62–64
  NIMH, 55, 64
  Ni-Zn, 64
Cells, PV
  back of, 394
  bypass diodes and, 52–53, 53f, 402
  cleaning of, 364, 441

configurations for, 47
crucible grown, 418
current; *See* Current, cell/module
description of, 49–52
development of, 14
dielectric coefficient of, 395
efficiency of
air mass and, 422–423
aSi, 409, 421–423
vs. bandgap energy, 440f
CdTe, 435, 437
CIGS, 433, 434
CIS, 433, 434
concentration and, 428, 441
crucible grown, 418
CSG, 419
EFG method, 418
first, 14
GaAs compounds, 427, 441
grounding and, 418
HIT, 439
intermediate band and, 442
material and, 396t
maximum, 394
MIS-IL, 438
M3EH-PPV, 443
multicrystalline, 409, 418–419
multijunctions, 402–403
next generation, 407–408
non-single-crystal, 408
purity and, 410
QW, 441
Schottky junctions, 402
series connection and, 53–54
single-crystal, 409
string ribbon, 419, 439
SunPower A-300 cell, 417
supertandem, 442
surface recombination and, 393
tandem silicon, 439
vs. wavelength, 394f
electrical field creation in, 49
encapsulants for, 416
failure of, 52, 53
fill factor; *See* Fill factor, cell
first, 14
geometry for, 396–398, 397f
heating of, 51–52
index of refraction in, 395
lifetime of, 416
manufacturing of, 16
maximizing performance of, 386–398
multicrystalline, 409, 418–419
net energy of, 14, 16
organic, 443
pn junction in; *See* pn junction
power; *See* Power, cell/module
to produce 1 kWh/day, 3
resistance; *See* Resistance, cell/module
on satellites, 14
series connection of, 52–54
silicon in; *See* Silicon
single-crystal; *See* Single-crystal silicon cells
standard calibration spectrum for, 26
SunPower A-300 cell, 417–418, 417f
surface of
coatings; *See* Coatings
lamination, 53, 416
polarization effect, 417–418
texture; *See* Textured surfaces
temperature, operating; *See* Temperature, cell/module
testing of, 427
thickness of, 52, 373
thin film; *See* Thin-film cells
voltage; *See* Voltage, cell/module
Ceramics, 419, 438
Cerium oxide, 441
CFCs, 353, 357t
CFR, 115
Chalcopyrite, 429
Chapin, Daryl, 14
Charge controllers
for 3.04 kW BBGC system, 224–225, 228t, 229, 232–234, 234t
for 7.20 kW BBGC system, 236–238, 241t
for 18 kW BBGC system, 246, 249, 252t, 254
for 120 V BBGC system, 224–225, 228t, 229, 232–234, 234t
for 120/240 V BBGC system, 236–238, 241t
in absorption stage, 233
for battery backup, 47, 89, 219
for cathodic protection system, 289–290
classification of, 72
current
for 3.04 kW, 120 V BBGC system, 224, 232
for highway signs, 292–293
for hybrid home, 323
for mountain cabin, 309t
for refrigeration system, 296, 298
description of, 72–76
efficiency of, 76, 80, 238
equalization mode, 75–76
in float stage, 233
function of, 47

generators and, 320–321
grounding of, 224
for highway signs, 292–293, 337, 338t
for hybrid home, 320–323, 339t
hysteresis and, 74, 74f, 76
Internet advice on, 232
inverters and, 246
lead-acid batteries and, 52
losses in, 219, 223
for mountain cabin, 306–308
MPPT and, 52, 79–80, 80f
Ni-Cd batteries and, 63
for outdoor lights, 285, 286
power, 224, 296
programming of, 232–233
for refrigeration system, 296–297
series/shunt, 76
stages in charging, 75, 75f
temperature and, 75, 76
UL 1741 standard, 103
voltage, 224, 233–234, 234t, 297, 298t
Chernobyl, 362
Chile, molybdenum deposits in, 432
China
Beijing, 2, 356
Chongqing, 356
energy consumption in, 5, 5f
GDP vs. kW per capita, 6f
Guiyang, 356
power plants in, 2
Shanghai, 456t
Tianjin, 356
Chloride, 65
Chlorine, 410, 429
Chlorofluorocarbons (CFCs), 353, 357t
Chongqing, particulates in, 356
Chromium, 174, 416
CIGS, 408, 428, 432–434, 432f
Circuit breakers
for 2.45 kW 1-phase system, 140–141
for 2.85 kW 1-phase system, 127, 130–132
for 3.04 kW BBGC system, 221, 227–230
for 3.20 kW 1-phase system, 138
for 7.20 kW BBGC system, 237–243
for 18 kW BBGC system, 253–256
for 21 kW 3-phase system, 146–149
for 120 V BBGC system, 221, 227–230
for 120/240 V BBGC system, 237–243
for 250 kW 3-phase system, 150, 155–156
in ac microinverter systems, 140–141
AIC rating of, 256
arc fault circuit interrupters and, 230
for battery strings, 266
busbars and, 131, 148
in combiner boxes, 98
grounding conductors and, 132
household, 230
for hybrid home, 322, 323t, 324
main, 138
for motors, 96
for mountain cabin, 309–310, 309t
NEC on, 96, 127, 130–131
for outdoor lights, 286
at PUC, 130–131
ratings for, 96
sizing of, 96
wiring and, 96
CIS; *See* Copper indium diselenide
Clean Air Act, 353–354, 361
Clean Power Estimator, 360–361
Clocks
FCC on, 115
generator start and, 89, 256, 322
midnight on, 30
noon on, 28–29, 29f
Close-spaced sublimation (CSS), 437
Clouds, 23, 69, 127, 308
Coal
acid rain and, 347
carbon dioxide from, 17p
environmental effects of, 354t
gallium and, 423
net energy of, 13
oil embargoes and, 4
production of, global, 4f, 5f
vs. PV, 363t
satellites and, 14
Coatings
antireflective, 52, 53, 388, 388t
corrosion and, 172, 173
dielectric constant of, 388
on edges of cells, 398
galvanized, 175, 358
pollution and, 358
snow load and, 191
thickness of, 388, 416
ultraviolet radiation and, 174
Cobalt, 62, 174
Code of Federal Regulations (CFR), 115
Codes
building, 96, 116, 179, 181–182
CFR, 115
definition of, 180
electrical, 181
NEC; *See National Electrical Code*
Coefficient of absorption, 388, 394, 420, 428
Coefficient of force, 190
Coefficient of linear expansion, 170, 172, 212p

Coefficient of thermal expansion, 170, 172
Coefficient of utilization, 282
Coffee maker, 245t
Collectors
concentrating, 52, 178, 195–196, 423, 427–428, 441
fixed; *See* Fixed collectors
mounts for; *See* Mounts
shade on; *See* Shade
structural loading by, 182–183
tracking; *See* Trackers
Colorado
21 kW 3-phase system in, 143–149
250 kW 3-phase system in, 150–156
air temperature range in, 143
CdTe arrays in, 437
Denver; *See* Denver
Columns, 170, 188, 203–205
Combiner boxes
for 2.85 kW 1-phase system, 125
for 3.04 kW BBGC system, 227, 229
for 7.20 kW BBGC system, 236–237, 239–240
for 18 kW BBGC system, 249, 251
for 120 V BBGC system, 227, 229
for 120/240 V BBGC system, 236–237, 239–240
for 250 kW 3-phase system, 151–154
circuit breakers in, 98
current in, 154–155
for hybrid home, 339t
Compact disc (CD) players, 88
Compass angles, 29
Complementary error function, 411
Compound interest, 7–8
Compression
air temperature and, 172
in bending, 169
buckling under, 170, 188, 203–205
crushing under, 170
Euler's formula for, 170
fatigue failure from, 165
in Hughes modular array, 188, 188f
modulus of elasticity for, 168
in rack mounts, 204–205, 204f
tensile strength and, 172
wind direction and, 165
Computers, 220t, 238t, 245t
Concentrating collectors, 52, 178, 195–196, 423, 427–428, 441
Concentration gradient, 379, 385, 397
Concrete, 363t, 364
Conducted EMI, 115
Conduction band
at absolute zero, 370, 370f
carrier concentration in, 374
curvature of, 387
description of, 370–372
direct vs. indirect bandgap and, 372
donor atoms and, 390
energy, momentum, and, 371–372
Fermi level and, 400–402
for GaAs compounds, 372, 372f
heat and, 371, 442–443
intermediate band and, 442
phosphorus and, 376
in QW cells, 441
for silicon, 372, 372f, 376, 377f
temperature and, 370, 370f
Conductivity, 375, 377, 414
Conductors; *See also* Wiring
air temperature and, 106
in conduit, 106
grounded; *See* Grounded conductors
grounding; *See* Grounding conductors
in junction boxes, 227
NEC on, 104, 108
negative, 78–79, 97, 224, 299
neutral; *See* Neutral conductors
phase conductors, 112, 138, 149, 267p
positive, 78–79, 97, 155
resistivity of, 156
sizing of, 127
ungrounded, 97, 128
Conduit, electrical
for 2.85 kW 1-phase system, 127–128, 128t, 129t, 132, 137
for 3.04 kW BBGC system, 227–229, 228t
for 7.20 kW BBGC system, 239–240
for 18 kW BBGC system, 249, 251–253
for 120 V BBGC system, 227–229, 228t
for 120/240 V BBGC system, 239–240
air temperature and, 106, 127–128, 128t, 137
conductors in, number of, 106
factory-installed, 194–195
grounding of, 137
height above roof, 128, 128t, 137
for hybrid home, 323
inside structures, 137
metallic interior of, 137
NEC on, 104, 127, 132
ultraviolet radiation and, 174
Conference on the Physical Basis of Climate and Climate Modeling, 353
Connected load, 294
Connecticut, electricity generation in, 359
Conservation of energy, 369, 372

Conservation of momentum, 369, 372
Consumer price index, 333, 334f
Consumption rates, 10–13, 12f, 91, 118, 121
Contacts
affixing, 413
aluminum, 413, 415–416, 420
for aSi cells, 420, 422f
back, coverage of, 394, 413
for CdTe cells, 436, 437
for CIS cells, 428, 432–433, 432f
for CSG cells, 419
current carried by, 415–416
diffusion of, 397, 399, 413
doping near, 393–394, 397–398, 413
EWT, 439
excess carriers and, 382, 385, 394–395, 397
fabrication of, 416
front, design of, 398, 413–414
for GaAs cells, 425f, 426t
graded junction and, 399
integrity test for, 427
lamination for, 53
LGBG process for, 439
metal, 400–402
for MIS-IL cells, 438–439
module connection to, 416
molybdenum for, 432
MWA, 439
MWT, 439
passivation at, 393
power loss in, 415
recombination centers and, 393
reflection by, 398
resistance between junction and, 396, 397f, 413–415, 414f
resistance of, 415
shadowing by, 439
for single-crystal silicon cells, 413–417
for SunPower A-300 cells, 417
thickness of, 415
width of, 415
zinc oxide for, 433
Control panels, 256
Conversion efficiency
in 250 kW 3-phase system, 150
ampere hours and, 303
of aSi cells, 409
in battery charger, 319
for hybrid home, 313
of lab thin films, 408
in microinverter systems, 141–142
overall, 117
payback and, 345
of polycrystalline cells, 409
PVWatts default, 121
of single-crystal cells, 409
worst-case, 219
Converters, 77–79, 78f
Copper
vs. aluminum, 106
from dezincification, 174
environmental concerns, 365
in galvanic series, 173t
generator losses from, 94–95
germanium and, 424
indium and, 429
RCRA on, 365
refining of, 429
selenium and, 430
tellurium and, 435
wiring, 104–106, 106t
Copper aluminum alloy, 176, 177t
Copper indium diselenide (CIS)
availability of, 408
bandgap energy of, 396t, 408f, 428, 440t
bandgap type, 372
cadmium and, 365, 400, 428
in cells
CdS and, 400
diagram of, 434f
discussion of, 428, 432–434
efficiency of, 396t
environmental concerns, 365
first, 428
indium availability and, 408
material usage, 441
production cost, 441
environmental concerns, 363t, 364, 365
photon current density vs. optical path length in, 408f
Copper indium (gallium) diselenide (CIGS), 408, 428, 432–434, 432f
Copper-zinc (Cu-Zn) batteries, 64
Corona discharge, 97
Corrected load, 294–295
Corrosion resistance
of aluminum, 176, 177, 179
of BOS components, 96
designing for, 117, 164, 172–174, 180, 182
of large-area arrays, 116
of molybdenum, 432
of steel, 175
Costa Rica, 359
Costs
availability and, 71–72, 71f, 311
of batteries, 61, 218, 272
for CdTe, 366, 435, 438
for CIS cells, 366

from Clean Power Estimator, 360
of collectors, 36
for CSG cells, 419
demand and, 103
design and, 179
externalities; *See* Externalities
of generators, 89–90, 90t, 95t, 319
installation, 193–195
labor, 193
of lead-acid batteries, 61
life-cycle, 195, 331–341, 343–344
of loads, 68–69
of modules, 14–15, 15f, 193, 280
mount design and, 192–193
for new power plants, 444
for next generation cells, 407
payback of, 217, 331, 344–345, 346t
recycling, 366
transmission congestion and, 444
types of, 331
per watt, 15f
Crevice corrosion, 173
Crucible growth of silicon, 418
Crystal defects, 389, 410
Crystalline silicon on glass (CSG), 409, 419
CSS, 437
Cuba, Havana, 356
Cuprite, 429
Cupronickel, 173t
Current, cell/module
for 2.85 kW 1-phase system, 123t, 127
for 3.04 kW BBGC system, 224t, 225t, 228t
for 3.20 kW 1-phase system, 137
for 7.20 kW BBGC system, 236t
for 18 kW BBGC system, 247, 252t
for 120 V BBGC system, 224t, 225t, 228t
for 120/240 V BBGC system, 236t
for 250 kW 3-phase system, 151–155
basics, 7, 49
for CdTe, 437
in charge control, 75, 232–233
in contacts, 415–416
exponential change in, 391
fan and, 67f, 273f
fill factor and, 50–51
formula for, 49, 386
ground fault, 97
for hybrid home, 316–317, 322–323
ideal vs. real, 49f
irradiance and, 50, 52
LCBs and, 77–78, 78f
limiting, 49
load and, 67–68, 67f, 68f
for mountain cabin, 304–307, 305t, 309t
MPPT and, 77–79, 78f
optimizing, 387–396
phase of, 445
for power maximum, 50
PWM inverters and, 86–88, 87f
for refrigeration system, 296–297
as selection criteria, 123
series connection, 52–55, 54f
short circuit, 49–50, 54, 386
square wave inverter and, 82
surges and, 97
temperature and, 51, 53
voltage drop and, 298
wavelength and, 52
zero, 50
Current density, 378
Curtain-wall glazing, 199
Cu-Zn batteries, 64

D

Dark current, 441
Day, Richard, 14
Daylight hours, 30
dc splicing mechanism, 227
dc-coupled BBGC systems
7.20 kW, 235–243
18 kW, 244–257
100 kW, 244
120 V, 220–235
120/240 V
7.2 kW, 235–243
18 kW, 244–257
vs. ac, 215
inverter efficiency in, 257
losses in, 257, 445
schematic diagram of, 216f
De Boisbaudran, Lecoq, 423
Dead-weight loads, 116, 182, 183, 188f, 249
Declination
angular displacement and, 31f
azimuth angles and, 31
collector orientation and, 36–37, 42
daylight hours and, 30
definition of, 27
formula for, 27–28
latitude and, 296
measurements of, 35
mounting angle and, 37f
row spacing and, 39–40
seasons and, 27f, 31f, 37
shadow band stands and, 35
solar altitude and, 30
solar noon and, 28, 31, 31f, 36

sunrise/sunset angles and, 30
zenith angle and, 31, 31f
Decommissioning, 365–366
Deflection, structural, 182
Delhi, particulates in, 356
Dendritic web, 419, 439
Denver
21 kW 3-phase system in, 143–149
250 kW 3-phase system in, 150–156
air temperature range in, 143
atmospheric density in, 183–184
average daily irradiation, 451t
mountain cabin stand-alone system near, 300–310, 311f
Department of Energy (DOE)
Energy Information Agency, 13
on global energy production, 4f, 5f
on grid-connected systems, xix
Zero Energy Homes Program, 194
Depletion, 9–10
Depletion allowance, 362
Depletion region, 381, 432
Depreciation, 345, 346t
Dezincification, 174
Diaspore, 423
Diborane, 363t
Dichlomethane, 357t
Dielectric coefficient, 395
Diesel, 90, 92–93, 334–335
Diffuse radiation, 23, 35, 36
Diffused junction, 379–380, 390, 411–412, 413f
Diffusion
absorption constant and, 383
definition of, 378
drift current and, 380, 397
drive-in, 412, 413f
excess carrier concentrations and, 385f
externally biased pn junction and, 384f
graded junction and, 399, 399f
illumination and, 383
impurity density and, 411–412
Kirchhoff's current law and, 380
metal junctions and, 400–402
pn junction and, 380, 380f
random thermal motion and, 378f
surface recombination and, 393
Diffusion constant, 379, 383, 389, 389f, 411
Digestion, 351
Diodes
bypass, 52–53, 53f, 402
current, equation for, 7, 385
theory of solid-state, 14
zener, 97, 116
Direct bandgap
of aSi cells, 372, 409, 420
bandgap diagram for, 372, 372f
of CdTe, 372, 435
of CIS, 372, 428
conduction band in, 372, 372f
of GaAs compounds, 372, 372f, 423
in next generation cells, 408
photon absorption and, 372, 373, 394
valence band in, 372, 372f
Direct mounts, 196, 199, 199f
Direct radiation, 23, 24, 281; *See also* Beam radiation
Direct subsidy, 362
Discount rate, 332–333
Distribution panel
for 2.45 kW 1-phase system, 142
for 21 kW 3-phase system, 146–147
for 120 V BBGC system, 230, 242, 261
in BBGC systems, 216f
branch circuits to, 107
busbars; *See* Busbars
feeder circuits to, 106–107
for mountain cabin, 303, 309, 310
NEC on, 322
PUC circuit breaker in, 130, 131f
for standby loads, 226
DOE; *See* Department of Energy
Donors
in aluminum, 413
conductivity and, 377
covalent bonding of, 379
description of, 377
energy of, 390
graded junction and, 399f
in indium phosphide, 427
from phosphorus, 376
pn junction and, 380, 380f
vs. recombination centers, 390
Schottky junctions and, 400–401
in silicon, 377f
space charge layer and, 381, 391
voltage and, 391
Double-axis trackers, 178, 297
Doubling time, 8–10, 9f
Douglas fir, 203t
Dow Jones Industrial Average, 333, 334f
Drag, 185–186, 186f
Drift current
definition of, 378
graded junction and, 399, 399f
metal junctions and, 401–402
pn junction and, 380, 380f, 384f
resistance and, 397

Drift velocity, 378, 393, 405p
Drive-in diffusion, 412, 413f
Dry corrosion, 174

E

Earth
  diameter of, 16p
  orbit of, 27–28, 27f
  polar axis inclination of, 27, 27f
  rotation of, 27–28
  seasons on, 25, 27, 33f–34f
  solar constant for, 22, 26
  to Sun distance, 22, 27, 43p
  surface of
    atmospheric weight on, 17p
    diffuse radiation at, 23
    heat reradiated from, 24
    radiation at, 23, 24
    solar energy density of, 22
    temperature of; *See* Air temperature
    water on, 16p
  Tropic of Cancer/Capricorn, 27
Earth Policy Institute, 17p
Earthquakes, 183, 192
Economic standards, 13
ED, 437
Eddy current, 82
Edge-defined film-fed growth (EFG), 418
Effective mass, 387
Effective wind area (EWA), 207t
EFG, 418
Egypt, Cairo, 356
Einstein relationship, 381, 389
Elastic limit, 166–167, 167f
Electric fields, 43p
Electrical codes, 181
Electrodeposition (ED), 437
Electromagnetic fields, 380–381
Electromagnetic interference (EMI), 109, 115
Electrometallic tubing (EMT), 267p
electromotive force (emf), 113, 273
Electron volts (eV), 371
Electron–hole pairs (EHPs)
  in aSi cells, 420
  bandgap energy and, 371
  bound, 443
  contacts and, 397f, 417
  creation of, 49, 371, 374
  EVA and, 418
  generation rate of, 375
  heterojunctions and, 400
  intermediate band and, 442
  MIS-IL and, 439
  photon path length and, 373
  pn junction and, 379, 382–383, 382f, 393, 402
  population of, 374
  recombination of, 375, 399, 441
  in silicon, 417
  in stacked cells, 422
  surface diffusion of, 393
  TCO and, 434
  temperature and, 376
  textured surfaces and, 388
  thermal velocity of, 375
Electrons, 369–371, 387, 427; *See also* Donors
Electroplating, 287, 288f
Electrostatic force, 377
Elongation
  in bar pull test, 166, 166f, 175
  formula for, 168
  strain and, 166
  temperature and, 171–172, 171f
  weight and, 211p
  yield point and, 167
emf, 113, 273
EMI, 109, 115
Emitter wrap-through (EWT), 439
EMT, 267p
Energy
  bandgap; *See* Bandgap energy
  in bar pull test, 167f, 168
  conservation of, 369, 372
  consumption of, global, 3
  mass conversion to, 21
  net, 13–14, 16
  in photons, 371
  power and, 26
  production of, global, 3, 4f, 5f
  renewable energy; *See* Renewable energy
  second law of thermodynamics on, 7, 14
  solar, 21–22
  units of measurement of, 2
Energy density
  blackbody, 22
  distance and, 43p
  of Ni-Cd batteries, 55
  at Sun surface, 22
  of sunlight; *See* Irradiation
Energy equity, 4–5
Energy Information Agency, 13
Energy payback time (EPBT), 446p
Energy storage
  alternatives for, 55, 67
  in batteries; *See* Batteries
  clouds and, 69

by consumers, 444
for critical vs. noncritical systems, 70–71, 70f, 272
in fuel cells, 65, 66–67
hydrogen for, 55, 65–67
PWM inverters and, 87
in water tanks, 274
England
London, 25
Manchester, 356
Environmental Protection Agency (EPA), 347, 353–357, 361, 365
EPBT, 446p
Epitaxial-growth junction, 390
Erosion corrosion, 174
Error function, 11
Erythema, 174
Ethylene vinyl acetate (EVA), 416, 417f, 418
Euler's formula, 170
Europe; *See also specific countries in*
ACE Designs project in, 439
energy consumption in, 5f
irradiation of, 33f–34f
PV manufacturing in, 16
EVA, 416, 417f, 418
EWA, 207t
EWT, 439
Excel, doubling time charts from, 9
Excess majority carriers, 375t, 382
Excess minority carriers, 375t, 384–385, 390, 393
Excitons, 443
Exponential growth, 7–11, 9f
Externalities
consideration of, 352–353
definition of, 331
greenhouse gases, 358–361
health and safety concerns, 362
of PV systems, 363–366
system life-cycle cost and, 347–348
toxic cleanup costs, 357

F

Fairbanks, average daily irradiation, 450t
Fans
120 V BBGC system for, 220t, 238t
air velocity moved by, 274
for hybrid home, 313t, 314
in inverters, 119
I-V curve and, 67f, 273f
for mountain cabin, 302, 302t, 307
power consumed by, 274
shading patterns, 250f
stand-alone system for, 272–274
switches for, 310
voltage requirements of, 217
Fatigue failure, 165, 165f, 182
Federal Communications Commission (FCC), 115
Feeder circuits, 106–107, 156
Feed-in Tariff (FIT), 349p
Fermi levels, 376, 386, 390, 400–402
Fill factor, cell
of CdTe, 437
of CIS, 433, 434
definition of, 50
for GaAs compounds, 427, 441
ideal vs. real, 50–51
I-V curve and, 50–51
maximizing, 51, 386
maximum power and, 50
resistance and, 50, 51, 433
reverse saturation current and, 51
of silicon in tandem, 439
Filtering, 86
Finger Lakes region, 357
Finite energy sources, 13
FIT, 349p
Fixed collectors
cost of, 36
direction facing, 40–43, 119
for highway signs, 292–293
for max irradiation, 35–37
for refrigeration system, 295–299
seasons and, 36f
tilt of, 35f, 37f, 38f, 119, 327
time of day and, 36f
vs. trackers, 201
utility peaking and, 41–43
Flat roofs
18 kW BBGC system on, 244, 245f, 246–249, 250f
21 kW 3-phase system on, 143–144
membrane, 244
rack mounts for, 198
zones on, 206f
Flicker, 110
Float stage, 75, 75f, 233
Flood loads, 182
Florida
atmospheric density in, 184
Boca Raton, 29f, 44p
CdTe arrays in, 437
Miami, 283–287, 451t
solar-tiled home in, 199f
Tampa, 220–235
West Palm Beach; *See* West Palm Beach

Florida Solar Energy Center, 461
Fluorescent lamps
ballasts, 82, 93
cost of, 68, 272
designing for, 272
efficiency of, 68
lifetime of, 281t
luminous efficacy of, 281t
noise from, 119
outlet rating and, 303
spectrum of, 23
use of, 282
wiring for, 107–108
Foot-candle (f-c), 280–281
Foot-pound (ft-lb), 2
Force; *See also* Loads
on arrays, 182–192
in bar pull test, 166
compressive; *See* Compression
designing for, 171, 178
electrostatic, 377
shear, 168, 168f, 184
stress; *See* Stress
tensile; *See* Tension
uplifting; *See* Uplifting force
weight, 116, 164
Fourier, J. B., 352
France
GDP vs. kW per capita, 6f
Paris, 457t
Free trade agreements, 6
Fuel cells, 65, 66–67, 355
Fuller, Calvin, 14
Functional requirements, 178
Furnace, 238t, 245t, 313t, 314
Fuses
for 2.85 kW 1-phase system, 127, 130
for 3.20 kW 1-phase system, 138
for 21 kW 3-phase system, 146
for 250 kW 3-phase system, 150–151, 155
for battery strings, 266
in combiner boxes, 98
for motors, 96
for mountain cabin, 309–310
NEC on, 104, 127
for outdoor lights, 286
ratings for, 96
for refrigeration system, 298t
wiring and, 96

G

GaAs; *See* Gallium arsenide
Gable roofs, 190, 206f
Galfan, 175
Gallium, 423–424
Gallium arsenide (GaAs)
absorption constant vs. photon wavelength for, 373f
bandgap energy of, 372f, 375t, 396t, 423, 425
bandgap of, 372
in cells
cascaded, 426f
composition of, 426t
discussion of, 423, 425–428
next generation, 441
performance of, 408
structure of, 425f
in compound tandem cells, 441
conduction band of, 372f
efficiency of, 396t
in heterojunctions, 400
intrinsic carrier concentration in, 374f, 375t
LEC method for, 425
melting point of, 375t
mobility of EHPs in, 375t
photon absorption and, 372
relative dielectric constant of, 375t
temperature and, 374f
thickness of, 372
valence band of, 372, 372f
Gallons per minute (GPM), 277–278, 278t
Galvalume, 175
Galvanic corrosion, 172–173, 173t
Galvanized steel, 358
Gas discharge lamps, 23, 281t, 282, 283
Gasoline
altitude and, 93
for automobiles, 359
availability of, 93
carbon dioxide and, 65, 366p
energy density of, 65
for generators
ALCC for, 341
combustion efficiency of, 319
cost of, 95t
life-cycle cost of, 337
maintenance and, 95t, 256
selection of, 90
for standby loads, 89
vs. hydrogen, 65
Gassing, 57, 63–64, 75–76, 233–234
Gauss' law, 380, 390–391
Gaussian curve, 11–12, 12f, 17p
GDP, 5, 5f, 6f, 358
Gel batteries, 61, 223, 350p
Generators
18 kW BBGC system and, 244, 254–255

120/240 V BBGC with 240 V SGCS and, 259
ac vs. dc, 90
altitude and, 93
availability and, 311
batteries and, 48f, 89–93, 234–235, 312
cost of, 89–90, 90t, 95t, 319
current from, 112, 324
derating factors, 93
diesel for
altitude and, 93
availability of, 92
combustion efficiency of, 319
contamination of, 92
cost of, 95t, 334
efficiency of, 92
maintenance for, 95t
selection of, 90, 92
size of, 95
for stand-alone systems, 89
temperature and, 92
efficiency of, 91–92, 92f, 312
failure of, 69
frequency, output, 93, 94, 113, 234–235
fuel consumption by, 91, 92f, 319, 320t
fuel for, 90–93, 334–335
gasoline for
ALCC for, 341
combustion efficiency of, 319
cost of, 95t
life-cycle cost of, 337
maintenance and, 95t, 256
selection of, 90
for stand-alone systems, 89
harmonics, 93
for highway signs, 337, 338t, 341
hours of operation, 319–320, 320t
for hybrid home, 312–314, 317–322, 320t, 324, 338–339, 339t
for hydroelectric power, 274
inverters and, 89, 91–92, 92f, 112, 234–235, 255, 312, 321–322
maintenance of
frequency of, 320, 320t
gas vs. diesel, 95, 95t
highway sign, 338, 338t, 339t
scheduled, 256
methane for, 90
modeling of, 112
natural gas for, 93, 95t
noise level of, 90t, 94
output stability of, 93–94
overloading of, 94, 255
payback from, 217
power, 94–95, 318–319
programming of, 322
propane for
altitude and, 93
availability of, 93
combustion efficiency of, 319
cost of, 95t, 338, 340f
for hybrid home, 312–313, 316, 324, 325f
maintenance for, 95t
selection of, 90
for stand-alone systems, 89
rotation speed, 91
selection of, 95
short-circuit faults, 112
size of, 90t, 91–92, 92f, 95t
starters for, 94, 256, 324
time of day and, 255
transformers and, 93
voltage, 93–94
Georgia, Atlanta; *See* Atlanta
Geothermal energy, 4, 354t
Germanite, 423
Germanium, 374f, 375t, 396t, 421, 423, 424–425, 442
Germany
FIT in, 349p
GDP vs. kW per capita, 6f
solar/wind systems in, 1
GFDI; *See* Ground fault detection and interruption device
Glass, 363t, 388t, 417f, 441
Global radiation, 23, 24, 32
Glow discharge, 97
Gold, 173t, 402, 429, 435
Google Earth, 39
Google Sketch-Up, 39
GPM, 277–278, 278t
Graded junctions, 398–399, 399f, 417
Grants, 348
Graphite, 62, 287, 438
Green pricing, 348
"Green tag," 347, 360
Greenhouse gases
carbon dioxide; *See* Carbon dioxide
carbon tetrachloride, 357t
CFCs, 353, 357t
characteristics of, 356–357
dichlomethane, 357t
earth surface temperature and, 25
first formal conference on, 353
halon, 357t
IPCC on, 353
methane; *See* Methane
nitrogen oxide; *See* Nitrogen oxide

pentafluoroethane, 357t
policies on, 352
REC and, 347
Gross domestic product (GDP), 5, 5f, 6f, 358
Ground fault detection and interruption device (GFDI)
for 2.85 kW 1-phase system, 126–127
for 3.04 kW BBGC system, 229
in 3.04 kW BBGC system, 230
for 120 V BBGC system, 229, 230
for 250 kW 3-phase system, 155
for mountain cabin, 310
for refrigeration system, 299
safety and, 365
Grounded conductors
for 2.85 kW 1-phase system, 128
for 3.04 kW BBGC system, 224
for 120 V BBGC system, 224
for 250 kW 3-phase system, 155–156
converters and, 78
current in, 97, 299
definition of, 97
fuses and, 155
GFDI and, 299
LCBs and, 78
MPPT and, 78
neutral conductor size and, 156
for outdoor lights, 286
purpose of, 128
for refrigeration system, 299
voltage drop on, 128
Grounding
for 2.85 kW 1-phase system, 126
of battery cases, 256–257
of conduit, 137
equipment, 98
EVA and, 418
IEEE 1547 on, 109
of modules, 155, 418
of mounts, 286
MPPT and, 78–79
NEC on, 97, 98, 108
for utility interactive system, 97
Grounding conductors
for 2.85 kW 1-phase system, 126–128, 132
in 3.04 kW BBGC system, 231–232
for 3.20 kW 1-phase system, 138
for 7.20 kW BBGC system, 243
for 18 kW BBGC system, 256
for 21 kW 3-phase system, 147
in 120 V BBGC system, 231–232
for 120/240 V BBGC system, 243
for 250 kW 3-phase system, 156
circuit breakers and, 132
current in, 97, 128, 299
definition of, 97
GFDI and, 299
for mountain cabin, 310
NEC on, 126–127
purpose of, 128, 299
for refrigeration system, 299
Ground-mounted arrays
250 kW 3-phase system, 150–156
cell temperature and, 199
components for, 199
Hughes, 187–188, 187f, 188f, 189f
photograph of, 200f
rack mounts for, 150, 200
shade and, 39–40
Grove, William, 66
Guiyang, sulfur oxide in, 356
Gust effect factors, 164, 165, 169, 189, 190
Gyration, radius of, 204

H

Half-life, 16p
Halogens, 432
Halon, 357t
Havana, air quality in, 356
Health, public, 362
Heat, 369, 371, 442–443
Heisenberg's uncertainty principle, 403–404
Hermann, K. S. L., 431
Heterojunction with Intrinsic Thin Layer (HIT), 439
Heterojunctions, 399–400, 420
High-pressure sodium lamps, 23, 281t, 282
Highway signs, portable, 291–293, 337, 338t, 341
Hip roofs, 190, 206f
HIT, 439
Hjelm, P. J., 432
Hodas, D. R., 359, 360
Holes, 371, 387; *See also* Acceptors; Electron–hole pairs
Hooke's law, 167–168
Horsepower, 275–276
Hot carrier cells, 442–443
Hot-dip galvanizing, 175
Hour angle, 30, 30f, 31, 39–40, 43
Hubbert, M. King, 11
Hubbert's Gaussian model, 10, 11–12, 12f
Hughes modular array, 187–188, 187f, 188f, 189f
Hurricane-prone regions
3.20 kW 1-phase system in, 133–138, 205–209
120 V BBGC system for, 220–235
fatigue limit and, 165–166

Florida; *See* Florida
power outages in, 69, 220
roughness categories for, 205
Hybrid home, 311–324, 325f, 338–339
Hydraulic power, 277
Hydroelectric power
BBGC systems and, 215, 216f, 257
classification of, 4
environmental effects of, 354t
generators for, 274
IEEE 1547 on, 109
Hydrogen
aSi and, 420
energy density of, 65
for energy storage, 55, 65–67
in fuel cells, 65, 66–67
in lead-acid batteries, 56–58, 60, 62
in Ni-Cd batteries, 64
PV-generated, 355
silicon and, 410
in Sun, 21, 43p
from water electrolysis, 66
Hydrogen selenide, 363t, 364
Hydrostatic pressure, 182, 192
Hysteresis, 74, 74f, 76, 82, 273

I

Ice, 164, 183, 192
Idaho, PM-10 particles in, 355
IEC, 105t
IEEE; *See* Institute of Electrical and Electronics Engineers
IEEE 928 standard, 105t
IEEE 929 standard, 103, 105t, 109, 112
IEEE 937 standard, 105t
IEEE 1013 standard, 105t
IEEE 1144 standard, 105t
IEEE 1145 standard, 105t
IEEE 1187 standard, 61–62, 105t
IEEE 1262 standard, 105t
IEEE 1361 standard, 105t
IEEE 1373 standard, 105t
IEEE 1374 standard, 105t
IEEE 1479 standard, 105t
IEEE 1513 standard, 105t
IEEE 1526 standard, 105t
IEEE 1547 standard
on clearing times, 109–110, 110t
on harmonics, 110–111, 111t
IEEE 929 and, 103
on islanding, 111
on power quality, 110–111
on THD, 84
title of, 105t
UL 1741 and, 109, 119
Illumination Engineering Society (IES), 281
Illumination level
current and, 49–50, 49f
fan output and, 273, 274
guidelines for, 281
of ideal vs. real cells, 49f
ideal vs. real loads and, 67f
I-V curve and, 49f, 67f, 77f
MPPT and, 50, 77, 274
of outdoor lights, 280
power and, 50, 50f
units of measurement of, 280–282
voltage and, 49–50, 49f, 50f
water pump system and, 77, 77f
Importance factor, 190, 211p
Incandescent lamps
blackbody spectrum of, 22, 281
cost of, 68, 272
filament temperature in, 22, 88
inverters and, 88
lifetime of, 281t
loading from, 88
luminous efficacy of, 281t
radiation from, 442
use of, 282
Incentive programs, 331, 349p, 444
Inclinometers, 39
India
Bombay, 356
Delhi, 356
energy consumption in, 5, 5f
Kolkata, 356
New Delhi, 457t
time zones in, 28
Indirect bandgap, 372, 372f, 373, 388, 394, 419
Indirect subsidy, 362
Indium, 408, 410, 429–430
Indium phosphide, 396t, 425–427, 426f, 426t, 441, 442
Induction machines, 109, 113
Inertia, moment of, 170, 204
Infinite energy sources, 13
Inflation rate, 332–333, 335
Infrared radiation
in blackbody spectra, 22f
carbon dioxide and, 23, 25
germanium and, 421
greenhouse gases and, 356–357
methane and, 25
temperature and, 22
thermophotovoltaic cells and, 442
water vapor and, 23

Infrastructure, electricity, 444
Inspections, 183, 230–231
Installation, 183, 193–195, 232, 364–365, 461–462
Institute of Electrical and Electronics Engineers (IEEE), 61, 84, 103; *See also specific IEEE standards*
Insulation
  galvanic corrosion and, 173
  in homes, 4, 209, 244, 348
  NEC on, 104, 106
  operating temperature and, 164
  resin, 419
  semiconductor material for, 370
  on wiring, 106, 126, 127, 128t, 174
Integrated mounts, 196, 198–199, 199f; *See also* Building-integrated PV
Interest rates, 8, 332–333, 335
Intergovernmental Panel on Climate Change (IPCC), 353, 356
Intergranular corrosion, 173–174
Intermediate band solar cells, 442
International Electrotechnical Commission (IEC), 105t
International Meteorological Organization, 352
International Organization for Standardization (ISO), 105t
Intrinsic carrier concentration, 374, 374f, 375t, 376, 386
Inversion layer, 438f, 439
Inverter bypass assembly, 98, 230, 242, 256
Inverters
  for 2.45 kW 1-phase system, 141–142
  for 2.85 kW 1-phase system, 122–125, 122t, 127–131
  for 3.04 kW BBGC system, 221, 228t, 229, 232–234, 234t
  for 3.20 kW 1-phase system, 136
  for 7.20 kW BBGC system, 238–242, 241t
  for 18 kW BBGC system, 244–246, 252t, 255
  for 21 kW 3-phase system, 143, 145–147, 149
  for 120 V BBGC system
    ac-coupled, 260–261
    dc-coupled, 221, 228t, 229, 232–234, 234t
  for 120/208 V 3-phase BBGC system, 261–263
  for 120/240 V BBGC system
    ac-coupled, 258–260, 258f
    dc-coupled, 238–242, 241t
  for 250 kW 3-phase system, 150–153, 155
  ac/dc disconnect for; *See* ac/dc disconnects
  batteries and, 89, 233, 312, 321
  for battery backup, 215, 217–218
  charge controllers and, 246
  "classic stacked" configuration, 238
  classification of, 72
  clearing times, 110t
  clocks of, 115
  collector orientation and, 119
  cost of, 81
  current
    for 2.45 kW 1-phase system, 141
    for 2.85 kW 1-phase system, 130
    for 21 kW 3-phase system, 147, 149
    in ac microinverter systems, 141
    ac/dc disconnect and, 130
    busbar rating and, 147
    conduction angle of, 114
    dc component in voltage and, 110
    filtering of, 115
    frequency adjustments and, 111–112, 111f
    harmonic content and, 111t
    for hybrid home, 323–324
    maximum, 122t
    for mountain cabin, 309–310, 309t
    for refrigeration system, 298
    search mode and, 88
    short-circuit faults, 112
    for utility interactive system, 88–89, 110–112, 221
  description of, 80–89
  efficiency of
    for battery backup, 218, 257
    conversion; *See* Conversion efficiency
    maximum, 80, 84
    in microinverter systems, 140
    modified sine wave, 81, 81t
    peak, 120
    power output and, 80, 294, 321, 322f
    pure sine, 81, 81t
    PVWatts default, 142
    PWM, 81, 81t, 88
    rated output and, 80
    SGCS, 120, 257
    square-wave, 81, 81t
    switches and, 88
  fans in, 119
  FCC on, 115
  frequency
    abnormal conditions in, 109, 110
    adjustment of, 111–112, 111f
    clearing times, 110, 110t
    eddy current losses and, 82
    hysteresis and, 82
    PWM, 85–88, 85f
    SFS for, 112, 114

THD of, 84
utility interactivity and, 109–114
voltage and, 86
function of, 47
generators and, 89, 91–92, 92f, 112, 234–235, 255, 312, 321–322
GFDI, 127, 155
harmonics
current and, 111t
FCC on, 115
fluorescent lamps and, 82
IEEE on, 111, 111t
modified sine wave, 81t, 83
motors and, 82
noise and, 80, 82
overheating and, 80
pure sine, 81t
PWM, 81t, 86
square-wave, 81, 81t, 82
housing for, 115
for hurricane-prone region, 136
for hybrid home, 313, 321–324, 339t
IEEE 929 standard on, 103
Internet advice on, 232
islanding, 89, 111, 112–114, 443–444
labels on, 115
losses in, 219
modeling of, 112
modified sine, 81, 81t, 82–84, 83f, 308
for each module, 140–142
monitoring by, 112, 120, 215
for mountain cabin, 300, 303, 308
mounting of, 119
MPPT and, 52, 80, 84–89, 120, 122
multilevel H-bridge for, 81t, 83–84, 83f
NEC on, 103, 120
noise from, 119
for outdoor lights, 285
PCC of, 112
power
for 2.85 kW 1-phase system, 122
for 3.04 kW BBGC system, 221
for 120 V BBGC system, 221
for 120/240 V BBGC system, 258–260
for 250 kW 3-phase system, 150–151
efficiency and, 80, 294, 321, 322f
flow of, 258
for hybrid home, 323–324
maximum, 119, 122t
for mountain cabin, 302t
MPPT and, 52, 80, 84
no-load consumption, 120
performance range, 81t
quality requirements for, 110–111
ratings for, 122t
for refrigeration system, 298
surge rating, 119–120
power factor for, 445
programming of, 232–235, 322, 445
pure sine, 81, 81t, 84
PWM; *See* Pulse-width-modulated inverters
for refrigeration system, 293–294
reliability of, 140
search mode for, 88, 322
selection considerations, 122
sell mode, 233–235, 259, 261, 312
for SGCS, 119–120, 257–258
short-circuit faults, 112
sleep mode, 120, 322
square wave, 81–82, 81f, 81t, 82f
stability of output from, 445
for stand-alone systems, 88, 271
standby loads and, 217–218
start-up of, 110
surges and, 80–81, 81t, 115–116, 119–120
synchronization of, 111, 258
transformers for; *See* Transformers
trip points of, 113, 119
UL on; *See* UL 1741 standard
utility interactive, 88–89, 109–116, 119–120, 215
voltage
for 2.85 kW 1-phase system, 123–124, 130–131
for 120/240 V BBGC system, 258
ac/dc disconnect and, 130–131
bulk, 233, 234t
clearing times, 109–110, 110t
dc component in output, 85–86, 110
distance to PCC, 112
float, 233, 234t
frequency and, 86
for hybrid home, 323
LBCI, 234, 234t
LBCO, 234, 234t
load and, 88
maximum, 122t
modified sine, 83–84
motor excitation frequency and, 86
MPPT and, 120
phase of, 131
PWM, 86–88
range of, 109, 110t
refloat, 233, 234t
for refrigeration system, 298t
search mode and, 88
sell, 233–235, 234t
square wave, 82

for standby loads, 89, 98, 215
SVS for, 112, 114
in utility interactive system, 88–89, 112
warranty on, 80, 120
weather rating, 122
Investment tax credits, 345, 346t
Ion implantation, 390
IPCC, 353, 356
Iran, Tehran, 356
Iron
arsenic and, 424
in batteries, 62, 65
carbon in, 174
cast, 169t, 170t, 173t
from ceramic substrates, 438
corrosion protection for, 175
fatigue failure in, 165, 165f
germanium and, 424
indium and, 429
oxygen and, 176
selective leaching of, 174
selenium and, 430
in steel, 174
tellurium and, 435
tensile strength of, 174–175
Iron oxides, 355, 364
Irradiance
air mass and, 24, 26
in Atlanta, 292t
cell current and, 50, 52
cell temperature and, 53
conversion to electricity, 51
definition of, 26
enhancement of, 195–196
fan performance and, 273–274, 273f
full sun, clear skies, 236
MPPT and, 274
for NOCT, 53
PV standard calibration spectrum, 26
at sunrise, 123
time of day and, 26, 26f
Irradiation
average daily, 33f–34f, 449
definition of, 26
maximizing, 35–37, 35f
during peak sun hours, 26, 26f
seasons and, 36f
time of day and, 36f
units of measurement of, 26
Islanding, 89, 111, 112–114, 443–444
ISO 9001 standard, 105t
Italy
GDP vs. kW per capita, 6f
Milan, 356

J

Japan
GDP vs. kW per capita, 6f
PV manufacturing in, 16
renewable energy programs in, 1
silicon tandem cells from, 439
Tokyo, 458t
J-bolts, 203
Junction boxes
for 2.85 kW 1-phase system, 125–130
for 3.04 kW BBGC system, 227
for 7.20 kW BBGC system, 239–240
for 18 kW BBGC system, 249, 252–253
for 120 V BBGC system, 227
for 120/240 V BBGC system, 239–240
NEC on, 227
series connection to, 126f
volume for, 227

K

Kentucky, Lexington, 244–257
Kenya, Nairobi, 458t
Kilowatt hour (kWh), 2–3
Kirchhoff's current law, 97, 131, 380
Kits, 193
Klaproth, M. H., 435
KOH; *See* Potassium hydroxide
Kolkata, particulates in, 356
Korea, GDP vs. kW per capita, 6f
Kyocera, 360
Kyoto Conference/Treaty, 6, 356

L

Labor costs, 193
Lag bolts/screws, 197, 202–203, 208–209, 210, 226–227
Laser ablation, 437
Laser-grooved, buried grid (LGBG) process, 439
Lasers, 282
Lateral drift, 182
Law of detailed balance, 380
Lead
calcium and, 59
in copper refining, 429
EPA on, 354, 355t
gallium and, 423
germanium and, 424
indium and, 429
RCRA on, 365
selenium and, 430

tellurium and, 435
zinc and, 175
Lead-acid batteries
AGM, 61, 223, 292, 350p
for backup systems, 218
bulk voltage, 75, 75f
capacity of, 58, 59f
cell caps for, 60–61
charging of, 56f, 57–61, 61f, 264–266, 265t, 295
chemistry of, 56–57, 56f
cost of, 61, 350p
current in, 56–60, 56f, 264–266
discharge of
current and, 264–266, 265t
discussion of, 56–61
lifetime and daily, 60f
resistance and, 265
schematic diagram of, 56f
temperature and, 59f, 294
voltage and, 58f, 61f, 265
efficiency of, 58
float voltage, 75, 75f
freezing point of, 62, 62t
gassing in, 57
with gel cells, 61
for hybrid home, 314–315
IEEE 1187 standard on, 61–62
lifetime of, 60, 60f, 61–62, 350p
maintenance on, 61
NEC on, 61
properties of, 57–62
rated capacity of, 294–295
resistance, 58–59, 61, 265
series-parallel connections of, 263–266, 264f, 265t
specific gravity of electrolyte, 61, 61f, 62t, 63
temperature of, 57–59, 59f, 61–62, 294
terminal voltage, 56–61, 58f, 61f, 62t, 265
venting of, 60
voltage of, 52, 57–59, 58f, 61, 61f, 62t
weight of, 350p
Least mean square curve, 9, 71
LEC method, 425
LEDs, 23, 68, 281t, 282
Lexington
18 kW BBGC system in, 244–257
air temperature range in, 244
latitude of, 244
LGBG process, 439
Life-cycle costing, 195, 331–341, 343–344
Lifetime
of cells, 416
consumption rates and, 10–11
contacts and, 413
excess carrier, 375t
exponential change in, 10–12
of fluorescent lamps, 281t
of incandescent lamps, 281t
of lead-acid batteries, 60, 60f, 61–62, 350p
of LED lamps, 68, 281t
minority carrier, 383, 389, 390
of modules, 163
net energy and, 14
of Ni-Cd batteries, 63
of NIMH batteries, 64
Lift
in water pump systems, 275–278, 278t
from wind; *See* Uplifting force
Light intensity, 373
Light-emitting diodes (LEDs), 23, 68, 281t, 282
Lighting systems
120 V BBGC system for, 220t
120/240 V BBGC system for, 238t, 245t
coefficient of utilization, 282
color temperature of, 281–282
fluorescent lamps in; *See* Fluorescent lamps
gas discharge lamps in, 23, 281t, 282, 283
high-pressure sodium lamps in, 23, 281t, 282
for hybrid home, 313t
illumination level; *See* Illumination level
incandescent lamps in; *See* Incandescent lamps
LED lamps in, 68, 281t, 282
low-pressure sodium lamps in, 282
lumen formula for, 282
maintenance factor, 282
metal halide lamps in, 281t, 282, 283
for mountain cabin, 302–303, 302t
outdoor; *See* Outdoor lights
power loss in ballasts, 283
room cavity ratio, 282
xenon lamps in, 282
Lightning, 97, 116, 126
Lightning arresters, 126, 310
Lightning rods, 97
Linear current boosters (LCBs), 72, 76–78, 78f, 275, 276, 279
Linear momentum for fluid flow, 184
Liquid-encapsulated Czochralski (LEC) method, 425
Lithium batteries, 55, 65
Lithium carbonate, 66
Live-weight loads, 116, 182, 183
Loads
adjustment of, 67
advanced management of, 444
battery discharge rate and, 58
for branch circuit taps, 156

to capacity matching, 444
charge controllers and, 72
connected, 294
corrected, 294–295
cost of, 68–69
critical, 69–70, 70f
current and, 54, 67–68, 67f
dead-weight, 116, 182, 183, 188f, 249
designing for, 182–192, 202–205, 272
energy storage and, 47
flood, 182
from hydrostatic pressure, 183, 192
ice, 164, 183, 192
live-weight, 116, 182, 183
MPPT and, 68, 68f, 77f
noncritical, 69–70, 70f, 96
power and, 67–68, 68f
procedures for determining, 326
rain, 183, 192
schematic diagram of, 48f, 53f
seismic, 183, 192
snow, 191–192
standby; *See* Standby loads
surges and, 80
voltage and, 54, 67–68, 67f
wind; *See* Wind load
worst-case conditions for, 113–114
Loans, 331, 341–345, 342t, 346t
London, sunlight in, 25
Los Angeles
air quality in, 356
FIT in, 349p
Love Canal, 357
Low battery cut in (LBCI) voltage, 234, 234t
Low battery cut out (LBCO) voltage, 234, 234t, 260
Low-carbon steel, 175
Low-pressure sodium lamps, 282
Luanda
average daily irradiation, 455t
latitude of, 296
stand-alone refrigeration system for, 293–299
Lumber, specific gravity of, 203t
Lumens, 280–281, 282
Luminous efficacy, 281, 281t

M

Magnesium, 173t, 176, 177, 177t, 438
Magnesium oxide, 388t, 425
Maintenance
ac/dc disconnects and, 155
designing for, 193
of generators
frequency of, 320, 320t
gas vs. diesel, 95, 95t
highway sign, 337–338, 338t, 339t
scheduled, 256
of highway sign system, 337–338, 338t, 339t
of lead-acid batteries, 61
live loads from, 183
manuals for, 232
operational requirements on, 179
Maintenance factor, 282
Majority carrier, 377, 382–383, 394, 398
Malachite, 429
Malthus, Thomas, 7
Manchester, particulates in, 356
Manganese, 176, 177t, 438
Manufacturing
cost of, 407
environmental concerns in, 363–364, 363t
location of, 16
Massachusetts, Boston, 452t
Material properties, 49, 164–172
MATLAB, 9
Maximum power point tracking (MPPT); *See also* Linear current boosters
by charge controllers, 52, 79–80, 80f
classification of, 72
description of, 68, 77
efficiency of, 77, 79
by inverters
design consideration, 122
minimum voltage, 123–124
modified sine wave, 84
monitoring of, 158
PWM, 85–88
range of, 120, 122t
in SGCS, 52
irradiance level and, 274
load and, 68, 68f, 77f
pumps and, 76–77, 275, 276
schematic diagram of, 78–79, 78f
for utility interactive system, 52
wiring and, 79
MBE, 425–427
Mechanical design
codes for; *See* Codes
constraints, 179
forces and, 182–192
functional requirements, 178
guidelines for, 180
loads and, 182–192, 202–205
of mounts, 192–202
operational requirements, 178–179
process of, 177–178

standards for; *See* Standards
steps in, 163
trade-offs, 180
Melbourne, average daily irradiation, 456t
Mendeleev, Dmitri, 423, 424
Metal halide lamps, 281t, 282, 283
Metal insulator semiconductor inversion layer (MIS-IL), 438–439, 438f
Metal organic chemical vapor deposition (MOCVD), 436, 437
Metal oxide varistor (MOV), 97, 116
Metallization wraparound (MWA), 439
Metallization wrap-through (MWT), 439
Methane
carbon dioxide and, 351, 357t
energy sources and, 354t
for generators, 90
IPCC on, 353
silicon cells and, 363t
sunlight and, 25
Mexico City
average daily irradiation, 459t
nitrogen oxide in, 356
Miami
average daily irradiation, 451t
parking lot lights for, 283–287
Micro grids, 443–444
Microwave oven, 220t, 238t, 245t, 313t
Milan, nitrogen oxide in, 356
Million Solar Roofs Initiative (MSRI), 1
Mineral depletion allowances, 348
Minimum Design Loads for Buildings; *See* ASCE 7-05 standard
Mining, 363t
Minority carrier
definition of, 377
diffusion length; *See* Minority carrier diffusion length
EVA and, 418
externally biased pn junction and, 383–385, 385f
illuminated pn junction and, 382–383
lifetime of, 383, 389, 390
reverse saturation current and, 386
surface recombination of, 393–394
Minority carrier diffusion length
absorption constant and, 383, 389
contact location and, 385
definition of, 383
diffusion constant and, 383
intrinsic region and, 393
lifetime and, 383
maximizing, 387, 388–389
metal junctions and, 402
photon absorption and, 383, 394, 397
predeposition and, 412
from surface, 383, 393
MIS-IL, 438–439, 438f
MOCVD, 436, 437
Modified sine inverter, 81, 81t, 82–84, 83f, 308
Modules
for 2.45 kW 1-phase system, 140–142, 141t
for 2.85 kW 1-phase system, 123–125, 123t
for 3.04 kW BBGC system, 224–226, 224t, 225t, 226f
for 3.20 kW 1-phase system, 134–135, 135f
for 6 kW 1-phase system, 139, 139f, 209, 209f
for 7.20 kW BBGC system, 235–236, 236t
for 18 kW BBGC system, 247–249
for 21 kW 3-phase system, 143, 144t
for 120 V BBGC system, 224–226, 224t, 225t, 226f
for 120/240 V BBGC system, 235–236, 236t
for 250 kW 3-phase system, 150–153
with bypass diodes, 52–53, 53f
for cathodic protection system, 289–290
cells in; *See* Cells, PV
cleaning of, 364, 441
connections to, 126, 416
cost of, 14, 15f, 193
with CSG cells, 419
current from; *See* Current, cell/module
description of, 52–54
efficiency of, 53–54, 117
EVA and, 418
framed, 115
grounding of, 155, 418
for highway signs, 292, 337, 338t
housing for, 53, 98
in Hughes modular array, 187f
for hurricane-prone region, 134–135, 135f
for hybrid home
discussion of, 315–317
excess energy produced, 317t
generator battery charger, 318t
generator hours and fuel, 320t
generator maintenance, 320t
life-cycle cost of, 338–339, 339t, 340f
for integrated mounts, 199
lead-acid batteries and, 52
lifetime of, 163
for microinverter system, 140–142, 141t
for mountain cabin, 306–307
number of, 124
for outdoor lights, 284, 285–286
parallel connection of, 52, 55, 117
power from; *See* Power, cell/module
procedure for estimating, 327

for refrigeration system, 296–298
scale drawings of, 134, 134f, 135f
series connection of, 55, 117, 126f
temperature, operating; *See* Temperature, cell/module
for utility interactive system, 52
voltage from; *See* Voltage, cell/module
for water pump system, 278–279, 279f
weight of, 164, 183
Modulus of elasticity, 167–168, 169t
Modulus of rigidity, 168, 169t
Molecular beam epitaxy (MBE), 425–427
Molybdenum, 402, 432, 434, 441
Momentum, conservation of, 369, 372
Monel, 173t
Monitoring
communication protocols for, 158
data analyzed, 158
IEEE 1547 on, 109, 110
by inverters, 112, 120, 215
in microinverter systems, 140, 142
for performance-based contract, 158
Montana, PM-10 particles in, 355
Moore, T. G., 352
MOSFET, 77–79, 82, 418
Motors
circuit breakers for, 96
current in, 96
damping of, 113
emf production by, 113, 273
excitation frequency of, 86
fuses for, 96
generators and, 93, 94
inverters and, 85–86
LCBs and, 76–77
NEC on, 96
overload protection for, 96
power factor for, 238–239
saturation flux of, 86
surges from, 116
switches for, 96
Mott, Nevill, 14
Mountain cabin stand-alone system, 300–310, 311f
Mounts
for 3.04 kW BBGC system, 226
for 3.20 kW 1-phase system, 137
for 120 V BBGC system, 226
angular displacement and, 36–37
ballasted, 198, 248, 249
BIPV; *See* Building-integrated PV
building codes on, 96, 116
cell temperature and, 196
declination and, 36–37
design of, 192–202
direct, 196, 199, 199f
direction, 40–43, 42f, 119
electrical grounding of, 98
ground; *See* Ground-mounted arrays
ground clearance of, 188
horizontal, 35, 41, 291–292
in Hughes modular array, 187f
for hurricane-prone region, 137, 207–208
for hybrid home, 315, 339t
integrated, 196, 198–199, 199f
inverters and, 119
for max irradiation, 35–37
pole, 200, 201f, 202f
rack; *See* Rack mounts
seasons and, 36, 36f
stand-off; *See* Stand-off mounts
tilt of
adjustment of, 179
drag, uplift, and, 186–187, 186f, 188, 191
in Hughes modular array, 187f, 188f
for hybrid home, 315
latitude and, 119, 248
for mountain cabin, 304–305, 305t
optimizing, 35f, 36–37, 195
performance and, 38f
procedure for estimating, 327
psh and, 296
roof slope and, 121
seasons and, 271
snow and, 143, 191–192
wind load and, 188, 191
zenith angle and, 37f
time of day and, 36f
tracker; *See* Trackers
for utility interactive system, 41–43, 119
weight of, 164
zenith angle and, 36, 37f
MOV, 97, 116
MPPT; *See* Maximum power point tracking
MSRI, 1
M3EH-PPV, 443
Muller, Franz, 435
Multicrystalline silicon cells, 409, 418–419
Multijunction cells, 402–403, 403f, 409
Multilevel H-bridge, 81t, 83–84, 83f
MWA, 439
MWT, 439

N

Nairobi, average daily irradiation, 458t
Narrow beam solid angle, 32
National Electrical Code (*NEC*)

about, 104
articles; *See specific NEC Articles*
on battery compartments, 96
on BOS components, 96
on branch circuits, 106–107
on circuit breakers, 96, 127, 130–131
on current, 127
design and, 104
on feeder circuits, 106–107
on fuses, 104, 127
on grounding, 97, 126–127
on household circuits, 230
on inverters, 103, 120
on lead-acid batteries, 61
on motors, 96
updates to, 104
on utility interactivity, 103
on wiring, 104–107, 106t, 127–128
National Energy Conservation and Policy Act, 4
National Fire Protection Association (NFPA); *See National Electrical Code*
National parks, U.S., 94
National Renewable Energy Laboratory (NREL), 37, 219, 449
Natural frequency, 113–114
Natural gas
altitude and, 93
environmental effects of, 354t
for generators, 93
oil embargoes and, 4
production of, global, 4f, 5f
NEC Article 110 (*NEC*), 108t, 115
NEC Article 210 (*NEC*), 108t, 303
NEC Article 220 (*NEC*), 220, 221, 303
NEC Article 230 (*NEC*), 108t, 266
NEC Article 240 (*NEC*), 108, 108t, 138, 156, 229
NEC Article 250 (*NEC*)
on 2.85 kW 1-phase system, 127, 132
on 18 kW BBGC system, 256
on 21 kW 3-phase system, 148
on 120 V BBGC system, 231–232
on 120/240 V BBGC system, 243
on 250 kW 3-phase system, 156
Article 690 and, 108, 108t
GFDI, 126–127
NEC Article 280 (*NEC*), 115
NEC Article 310 (*NEC*), 106, 108, 108t, 127–128, 156
NEC Article 314 (*NEC*), 227, 267p
NEC Article 339 (*NEC*), 108t
NEC Article 400 (*NEC*), 108t
NEC Article 430 (*NEC*), 96
NEC Article 480 (*NEC*), 108t
NEC Article 490 (*NEC*), 108t
NEC Article 690 (*NEC*)
acceptance of, 120
on branch circuit loads, 302
on busbars, 147, 242
on charge controllers, 290
on circuit breakers, 130
on conduit, 132
contents of, 108t
cross references, 108, 108t
on current, 127
on grounding conductors, 126
on inverters, 221, 239
NEC Article 240 and, 108
on PUC, 130, 242
on standby loads, 239
topic of, 104
on voltage, 123
on wiring, 127, 130
Negative conductors, 78–79, 97, 224, 299
Net energy, 13–14, 16
Net impurity concentration, 411–412
Net metering, 349p
Neutral conductors
for 3.20 kW 1-phase system, 138
for 21 kW 3-phase system, 147–149
for 120 V BBGC system, 260
for 250 kW 3-phase system, 156
autotransformers and, 261f
current in, 230–231, 267p
NEC on, 108t
New Delhi, average daily irradiation, 457t
New Mexico, Albuquerque, 71f, 453t
New York City, nitrogen oxide in, 356
New York State
Albany, 140–142, 141t
Finger Lakes region, 357
Love Canal, 357
New York City, 356
Newfoundland, 28
Ni-Cd batteries, 55, 62–64
Nickel, 173t, 358, 429
Nickel-cadmium (Ni-Cd) batteries, 55, 62–64
Nickel-metal hydride (NIMH), 55, 64
Nickel-zinc (Ni-Zn) batteries, 64
NIMH, 55, 64
Nitrogen oxides
acid rain and, 352, 357, 359
allowed emissions of, 361
carbon dioxide and, 357t
credits for, 360
energy sources and, 354t
EPA on, 354, 355t

IPCC on, 353
World Bank on, 356
Ni-Zn batteries, 64
Nominal operating cell temperature (NOCT), 53, 123, 123t, 196
Nonrenewable energy sources, 13
Normal distribution, 11–12, 12f, 17p
Normal incidence pyrheliometers, 32–35
North America
Canada; *See* Canada
energy consumption in, 5, 5f
irradiation of, 33f–34f
population of, 5
U.S.; *See* United States
North Dakota, Bismarck, 311–324, 325f, 338–339, 453t
NREL, 37, 219, 449
n-type impurities, 376–377, 379, 390, 411, 439
Nuclear energy, 4, 4f, 5f, 354t, 362

O

Occupational Safety and Health Administration (OSHA), 364
OECD, 5f
Ohio, CdTe arrays in, 437
Ohio River region, 354, 357
Ohm's law, 378
Ohms-per-square resistance, 414–415, 414f
Oil, 3–4, 10, 13, 14, 354t
Oklahoma City
2.85 kW 1-phase system in, 121–132, 132f
air temperature range in, 123
OMVPE, 425–427
One-axis trackers, 36–37, 327
Operational requirements, 178–179
Optical up-/down-converters, 443
Oregon, PM-10 particles in, 355
Organization for Economic Cooperation and Development (OECD), 5f
Organometallic vapor phase epitaxy process (OMVPE), 425–427
OSHA, 364
Outdoor lights
coverage pattern for, 283f
fixture spacing, 283
IES on, 281
illumination level of, 280–281
for mountain cabin, 302t
mounts for, 200, 201f
pole for, 283, 286–287
protection for, 179, 196
RCR for, 282
seasons and, 271
stand-alone system for, 283–287
Overcurrent protection
for 2.85 kW 1-phase system, 130
for 3.20 kW 1-phase system, 138
for 250 kW 3-phase system, 150, 155–156
AIC rating and, 256
busbars; *See* Busbars
circuit breakers; *See* Circuit breakers
classification of, 95
fuses; *See* Fuses
interrupting capacity of, 266
NEC on, 108
specified maximum, 155
Oxygen
aluminum and, 176
in copper refining, 429
in dry corrosion, 174
iron and, 176
photosynthesis and, 25, 351
silicon and, 409
sodium and, 176
Ozone, 23, 354–355

P

Paris, average daily irradiation, 457t
Particulates, 354–356, 354t, 355t
Passive solar energy, 312, 354t
Payback, 217, 331, 344–345, 346t
PCC, 112
Peak sun hours (psh)
current and, 289
definition of, 26
irradiation and, 26, 26f
load and, 315
mount tilt and, 295–296, 305t
payback and, 345
power and, 284, 293t
tracking and, 297
Pearson, Gerald, 14
Pechiney process, 424
Pennsylvania
cadmium deposits in, 431
Three Mile Island, 362
Pentafluoroethane, 357t
Performance-based contract, 158
Periodic inspections, 179
Permits, 120–121, 181, 461–462
Petroleum
acid rain and, 347
carbon dioxide from burning, 17p
consumption of, global, 17p
gallons per barrel of, 17p
to generate 1kWh, 3

production of
global rate of, 3–4, 4f, 5f, 12, 12f
Hubbert on, 12–13, 12f
U.S., 12–13, 12f
underground storage of, 287
pH levels, 352, 357
Phase conductors, 112, 138, 149, 267p
Phoenix, tracking collectors in, 36
Phosphine, 363t
Phosphoric acid, 66
Phosphorus, 376–377, 377f, 410, 438
Photogalvanic effect, 14
Photoluminescence, 443
Photons
absorption of, 369, 371–375, 443
cell interaction with, 49
collection efficiency, 394, 394f
diffusion length and, 389
electrons and, 369, 372
emission of, 404p, 443
energy in, 371
excitons and, 443
flux, 375, 443
frequency of, 371
heat and, 371, 442
mass of, 369
pn junction and, 382–383, 382f, 391, 394f
reflection of, 23, 24, 369, 387–388
transmission of, 369
wavelength of, 371, 373f
Wilson on, 14
Photosynthesis, 25, 351, 359–360
Photovoltaic (PV) systems
BOS components, 95–98
classification of energy from, 4
data package for, 461–462
demand for, 103
development of, 14
environmental effects of, 354t
factory-installed, 194
function of, 163
medium-sized, 143
power production, 10
safety factor for, 169
standard calibration spectrum for, 26
Physical vapor deposition (PVD), 433, 437
Pigmentation, 174
Pin junction, 392–393, 392f
Pipes, 275, 276, 276f, 277f, 301
Pitting, 173
Plan view, 133
Planck, Max, 14, 22
Planck's constant, 22, 371, 403–404
Plants, 25, 351, 359–360, 415, 430
Plasma decomposition, 420
Platinum, 173t
PM-10 particles, 355
pn junction
depth of, 412
efficiency and, 394–396
EHPs at, 49
epitaxial growth of, 439
externally biased, 383–386, 384f, 385f
formation of, 379–380, 380f, 390, 411–412
graded, 398–399, 399f, 417
with heterojunctions, 399–400, 420
low injection levels, 384
maximizing width of, 390–393
photons and, 382–383, 382f
resistance and, 396–398, 397f
reverse saturation current and, 386–387
surface recombination and, 393–394
in tandem, 402–403, 403f
textured surfaces and, 395f
tunneling in, 403–404, 403f, 404f, 421, 439
voltage across, 380–381, 384, 391–392
Point of common coupling (PCC), 112
Point of utility connection (PUC)
for 2.85 kW 1-phase system, 130–131, 131f
for 3.04 kW BBGC system, 221, 232
for 3.20 kW 1-phase system, 138
for 7.20 kW BBGC system, 242
for 120 V BBGC system, 221, 232
for 120/240 V BBGC system, 242
for 250 kW 3-phase system, 155
circuit breakers at, 130–131
frequency adjustments and, 111f
Pole mounts, 200, 201f, 202f
Polycrystalline silicon cells, 409
Population, global, 2, 7, 16p
Positive conductors, 78–79, 97, 155
Potassium carbonate, 66
Potassium hydroxide, 62–63, 64, 66
Power, cell/module
for 2.45 kW 1-phase system, 142
for 2.85 kW 1-phase system, 125
for 3.04 kW BBGC system, 224t, 225t
for 7.20 kW BBGC system, 236, 236t
for 18 kW BBGC system, 250–251
for 120 V BBGC system, 224t, 225t
for 120/240 V BBGC system, 236t
for 250 kW 3-phase system, 150–151
birds and, 54
charge control and, 80, 232–233, 323
contacts and, 415
controller and, 224
for highway signs, 292–293, 293t
for hybrid home, 315–317, 316t, 317t, 318t

illumination level and, 50f
LCBs and, 77
load and, 67–68, 68f
maximum, 50–52, 386
in microinverter systems, 142
for mountain cabin, 306–308
MPPT and, 68, 68f, 77, 79, 80f
for outdoor lights, 284–285, 285t
parallel connection, 55
for refrigeration system, 296–297
series connection, 52–55
temperature and, 51, 51f
typical, 47
for utility interactive system, 52, 117–118
Power factor, 220, 238–239, 445
Power plants, 118, 444
Predeposition, 411–412
Present worth, 333–336, 344
Present worth factor, 333, 335–336, 340–341
Price supports, 348
Prime lending rate, U.S., 333, 334f
Propane, 90, 93, 312
Proportional limit, 166, 167f, 169
Protons, 427
psh; *See* Peak sun hours
PSI Process, 419
PSPICE, 100p
p-type impurities, 377, 379, 390, 439
PUC; *See* Point of utility connection
Pulse-width-modulated (PWM) inverters
amplitude control by, 85, 85f
cost of, 81
description of, 85–88
efficiency of, 81, 81t, 86, 88
energy storage and, 87
frequency control by, 85, 85f
full-bridge, 87–88, 88f
as motor controller, 86
MPPT and, 86–88
output power range, 81t
surges and, 81t
three-level, 86–87, 87f
two-level, 86f
waveform control by, 85, 85f
Pump systems
horsepower of, 275–276
LCBs and, 76–77, 275, 276
pipes in, 275, 276, 276f, 277f, 301
power factor for, 238–239
pumping height for, 275–277
pumping time of, 275–277
speed of, 277
tracking collectors for, 201
water; *See* Water pump systems
Pure sine inverter, 81, 81t, 84
PVD, 433
PVWatts
on 18 kW BBGC system, 250–251
on 2.85 kW 1-phase system, 121–122, 138
for battery backup, 219
Clean Power Estimator, 360–361
cost of, 118
default values in, 121
for orientation data, 41
for performance data, 119, 138
Pyranometers, 32, 35
Pyrheliometers, 32–35

Q

Q, 59, 113–114, 114t
Quad, 2
Quantum mechanical tunneling, 404
Quantum well (QW) cells, 441
Quartz, 425

R

Rack mounts
for 250 kW 3-phase system, 150
ballasted, 198
buckling of, 204
cell temperature and, 196, 198
compression in, 204–205, 204f
description of, 198
for ground-based arrays, 150, 200
photograph of, 198f
Radiated EMI, 115
Radio frequency magnetron sputtering, 436
Rain, acid, 331–332, 347, 352, 357, 359
Rain load, 183, 192
Random thermal motion, 378, 378f, 380
RCR, 282
RCRA, 365
Rebates, 348
REC, 347, 360
Receptacles, 96, 302t, 303, 310
Recombination centers, 389–390, 393, 441
Recombination rate, 393
Recycling costs, 366
Refrigerators
120 V BB stand-alone system for, 293–299
120 V BBGC system for, 220, 220t
120/240 V BBGC system for, 238t, 245t
ac vs. dc, 69
annualized life-cycle cost for, 341
current for, 220
for hybrid home, 313t
life-cycle cost of, 332, 336–337, 337t

for mountain cabin, 300, 302t
power consumption of, 220
power factor for, 220, 239
specifications for, 272
Regenerative comparator circuit, 74, 76
Regression analysis, 9
Regulations, 444
Reliability, 103, 193; *See also* Availability
Renewable energy
definition of, 13
geothermal, 4, 354t
goals of, 444
health and safety concerns, 362
hydroelectric power; *See* Hydroelectric power
production of, global, 4f, 5f
PV; *See* Photovoltaic systems
wind; *See* Wind turbines
Renewable Energy Credit (REC), 347, 360
Renewable portfolio standards (RPS), 1
Resistance, cell/module
fill factor and, 51
photon absorption and, 375
series, 386, 393–394, 396–398, 397f
shunt, 396–398, 397f
surface recombination and, 393–394
Resonant frequency, 113–114
Resource and Conservation Recovery Act (RCRA), 365
Reverse fatigue limit, 165
Reverse saturation current
diode current and, 7, 385
fill factor and, 51
metal junctions and, 402
minimizing, 386–387
open circuit voltage and, 50, 386
temperature and, 51
Rio de Janeiro, sulfur oxide in, 356
RLC circuits, 113–114, 159p
"Romex" cable, 230
Roofs
actual dimension of, 133
area of, 117
asphalt shingles on, 195
attachment points on, 194
BIPV, 105t, 115, 194, 195
corners and edges of, 133
dead loads on, 183
diagrams of, 133
exposure of, 119, 205, 210, 210t
flat; *See* Flat roofs
force coefficient for, 190
gable, 190, 206f
height of, 207, 210, 210t
hip, 190, 206f
in hurricane-prone region, 133–134, 134f
live loads on, 183
MSRI, 1
penetrations into, 117, 182, 197
shapes of, 132–133
slope of
for rack mounts, 198
rise:run ratio, 133
vertical:horizontal units, 121
wind speed and, 207, 207t
snow load on, 191
spanners for, 197, 197f
trusses for, 134, 194, 197f, 208
wind and, 117, 133, 186–187, 190–191, 190f
wiring and, 126f
withdrawal load on, 202–203
zones on, 134, 134f, 205, 206f, 207–209, 207t
Room cavity ratio (RCR), 282
Roughness, surface, 205; *See also* Topography
Row spacing
for 18 kW BBGC system, 248–249, 248f
for 21 kW 3-phase system, 144, 144f
for 250 kW 3-phase system, 153
angular displacement and, 39–40
azimuth angle and, 39–40, 144, 144f
declination and, 39–40
hour angle and, 39–40
latitude and, 39–40, 153
mount height and, 40f
shade and, 39–40, 40f
solar altitude and, 39–40, 40f, 144, 144f
time of day and, 39–40
wind loading and, 188
RPS, 1
Rupture strength, 167, 167f
Russia
Chernobyl, 362
GDP vs. kW per capita, 6f

S

Sacramento
average daily irradiation, 450t
south- vs. west-facing systems in, 42f
SAE, 181
Safety factors, 169–170, 203t
Sahara Desert, 25
Salt domes, 431
Sandia frequency shift (SFS), 112, 114
Sandia voltage shift (SVS), 112, 114
Saudi Arabia, GDP vs. kW per capita, 6f
Scatter plots, 9

Schematic diagrams
2.45 kW 1-phase system, 142f
2.85 kW 1-phase system, 132f
3.04 kW BBGC system, 228f
3.20 kW 1-phase system, 139f
7.20 kW BBGC system, 237f, 243f
18 kW BBGC system, 254f, 255f
21 kW 3-phase system, 147
120 V BBGC system
with 240 V SGCS, 260f
dc-coupled, 228f
120/208 V 3-phase BBGC system, 262f
120/240 V BBGC system
with 240 V SGCS, 258f
dc-coupled, 237f, 243f
250 kW 3-phase, 157f
ac-coupled BB, 216f
basic, 48f
for cathodic protection system, 291f
dc-coupled BBGC systems, 216f
for fan, 273f
full-bridge PWM, 88f
LCB, 77–78, 78f
microinverter, 142f
MPPT, 78f
for outdoor lights, 286f
for refrigeration system, 299f
Schottky, Walter, 14
Schottky barrier device, 49
Schottky junctions, 400–402
Schroeder, Johann, 424
Scotland, 431
Screen printing, 437
Seattle
availability in, 72
average daily irradiation, 454t
tracking collectors in, 36
Second law of thermodynamics, 7, 14
Secondary spot networks, 109
Security, 179, 196
Seismic loads, 183, 192
Selective leaching, 174
Selenium, 14, 364–365, 429, 430–431, 433, 435
Semiconductors
at absolute zero, 370, 370f, 376
absorption constant for, 373, 373f
bandgap energy in; *See* Bandgap energy
conduction band in; *See* Conduction band
conductivity of, 375
EHPs in; *See* Electron–hole pairs
extrinsic, 376, 377
graded junctions in, 398–399, 399f, 417
heterojunctions, 399–400, 420
intermediate band in, 442
intrinsic, 376
metal junctions with, 400–402, 401f, 402f
with multijunctions, 402–403, 403f
pin junction, 392–393, 392f
pn junction of; *See* pn junction
Schottky junctions, 400–402
temperature and, 374, 374f
thermal equilibrium in, 374
with tunnel junctions, 403–404, 403f, 404f, 421, 439
valence band in; *See* Valence band
Series controllers, 76
SFS, 112, 114
SGCS; *See* Straight grid-connected systems
Shade
18 kW BBGC system and, 250f, 252t
bypass diodes and, 52–53, 53f
computation of, 39–40
from contacts, 439
designing for, 179
field measurement of, 38–39, 39f
row spacing and, 39–40, 40f
site selection and, 38, 196
solar access rights, 362
Shadow-band stands, 35
Shanghai, average daily irradiation, 456t
Shear, 168, 168f, 184
Sheet steel, 175
Shingles, asphalt, 195
Shingles, solar, 194, 199
Shipments, 15–16, 15f
Shockley, William, 14
Shunt controllers, 76
Silane, 410, 420
Silica dust, 363t
Silicon
absorption constant for, 373f
aluminum and, 176, 177, 177t, 413
amorphous; *See* Amorphous silicon
bandgap energy of, 372, 372f, 375t, 377f, 396t, 408f
bandgap of, 372
bandgap type, 372f
bonding of, 376–377
boron and, 376–377, 377f
in cells, 409, 419, 420, 422
on ceramics, 419, 438
conductivity of, 377
crucible growth of, 418
CSG, 409, 419
diffusion constant for, 389f
disposal of, 365
EFG of, 418
environmental concerns, 364, 365, 409

epitaxial growth of, 419, 439
HIT cells, 439
intrinsic carrier concentration in, 374f, 375t, 376
melting point of, 375t
MIS-IL cells, 438–439, 438f
mobility of EHPs in, 375t, 376, 389, 389f
MPPT and, 391
multicrystalline, 409, 418–419
passivation of surfaces, 393
phosphorus and, 376–377, 377f, 410
photon current density vs. optical path length in, 408f
polycrystalline, 409
processing of, 438
production of, 409
purity of, 410
relative dielectric constant of, 375t
shortage of refined, 15
single-crystal; *See* Single-crystal silicon cells
string ribbon technology for, 419, 439
tandem cells, 439
temperature and, 374f, 376, 389, 389f
Silicon oxide varistor (SOV), 97, 116
Silicon oxides, 388t
Silicon-controlled rectifiers (SCRs), 82, 83
Silver, 173t, 365, 424, 429, 435
Silver oxide, 65
Single-axis trackers, 36–37, 327
Single-crystal silicon cells
aSi in tandem with, 439
contacts for, 413–416
efficiency of, 409, 417
junction fabrication, 411–412
resistivity of, 415
SunPower A-300 cell, 417–418, 417f
thickness of, 408, 438
wafer fabrication, 410–411
Skin cancer, 174
Skin friction forces, 184
Sky; *See also* Atmosphere
clear
color of, 23
irradiance in, 236
sunlight and, 23
clouds in, 23, 69, 127, 308
position of sun in
air temperature and, 25
angular displacement, 30–31, 31f, 36–37, 37f, 39–40
azimuth angle; *See* Azimuth angle
declination; *See* Declination
hour angle, 30, 30f, 31, 39–40, 43
solar altitude; *See* Solar altitude
zenith angle and; *See* Zenith angle
Slenderness ratio, 204
"Smart garage" concept, 445
Smart grids, 444–445
Smokestack scrubbers, 347
Snow
array surfaces and, 191
ASCE on, 181
density of "wet," 244
loading from, 191–192
rack tilt and, 143, 191
stress from, 164
Society of Automotive Engineers (SAE), 181
Sodium, 65, 176, 441
Sodium aluminate, 423–424
Sofia, nitrogen oxide in, 356
Soil
acid rain and, 352
acidification of, 357–358
cathodic protection system design and, 288–291
dry vs. wet, 290
electroplating in, 287
expansive, 192
hydrostatic pressure in, 192
loading from, 182
resistivity of, 289
sunlight reflected from, 23
Solar altitude
angular displacement and, 30
azimuth angle and, 31, 32f
collector orientation and, 43
declination and, 30
definition of, 29
diagram of, 30f
double-axis trackers and, 297
hour angle and, 30
row spacing and, 39–40, 40f, 144, 144f
shade and, 38, 38–40
Solar constant, 22, 26
Solar midnight, 30
Solar noon
air mass at, 36
analemma, 28–29
azimuth angle and, 29
clock noon and, 28–29, 29f
collector orientation and, 36–37
declination and, 28, 31, 31f
definition of, 28
earth orbit and, 28
longitude and, 28
mounting angle and, 37f
zenith angle at, 31, 31f

Solar Pathfinder, 38–39, 39f
Solar radiation; *See* Sunlight
Solar thermal collectors, 196
Soluble Threshold Limit Concentration (STLC), 365
Solvents, 355t, 357, 363t, 364
Source circuit combiner boxes, 98
South America, 33f–34f; *See also specific countries in*
Southern California Gas Company, 66
Southern yellow pine, 203t, 208, 226
SOV, 97, 116
Space charge layer, 380f, 381, 391, 405p
Spain
  FIT in, 349p
  GDP vs. kW per capita, 6f
  renewable energy programs in, 1
Spangle, 175
Speed of light, 22, 371
Sphalerite, 423
Spring constant, 167
Square wave inverters, 81–82, 81f, 81t, 82f
Stable processes, 351
Staebler–Wronski effect, 420–421
Stagnation point, 184
Stainless steel
  aSi and, 421, 422f
  crevice corrosion and, 173
  in galvanic series, 173t
  modulus of elasticity for, 169t
  modulus of rigidity for, 169t
  molybdenum and, 432
  tensile strength of, 170t
  yield strength of, 170t
Standard candle, 280–281
Standard deviation, 11–12, 17p
Standard test conditions (STC), 123
Standard time, 28
Standards, 180, 444; *See also specific documents*
Standby loads
  for 3.04 kW BBGC system, 220–221, 220t
  for 7.20 kW BBGC system, 238–239, 238t
  for 120 V BBGC system, 220–221, 220t
  for 120/240 V BBGC system, 238–239, 238t
  battery backup for, 215, 217–218
  bypass switches for, 98
  synchronization for, 89
  wiring of, 230–231
Stand-off mounts
  for 2.85 kW 1-phase system, 121
  for 3.20 kW 1-phase system, 205–209
  aspect ratios of, 197
  cell temperature and, 196, 197
  description of, 196–197
  height above roof, 196, 197
  for hurricane-prone region, 205–209
  lag bolts/screws for, 202–203
  photograph of, 197f
  slope of roof and, 121
  uplift pressure on, 190f
  wind loads on, 190
STC, 123
Steel
  vs. aluminum, 169–170, 176, 211p
  bar pull test, 166–167, 166f, 175
  cathodic protection of, 287–288
  coefficient of linear expansion for, 172
  coefficient of thermal expansion for, 172
  composition of, 175
  corrosion protection for, 175
  cost of, 179
  fatigue failure in, 165, 165f
  in galvanic series, 173t
  grades of, 175
  heat-treatment of, 174
  intergranular corrosion and, 173–174
  in manufacturing, 363t
  modulus of elasticity for, 169t, 170
  modulus of rigidity for, 169t
  in mounts, 364
  pollution and, 358
  production of, 354–355
  stainless; *See* Stainless steel
  stress corrosion cracking of, 174
  stress limits of, 164
  stress-strain curve for, 166–167, 167f
  tensile strength of, 170t, 212p
  thermal stress on, 171–172, 171f
  vs. titanium, 170
  uniform attack of, 172
  weight of, 211p
  welding of, 174
  yield strength of, 170t
  zinc and, 173, 175
STLC, 365
Stockholm
  average daily irradiation, 459t
  particulates in, 356
Straight grid-connected systems (SGCS)
  2.45 kW 1-phase, 140–142
  2.85 kW 1-phase, 121–132
  3.20 kW 1-phase, 133–138
  21 kW 3-phase, 143–149
  120/208 V 3-phase, 261–263
  240 V
    120 V BBGC system, 260–261
    120/240 V BBGC system, 258–260

250 kW 3-phase, 150–156
aesthetics of, 115
cell/module voltage in, 98
conversion efficiency of, 117
definition of, 103
EMI with, 115
energy output of, 117–118
generators and, 312
IEEE 1547 standard on, 109–112
installation of, 119, 120–121
inverters for, 119–120, 257
islanding in, 112–114
losses from, 117–118, 118f
losses in, 445
mechanical considerations, 116–117
modules for, 52
MPPT and, 52
NEC on, 104–106, 108, 108t
performance monitoring, 158
roof space for, 117
surge protection for, 115–116
voltage, maximum open-circuit, 52
wiring for, 106–108, 106t
Strain, 166–169, 167f
Streamlining, 185, 186f
Stress
in bar pull test, 166–168, 167f
bending, 164, 169, 427
definition of, 164
discussion of, 164–172
elastic limit of, 166–167, 167f
limits of, 165, 165f
proportional limit of, 166, 167f, 169
shear, 168, 168f, 184
yield strength, 167, 169–170, 170t
Stress corrosion cracking, 174
String ribbon technology, 419, 439
Strip steel, 175
Stromeyer, Friedrich, 431
Structural design, 116–117
Subsidies, 348, 361–362
Sulfur, 65, 424, 429, 430, 431, 434
Sulfur oxides
acid rain and, 331, 347, 352, 357, 359
allowed emissions of, 361
credits for, 360
energy sources and, 354t
EPA on, 354, 355, 355t
structures and, 358
World Bank on, 356
Sulfuric acid, 56–57, 56f, 429
Sun
composition of, 21
declination of; *See* Declination
density of, average, 43p
diameter of, 43p
to Earth distance, 22, 27, 43p
energy radiated by, 21–22, 25
hydrogen consumption by, 43p
surface of
energy density of, 22
temperature of, 21, 281
Sunburn, 174
Sunlight
atmosphere and, 23–25, 422–423
carbon dioxide and, 23, 25
cell interaction with, 49
clouds and, 23, 127
color of, 281
diffused, 23
direct, 23
energy density of; *See* Irradiation
hours of, 30
illumination level of, 281
intensity of, 24, 26
latitude and, 25
measurements of, 32–35
methane and, 25
peak energy, 408, 442
photons in; *See* Photons
photosynthesis and, 25
power density of, 26
reflected, 23, 24
seasons and, 25
spectrum of, 22–23, 22f, 24f, 409
wavelength of, 371
SunPower A-300 cell, 417–418, 417f
Sunrise angle, 30
Sunset angle, 30
Supertandem cells, 442
Surface recombination, 393–394
Surge protection
for 3.04 kW BBGC system, 227
for 7.20 kW BBGC system, 240
for 120 V BBGC system, 227
for 120/240 V BBGC system, 240
description of, 97
IEEE on, 109
for inverters, 80–81, 81t, 97, 116
from loads, 80, 116
MOVs, 97, 116
NEC on, 115
SOVs, 97, 116
Sustainability, tests for, 13–14
SVS, 112, 114
Sweden, Stockholm; *See* Stockholm
Switches, 84, 96, 104, 310
Synchronous machines, 109, 113

T

Tampa, 120 V BBGC system for, 220–235
Tangential stress, 168, 168f, 184
Tantalum pentoxide, 388t
Tap conductors, 138
Tariffs, 348, 349p, 362
Taxes, 348, 362
TC-82 standards, 105t
TCLP, 365
TCO, 419, 420, 433–434, 436
Tehran, sulfur oxide in, 356
Televisions (TVs), 88, 238t, 313t
Tellurium, 363t, 364–365, 429, 430–431, 435–437
Temperature
  air; *See* Air temperature
  blackbody radiation and, 22f
  at Earth surface, 25
  at sun surface, 21
Temperature, cell/module
  2.85 kW 1-phase system and, 123, 123t, 124t
  3.04 kW BBGC system and, 224t, 225, 228t
  7.20 kW BBGC system and, 236, 236t
  18 kW BBGC system and, 247
  120 V BBGC system and, 224t, 225, 228t
  120/240 V BBGC system and, 236, 236t
  current and, 51
  formula for, 53
  high/low limit, 123
  for hybrid home, 316–317
  lowering, 196
  at mountain cabin, 306
  NOCT, 53, 123, 123t, 196
  power and, 51, 51f
  testing for, 427
  voltage and, 51, 51f, 53
  water pump system and, 279, 279f
Tensile strength
  of aluminum, 170t, 176, 177
  of cast iron, 170t
  compression and, 172
  designing for, 165
  fatigue limit and, 165
  of steel, 167f, 170t, 174–175
  stress corrosion cracking and, 174
  stress-strain curve for, 167f
  of titanium, 170t
  uplift and, 190, 203
  of wrought iron, 170t
  yield strength and, 170t
Tension, 165–169, 172, 188
Terminal blocks, 227, 240, 299
Testing, spectral conditions for, 23
Texas, Austin, 454t
Textured surfaces
  diagram of, 395f
  etching process for, 411
  for MIS-IL cells, 439
  pn junction and, 395f
  purpose of, 388, 394–395
  reflection and, 52
  sunlight spectrum and, 388, 416
  in SunPower A-300 cell, 417, 417f
THD, 84, 101p, 111
Thermal velocity, 378, 404p–405p
Thermal voltage, 50
Thermodynamics, second law of, 7, 14
Thermoelectronics, 443
Thermophotovoltaic cells, 442
THHN wire, 106, 106t, 127, 128t
Thin-film cells
  aSi in, 409, 420, 422
  carbon dioxide and, 363t
  CdS and, 400, 436
  challenges of, 428
  CIS for, 428, 432
  conversion efficiency of, 408
  CSG, 409, 419
  direct mounting of, 199
  environmental concerns, 363t, 365
  GaAs compounds in, 423
  manufacturing of, 363t
  mounting of, 199
  next generation, 407–408, 419
  photon current density vs. optical path length in, 408, 408f
  power from, 196
  PSI Process for, 419
  silicon in, 409, 419, 420, 422
  temperature and, 196
  thickness of, 408
Thin-Film Photovoltaics Partnership Program, 432
Third-party ownership, 331
Three Mile Island, 362
THWN-2 wire, 106, 106t, 127, 128t
Tianjin, particulates in, 356
Tidal energy, 354t
Tiles, solar, 194, 199f
Time value of money, 332
Time zones, 28
Tin, 424, 435, 436
Tin oxide, 420
Titanium, 169t, 170t, 173t, 416
Titanium oxide, 388t
Tobacco industry, 347, 362
Tokyo, average daily irradiation, 458t
Topographical factor, 190

Topography, 187, 191; *See also* Roughness, surface
Torsional shear stress, 169
TOT, 11
Total harmonic distortion (THD), 84, 101p, 111
Total Threshold Limit Concentration (TTLC), 365
Toughness of material, 168
Toxic material storage, 287
Toxicity Characteristic Leaching Procedure (TCLP), 365
Trackers
  active, 201
  cost of, 36
  directionality of, 40–43
  double-axis, 178, 297
  vs. fixed collectors, 201
  functional requirements for, 178
  irradiation and, 35–37
  optimizing, 195–196
  output from, 200–201
  passive, 201
  photograph of, 202f
  seasons and, 36f
  single-axis, 36–37, 327
  tilt of, 35f, 38f
  time of day and, 36f
  for utility interactive system, 41–43
Transformers
  for ac-coupled BBGC systems, 260–262, 261f, 263f
  autotransformers, 260–261, 261f
  generators and, 93
  inverters and, 82, 85–86
Transistors, bipolar, 14, 82, 83
Transistors, insulated gate, 83
Transistors, MOS, 83
Transparent conducting oxide (TCO), 419, 420, 433–434, 436
Trapping centers, 390, 410, 421
Trichlorosilane, 410
Tropic of Cancer, 27
Tropic of Capricorn, 27
TTLC, 365
Tungsten, 281
Tunisia, CdTe arrays in, 437
Tunnel junctions, 403–404, 403f, 404f, 421, 439
Two-axis trackers, 178, 297

U

UL 1741 standard
  on battery-backup, 215
  generators and, 234–235
  GFDI, 127
  IEEE 929 and, 103
  IEEE 1547 and, 109, 112, 119, 120
  NEC and, 120, 127
  selection requirement, 122
  on SGCS, 127
  stand-alone systems and, 271
  title of, 105t
  on utility stability, 259
Ultimate strength; *See* Tensile strength
Ultrasonic bonding, 416
Ultraviolet radiation
  in blackbody spectra, 22f
  degradation from, 116–117, 174
  encapsulant and, 416
  EVA and, 416
  ozone and, 23
  regions of, 174
  wavelength range of, 174
Underwriters Laboratories (UL), 103; *See also* UL 1741 standard
Ungrounded conductors, 97, 128
Uniform attack, 172
Uninterruptible power supplies (UPS), 84, 289
Unit electrical cost, 341
United Kingdom (U.K.)
  England; *See* England
  GDP vs. kW per capita, 6f
  Scotland, 431
United Nations, Special Session on Environment and Development, 1
United States (U.S.); *See also specific states*
  DOE; *See* Department of Energy
  eastern seaboard ozone, 355
  GDP vs. kW per capita, 6f
  Kyoto Conference/Treaty and, 6, 356
  land area of, 359
  manufactured housing in, 194
  molybdenum deposits in, 432
  National Energy Conservation and Policy Act, 4
  Naval Observatory, 29, 285
  oil reserves in, 10
  petroleum production in, 12–13, 12f
  population of, 17p
  prime lending rate of, 333, 334f
  PV manufacturing in, 16
  solar systems in, 1
  wind speed in, 248
Unstable processes, 351
Uplifting force
  on 3.20 kW 1-phase system, 208–209
  on 6 kW 1-phase system, 210
  on 120 V BBGC system, 226

ballasted systems and, 249
maximum, 184
mount tilt and, 186–187, 186f, 188, 191
roof slope and, 207t
roof zone and, 207t, 210
on stand-off mount, 190f
tensile strength and, 190, 203
wind speed and, 207t
withdrawal load and, 190
UPS, 84, 289
USE-2 wire, 106t, 126, 127, 128t
Utah, sulfur oxide in, 355
Utility escalation rates, 345, 346t
Utility interactive systems
abnormal conditions in, 109
battery backup for; *See* Battery-backup grid-connected (BBGC) systems
classes of, 103
cost of, 103
DOE on, xix
EMI and, 109
excess energy in, 47
frequency in, 109
grid failure and, 47, 109
IEEE 1547 on; *See* IEEE 1547 standard
interfacing circuitry for, 47
inverters and, 88–89
islanding in, 89, 111, 112–114, 443–444
isolation of, 109
micro grids, 443–444
MPPT and, 80
orientation for peak loads, 41–43, 195
schematic diagram of, 48f
smart grids, 444–445
straight; *See* Straight grid-connected systems
surges and, 109
synchronization of, 109
THD of, 84
voltage to/from, 52, 98, 109

V

Vacuum level, 400, 401f
Vacuum tubes, 14
Valence band
at absolute zero, 370, 370f
acceptor atoms and, 390
carrier concentration in, 374
curvature of, 387
description of, 370–372
direct vs. indirect bandgap and, 372
energy, momentum, and, 371–372
for GaAs compounds, 372, 372f
intermediate band and, 442
phosphorus and, 376
in QW cells, 441
for silicon, 372, 372f, 376, 377f
temperature and, 370, 370f
Vanguard I satellite, 14
VB Engineering, 461
Velocity pressure exposure coefficient, 190
Vermont, 71f
Vermont, Burlington; *See* Burlington
Video cassette recorders (VCRs), 88
Visible light, 22f, 23, 25
Vitamin D, 174
Volcanoes, 25
Voltage, cell/module
for 2.45 kW 1-phase system, 140
for 2.85 kW 1-phase system, 123–124, 128
for 3.04 kW BBGC system, 224t, 225, 225t
for 3.20 kW 1-phase system, 134–135
for 7.20 kW BBGC system, 236, 236t
for 18 kW BBGC system, 247, 252t
for 120 V BBGC system, 224t, 225, 225t
for 120/240 V BBGC system, 236, 236t
for 250 kW 3-phase system, 150, 151
for CdTe, 437
CIS, 434
controller and, 75, 224
fan and, 67f, 273f
fill factor and, 50–51
grounded conductors and, 128
for hybrid home, 315–317, 323
ideal vs. real, 49f
illumination level and, 50f
inverters and, 123
LCBs and, 77–78, 78f
limiting, 49
load and, 67–68, 67f, 68f
maximum power, 50–52
for mountain cabin, 306
MPPT and
for 2.85 kW 1-phase system, 123–124
controllers and, 80f
converters and, 77–79
load and, 68, 68f
for silicon, 391
water pump system and, 77f
NEC on, 123
open circuit, 50–53, 386, 402
parallel connection, 55
pn junction voltage and, 391–392
PWM inverters and, 86–88, 87f
for refrigeration system, 297, 298t
series connection, 52–55, 54f
for silicon, 391

square wave inverter and, 82
tandem silicon, 439
temperature and, 51, 51f, 53
typical, 47
in utility interactive system, 52, 98
wiring and, 128–130
zero, 49–50
Voltage drop
in 2.85 kW 1-phase system, 129–130
in 3.04 kW BBGC system, 228t, 229
in 3.20 kW 1-phase system, 137
in 7.20 kW BBGC system, 241t
in 18 kW BBGC system, 252
in 21 kW 3-phase system, 145–149, 146t, 148f
in 120 V BBGC system, 228t, 229
in 120/208 V 3-phase BBGC system, 241t, 262
in 250 kW 3-phase system, 153–154, 153t, 154t, 156
basics, 107, 128
calculation of, 298
compensation for, 112
in hybrid home, 322–323
module choice and, 143
NEC on, 106–107
for outdoor lights, 285–286
in refrigerator system, 298t
from source to load, 107–108
worst-case conditions for, 137

W

Wafer fabrication, 410–411
Warranty, 69
Washing machines, 220t, 238t, 313t
Washington, Seattle; *See* Seattle
Waste Extraction Test (WET), 365
Water, weight of, 275
Water heater
excess energy for, 307
fuel cells for, 66
in MSRI, 1
power for, 217
size of solar, 235
subsidies for, 348
in water pump sizing, 301
Water pollution, 357
Water pump systems
ac vs. dc, 301
current for, 277–280, 278t
GPM of, 277–278, 278t
high- vs. medium head, 276–277, 277f
horsepower of, 275–276, 301
for hybrid home, 313t
I-V curve and, 77f
LCB for, 76–77, 275, 276, 279
lift in, 275–278, 278t
for mountain cabin, 300, 302t, 303
noise from, 119
performance specifications for, 274–277
pipes in, 275, 276, 276f, 277f, 301
power factor for, 239
power for, 277–280, 278t
pumping height for, 275–277
pumping time of, 275–277, 301
speed of, 277
stand-alone system for, 274–280
switches for, 310
tracking collectors for, 201
voltage for, 239, 277–280, 278t
wire-to-water efficiency, 275–276
work of, 275
Water storage tanks, 274, 301, 303, 307
Water vapor, 23
Water wave energy, 354t
Watercraft, 41
Watt hours (Wh), 223, 293, 307
Weather-sealing materials, 164, 174
Weld decay, 174
West Palm Beach
3.2 kW 1-phase system in, 133–138, 135f, 139f, 205–209, 205f
3.20 kW 1-phase system in, 134f
air temperature range in, 135
WET, 365
White oak, 203t
White spruce, 203t
Wilson, Alan, 14
Wind direction, 164–165, 190, 358
Wind load
on 3.04 kW BBGC system, 226
on 18 kW BBGC system, 249–250
on 3.20 kW 1-phase system, 133, 134, 205–209
on 120 V BBGC system, 226
analysis of, 187–188
array area and, 190–191
ASCE on, 134, 189–191, 205–210
buckling and, 204
building codes on, 96
classification of, 183
designing for, 116, 164–165, 182, 365
on Hughes modular array, 187–188, 188f, 189f
in hurricane-prone region, 133, 134, 205–209
interior vs. exterior rows, 188
range of, 116
roof zones, 134, 134f, 205

row configuration and, 208–209
topography and, 187, 191
worst-case conditions for, 191
Wind speed
air pressure and, 184–191
ASCE on, 189–191
designing for, 164–165, 181, 186–191
high (>125 mph)
3.20 kW design for, 133–138, 135f, 139f, 205–209, 205f
ASCE on, 190
outdoor light pole design for, 286–287
possibility of, 181
rack mounts for, 198f
solar tiles for, 199f
worst-case, 191
in hurricanes, 133
low (<90 mph)
6 kW system, 139, 209–210, 209f
designing for, 133
medium (90 to 125 mph)
18 kW BBGC system, 248
120 V backup system, 226
designing for, 181
Hughes modular array, 188, 188f, 189f
for NOCT, 53
roof slope and, 207t
roof zone and, 207t
snow load and, 191
stagnation point, 184
U.S. typical, 248
velocity profiles, 185f
Wind tunnel testing, 188, 206, 207, 210–211
Wind turbines
BBGC systems and, 215, 216f, 257
classification of energy from, 4
environmental effects of, 354t
IEEE 1547 on, 109
motors attached to, 113
Winkler, Clemens, 424
Wiring; *See also* Conductors
for 2.45 kW 1-phase system, 141
for 2.85 kW 1-phase system, 126–131, 126f, 128t, 129t
for 3.04 kW BBGC system, 222, 227–230, 228t
for 3.20 kW 1-phase system, 136–138
for 7.20 kW BBGC system, 239–243, 241t
for 18 kW BBGC system, 251–256, 252t, 253t
for 21 kW 3-phase system, 145–149, 146t
for 120 V BBGC system, 222, 227–230, 228t
for 120 V stand-alone system, 298
for 120/240 V BBGC system, 239–243, 241t
for 250 kW 3-phase system, 153–156, 153t, 154t
in ac microinverter systems, 141
air temperature and, 127, 128t
between batteries, 263–264, 266
for branch circuit taps, 156
for cathodic protection system, 289
circuit breakers and, 96
cost of, 79
current in, 96, 126–129, 128t, 129t
efficiency of, 218
fuses and, 96
household, 230–231
for hurricane-prone region, 136–138
for hybrid home, 313, 322–324, 323t
losses in, 218, 219, 223
for microinverter, 141
for mountain cabin, 309–310, 309t
MPPT and, 79
NEC on, 104–107, 106t, 127, 300
for outdoor lights, 284–286
overcurrent protection for, 130
power loss in, 79
for refrigeration system, 298, 298t
resistance of, 263–264, 298, 298t
size of, 106–108, 106t
of standby loads, 230–231
voltage drop across; *See* Voltage drop
Withdrawal loads, 202–203, 203t
Wood Engineering and Construction Handbook (Faherty and Williamson), 203
Work function of metal, 400
Working stress, 169–170
World Bank, 356
World Resources Institute (WRI), 356, 358
WRI, 356, 358
Wrought iron, 170t

X

Xenon lamps, 282

Y

Yield point, 167, 167f
Yield strength, 167, 169–170, 170t
Young's modulus, 167–168, 169t

Z

Zener diode, 97, 116
Zenith, 29
Zenith angle
angular displacement and, 31, 31f
azimuth angles and, 31

collector orientation and, 36–37
daylight hours and, 30
declination and, 31, 31f
definition of, 24
mounting angle and, 37f
solar altitude and, 30, 31
at solar noon, 31, 31f
sunrise/sunset angles and, 30
Zero Energy Homes Program, 194
Zinc
aluminum and, 173, 175, 176, 177t
in batteries, 64–65
cadmium and, 431
gallium and, 423
in galvanic series, 173t
germanium and, 424–425
in hot-dip galvanizing, 175
indium and, 429
indium phosphide and, 427
pollution and, 358
selective leaching of, 174
steel and, 173, 175
Zinc oxide, 433, 441
Zirconium oxide, 66